Selected Titles in This Series

(Continued in the back of this publication)

Structured Matrices in Mathematics, Computer Science, and Engineering I

CONTEMPORARY MATHEMATICS

280

Structured Matrices in Mathematics, Computer Science, and Engineering I

Proceedings of an AMS-IMS-SIAM
Joint Summer Research Conference
University of Colorado, Boulder
June 27–July 1, 1999

Vadim Olshevsky
Editor

American Mathematical Society
Providence, Rhode Island

This volume contains the proceedings of an AMS-IMS-SIAM Joint Summer Research Conference held at the University of Colorado, Boulder, Colorado on June 27–July 1, 1999, with support from the National Science Foundation, grant DMS-9618514.

2000 *Mathematics Subject Classification.* Primary 15–XX, 47–XX, 65–XX, 93–XX.

Any opinions, findings, and conclusions or recommendations expressed in this material are those of the authors and do not necessarily reflect the views of the National Science Foundation.

Library of Congress Cataloging-in-Publication Data

Structured matrices in mathematics, computer science, and engineering : proceedings of an AMS-IMS-SIAM joint summer research conference, University of Colorado, Boulder, June 27–July 1, 1999 / Vadim Olshevsky, editor.

p. cm. — (Contemporary mathematics, ISSN 0271-4132 ; 280–281)

Includes bibliographical references.

ISBN 0-8218-1921-6 (v. 1 : alk. paper)–ISBN 0-8218-2092-3 (v. 2 : alk. paper)

1. Matrices—Congresses. I. Olshevsky, Vadim, 1961– II. Contemporary mathematics (American Mathematical Society) ; v. 280–281.

QA188.S764 2001
512.9′434—dc21 2001041241

∞ The paper used in this book is acid-free and falls within the guidelines
established to ensure permanence and durability.
Visit the AMS home page at URL: `http://www.ams.org/`

10 9 8 7 6 5 4 3 2 1 06 05 04 03 02 01

Contents

Structured Matrices in Mathematics, Computer Science, and Engineering I

Part III. Control Theory

Part IV. Spectral Properties. Conditioning

Structured Matrices in Mathematics, Computer Science, and Engineering II

Part V. Fast Algorithms

Part VI. Numerical Issues

Part VII. Iterative Methods. Preconditioners

Part VIII. Linear Algebra and Various Applications

Foreword

Many important problems in applied sciences, mathematics, and engineering can be reduced to matrix problems. Moreover, various applications often introduce a special structure into the corresponding matrices, so that their entries can be described by a certain compact formula. Among classical examples are Toeplitz matrices $[a_{i-j}]$, Hankel matrices $[a_{i+j}]$, Toeplitz-plus-Hankel matrices, Vandermonde matrices $[a_i^{j-1}]$, Cauchy matrices $[\frac{1}{a_i - b_j}]$, Pick matrices $[\frac{1-a_i a_j^*}{b_i + b_j^*}]$, and also Bezoutians, controllability and observability matrices and others. Though standard linear algebra methods are, of course, readily available, there are several reasons why they can be unattractive in many instances. Along with just the desire to find elegant structure-exploiting solutions, there are also practical computational considerations such as the storage limitations, the need in reducing computational complexity as well as in obtaining a better numerical accuracy. In many cases these goals can be achieved by solving the underlying problem in terms of only $O(n)$ of parameters defining a structured $n \times n$ matrix via a compact formula as in the above examples.

Structured matrices have been under close study for a long time, and in quite diverse (and seemingly unrelated) areas. Typically, not only an area of application gives rise to certain patterns of structure, but it often provides a technique to solve the associated matrix problems. As an illustration of this principle we mention the classical interpolation problems of Caratheodory-Toeplitz and of Nevanlinna-Pick. Not only are they related to positive definite Toeplitz and Pick matrices, resp., but the recursive algorithms of Schur and Nevanlinna for computing their solutions admit nice matrix interpretations. In fact, these classical interpolation algorithms can be understood as efficient structure-exploiting ways to compute the Cholesky decompositions for these matrices. As was mentioned above, structured matrices were studied from different points of view, in mathematics, computer science and engineering. For example, the same Toeplitz and Pick matrices were closely looked at using the methods of reproducing kernel Hilbert spaces, lifting-of-commutants, state-space methods, as well as the methods of system theory and signal processing, network theory, linear prediction, to mention just a few mathematical and engineering fields. Interestingly, in the latter a physical intuition often provides deep insights into structured matrix problems. An interplay between the techniques of engineers and mathematicians is reflected in these volumes. There are several other areas providing their own applications and their own languages to attack structured matrix problems. It can be quite difficult to survey all such connections; in fact, browsing through the papers of these volumes can give a flavor of the plethora of different techniques and approaches. It appears that the theory of structured matrices is positioned to bridge the gaps between these diverse areas.

Significant progress has been recently made in studying relevant numerical issues. It was quite well-understood for a long time that structure can be exploited to speed-up computations and to design fast algorithms. However, many of those suffered from the loss of numerical accuracy. In the past few years a number of algorithms blending speed and accuracy has been developed. This progress is fully reflected in these volumes.

Though structured matrices have been under close study for a long time, in the past decade they have enjoyed a significant growth in popularity. One reason for this is in that the theory of structured matrices is poised to bridge diverse applications in the sciences and engineering, deep mathematical theories and computational and numerical issues. Hence, it is not surprising that the number of researchers in our scientific community is rapidly increasing. Special sessions and minisymposia devoted to structured matrices were included in the programs of various general mathematical and engineering conferences, among which we mention various SIAM meetings, ILAS, IWOTA, SPIE, and MTNS. Moreover, in the past few years several special international conferences focusing solely on different aspects of structured matrices were held in Santa Barbara (USA, Aug. 1996), Cortona (Italy, Sept. 1996), Boulder (USA, July 1999), Chemnitz(Germany, Jan. 2000), Cortona (Italy, Sep. 2000), and we are looking forward to the next meeting in South Hadley (USA, Aug. 2001).

These meetings, especially the one in Boulder (July 1999), brought together quite diverse audiences of participants, many of which have never actually met earlier in the framework of such a comprehensive cross-disciplinary conference. In fact, it was a unique "cross-fertilization" atmosphere of the Boulder meeting that suggested the idea to pursue this publishing project. The detailed table of contents will provide a general idea of these volumes. Thirty eight papers devoted to the different aspects of the theory of structured matrices and using different techniques are collected under one cover. We hope that the reader will enjoy a plethora of different problems, different focuses, and different methods that all contribute to one unified theory of structured matrices.

Part I is devoted to a connection of structured matrices to several problems in interpolation and approximation. In the first paper the reader will see a natural connection of our subject to reproducing kernels. Other papers emphasize algorithmical aspects of tangential interpolation problems, exploit the maximum-volume concept, and present relations to Pade approximations and to the extended Euclidean algorithm.

Part II provides the perspective of engineers and presents their insights into the subject. It starts with a paper on systems on low Hankel rank. The reader will notice that the other parts of these volumes contain several other papers that discuss some of the related issues but from different points of view. Part 2 also contains an application of the "language of signal flow graphs" to study structured matrices, an exposition of recent advances in tensor analysis, and several applications.

Part III is dedicated to recent relevant developments in control theory.

Part IV contains several papers discussing spectral properties of Toeplitz, block Toeplitz, Pick and Hankel matrices.

PART V presents several fast algorithms for various classes of structured matrices.

Part VI provides a snapshot of the current state-of-art in numerical issues related to structured matrices. It starts with a paper containing a "quite heavy" error analysis leading to the first provably stable variant of the unitary Hessenberg QR algorithm. The other papers cover stability issues of fast algorithms and inversion formulas.

Part VII contains two papers devoted to preconditioners design.

Finally, part VIII contains a discussion of the concept of approximate displacement rank, new inversion formulas for structured matrices, a survey on completion problems and several related topics.

Vadim Olshevsky

Department of Mathematics and Statistics
Georgia State University
Atlanta, GA 30303, USA

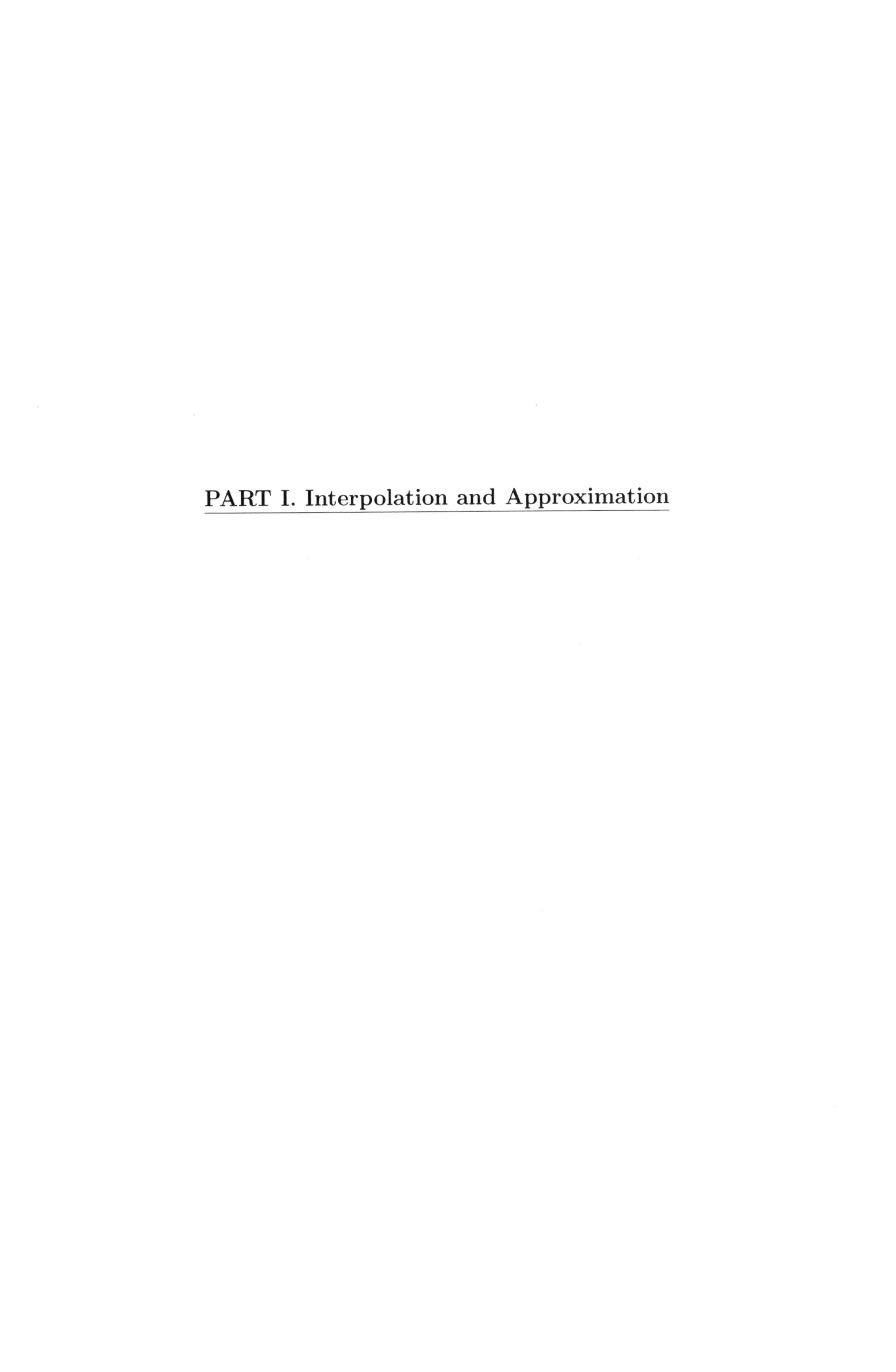

PART I. Interpolation and Approximation

Contemporary Mathematics
Volume **280**, 2001

Structured Matrices, Reproducing Kernels and Interpolation

Harry Dym

Dedicated to the memory of Naftali Kravitsky (ז"ל)

1. Introduction

The purpose of this paper is to discuss some natural connections between structured matrices and a class of interpolation problems. The link rests on interpreting the structured matrix as the Gram matrix of a reproducing kernel space of a special kind. If the matrix is positive, then the space is a finite dimensional Hilbert space. If the matrix is Hermitian and invertible, then the space is a finite dimensional Krein space. If the matrix is invertible but not Hermitian, then pairs of reproducing kernel spaces come into play. We shall not consider that case here; see [AD4] for additional information . There are also connections when P is Hermitian and singular, but we shall not consider that case in any detail here either; see [D6] for an elementary approach to this case and references to the literature.

Most of the material that will be presented here has been published in one form or another. However, it is often imbedded as part of an objective of wider scope that on the one hand serves to illustrate its usefulness, but on the other hand makes it more difficult to extract. The main objective of the present paper is make this circle of ideas accessible (and hopefully attractive) to a wider audience. Accordingly, we shall not dwell excessively on proofs that are available elsewhere unless they are both short and particularly instructive. However, in order to make the main ideas plausible we shall discuss the strategy underlying some of the proofs and shall supplement the discussion with a number of illustrative examples. In marked contrast, the full details of the material in Section 13 is only available at the moment in preprint form [BoD2]. Nevertheless it is included because it describes a rather interesting class of structured matrices and their role in a perhaps less familiar type of interpolation problem.

In the sequel we shall discuss a number of spaces of vector valued functions (vvf's). Some of these spaces are most naturally viewed as subspaces of the Hardy space $H_2^m(\Omega_+)$ of $m \times 1$ vvf's that are analytic and square integrable in an appropriate sense in the region Ω_+, where Ω_+ is either the open unit disc $\mathbb{D}$, the open upper

---

1991 *Mathematics Subject Classification.* Primary 54C40, 14E20; Secondary 46E25, 20C20.
H. Dym wishes to thank Renee and Jay Weiss for endowing the chair which supports his research.

half plane $\mathbb{C}_+$ or the open right half plane Π_+. All three of these cases (and more) can be treated uniformly by introducing the appropriate notation which emerges upon reexpressing the Cauchy formula (for scalar f)

$$f(\omega) = \frac{1}{2\pi} \oint \frac{f(\lambda)}{\lambda - \omega} d\lambda \qquad (\omega \in \Omega_+),$$

as an inner product

$$f(\omega) = \langle f, \frac{1}{\rho_\omega(\lambda)} \rangle$$

with respect to the standard inner product on the boundary Ω_0 of Ω_+ that is spelled out in the following table:

Ω_+	$\mathbb{D}$	$\mathbb{C}_+$	Π_+
$a(\lambda)$	1	$\sqrt{\pi}(1 - i\lambda)$	$\sqrt{\pi}(1 + \lambda)$
$b(\lambda)$	λ	$\sqrt{\pi}(1 + i\lambda)$	$\sqrt{\pi}(1 - \lambda)$
$\rho_\omega(\lambda)$	$1 - \lambda\omega^*$	$-2\pi i(\lambda - \omega^*)$	$2\pi(\lambda + \omega^*)$
Ω_0	$\mathbb{T}$	$\mathbb{R}$	$i\mathbb{R}$
$\langle f, g \rangle$	$\frac{1}{2\pi} \int_0^{2\pi} g(e^{i\theta})^* f(e^{i\theta}) d\theta$	$\int_{-\infty}^{\infty} g(x)^* f(x) dx$	$\int_{-\infty}^{\infty} g(iy)^* f(iy) dy$
λ°	$1/\lambda^*$ if $\lambda \neq 0$	λ^*	$-\lambda^*$
$f^\#(\lambda)$	$f(\lambda^\circ)^*$	$f(\lambda^\circ)^*$	$f(\lambda^\circ)^*$
$\delta_\omega(\lambda)$	$\lambda - \omega$	$2\pi i(\lambda - \omega)$	$-2\pi(\lambda - \omega)$
$ab' - ba'$	1	$2\pi i$	-2π
$\varphi_{j,\omega}(\lambda)$	$\lambda^j/(1 - \lambda\omega^*)^{j+1}$	$-1/2\pi i(\lambda - \omega^*)^{j+1}$	$(-1)^j/2\pi(\lambda + \omega^*)^{j+1}$
$(R_\alpha \rho_\omega^{-1})(\lambda)$	$\omega^*/\rho_\omega(\alpha)\rho_\omega(\lambda)$	$2\pi i/\rho_\omega(\alpha)\rho_\omega(\lambda)$	$-2\pi/\rho_\omega(\alpha)\rho_\omega(\lambda)$

Table 1.1

This strategy was adopted in [D1] and [AD1].The observation that the kernels $\rho_\omega(\lambda)$ which intervene in these problems can be expressed in terms of a pair of polynomials $a(\lambda)$ and $b(\lambda)$ as

$$\rho_\omega(\lambda) = a(\lambda)a(\omega)^* - b(\lambda)b(\omega)^* \tag{1.1}$$

is due Lev-Ari and Kailath [LAK]. They noticed that certain fast algorithms in which the term $\rho_\omega(\lambda)$ intervenes will work if and only if $\rho_\omega(\lambda)$ can be expressed in the form (1.1). A general theory of reproducing kernels with denominators of this form and their applications was developed in [AD3]–[AD5]; for related developments that were obtained independently by a number of mathematicians in the former Soviet Union, see Nudelman [Nu]. The polynomials correspond to a distributed form of the conformal map b/a which sends $\mathbb{D}$ into Ω_+. Explicit formulas for them are given in Table 1.1. For each of the three listed choices of the kernel function $\rho_\omega(\lambda)$, the region

$$\Omega_+ = \{\omega \in \mathbb{C} : \rho_\omega(\omega) > 0\}$$

and its boundary

$$\Omega_0 = \{\omega \in \mathbb{C} : \rho_\omega(\omega) = 0\} \ .$$

The rest of the notation is fairly standard: The symbol A^* denotes the adjoint of an operator A on a Hilbert space, with respect to the inner product of the space. If A is a finite matrix, then the adjoint will always be computed with respect to the standard inner product so that, in this case, A^* will be the Hermitian transpose, or just the complex conjugate if A is a number. However, the complex conjugate of a complex number λ will also be designated by $\overline{\lambda}$. The symbol $\sigma(A)$ denotes the spectrum of a matrix A and J stands for the $m \times m$ signature matrix

$$J = \begin{bmatrix} I_p & 0 \\ 0 & -I_q \end{bmatrix}$$

with $p \geq 1$, $q \geq 1$ and $p + q = m$. If $F(\lambda)$ is a matrix valued function (mvf), then $\mathcal{A}_F$ denotes its domain of analyticity. The following acronyms will be used: mvf = matrix valued function, vvf = vector valued function, RKHS = reproducing kernel Hilbert space, RKKS = reproducing kernel Krein space and RK = reproducing kernel.

2. Structured Matrices

We shall introduce three different measures of structure in terms of a given pair of $n \times n$ complex matrices M and N:

$$\nabla_{\Omega_+} P = M^*PM - N^*PN \text{ when } \Omega_+ = \mathbb{D} \ . \tag{2.1}$$

$$\nabla_{\Omega_+} P = M^*PN - N^*PM \text{ when } \Omega_+ = \mathbb{C}_+ \ . \tag{2.2}$$

$$\nabla_{\Omega_+} P = M^*PN + N^*PM \text{ when } \Omega_+ = \Pi_+ \ . \tag{2.3}$$

The reason for this notation will become apparent later. Throughout this paper we shall always assume that the matrices M and N define a regular matrix pencil, i.e., that the matrix polynomial

$$G(\lambda) = M - \lambda N \tag{2.4}$$

is invertible except for at most n points in the complex plane. Then the matrix polynomial

$$H(\lambda) = \lambda M^* - N^* \tag{2.5}$$

enjoys the same property .

An $n \times n$ Hermitian matrix P is said to be structured if the rank of

$$\nabla_{\Omega_+} P$$

is small with respect to n. Sometimes, it is also required that this number (which is often referred to as the displacement rank of P) is independent of n . But that may not always be a meaningful condition. For a comprehensive introduction to structured matrices, the monograph [HR] by Heinig and Rost and the survey articles [KaSa] by Kailath and Sayed and [Ol] by Olshevky are recommended.

3. Reproducing Kernel Spaces

A Hilbert space $\mathcal{H}$ of $m \times 1$ vector valued functions that are defined on some subset Δ of $\mathbb{C}$ is said to be a RKHS (reproducing kernel Hilbert space) if there exists an $m \times m$ mvf $K_\omega(\lambda)$ on $\Delta \times \Delta$ such that for every choice of $\omega \in \Delta$, $u \in \mathbb{C}^m$ and $f \in \mathcal{H}$:

(1) $K_\omega u \in \mathcal{H}$ (as a function of λ).

(2) $\langle f, K_\omega u\rangle_{\mathcal{H}} = u^* f(\omega)$. (3.1)

The main facts are:

(1) The RK (reproducing kernel) is unique, i.e., if $K_\omega(\lambda)$ and $L_\omega(\lambda)$ are both RK's for the same RKHS, then $K_\omega(\lambda) = L_\omega(\lambda)$ for every choice of ω and λ in Δ.

(2) $K_\alpha(\beta)^* = K_\beta(\alpha)$. (3.2)

(3) $\sum_{i,j=1}^n u_j^* K_i(\omega_j) u_i \geq 0$ (3.3)

for every choice of $\omega_1, \ldots, \omega_n$ in Δ and $u_1, \ldots, u_n$ in $\mathbb{C}^m$.

Example 3.1. The Hardy space $H_2^m(\Omega_+)$ is a RKHS with RK

$$K_\omega(\lambda) = I_m/\rho_\omega(\lambda) \tag{3.4}$$

for each of the classical choices of Ω_+, where $\rho_\omega(\lambda)$ and the corresponding inner product are specified in Table 1.1. Basically this is just Cauchy's theorem for $H_2(\Omega_+)$.

Example 3.2. Let

$$\mathcal{M} = \{F(\lambda)u : u \in \mathbb{C}^n\} \ , \tag{3.5}$$

where $F(\lambda)$ is an $m \times n$ mvf that is meromorphic in some open nonempty subset Δ of $\mathbb{C}$ and has n linearly independent columns $f_1(\lambda), \ldots, f_n(\lambda)$ (in the sense of Subsection 4.2) and let P be any $n \times n$ positive definite matrix (i.e., $P > 0$). Then the space $\mathcal{M}$ endowed with the inner product

$$\langle F(\lambda)u, F(\lambda)v\rangle_{\mathcal{M}} = v^* P u \tag{3.6}$$

for every choice of u and v in $\mathbb{C}^n$, is an n dimensional RKHS with RK

$$K_\omega(\lambda) = F(\lambda)P^{-1}F(\omega)^* \tag{3.7}$$

(at the points of analyticity). The verification is by direct computation

Formulas (3.6) and (3.7) In Example 3.2 remain valid if the matrix P is Hermitian and invertible rather than positive definite. In this case, the space $\mathcal{M}$ is a reproducing kernel Krein space (RKKS) with respect to the indefinite inner product (3.6). That is to say, the space $\mathcal{M}$ admits a direct sum decomposition

$$\mathcal{M} = \mathcal{M}_+ + \mathcal{M}_- \quad \text{with} \quad \mathcal{M}_+ \cap \mathcal{M}_- = \{0\}$$

such that:

(1) $\mathcal{M}_+$ is a Hilbert space.

(2) $\mathcal{M}_-$ is a Hilbert space with respect the negative of the indefinite inner product (3.6).

(3) $\mathcal{M}_+$ is orthogonal to $\mathcal{M}_-$ with respect to the indefinite inner product (3.6).

This is easily verified by setting

$$\mathcal{M}_\pm = \{F(\lambda)\Pi_\pm u : u \in \mathbb{C}^n\},$$

where $\Pi_\pm$ denotes the orthogonal projection of $\mathbb{C}^n$ onto the span of the eigenvectors of P corresponding to the eigenvalues that fall in the interval between 0 and $\pm\infty$.

For ease of future reference, we shall summarize this more general setting in the next example.

EXAMPLE 3.3. Let $\mathcal{M}$ be the space defined in Example 3.2 endowed with the now indefinite inner product (3.6) that is defined in terms of an invertible Hermitian matrix P. Then $\mathcal{M}$ is an n dimensional RKKS with RK given by formula (3.7).

4. Detour on R_α Invariance

A major role in this subject is played by the generalized backwards shift operator R_α that acts on matrix valued meromorphic functions by the rule

$$R_\alpha F(\lambda) = \frac{F(\lambda) - F(\alpha)}{\lambda - \alpha}$$

for every point α in the domain of analyticity of F. In this section we shall study finite dimensional spaces of vector valued functions which are invariant under the action of R_α for at least one appropriately chosen point $\alpha \in \mathbb{C}$. The contents are taken largely from Section 3 of [D3] and Section 4 of [D5] .

4.1. The main observation.

THEOREM 4.1. *Let $\mathcal{M}$ be an n dimensional vector space of $m \times 1$ vector valued functions which are meromorphic in some open nonempty set $\Delta \subset \mathbb{C}$ and suppose further that $\mathcal{M}$ is R_α invariant for some point $\alpha \in \Delta$ in the domain of analyticity of $\mathcal{M}$. Then $\mathcal{M}$ is spanned by the columns of a rational $m \times n$ matrix valued function of the form*

$$F(\lambda) = V\{M - \lambda N\}^{-1} , \tag{4.1}$$

where $V \in \mathbb{C}^{m\times n}$, $M, N \in \mathbb{C}^{n\times n}$,

$$MN = NM \text{ and } M - \alpha N = I_n . \tag{4.2}$$

Moreover, $\lambda \in \Delta$ is a point of analyticity of F if and only if the $n \times n$ matrix $M - \lambda N$ is invertible.

PROOF. Let $f_1, \ldots, f_n$ be a basis for $\mathcal{M}$ and let

$$F(\lambda) = [f_1(\lambda) \cdots f_n(\lambda)]$$

be the $m \times n$ matrix valued function with columns $f_1(\lambda), \ldots, f_n(\lambda)$. Then, because of the presumed R_α invariance of the columns of F,

$$R_\alpha F(\lambda) = \frac{F(\lambda) - F(\alpha)}{\lambda - \alpha} = F(\lambda)E_\alpha$$

for some $n \times n$ matrix E_α which is independent of λ. Thus

$$F(\lambda)\{I_n - (\lambda - \alpha)E_\alpha\} = F(\alpha) ,$$

and hence, since $\det\{I_n - (\lambda - \alpha)E_\alpha\} \not\equiv 0$,

$$F(\lambda) = F(\alpha)\{I_n + \alpha E_\alpha - \lambda E_\alpha\}^{-1} ,$$

which is of the form (4.1) with $V = F(\alpha)$, $M = I_n + \alpha E_\alpha$ and $N = E_\alpha$.

Suppose next that F is analytic at a point $\omega \in \Delta$ and that $u \in \ker(M - \omega N)$. Then

$$F(\lambda)(M - \lambda N)u = Vu = 0 ,$$

first for $\lambda = \omega$, and then for every $\lambda \in \Delta$ in the domain of analyticity of F. Thus, for all such λ,

$$(\omega - \lambda)F(\lambda)Nu = F(\lambda)\{M - \lambda N - (M - \omega N)\}u = 0 .$$

Therefore, since the columns of $F(\lambda)$ are linearly independent functions of λ, $Nu = 0$. But this in conjunction with the prevailing assumption $(M - \omega N)u = 0$ implies that

$$u \in \ker M \cap \ker N \Longrightarrow u = 0 \Longrightarrow M - \omega N \text{ is invertible.}$$

Thus we have shown that if F is analytic at ω, then $M - \omega N$ is invertible. Since the opposite implication is easy, this serves to complete the proof. □

COROLLARY 4.2. *If* $\det(M - \lambda N) \not\equiv 0$ *and* $F(\lambda) = V(M - \lambda N)^{-1}$ *is a rational* $m \times n$ *matrix valued function with* n *linearly independent columns, then:*

(1) *M is invertible if and only if F is analytic at zero.*
(2) *N is invertible if and only if F is analytic at infinity and $F(\infty) = 0$.*

Moreover, in case (1) F can be expressed in the form

$$F(\lambda) = C(I_n - \lambda A)^{-1} , \tag{4.3}$$

whereas in case (2) F can be expressed in the form

$$F(\lambda) = C(A - \lambda I_n)^{-1} . \tag{4.4}$$

PROOF. The first assertion is contained in the theorem, the second is obtained in much the same way. More precisely, if $\lim_{\lambda \to \infty} F(\lambda) = 0$ and $u \in \ker N$, then

$$F(\lambda)Mu = F(\lambda)(M - \lambda N)u = Vu .$$

But now, upon letting $\lambda \to \infty$ it follows that

$$Vu = 0 \Longrightarrow F(\lambda)Mu = 0 \Longrightarrow Mu = 0 \Longrightarrow u \in \ker M \cap \ker N \Longrightarrow u = 0 .$$

Thus N is invertible. The other direction is easy, as are formulas (4.3) and (4.4). Just take $C = VM^{-1}$ and $A = NM^{-1}$ in the first case, and $C = VN^{-1}$ and $A = MN^{-1}$ in the second. □

COROLLARY 4.3. *Let f be an $m \times 1$ vector valued function which is meromorphic in some open nonempty set $\Delta \subset \mathbb{C}$ and let $\alpha \in \Delta$ be a point of analyticity of f. Then f is an eigenfunction of R_α if and only if it can be expressed in the form*

$$f(\lambda) = \frac{v}{\rho_\omega(\lambda)}$$

for one or more choices of $\rho_\omega(\lambda)$ in Table 1.1 with $\rho_\omega(\alpha) \neq 0$ and some nonzero constant vector $v \in \mathbb{C}^m$.

4.2. Linear independence. It seems worthwhile to emphasize that herein the n columns of an $m \times n$ mvf $F(\lambda)$ are said to be linearly independent if they are linearly independent in the vector space of continuous $m \times 1$ vector valued functions on the domain of analyticity of F. Thus, for example, if $F(\lambda)$ is meromorphic and $F(\lambda)u = 0$ for some $u \in \mathbb{C}^n$ and all points λ in the domain of analyticity of F, then $u = 0$. If

$$F(\lambda) = C(I_n - \lambda A)^{-1} \text{ or } F(\lambda) = C(A - \lambda I_n)^{-1} ,$$

this is easily seen to be equivalent to the statement that

$$\bigcap_{j=0}^{n-1} \ker CA^j = 0 ,$$

i.e., that the pair (C, A) is observable. Such a realization for F is minimal in the sense of Kalman because (in the usual terminology, see e.g., Zhou, Doyle and Glover [ZDG]) the pair (A, B) is automatically controllable:

$$\bigcap_{j=0}^{n-1} \ker B^* A^{*j} = \{0\} \text{ (equivalently, } \operatorname{rank}[B\ AB \cdots A^{n-1}B] = n) ,$$

since $B = I_n$.

5. A Special Class of Reproducing Kernel Spaces

We shall be particularly interested in RKKS's of $m \times 1$ vector valued meromorphic functions in $\mathbb{C}$ with RK's of a special form, which will be described below in the statement of Theorem 5.1. The theorem is an elaboration of a fundamental result that is due to de Branges [dB]. It is formulated in terms of the polynomials $a(\lambda)$ and $b(\lambda)$ which are given in Table 1.1 in order to obtain a statement which is applicable to each of the three classical choices of Ω_+.

A set Δ is said to be symmetric with respect to Ω_0 (or $\rho_\omega(\lambda)$) if for every $\lambda \in \Delta$ (except 0 for $\Omega_0 = \mathbb{T}$) the point $\lambda^o \in \Delta$; note that $\rho_\omega(\omega^o) = 0$.

THEOREM 5.1. *Let $\mathcal{K}$ be a RKKS of $m \times 1$ vector valued functions that are analytic in an open subset Δ of $\mathbb{C}$ that is symmetric with respect to Ω_0 and assume that $\Delta \cap \Omega_0 \neq \emptyset$. Then the reproducing kernel $K_\omega(\lambda)$ of $\mathcal{K}$ can be expressed in the form*

$$K_\omega(\lambda) = \frac{J - \Theta(\lambda) J \Theta(\omega)^*}{\rho_\omega(\lambda)} , \tag{5.1}$$

for some choice of $m \times m$ matrix valued function $\Theta(\lambda)$ which is analytic in Δ and some signature matrix J, if and only if the following two conditions hold:

(1) *$\mathcal{K}$ is R_α invariant for every $\alpha \in \Delta$.*

(2) *The structural identity*

$$\langle R_\alpha(bf), R_\beta(bg)\rangle_{\mathcal{K}} - \langle R_\alpha(af), R_\beta(ag)\rangle_{\mathcal{K}} = |ab' - ba'|^2 g(\beta)^* J f(\alpha) \tag{5.2}$$

holds for every choice of α, β in Δ and f, g in $\mathcal{K}$.

Moreover, in this case, the function $\Theta(\lambda)$ which appears in (5.1) is unique up to a J unitary constant factor on the right; it can be taken equal to

$$\Theta(\lambda) = I_m - \rho_\mu(\lambda) K_\mu(\lambda) J \tag{5.3}$$

for any point $\mu \in \Delta \cap \Omega_0$.

This formulation is adapted from [AD4]; see especially Theorems 4.1, 4.3 and 4.4. The restriction to the three choices of $a(\lambda)$ and $b(\lambda)$ specified earlier, permits some simplification in the presentation, because the terms $r(a,b;\alpha)f$ and $r(b,a;\alpha)f$ which intervene there are constant multiples of $R_\alpha(af)$ and $R_\alpha(bf)$, respectively.

The restriction $\Delta\cap\Omega_0=\emptyset$ can be relaxed at the expense of a more sophisticated formulation. However, since we shall be dealing with finite dimensional spaces and rational functions, there is no need for this extra complication. The interested reader can refer to [AD4] for more information.

For the three cases of interest, the structural identity (5.2) can be reexpressed as:

$$\langle (I+\alpha R_\alpha)f,(I+\beta R_\beta)g\rangle_{\mathcal{K}}-\langle R_\alpha f,R_\beta g\rangle_{\mathcal{K}}=g(\beta)^*Jf(\alpha) \tag{5.4}$$

if $\Omega_+=\mathbb{D}$,

$$\langle R_\alpha f,g\rangle_{\mathcal{K}}-\langle f,R_\beta g\rangle_{\mathcal{K}}-(\alpha-\beta^*)\langle R_\alpha f,R_\beta g\rangle_{\mathcal{K}}=2\pi i\; g(\beta)^*Jf(\alpha) \tag{5.5}$$

if $\Omega_+=\mathbb{C}_+$, and

$$\langle R_\alpha f,g\rangle_{\mathcal{K}}+\langle f,R_\beta g\rangle_{\mathcal{K}}+(\alpha+\beta^*)\langle R_\alpha f,R_\beta g\rangle_{\mathcal{K}}=-2\pi\; g(\beta)^*Jf(\alpha) \tag{5.6}$$

if $\Omega_+=\Pi_+$.

Formula (5.5) appears in de Branges [dB]; formula (5.4) is equivalent to a formula which appears in Ball [Ba], who adapted de Branges' work to the disc, including an important technical improvement due to Rovnyak [Rov]. All three of these references deal with the Hilbert space case only.

From time to time we shall refer to a RKKS with a RK of the form (5.1) as a dBK space $\mathcal{K}(\Theta)$ and to a RKHS with a RK of this form as a de Branges space $\mathcal{H}(\Theta)$.

6. An Important Conclusion

The role of the two conditions in Theorem 5.1 becomes particularly transparent when $\mathcal{K}$ is finite dimensional. Indeed, if the n dimensional space $\mathcal{M}$ considered in Example 3.3 is R_α invariant for some point α in the domain of analyticity of $F(\lambda)$, then, by Theorem 4.1, $F(\lambda)$ can be expressed in the form

$$F(\lambda)=V(M-\lambda N)^{-1} \tag{6.1}$$

with M and N satisfying (4.2). Thus R_α invariance forces the elements of $\mathcal{M}$ to be rational of the indicated form. Since

$$(R_\beta F)(\lambda)=F(\lambda)N(M-\beta N)^{-1}$$

for every point β at which the matrix $M-\beta N$ is invertible, i.e., for every $\beta\in\mathcal{A}_F$, the domain of analyticity of F, it is readily checked that

$$\begin{aligned}\langle R_\alpha Fu,Fv\rangle_{\mathcal{M}}&=\langle FN(M-\alpha N)^{-1}u,Fv\rangle_{\mathcal{M}}\\&=v^*PN(M-\alpha N)^{-1}u\ ,\end{aligned} \tag{6.2}$$

and similarly that

$$\langle Fu,R_\beta Fv\rangle_{\mathcal{M}}=v^*(M^*-\beta^*N^*)^{-1}N^*Pu\ , \tag{6.3}$$

and

$$\langle R_\alpha Fu,R_\beta v\rangle_{\mathcal{M}}=v^*(M^*-\beta^*N^*)^{-1}N^*PN(M-\alpha N)^{-1}u \tag{6.4}$$

for every choice of α, β in $\mathcal{A}_F$ and u, v in $\mathbb{C}^n$. For each of the three special choices of Ω_+ under consideration, it is now readily checked that the structural identity (5.2) reduces to a matrix equation for P by working out (5.4)–(5.6) with the aid of (6.2)–(6.4). In other words:

In a finite dimensional R_α invariant RKKS $\mathcal{M}$ *with Gram matrix P, the structural identity (5.2) is equivalent to a Lyapunov-Stein equation for P.*

This last conclusion seems to have been first established explicitly in [D2] by a considerably lengthier calculation. The present more appealing argument is adapted from [D3] and [D5].

If F is analytic at zero, then we may presume that $M = I_n$ in (6.1) and take $\alpha = \beta = 0$ in the structural identity (5.2).

THEOREM 6.1. *Let $\mathcal{M}$ denote the finite dimensional RKHS that was introduced in Example 3.3 and let $F(\lambda)$ be given by (6.1). Then the RK of $\mathcal{M}$ can be expressed in the form*

$$K_\omega(\lambda) = \frac{J - \Theta(\lambda) J \Theta(\omega)^*}{\rho_\omega(\lambda)} \tag{6.5}$$

with $\rho_\omega(\lambda)$ as in Table 1.1 if and only if P is a solution of the equation

$$M^*PM - N^*PN = V^*JV \text{ when } \Omega_+ = \mathbb{D} \,, \tag{6.6}$$

$$M^*PN - N^*PM = 2\pi i V^*JV \text{ when } \Omega_+ = \mathbb{C}_+ \,, \tag{6.7}$$

$$M^*PN + N^*PM = -2\pi V^*JV \text{ when } \Omega_+ = \Pi_+ \,. \tag{6.8}$$

Moreover, in each of these cases $\Theta(\lambda)$ is uniquely specified up to a J unitary constant multiplier on the right by the formula

$$\Theta(\lambda) = I_m - \rho_\mu(\lambda) F(\lambda) P^{-1} F(\mu)^* J \tag{6.9}$$

for any choice of the point $\mu \in \Omega_0 \cap \mathcal{A}_F$.

It is well to note that formula (6.9) is a realization formula for $\Theta(\lambda)$, and that in the usual notation of (4.3) and (4.4) it depends only upon A, C and P. It can be reexpressed in one of the standard A, B, C, D forms by elementary manipulations. A very general class of realization formulas of the form (6.9) and extensions thereof may be found in [AD6].

7. Factorization and Recursive Methods

In this section we shall exhibit the connection in the setting of Theorem 6.1 between:

(1) R_α-invariant subspaces of the finite dimensional RKKS $\mathcal{M}$.
(2) Subblocks of the invertible structured Hermitian matrix P that serves to define its indefinite inner product via formula (3.6).
(3) Factors of the mvf $\Theta(\lambda)$.

THEOREM 7.1. *Let*

$$P = \begin{bmatrix} P_{11} & P_{12} \\ P_{21} & P_{22} \end{bmatrix}, \quad M = \begin{bmatrix} M_{11} & M_{12} \\ M_{21} & M_{22} \end{bmatrix}$$

and

$$N = \begin{bmatrix} N_{11} & N_{12} \\ N_{21} & N_{22} \end{bmatrix}$$

be conformable block decompositions, where the upper left hand block in each of these three matrices is $k \times k$, and suppose that P_{11} is invertible and that

$$M_{21} = N_{21} = 0 \ . \tag{7.1}$$

Let

$$\Pi = \begin{bmatrix} -P_{11}^{-1} P_{12} \\ I_{n-k} \end{bmatrix} \tag{7.2}$$

and let

$$Q = \Pi^* P \Pi = P_{22} - P_{21} P_{11}^{-1} P_{12} \tag{7.3}$$

be the Schur complement of P_{22} with respect to P. Then for any point $\mu \in \Omega_0$ at which $G(\mu) = M - \mu N$ is invertible, the mvf

$$\Theta(\lambda) = I_m - \rho_\mu(\lambda) F(\lambda) P^{-1} F(\mu)^* J \tag{7.4}$$

admits a factorization of the form

$$\Theta(\lambda) = \Theta_1(\lambda)\Theta_2(\lambda) \ , \tag{7.5}$$

where

$$\Theta_1(\lambda) = I_m - \rho_\mu(\lambda) F(\lambda) \begin{bmatrix} I_k \\ 0 \end{bmatrix} P_{11}^{-1} \begin{bmatrix} I_k \\ 0 \end{bmatrix}^* F(\mu)^* J \ , \tag{7.6}$$

$$\Theta_2(\lambda) = I_m - \rho_\mu(\lambda) W (M_{22} - \lambda N_{22})^{-1} Q^{-1} (M_{22}^* - \overline{\mu} N_{22}^*)^{-1} W^* J \tag{7.7}$$

and

$$W = F(\mu)\Pi(M_{22} - \mu N_{22}) \ . \tag{7.8}$$

Moreover,

$$M_{22}^* Q M_{22} - N_{22}^* Q N_{22} = W^* J W \qquad if \quad \Omega_+ = \mathbb{D} \ , \tag{7.9}$$

$$M_{22}^* Q N_{22} - N_{22}^* Q M_{22} = 2\pi i W^* J W \quad if \quad \Omega_+ = \mathbb{C}_+ \tag{7.10}$$

and

$$M_{22}^* Q N_{22} + N_{22}^* Q M_{22} = -2\pi W^* J W \qquad if \quad \Omega_+ = \Pi_+ \ . \tag{7.11}$$

PROOF. The assumptions guarantee that the subspace

$$\mathcal{M}_1 = \left\{ F(\lambda) \begin{bmatrix} I_k \\ 0 \end{bmatrix} u : u \in \mathbb{C}^k \right\}$$

of $\mathcal{M}$ that is spanned by the first k columns of $F(\lambda)$ is R_α-invariant. Therefore, by Theorem 6.1, its RK,

$$F(\lambda) \begin{bmatrix} I_k \\ 0 \end{bmatrix} P_{11}^{-1} \begin{bmatrix} I_k \\ 0 \end{bmatrix}^* F(\omega)^*$$

can be expressed in terms of the mvf $\Theta_1(\lambda)$.

Next, it is readily checked that $\mathcal{M}$ admits an orthogonal direct sum decomposition

$$\mathcal{M} = \mathcal{M}_1 \dot{+} \mathcal{M}_2 \ ,$$

where

$$\mathcal{M}_2 = \{F(\lambda)\Pi u : u \in \mathbb{C}^{n-k}\}$$

and

$$\langle F\Pi u, F\Pi v\rangle_{\mathcal{M}} = v^*\Pi^* P\Pi u = v^* Q u$$

for every choice of $u, v \in \mathbb{C}^{n-k}$. Thus $\mathcal{M}_2$ is also a RKKS, but in general, it is not R_α invariant. However, it turns out that the mvf

$$\Theta_1(\lambda)^{-1} F(\lambda)\Pi(M_{22} - \lambda N_{22})$$

is independent of λ. Thus, as $\Theta_1(\mu) = I_m$,

$$\widehat{F}(\lambda) = \Theta_1(\lambda)^{-1} F(\lambda)\Pi = W(M_{22} - \lambda N_{22})^{-1} . \tag{7.12}$$

This important fact is established in the proof of formula (4.23) in [AD5]. It serves to guarantee that the space

$$\widehat{\mathcal{M}}_2 = \{\widehat{F}(\lambda)u : u \in \mathbb{C}^{n-k}\}$$

endowed with the indefinite inner product

$$\langle \widehat{F}u, \widehat{F}v\rangle_{\widehat{\mathcal{M}}_2} = v^* Q u$$

is an R_α-invariant RKKS that satisfies the structural identity (5.2), since Q is a solution of one of the equations (7.9)-(7.11), depending on the choice of Ω_+. Therefore $\widehat{\mathcal{M}}_2$ has a reproducing kernel $\widehat{F}(\lambda)Q^{-1}\widehat{F}(\omega)^*$ of the form

$$\widehat{F}(\lambda)Q^{-1}\widehat{F}(\omega)^* = \frac{J - \Theta_2(\lambda)J\Theta_2(\omega)^*}{\rho_\omega(\lambda)} ,$$

where $\Theta_2(\lambda)$ is uniquely specified by formula (7.7) up to a J-unitary constant factor on the right. The factorization (7.5) emerges on theoretical grounds upon checking that both

$$\frac{J - \Theta(\lambda)J\Theta(\omega)^*}{\rho_\omega(\lambda)}$$

and

$$\frac{J - \Theta_1(\lambda)J(\Theta)_1(\omega)^*}{\rho_\omega(\lambda)} + \Theta_1(\lambda)\left\{\frac{J - \Theta_2(\lambda)J\Theta_2(\omega)^*}{\rho_\omega(\lambda)}\right\}\Theta_1(\omega)^*$$

are reproducing kernels for $\mathcal{M}$, or upon just multiplying out. □

8. More on Factorization

If M and N are upper triangular and P is positive definite, then the procedure described in Section 7 can be iterated n times in steps of size one. This is one way to obtain the factorization

$$\Theta(\lambda) = \Theta_1(\lambda)\cdots\Theta_n(\lambda) ,$$

where each term $\Theta_i(\lambda)$ is a J-inner mvf of McMillan degree one, i.e., a so called elementary section or Blaschke-Potapov factor. If P is just Hermitian and invertible, then one can no longer guarantee that all the subblocks

$$P_{[k]} = \begin{bmatrix} p_{11} & \cdots & p_{1k} \\ \vdots & & \vdots \\ p_{k1} & \cdots & p_{kk} \end{bmatrix} , \quad k = 1, \ldots, n ,$$

of P are invertible. However, since $p_{ij} \neq 0$ for at least one choice of i and j, one can permute the indices to form a new Hermitian matrix P' such that at least one of the matrices $(P')_{[1]}$, $(P')_{[2]}$ is invertible. Then, since the Schur complement of the invertible block with respect to P' is also an invertible Hermitian matrix, this procedure may be reiterated. Ultimately one obtains a new matrix P'' and a sequence of positive integers $i_1 < i_2 < \cdots < i_k$ with $|i_{j+1} - i_j| \leq 2$, for $j = 1, \ldots, k-1$, and $i_k = n$ such that each of the subblocks $(P'')_{[i_j]}$ is invertible. This amounts to reindexing the basis $f_1(\lambda), \ldots, f_n(\lambda)$ of the space $\mathcal{M}$. However, if $\tilde{f}_1(\lambda), \ldots, \tilde{f}_n(\lambda)$ denote the basis corresponding to the shuffled indices, there is no longer any guarantee that the spaces spanned by $\{\tilde{f}_1, \ldots, \tilde{f}_j\}$ (that correspond to $(P'')_{[j]}$) are R_α invariant. In fact there are $\mathcal{K}(\Theta)$ spaces which have no nondegenerate R_α invariant subspaces and hence the corresponding $\Theta(\lambda)$ cannot be factored. Three simple examples (that are adapted from the thesis [A]) are furnished on pages 152-154 of [AD2]. However, if each of the original basis elements $f_1(\lambda), \ldots, f_n(\lambda)$ of the space $\mathcal{M}$ is an eigenfunction of the operator R_α, then the spaces spanned by any finite linear combination of them is R_α invariant. Thus in this case (i.e., when M and N are diagonal matrices) we can obtain a factorization of the form

$$\Theta(\lambda) = \Theta_1(\lambda) \cdots \Theta_k(\lambda)$$

with factors $\Theta_j(\lambda)$ that are of degree at most two. The "elementary factors" of degree 2 can be expressed in the form

$$(I_m + \{b_{\beta,\alpha}(\lambda) - 1\}W_{12})(I_m + \{b_{\alpha,\beta}(\lambda) - 1\}W_{21}) ,$$

where

$$b_{\alpha,\beta}(\lambda) = \begin{cases} (\lambda - \alpha)/(1 - \lambda\overline{\beta}) & \text{if} \quad \Omega_+ = \mathbb{D} \\ (\lambda - \alpha)/(\lambda - \overline{\beta}) & \text{if} \quad \Omega_+ = \mathbb{C}_+ \end{cases}$$

and

$$W_{ij} = u_i(u_j^* J u_i)^{-1} u_j^* J .$$

See Section 7 of [AD2] for additional information. The same reference discusses recursive methods; see also Section 8 of [D3] and the literature cited in both.

9. The Nevanlinna-Pick Problem

The de Branges spaces $\mathcal{H}(\Theta)$ are a useful tool in the resolution of a number of matrix valued interpolation problems. The bitangential Nevanlinna-Pick problem is a good starting point. It is both a significant problem in its own right and a good model for the general case that will be discussed in the next section. We shall formulate it for the Schur class

$$\mathcal{S}^{p\times q}(\Omega_+) = \{p \times q \text{ mvf's } S(\lambda) \text{ that are analytic and contractive in } \Omega_+\} . \quad (9.1)$$

There are a number of different ways of defining the term contractive for mvf's. Perhaps the simplest is that the inequality

$$\|S(\lambda)\eta\| \leq \|\eta\|$$

hold for the Euclidian lengths of the indicated vectors for every point $\lambda \in \Omega_+$ and every vector $\eta \in \mathbb{C}^q$. For a useful maximum principle and other properties, see e.g., Section 0.3 of [D1].

The data for the NP problem (in the class $\mathcal{S}^{p\times q}(\Omega_+)$) is a set of points $\omega_1, \ldots, \omega_n \in \Omega_+$, two sets of vectors $\xi_1, \ldots, \xi_n \in \mathbb{C}^p$ and $\eta_1, \ldots, \eta_n \in \mathbb{C}^q$ and an integer l between 0 and n. The problem is to:

(1) Formulate conditions on the data that insure the existence of at least one mvf $S \in \mathcal{S}^{p\times q}(\Omega_+)$ such that

$$\xi_j^* S(\omega_j) = \eta_j^* \quad \text{for} \quad j = 1, \ldots, \ell$$

and (9.2)

$$S(\omega_j)\eta_j = \xi_j \quad \text{for} \quad j = \ell+1, \ldots, n \ .$$

(2) Obtain a description of the set of all solutions when the conditions for existence are met.

There are a number of ways to solve problems of this sort, see e.g., [Ar], [BH], [BGR], [DFK], [FF], [IS], [KKY] and [D4]. However, the strategy that we shall adopt here seems to fit particularly well with the notion of structured matrices. We shall recast the problem by introducing the orthogonal projections $\mathfrak{p}$ of $L_2^k(\Omega_0)$ onto $H_2^k(\Omega_+)$ (viewed as a subspace of $L_2^k(\Omega_0)$) and

$$\mathfrak{q}' = I - \mathfrak{p}$$

and taking advantage of the evaluations

$$\mathfrak{p}S^* \frac{\xi}{\rho_\omega} = S(\omega)^* \frac{\xi}{\rho_\omega} \quad \text{and} \quad \mathfrak{q}' S \frac{\eta}{\delta_\omega} = \frac{S(\omega)}{\delta_\omega}\eta \tag{9.3}$$

that are valid for every choice of $S \in \mathcal{S}^{p\times q}(\Omega_+)$, $\xi \in \mathbb{C}^p$, $\eta \times \mathbb{C}^q$ and $\omega \in \Omega_+$. (In order to keep the notation simple, the "height" of the vvf's is not indicated explicitly in the symbols for the projections.) Then, upon setting

$$g_j = \frac{\xi_j}{\rho_{\omega_j}} \quad \text{and} \quad h_j = \frac{\eta_j}{\rho_{\omega_j}} \quad \text{for} \quad j = 1, \ldots, \ell$$

and

$$g_j = \frac{\xi_j}{\delta_{\omega_j}} \quad \text{and} \quad h_j = \frac{\eta_j}{\delta_{\omega_j}} \quad \text{for} \quad j = \ell+1, \ldots, n \ ,$$

it is readily seen that the stated conditions (9.2) are equivalent to requiring that

$$\mathfrak{p}S^* g_j = h_j \quad \text{for} \quad j = 1, \ldots, \ell$$

and (9.4)

$$\mathfrak{q}' S h_j = g_j \quad \text{for} \quad j = \ell+1, \ldots, n \ .$$

But this in turn is the same as requiring that

$$g_j - Sh_j \in H_2^p(\Omega_+) \quad \text{for} \quad j = \ell+1, \ldots, n$$

and (9.5)

$$-S^* g_j + h_j \in H_2^q(\Omega_+)^\perp \quad \text{for} \quad j = 1, \ldots, \ell \ .$$

In fact, in each of the preceding two constraints, the imposed condition is automatically fulfilled for the complementary set of integers between 1 and n. Thus, we can let j run from 1 to n in both cases and hence, upon introducing the notation

$$f_j(\lambda) = \begin{bmatrix} g_j(\lambda) \\ h_j(\lambda) \end{bmatrix}, \qquad F(\lambda) = [f_1(\lambda) \cdots f_n(\lambda)] \quad \text{and}$$

$$\Delta_S = \begin{bmatrix} I_p & -S \\ -S^* & I_q \end{bmatrix}, \tag{9.6}$$

we can reexpress the requirement that S be a solution of the Nevanlinna-Pick problem as

$$\Delta_S F u \in H_2^p(\Omega_+) \oplus H_2^q(\Omega_+)^\perp \tag{9.7}$$

for every $u \in \mathbb{C}^n$.

Notice that for the given data ,

$$F(\lambda) = V(M - \lambda N)^{-1} , \tag{9.8}$$

where e.g., for $\Omega_+ = \mathbb{D}$,

$$V = \begin{bmatrix} \xi_1 & \cdots & \xi_n \\ \eta_1 & \cdots & \eta_n \end{bmatrix} , \quad M - \lambda N = \begin{bmatrix} I_\ell - \lambda A_1 & 0 \\ 0 & \lambda I_r - A_2 \end{bmatrix} ,$$

$$A_1 = \operatorname{diag}\{\overline{\omega}_1, \ldots, \overline{\omega}_\ell\} \text{ and } A_2 = \operatorname{diag}\{\omega_{\ell+1}, \ldots, \omega_n\} .$$

10. A Hierarchy of Interpolation Problems

The Nevanlinna-Pick problem of Section 9 is a special case of a more general interpolation problem that we shall refer as the Basic Interpolation Problem (BIP). For ease of exposition it is convenient to formulate this problem in terms of three matrices: $V \in \mathbb{C}^{m\times n}$, $M, N \in \mathbb{C}^{m\times m}$ and the mvf's

$$G(\lambda) = M - \lambda N \quad \text{and} \quad F(\lambda) = VG(\lambda)^{-1}.$$

More general starting points are discussed in [D3] and [D5]. As usual, it is assumed that $G(\lambda)$ is invertible except for at most n values of λ. Let

$$\mathcal{S}_B(V, M, N) = \{S \in \mathcal{S}^{p\times q}(\Omega_+) : \Delta_S F u \in H_2^p(\Omega_+) \oplus H_2^q(\Omega_+)^\perp \text{ for every } u \in \mathbb{C}^n\} .$$

Then the BIP is to:

(1) Find conditions on V, M and N which insure that $\mathcal{S}_B(V, M, N)$ is not empty.

(2) Furnish a description of the set $\mathcal{S}_B(V, M, N)$ when the conditions in (1) are met.

Thus, we see that the BIP is exactly the same as the Nevanlinna-Pick problem except that the data $F(\lambda)$ is more general. There is in fact a hierarchy of interpolation problems in this framework that can be listed in increasing order of difficulty as follows:

(1) The mvf $G(\lambda)$ is invertible on Ω_0.
   (a) $\ell = 0$ or $\ell = n$ (that is a one sided problem).
   (b) The Lyapunov-Stein equation for the region Ω_+ with boundary equal to the given Ω_0 (which is taken from the list (6.6) - (6.8)) has exactly one positive semidefinite solution.
   (c) The Lyapunov-Stein equation referred to in (1b) has more than one positive semidefinite solution.

(2) The mvf $G(\lambda)$ is not invertible for at least one point $\lambda \in \Omega_0$.

It is not hard to establish necessary conditions for the BIP to be solvable when $G(\lambda)$ is invertible on Ω_0.

THEOREM 10.1. *Let $G(\lambda)$ be invertible on Ω_0. Then for every $S \in \mathcal{S}_B(V, M, N)$, the matrix P_S that is determined by the rule*

$$v^* P_S u = \langle \Delta_S F u, F v \rangle \tag{10.1}$$

for $u, v \in \mathbb{C}^n$ *and the inner product that is given in Table 1.1, is a positive semidefinite solution of the Lyapunov-Stein equation for the region* Ω_+ *corresponding to* Ω_0 *(see (6.6)-(6.8)).*

A proof of Theorem 10.1 is furnished in Section 7 of [D3] and Section 5 of [D5].

Theorem 10.1 admits a converse: If $G(\lambda)$ is invertible on Ω_0 and the corresponding Lyapunov-Stein equation admits a positive semidefinite solution P, then the BIP is solvable, i.e.,

$$\mathcal{S}_B(V, M, N) \neq \emptyset \ .$$

However, this is only part of the story because this equation may have many positive semidefinite solutions and this extra freedom can be used to impose more interpolation conditions. This leads to more refined interpolation problems which we have referred to elsewhere as the aBIP (augmented BIP) and the $\widehat{\text{aBIP}}$. To describe the salient features of these problems with a minimum of fuss and bother, it is convenient to introduce two subsets of the set of solutions of the BIP for given V, M and N that are defined in terms of a supplementary positive semidefinite $n \times n$ matrix P:

$$\mathcal{S}(V, M, N; P) = \{S \in \mathcal{S}_B(V, M, N) : \langle \Delta_S F u, F u \rangle = u^* P u \text{ for every } u \in \mathbb{C}^n\} \ .$$

$$\widehat{\mathcal{S}}(V, M, N; P) = \{S \in \mathcal{S}_B(V, M, N) : \langle \Delta_S F u, F u \rangle \leq u^* P u \text{ for every } u \in \mathbb{C}^n\} \ .$$

Clearly,

$$\mathcal{S}(V, M, N; P) \subseteq \widehat{\mathcal{S}}(V, M, N; P) \subseteq \mathcal{S}_B(V, M, N) \ . \tag{10.2}$$

THEOREM 10.2. *If P is a positive semidefinite solution of the Lyapunov-Stein equation for a given Ω_+ and* $\det\{G(\lambda)\} \not\equiv 0$, *then*

$$\widehat{\mathcal{S}}(V, M, N; P) \neq \emptyset \ . \tag{10.3}$$

If it is also assumed that $G(\lambda)$ is invertible on Ω_0, then

$$\mathcal{S}(V, M, N; P) = \widehat{\mathcal{S}}(V, M, N; P) \ . \tag{10.4}$$

A proof of statement (10.4) is furnished in [D5]. Connections with the abstract approach of [KKY], which leads to a proof of (10.3) and a representation formula for the solutions is also discussed there. A direct proof of (10.3) and a representation formula for the case $\Omega_+ = \mathbb{D}$ are given in [BoD1]. Examples for which (10.3) fails to hold may be found in [Sa] and [BoD2] . The next theorem , which is taken from [BoD1] , is cited to give some indication of the flavor.

THEOREM 10.3. *Let $\Omega_+ = \mathbb{D}$, let P be a positive semidefinite solution of the Lyapunov-Stein equation (6.6), let*

$$V_1 = [I_p \quad O_{p\times q}]V \ ,$$

$$k = \text{rank}\{N^*PN + V_1^*V_1\} - \text{rank } P$$

and assume that $G(\lambda)$ is invertible on $\mathbb{T}$. Then:

(1) $k \leq \min(p, q)$.
(2) *There exists a rational J-inner mvf*

$$\Theta(\lambda) = \begin{bmatrix} \Theta_{11}(\lambda) & \Theta_{12}(\lambda) \\ \Theta_{21}(\lambda) & \Theta_{22}(\lambda) \end{bmatrix}$$

of McMillan degree equal to the rank of P with diagonal blocks of size $p \times p$ and $q \times q$, respectively, such that

$$\widehat{\mathcal{S}}(V, M, N; P) = \{(\Theta_{11}\mathcal{E} + \Theta_{12})(\Theta_{21}\mathcal{E} + \Theta_{22})^{-1}\} ,$$

where

$$\mathcal{E} = U_1 \begin{bmatrix} \varphi & 0 \\ 0 & I_k \end{bmatrix} U_2 ,$$

U_1 and U_2 are constant unitary matrices that depend upon the data and $\varphi \in \mathcal{S}^{(p-k)\times(q-k)}(\mathbb{D})$.

11. Some More Structured Matrices

The off diagonal block in the Pick matrix associated with two sided tangential interpolation problem is typically a non Hermitian structured matrix. In this section we shall furnish a number of examples. For the sake of definiteness, we shall fix $\Omega_+ = \mathbb{D}$ and take

$$M - \lambda N = \begin{bmatrix} I_\ell - \lambda A_1 & 0 \\ 0 & \lambda I_r - A_2 \end{bmatrix}$$

with

$$\sigma(A_1) \subset \mathbb{D} \quad \text{and} \quad \sigma(A_2) \subset \mathbb{D}$$

for the first three cases. Then, upon invoking the block decompositions,

$$P = \begin{bmatrix} P_{11} & P_{12} \\ P_{21} & P_{22} \end{bmatrix} \quad \text{and} \quad V = \begin{bmatrix} V_{11} & V_{12} \\ V_{21} & V_{22} \end{bmatrix}$$

with $P_{11} \in \mathbb{C}^{\ell\times\ell}$, $P_{22} \in \mathbb{C}^{r\times r}$, $V_{11} \in \mathbb{C}^{p\times\ell}$ and $V_{22} \in \mathbb{C}^{q\times r}$, we can reexpress equation (6.6) as four separate equations, one for each block of P:

$$P_{11} - A_1^* P_{11} A_1 = V_{11}^* V_{11} - V_{21}^* V_{21} . \tag{11.1}$$

$$-P_{12}A_2 + A_1^* P_{12} = V_{11}^* V_{12} - V_{21}^* V_{22} . \tag{11.2}$$

$$A_2^* P_{22} A_2 - P_{22} = V_{12}^* V_{12} - V_{22}^* V_{22} . \tag{11.3}$$

We do not write the equation for P_{21} because it is the adjoint of the equation for P_{12}. Because of the assumptions on the spectrum of A_1 and A_2, the diagonal blocks P_{11} and P_{22} are uniquely determined by equations (11.1) and (11.3), respectively. However, the off diagonal block P_{12} is uniquely determined by equation (11.2) if and only if

$$\sigma(A_2) \cap \sigma(A_1^*) = \phi.$$

This last condition corresponds to the case (1b) in the hierarchical list of interpolation problems in the preceding section. Notice that even though we are interested in Hermitian solutions P of the full equation, the off diagonal block is a horse (actually a structured matrix) of a different color. The next several paragraphs will illustrate some of the possibilities.

Case 1. ($\Omega_+ = \mathbb{D}$) P_{12} is a Hankel matrix.

Choose

$$A_1 = \overline{\omega} I_\ell + T_\ell , \ A_2 = \omega I_r + T_r ,$$

$$\begin{bmatrix} V_{11} \\ V_{21} \end{bmatrix} = \begin{bmatrix} \xi_1 & 0 & \cdots & 0 \\ \eta_1 & \eta_2 & \cdots & \eta_\ell \end{bmatrix} \quad \text{and} \quad \begin{bmatrix} V_{12} \\ V_{22} \end{bmatrix} = \begin{bmatrix} \xi_{\ell+1} & \xi_{\ell+2} & \cdots & \xi_{\ell+r} \\ \eta_{\ell+1} & 0 & \cdots & 0 \end{bmatrix} ,$$

where $\omega \in \mathbb{D}$ and T_k denotes the $k\times k$ shift matrix with 1's on the first superdiagonal and 0's elsewhere. Then the entries p_{st} of the P_{12} block of P are subject to the recursion

$$p_{s,t-1} - p_{s-1,t} = v_s^* J v_t \tag{11.4}$$

for $s = 1, \dots, \ell$ and $t = \ell + 1, \dots, n$, where $v_s = \mathrm{col}[\xi_s, \eta_s]$ denotes the s'th column of V and it is understood that $p_{st} = 0$ if either $s = 0$ or $t = \ell$. Therefore, we must have

$$v_1^* J v_{\ell+1} = 0 \ . \tag{11.5}$$

Because of the special form of the columns of V we also have

$$v_s^* J v_t = 0$$

for $s = 2, \dots, \ell$ and $t = \ell + 2, \dots, n$. Thus, it is readily seen that

$$P_{12} = \begin{bmatrix} c_1 & c_2 & c_3 & \cdots & c_r \\ c_2 & c_3 & & & \\ c_3 & & & & \\ \vdots & & & & \vdots \\ c_\ell & & & & c_{\nu-1} \end{bmatrix} \qquad (\nu = \ell + r)$$

is an $\ell \times r$ Hankel matrix with

$$c_j = \begin{cases} -\xi_1^* \xi_{\ell+j+1} & \text{for} \quad j = 1, \dots, r-1 \\ -\eta_{j+1}^* \eta_{\ell+1} & \text{for} \quad j = 1, \dots, \ell - 1 \ . \end{cases}$$

These two recipes must agree on the overlap. This consistency condition together with (11.5) can be reexpressed as

$$v_1^* J v_{\ell+1+j} + v_{1+j}^* J v_{\ell+1} = 0 \quad \text{for} \quad j = 0, \dots, \gamma - 1 \ , \tag{11.6}$$

where $\gamma = \min(\ell, r)$. The remaining entries in the Hankel matrix c_j for $j = \max(\ell, r), \dots, \ell + r - 1$, are unspecified. This reflects the lack of uniqueness in the set of solutions to equation (11.2) for this choice of the data. For additional discussion of this example and its role in interpolation theory, see Section 10 of [D2].

Case 2. $(\Omega_+ = \mathbb{D})$ P_{12} is a Cauchy matrix.

Choose

$$A_1 = \mathrm{diag}(\overline{\alpha}_1, \dots, \overline{\alpha}_\ell) \quad \text{and} \quad A_2 = \mathrm{diag}(\overline{\beta}_{\ell+1}, \dots, \overline{\beta}_{\ell+r})$$

with $\sigma(A_1^*) \cap \sigma(A_2) = \phi$. Then the entries p_{st}, $s = 1, \dots, \ell$, $t = \ell + 1, \dots, n$, of P_{12} are subject to the recursion

$$-p_{st}\overline{\beta}_t + \alpha_s p_{st} = v_s^* J v_t \ ,$$

and hence, because of the spectral assumptions,

$$p_{st} = \frac{v_s^* J v_t}{\alpha_s - \overline{\beta}_t} \ ,$$

for $s = 1, \dots, \ell$ and $t = \ell + 1, \dots, n$. Thus, P_{12} is a Cauchy matrix, as claimed.

If $p = q = 1$ and we choose

$$v_s = \begin{bmatrix} \bar{\xi}_s \\ 1 \end{bmatrix} \quad \text{for} \quad s = 1, \ldots, \ell$$

and

$$v_t = \begin{bmatrix} 1 \\ \bar{\eta}_t \end{bmatrix} \quad \text{for} \quad t = \ell + 1, \ldots, n \ ,$$

then

$$p_{st} = \frac{\xi_s - \bar{\eta}_t}{\alpha_s - \bar{\beta}_t} \ ,$$

that is to say P_{12} is a Loewner matrix.

Case 3. $(\Omega_+ = \mathbb{D})$ P_{12} is a Vandermonde matrix.

Fix $\ell = r$ and choose

$$A_1 = \operatorname{diag}(\overline{\alpha}_1, \ldots, \overline{\alpha}_\ell) \ , \ A_2 = T_\ell$$

(the $(\ell \times \ell)$ shift matrix) and assume that $\sigma(A_1^*) \cap \sigma(A_2) = \phi$. This forces the α_j to be different from zero. Next choose the column vectors v_s in V in such a way that

$$v_s^* J v_t = \begin{cases} \alpha_s & \text{for} \quad s = 1, \ldots, \ell \quad \text{and} \quad t = \ell + 1 \\ 0 & \text{for} \quad s = 1, \ldots, \ell \quad \text{and} \quad t = \ell + 2, \ldots, n \ . \end{cases}$$

Then it is readily checked that

$$P_{12} = \begin{bmatrix} 1 & x_1 & \cdots & x_1^{\ell-1} \\ \vdots & \vdots & & \vdots \\ 1 & x_\ell & \cdots & x_\ell^{\ell-1} \end{bmatrix}$$

is a Vandermonde matrix with $x_j = 1/\alpha_j$.

One might ask whether or not this choice of data is compatible with a positive definite (or semidefinite) P. The answer is yes. For example if $\ell = r = 3$ and

$$V = \begin{bmatrix} \xi_1 & \xi_2 & \xi_3 & \xi_4 & 0 & 0 \\ 0 & 0 & 0 & \eta_4 & \eta_5 & \eta_6 \end{bmatrix}$$

(with $\xi_j \in \mathbb{C}^p$ and $\eta_j \in \mathbb{C}^q$, as usual), then

$$P_{11} = \sum_{k=0}^{\infty} (A_1^*)^k V_{11}^* V_{11} A_1^k$$

and

$$P_{22} = \sum_{k=0}^{\infty} (T_3^*)^k (V_{22}^* V_{22} - V_{12}^* V_{12}) T_3^k \ .$$

Choose

$$V_{11} = tI_3 \ , \ V_{22} = I_3$$

$$\xi_4 = \frac{1}{4t} \begin{bmatrix} 1 \\ 2 \\ 3 \end{bmatrix} \ .$$

Then clearly $P_{11} = t^2Q_{11}$ and $P_{22} = Q_{22} + o(1)$ as $t \to \infty$, where $Q_{11} > 0$ and $Q_{22} > 0$, while P_{12} is constant, since

$$\alpha_j = v_j^* J v_4 = \xi_j^* \xi_4 = \frac{j}{4} \quad \text{for} \quad j = 1, 2, 3 \; .$$

Therefore,

$$P_{11} - P_{12}P_{22}^{-1}P_{21} = t^2Q_{11} - P_{12}Q_{22}^{-1}P_{21} + o(1)$$

as $t \to \infty$. Thus P is positive definite for large enough t.

Case 4. ($\Omega_+ = \mathbb{D}$) P_{12} is a Toeplitz matrix.

Now let the data of Case 1, but with $\omega = 0$, be applied to the equation for the P_{12} block that is associated with equation (6.7). Then P_{12} is a solution of the equation

$$P_{12} - T_\ell^* P_{12} T_r = -2\pi i(V_{11}^* V_{12} - V_{21}^* V_{22}^*) \; .$$

It is readily checked that

$$P_{12} \;=\; \begin{bmatrix} a_1 & a_2 & \cdots & a_r \\ b_2 & a_1 & \cdots & a_{r-1} \\ \vdots & & \ddots & \\ b_\ell & b_{\ell-1} & & \end{bmatrix}$$

is a Toeplitz matrix with

$$\begin{aligned} a_j &= -2\pi i v_1^* J v_{\ell+j} \; , \quad \text{for} \quad j = 1, \ldots, r \\ b_j &= -2\pi i v_j^* J v_{\ell+1} \; , \quad \text{for} \quad j = 2, \ldots, \ell \; . \end{aligned}$$

Finally, we remark that non-Hermitian structured matrices also occur naturally in the theory of reproducing kernel pairs. For additional information on the latter, see [AD4] and, for applications to the extension of the Iohvidov laws, [AD5].

12. $G(\lambda)$ Invertible Off the Boundary

In previous sections we paid special attention to the case that $G(\lambda)$ is invertible on Ω_0. In that setting the columns of the mvf $F(\lambda) = VG(\lambda)^{-1}$ have finite norm with respect to the standard inner product. In the remaining sections of this paper we shall focus on the opposite extreme, i.e., on the case when $G(\lambda) = M - \lambda N$ is invertible off Ω_0.

THEOREM 12.1. *Let $G(\lambda)$ be invertible for every point $\lambda \notin \Omega_0$ and let P be any $n \times n$ positive semidefinite matrix. Then a mvf $S \in \mathcal{S}^{p\times q}(\Omega_+)$ belongs to the set $\widehat{\mathcal{S}}(V, M, N; P)$ if and only if*

$$\langle \Delta_S Fu, Fu \rangle \leq u^* P u \tag{12.1}$$

for every $u \in \mathbb{C}^n$.

PROOF. Assume that the inequality (12.1) is in force for every $u \in \mathbb{C}^n$. Then the decomposition

$$\Delta_S = \begin{bmatrix} I_p \\ -S^* \end{bmatrix} [I_p \quad -S] + \begin{bmatrix} 0 & 0 \\ 0 & I_q - S^*S \end{bmatrix}$$

insures that

$$[I_p \quad -S]Fu \in L_2^p(\Omega_0)$$

for every $u \in \mathbb{C}^n$. The condition on $G(\lambda)$ guarantees further that each such vvf belongs to the Smirnov class and hence, by the Smirnov maximum principle, to $H_2^p(\Omega_+)$.

A similar argument based on the decomposition

$$\Delta_S = \begin{bmatrix} -S \\ I_q \end{bmatrix} [-S^* \quad I_q] + \begin{bmatrix} I_p - SS^* & 0 \\ 0 & 0 \end{bmatrix}$$

permits one to conclude from (12.1) that

$$[-S^* \quad I_q]Fu \in H_2^q(\Omega_+)^\perp$$

for every $u \in \mathbb{C}^n$. Thus condition (12.1) insures that $S \in \widehat{\mathcal{S}}(V, M, N; P)$. The converse is selfevident. □

The interpolation problem in this setting is more complicated, because the interpolation conditions are nontangential limits, rather than explicit evaluation of mvf's and their derivatives inside Ω_+. Moreover, boundary interpolation problems are automatically bitangential problems. A full analysis of this setting is beyond the scope of this paper. However, in order to at least indicate some of the possibilities for the structured matrices that can arise, we shall sketch two examples for $\Omega_+ = \mathbb{D}$.

Example 1.

$$M = I_n \ , \ N = \operatorname{diag}\{\overline{\beta}_1, \ldots, \overline{\beta}_n\}$$

with $\beta_j \in \mathbb{T}$ for $j = 1, \ldots, n$.

In this setting the entries p_{st}, $s, t = 1, \ldots, n$, of any $n \times n$ matrix P that solves the Stein equation (6.6) must satisfy the conditions

$$p_{st}(1 - \beta_s \overline{\beta}_t) = v_s^* J v_t \ ,$$

where v_s, $s = 1, \ldots, n$, denote the columns of V. Notice that the columns

$$f_j(\lambda) = \frac{v_j}{\rho_{\beta_j}(\lambda)} \ , \ j = 1, \ldots, n \ ,$$

of $F(\lambda)$ now have infinite norm with respect to the standard inner product. Moreover, if $\beta_s \overline{\beta}_t = 1$, then the data is subject to the constraint $v_s^* J v_t = 0$, but p_{st} is arbitrary.

Example 2. $M = I_n$, $N = \overline{\beta} I_n + T$ with $\beta \in \mathbb{T}$, where $T = T_n$ denotes the $n \times n$ shift matrix.

In this setting the Stein equation (6.6) reduces to

$$-\overline{\beta} T^* P - P \beta T - T^* P T = V^* J V \ . \tag{12.2}$$

The solutions of this equation have some features in common with the solutions of the simpler equation (11.4) that was considered in the preceding section. For example, if $n = 4$, then

$$P_{12} = \begin{bmatrix} a_{11} & a_{12} & a_{13} & b_{13} \\ a_{21} & a_{22} & b_{23} & b_{24} \\ a_{31} & b_{32} & b_{33} & b_{34} \\ c_{31} & c_{32} & c_{33} & c_{34} \end{bmatrix} ,$$

where:

(1) The a_{st} are uniquely determined from the equation.
(2) The c_{st} (i.e., the bottom row of P) are completely arbitrary.
(3) The remaining entries (i.e., the b_{st}) are uniquely determined from the a_{st} and the c_{st}.

A complete analysis of equation (12.2) is furnished in [BoD2].

13. Another Boundary Interpolation Problem

In this section we shall sketch the main features of a boundary interpolation problem of a more sophisticated sort, which we shall refer to as the $\widehat{\text{CFFP}}$, an acronym for Carathéodory-Fejér Full Matrix Boundary Problem. The analysis of this problem leads naturally to the consideration of a more complex family of structured matrices than considered earlier.

The problem is formulated in terms of a point $\beta \in \mathbb{T}$ and a set of $p \times p$ matrices $\gamma_0, \ldots, \gamma_{2k+1}$, where γ_0 is unitary. The first objective is to find conditions on the data which guarantee the existence of a mvf $S \in \mathcal{S}^{p\times p}(\mathbb{D})$ such that:

(1) The nontangential limits

$$S_j := \angle \lim_{\lambda\to\beta} \frac{S^{(j)}(\lambda)}{j!} \tag{13.1}$$

exist for $j = 0, \ldots, 2k+1$.

(2) The limits

$$S_j = \gamma_j \quad \text{for} \quad j = 0, \ldots, 2k\ . \tag{13.2}$$

(3) The limit S_{2k+1} is subject to the constraint

$$(-1)^k \beta^{2k+1}(\gamma_{2k+1} - S_{2k+1})\gamma_0^* \geq 0\ . \tag{13.3}$$

The second objective is to furnish a description of the set of all solutions when the conditions for existence are fulfilled. The rather odd-looking constraint (13.3) is precisely the condition that is needed to fit this problem into the framework of the $\widehat{\text{aBIP}}$.

It turns out that the condition for the existence of a solution have a neat formulation in terms of three block matrices that are defined in terms of the data: The block Hankel matrix

$$H_\gamma = \begin{bmatrix} \gamma_1 & \cdots & \gamma_{k+1} \\ \vdots & & \vdots \\ \gamma_{k+1} & \cdots & \gamma_{2k+1} \end{bmatrix}, \tag{13.4}$$

the lower triangular block Toeplitz matrix

$$L_\gamma = \begin{bmatrix} \gamma_0 & & 0 \\ \vdots & \ddots & \\ \gamma_k & \cdots & \gamma_0 \end{bmatrix}, \tag{13.5}$$

and the block upper triangular matrix

$$D_\beta = [d_{ij}]\ ,\ i, j = 0, \ldots, k\ ,$$

where

$$d_{ij} = \begin{cases} 0_{p\times p} \ , & \text{if} \quad i > j \\ (-1)^j \binom{j}{i} \beta^{i+j+1} I_p \ , & \text{if} \quad i \leq j \ . \end{cases} \tag{13.6}$$

Kovalishina [Ko] showed that the condition

$$H_\gamma D_\beta L_\gamma^* > 0 \tag{13.7}$$

is sufficient to guarantee the existence of solutions to the $\widehat{\text{CFFP}}$ and gave a linear fractional description of the set of all solutions when this condition is in force. She also showed that the weaker constraint (13.8) that will be given below is necessary. The following theorem, which is established in [BoD2], eliminates the gap between these two sets of conditions. A description of the set of all solutions under the less restrictive condition (13.8) is also furnished there.

THEOREM 13.1. *The $\widehat{\text{CFFP}}$ based on $\gamma_0, \dots, \gamma_{2k+1}$ and β has a solution if and only if*

$$H_\gamma D_\beta L_\gamma^* \geq 0 \ . \tag{13.8}$$

The strategy adopted in [BoD2] to prove Theorem 13.1 is to identify the set of solutions of the $\widehat{\text{CFFP}}$ with the set of solutions of an associated $\widehat{\text{aBIP}}(V, M, N; P)$, i.e., with the set $\widehat{\mathcal{S}}(V, M, N; P)$ based on the matrices

$$M = I_n \ , \text{where } n = (k+1)p \ ,$$

$$N = \begin{bmatrix} \overline{\beta} I_p & I_p & 0 & 0 & \cdots & 0 \\ 0 & \overline{\beta} I_p & I_p & 0 & \cdots & 0 \\ \vdots & \ddots & \ddots & \ddots & & \vdots \\ 0 & 0 & & & & I_p \\ 0 & 0 & & & & \overline{\beta} I_p \end{bmatrix} , \tag{13.9}$$

$$V = \begin{bmatrix} V_1 \\ V_2 \end{bmatrix} = \begin{bmatrix} I_p & 0 & \cdots & 0 \\ \gamma_0^* & \gamma_1^* & \cdots & \gamma_k^* \end{bmatrix} \tag{13.10}$$

and

$$P = H_\gamma D_\beta L_\gamma^* \ . \tag{13.11}$$

The argument proceeds in steps.

Step 1. If $S(\lambda)$ is a solution of the $\widehat{\text{CFFP}}$, then P is a positive semidefinite solution of the Stein equation

$$P - N^* P N = V^* J V \tag{13.12}$$

and

$$S \in \widehat{\mathcal{S}}(V, M, N; P) \ .$$

Step 2. If $Q = [q_{ij}]$, $i, j = 0, \dots, k$, is a positive semidefinite solution of the homogeneous Stein equation

$$Q - N^* Q N = 0, \tag{13.13}$$

then all the block entries except possibly the lower right corner corner vanish, i.e.,

$$q_{ij} = 0 \quad \text{for} \quad \begin{cases} i = 0, \dots, k-1 & \text{and} \quad j = 0, \dots, k \\ i = k & \text{and} \quad j = 0, \dots, k-1 \,. \end{cases}$$

Step 3. If $S \in \widehat{\mathcal{S}}(V, M, N; P)$, then the $n \times n$ mvf

$$W(\lambda) = (\lambda I_n - N^*)^{-1}\{V_1^*(V_1 - S(\lambda)V_2)G(\lambda)^{-1} - P\} \tag{13.14}$$

admits a representation of the form

$$W(\lambda) = \frac{1}{\pi}\int_0^{2\pi} \frac{d\sigma(t)}{e^{it} - \lambda} \,, \; \lambda \in \mathbb{D} \,, \tag{13.15}$$

where $d\sigma$ is a finite $n \times n$ matrix valued measure on $\mathbb{T}$.

This representation is obtained by first showing that the mvf $\lambda W(\lambda) + P/2$ is of Carathéodory class and then invoking the Herglotz representation

$$\frac{P}{2} + \lambda W(\lambda) = C + \frac{1}{2\pi}\int_0^{2\pi} \frac{e^{it} + \lambda}{e^{it} - \lambda} d\sigma(t) \,.$$

Step 4. The nontangential limit

$$\angle \lim_{\lambda \to \beta} (\lambda - \beta) W(\lambda) = -\frac{1}{\pi}\sigma(\{t_0\}) \,, \tag{13.16}$$

exists, where $\beta = e^{it_0}$. This statement follows readily from (13.15).

Step 5. If $S \in \widehat{\mathcal{S}}(V, M, N; P)$, then the nontangential limit

$$\angle \lim_{\lambda \to \beta} \left\{ \frac{S(\lambda) - \gamma_0 - \gamma_1(\lambda - \beta) - \cdots - \gamma_{2k+1}(\lambda - \beta)^{2k+1}}{(\lambda - \beta)^{2k+1}} \right\}$$

exists.

This conclusion is obtained by identifying the limit with the lower right corner of the matrix

$$\angle \lim_{\lambda \to \beta} (\lambda - \beta) W(\lambda) (D_\beta L_\gamma^*)^{-1} \,,$$

which exists by Step 4. Notice that Step 5 justifies (13.1) and (13.2), but does not yield any information about the value of S_{2k+1}.

Step 6. If $S \in \widehat{\mathcal{S}}(V, M, N; P)$ then the matrix P_S that is defined by the rule (10.1) is given by the formula

$$P_S = H_S D_\beta L_\gamma^* \,, \tag{13.17}$$

where the matrix H_S is obtained from H_γ by replacing γ_{2m+1} by S_{2m+1}. Moreover, P_S is a positive semidefinite solution of the Stein equation (13.12).

Step 7. If $S \in \widehat{\mathcal{S}}(V, M, N; P)$, then S is a solution of the $\widehat{\text{CFFP}}$.

In view of Steps 2 and 6,

$$H_\gamma - H_S = (P - P_S)(D_\beta L_\gamma^*)^{-1} = \begin{bmatrix} 0 & \cdots & 0 \\ \vdots & & \vdots \\ 0 & \cdots & \delta \end{bmatrix} (D_\beta L_\gamma^*)^{-1} \,.$$

The bottom right hand corner of this last formula yields the identity

$$\gamma_{2k+1} - S_{2k+1} = \delta(-1)^k \beta^{-(2k+1)} \gamma_0 \,. \tag{13.18}$$

But this in turn leads easily to the inequality (13.3), since $\delta \geq 0$ and γ_0 is unitary.

Now, having identified $\widehat{S}(V, M, N; P)$ with the set of solutions of the $\widehat{\text{CFFP}}$, we can obtain a description of the set of all solutions to the latter on the basis of a general theorem which supplies a description of all solutions of the $\widehat{\text{aBIP}}$. A proof of this theorem too may be found in [BoD2]. It may also be obtained from the more general setting that was investigated by Katsnelson, Kheifetz and Yuditskii [KKY]; see also [KY].

There are a number of subtle points connected with this problem. In particular the data V yields only the coefficient $\gamma_0, \ldots, \gamma_k$. The coefficients $\gamma_{k+1}, \ldots, \gamma_{2k+1}$ enter via the last block column of $H_\gamma P(D_\beta L_\gamma^*)^{-1}$.

It is also worth noting that the structured matrix D_β may be characterized as the unique solution of the equation

$$D + \beta(I_n + \beta T^*)DT = \beta E^* E \ , \tag{13.19}$$

where

$$E = [I_p \ \ 0 \ \ \cdots \ \ 0] \tag{13.20}$$

is a block row matrix with $k+1$ blocks of size $p \times p$; see Section 10 of [BoD2] for the details.

I wish to thank Vladimir Bolotnikov for reading the final version of this paper. He reported that he did not find any misprints. I hope he is right. In any event, the responsibility is all mine.

References

[A] D. Alpay, *Reproducing kernel Krein spaces of analytic functions and inverse scattering*, Ph.D. Thesis, Department of Theoretical Mathematics, The Weizmann Institute of Science, Rehovot, Israel (submitted October 1985).

[AD1] D. Alpay and H. Dym, Hilbert spaces of analytic functions, inverse scattering, and operator models I, *Integral Equations Operator Theory* **7** (1984), 589–641.

[AD2] D. Alpay and H. Dym, On applications of reproducing kernel spaces to the Schur algorithm and rational J unitary factorization, in: *Schur Methods in Operator Theory and Signal Processing* **OT18**, Birkhäuser, 1986, pp. 89–159.

[AD3] D. Alpay and H. Dym, On reproducing kernel spaces, the Schur algorithm and interpolation in a general class of domains, in: *Operator Theory and Complex Analysis*, (T. Ando and I. Gohberg, eds.), *Oper. Theory: Adv. Appl.* **OT59**, Birkhäuser-Verlag, Basel, 1992, pp. 30–77.

[AD4] D. Alpay and H. Dym, On a new class of structured reproducing kernel spaces, *J. Funct. Anal.* **111** (1993), 1–28.

[AD5] D. Alpay and H. Dym, On a new class of reproducing kernel spaces and a new generalization of the Iohvidov laws, *Linear Algebra Appl.* **178** (1993), 109–183.

[AD6] D. Alpay and H. Dym, On a new class of realization formulas and their application, *Linear Algebra Appl.*, in press.

[Ar] D. Z. Arov, The generalized bitangent Carathéodory-Nevanlinna-Pick problem and (j, J_0) inner matrix valued functions, *Russian Acad. Sci. Izv. Math.* **42** (1994), No.1, 1–26.

[Ba] J.A. Ball, Models for non contractions, *J. Math. Anal. Appl.* **52** (1975), 240–254.

[BGR] J.A. Ball, I. Gohberg and I. Rodman, *Interpolation of Rational Matrix Functions*, Birkhäuser-Verlag, Basel, 1990.

[BH] J.A. Ball and J.W. Helton, Interpolation problems of Pick-Nevanlinna and Loewner types for meromorphic matrix functions: parametrization of the set of all solutions, *Integral Equations Operator Theory* **9** (1986), 155–203.

[BoD1] V. Bolotnikov and H.Dym, On degenerate interpolation maximum entropy and extremal problems for matrix Schur functions, *Integral Equations Operator Theory* **32** (1998), No. 4, 367–435.

[BoD2] V. Bolotnikov and H.Dym, *On boundary interpolation for matrix Schur functions*, Preprint MCS99-22, Department of Mathematics, The Weizmann Institute of Science, Rehovot, Israel
(available at http://www.wisdom.weizmann.ac.il).

[dB] L. de Branges, Some Hilbert spaces of analytic functions I, *Trans. Amer. Math. Soc.* **106** (1963), 445–468.

[D1] H. Dym, *J Contractive Matrix Functions, Reproducing Kernel Hilbert Spaces and Interpolation*, CBMS Reg. Conf., Ser. in Math. **71**, Amer. Math. Soc., Providence, RI, 1989.

[D2] H. Dym, *On reproducing kernel spaces, J unitary matrix functions, interpolation and displacement rank*, in: The Gohberg Anniversary Collection II (H. Dym, S. Goldberg, M.A. Kashoek and P. Lancaster, eds.), *Oper. Theory Adv. Appl.* **OT41**, Birkhäuser-Verlag, Basel, 1989, pp. 173–239.

[D3] H. Dym, *Shifts, realizations and interpolation, redux*, in: Operator Theory and its Applications (A. Feintuch and I. Gohberg, eds.), *Oper. Theory Adv. Appl.* **OT73**, Birkhäuser-Verlag, Basel, 1994, pp. 182–243.

[D4] H. Dym, Book Review, *Bulletin A.M.S.* **31** (1994), 125–140.

[D5] H. Dym, A basic interpolation problem, in: *Holomorphic Spaces*, (S. Axler, J. E. McCarthy and D. Sarason, eds.), Cambridge University Press, Cambridge, 1998, pp. 381–425.

[D6] H. Dym, Notes on Riccati equations and reproducing kernel spaces, *Oper. Theory Adv. Appl.*, to appear (also available at http://www.wisdom.weizmann.ac.il).

[DFK] V.K. Dubovoj, B. Fritzsche and B. Kirstein, *Matricial version of the classical Schur problem*, Teubner, Leipzig, 1992.

[FF] C. Foias and A. Frazho, *The Commutant Lifting Approach to Interpolation Problems*, Birkhäuser-Verlag, Basel, 1990.

[HR] G. Heinig and K. Rost, *Algebraic Methods for Toeplitz-like Matrices and Operators*, Birkhäuser, Basel, 1984.

[IS] T. S. Ivanchenko and L. A. Sakhnovich, *An operator approach to the Potapov scheme for the solution of interpolation problems*, in: Matrix and Operator Valued Functions (I. Gohberg and L.A. Sakhnovich, eds.), *Oper. Theory. Adv. Appl.* **OT72**, Birkhäuser-Verlag, Basel, 1994, pp. 48–86.

[KaSa] T. Kailath and A. H. Sayed, Displacement structure: theory and applications, *SIAM Review* **37** (1995), 297–386.

[KKY] V. Katsnelson, A. Kheifets and P. Yuditskii, An abstract interpolation problem and extension theory of isometric operators, in: *Operators in Spaces of Functions and Problems in Function Theory* (V. A. Marchenko, ed.) **146**, Baukova Dumka, Kiev, 1987, pp. 83–96. English transl. in: *Topics in Interpolation Theory* (H. Dym, B. Fritzche, V. Katsnelson and B. Kirstein, eds.), *Oper. Theory Adv. Appl.* **OT95**, Birkhäuser Verlag, Basel, 1997, pp. 283–298.

[KY] A. Kheifets and P. Yuditskii, An analysis and extension approach of V. P. Potapov's approach to scheme interpolation problems with applications to the generalized bitangential Schur-Nevanlinna-Pick problem and J-inner-outer factorization, in: *Matrix and Operator Valued Functions* (I.Gohberg and L. A. Sakhnovich, eds.), *Oper. Theory Adv. Appl.* **OT72**, Birkhäuser Verlag, Basel, 1994, pp. 133–161.

[Ko] I. V. Kovalishina, A multiple boundary problem for contracting matrix-valued functions in the unit circle, *Teoriya Funktsii, Funktsianal'nyi Analiz*

Ikh Prilozheniya **51** (1989), 38-55. English transl. in: *Journal of Soviet Mathematics* **52(6)** (1990), 3467–3481.

[LAK] H. Lev-Ari and T. Kailath, *Triangular factorization of structured Hermitian matrices*, in: I. Schur Methods in Operator Theory and Signal Processing (I. Gohberg, ed), *Oper. Theory Adv. Appl.* **OT18**, Birkhäuser-Verlag, Basel, 1986, pp. 301–324.

[Nu] A. A. Nudelman, *Some generalizations of classical interpolation problems*, in: Operator Extensions, Interpolation of Functions and Related Topics (A. Gheondea, D. Timotin and F.-H. Vascilescu, eds.), *Oper. Theory Adv. Appl.* **OT61**, Birkhäuser, Basel, 1993, pp. 312–322.

[Ol] V. Olshevsky, Pivoting for structured matrices with applications, *Linear Algebra Appl.*, in press
(also available at http://www.cs.gsu.edu/~matvro/papers.html).

[Rov] J. Rovnyak, *Characterization of spaces* $\mathcal{K}(M)$, unpublished manuscript, 1968.

[Sa] D. Sarason, Nevanlinna-Pick interpolation with boundary data, *Integral Equations Operator Theory* **30** (1998), 231–250.

[ZDL] K. Zhou, J. C. Doyle and K. Glover, *Robust and Optimal Control*, Prentice Hall, New Jersey, 1996.

Current address: Department of Mathematics, The Weizmann Institute of Science, Rehovot 76100, Israel

E-mail address: dym@wisdom.weizmann.ac.il

Contemporary Mathematics
Volume **280**, 2001

A Superfast Algorithm for Confluent Rational Tangential Interpolation Problem via Matrix-vector Multiplication for Confluent Cauchy-like Matrices

Vadim Olshevsky and Amin Shokrollahi

Abstract. Various problems in pure and applied mathematics and engineering can be reformulated as linear algebra problems involving dense structured matrices. The *structure* of these dense matrices is understood in the sense that their n^2 entries can be completeley described by a smaller number $O(n)$ of parameters. Manipulating directly on these parameters allows us to design efficient *fast algorithms*. One of the most fundamental matrix problems is that of multiplying a (structured) matrix with a vector. Many fundamental algorithms such as convolution, Fast Fourier Transform, Fast Cosine/Sine Transform, and polynomial and rational multipoint evaluation and interpolation can be seen as superfast multiplication of a vector by structured matrices (e.g., Toeplitz, DFT, Vandermonde, Cauchy). In this paper, we study a general class of structured matrices, which we suggest to call *confluent Cauchy-like matrices*, that contains all the above classes as a special case. We design a new superfast algorithm for multiplication of matrices from our class with vectors. Our algorithm can be regarded as a generalization of all the above mentioned fast matrix-vector multiplication algorithms. Though this result is of interest by itself, its study was motivated by the following application. In a recent paper [**18**] the authors derived a superfast algorithm for solving the classical tangential Nevanlinna-Pick problem (rational matrix interpolation with norm constrains). Interpolation problems of Nevanlinna-Pick type appear in several important applications (see, e.g., [**4**]), and it is desirable to derive efficient algorithms for several similar problems. Though the method of [**18**] can be applied to compute solutions for certain other important interpolation problems (e.g., of Caratheodory-Fejer), the solution for the most general *confluent tangential* interpolation problems cannot be easily derived from [**18**]. Deriving new algorithms requires to design a special fast algorithm to multiply a confluent Cauchy-like matrix by a vector. This is precisely what has been done in this paper.

1991 *Mathematics Subject Classification.* Primary: 15A06 Secondary: 47N70, 42A70.

Key words and phrases. Rational matrix tangential interpolation. Nevanlinna-Pick problem. Caratheodory-Fejer problem. Cauchy matrices. Superfast algorithms.

The work of the first author was supported by NSF grants CCR 9732355 and 0098222.

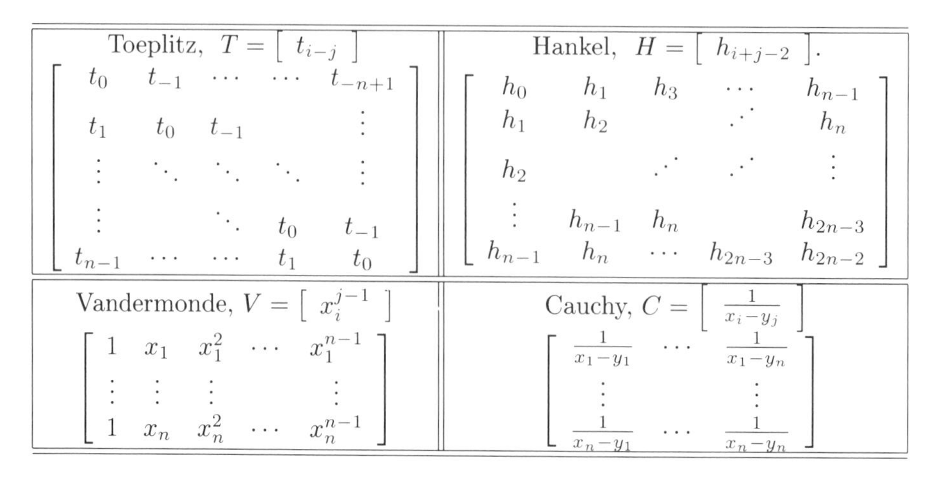

FIGURE 1. *Some examples of matrices with structure.*

Toeplitz matrices	convolution	$M(n)$
Hankel matrices	convolution	$M(n)$
Vandermonde matrices	multipoint polynomial evaluation	$M(n)\log n$
DFT matrices (i.e., Vandermonde matrices with special nodes)	discrete Fourier transform	$M(n)$
inverse Vandermonde matrices	polynomial interpolation	$M(n)\log n$
Cauchy matrices	multipoint rational evaluation	$M(n)\log n$
inverse Cauchy matrices	rational interpolation	$M(n)\log n$

FIGURE 2. *Connection between fundamental algorithms and structured matrix-vector multiplication*

1. Introduction

1.1. Several examples of matrices with structure. Structured matrices are encountered in a surprising variety of areas (e.g., signal and image processing, linear prediction, coding theory, oil exploration, to mention just a few), and algorithms (e.g., for Pade approximations, continuous fractions, classical algorithms of Euclid, Schur, Nevanlinna, Lanzcos, Levinson). There is an extensive literature on structured matrices, we mention here only several large surveys [**13**], [**10**], [**3**], [**17**] and two recent representative papers on rational interpolation [**7, 18**] and on list decoding of algebraic codes [**19**].

Many fundamental algorithms for polynomial and rational computations can be seen as algorithms for matrices with structure. Examples include Toeplitz $[t_{i-j}]$, Hankel $[h_{i+j-2}]$, Vandermonde $[x_i^{j-1}]$, and Cauchy matrices $[1/(x_i - y_j)]$, shown in Table 1.

Multiplication of these matrices with vectors often has an analytic interpretation. For instance, the problem of *multipoint evaluation* of a rational function

$$a(x) = \sum_{k=1}^{n} \frac{a_k}{x - y_k}$$

at the points $x_1, \ldots, x_n$ is clearly equivalent to computing the product of a Cauchy matrix by a vector:

$$\begin{bmatrix} a(x_1) \\ a(x_2) \\ \vdots \\ a(x_n) \end{bmatrix} = \begin{bmatrix} \frac{1}{x_1-y_1} & \frac{1}{x_1-y_2} & \cdots & \frac{1}{x_1-y_n} \\ \frac{1}{x_2-y_1} & \frac{1}{x_2-y_2} & \cdots & \frac{1}{x_2-y_n} \\ \vdots & \vdots & & \vdots \\ \frac{1}{x_n-y_1} & \frac{1}{x_n-y_2} & \cdots & \frac{1}{x_n-y_n} \end{bmatrix} \cdot \begin{bmatrix} a_1 \\ a_2 \\ \vdots \\ a_n \end{bmatrix}.$$

Table 2 lists some further analytic interpretations for various matrices with structure relating them to several fundamental algorithms. Its last column lists running times of the corresponding algorithms. Here, we denote by

$$(1.1) \qquad M(n) = \begin{cases} n \log n & \text{if the field } K \text{ supports FFT's of length } n \\ n \log n \log \log n & \text{otherwise} \end{cases}$$

the running time of basic polynomial manipulation algorithms such as multiplication and division with remainder of polynomials of degree $< n$, cf. [**1**, Th. (2.8), Th. (2.13), Cor. (2.26)].

The running times in Table 2 are well known. We only mention that the problem of fast multiplying a Cauchy matrix by a vector is known in the numerical analysis community as the *Trummer problem.* It was posed by G. Golub and solved by Gerasoulis in [**6**]. We also mention the celebrated *fast multipole method* (FMM) of [**20**] that computes the approximate product of a Cauchy matrix by a vector (the FFM method is important in computaional potential theory).

In this extended abstract we continue the work of our colleagues and propose a new superfast algorithm to multiply by a vector a confuent Cauchy-like matrix (a far reaching generalization of a Cauchy matrix). To introduce confluent Cauchy-like matrices we need a concept of *displacement* recalled next.

1.2. More general matrices, displacement structure. Many applications give rise to the more general classes of structured matrices defined here. We start with a simple clarifying example. Let us define two auxiliary diagonal matrices $A_\varsigma = \text{diag}\{x_1, \ldots, x_n\}$, and $A_\pi = \text{diag}\{y_1, \ldots, y_n\}$. It is immediate to see that for a Cauchy matrix $[1/(x_i - y_j)]$, the matrix

$$A_\varsigma C - C A_\pi = \left[\, \tfrac{x_i - y_j}{x_i - y_j} \,\right] = \begin{bmatrix} 1 & 1 & \cdots & 1 \\ 1 & 1 & \cdots & 1 \\ \vdots & \vdots & & \vdots \\ 1 & 1 & \cdots & 1 \end{bmatrix}$$

is the all-one matrix, and hence $\text{rank}(A_\varsigma C - C A_\pi) = 1$.

This observation is used to define a more general class of matrices which have appeared in many applications, e.g., [**4, 18**] and which have attracted much attention recently (see, e.g., [**17**] and the references therein). For these matrices the

Toeplitz-like matrices R	rank $(ZR - RZ) << n$
Hankel-like matrices R	rank $(ZR - RZ^T) << n$
Vandermonde-like matrices R	rank $(D_x^{-1}R - RZ^T) << n$
Cauchy-like matrices R	rank $(D_xR - RD_y) << n$

TABLE 1. Definitions of basic classes of structured matrices

parameter

$$\alpha = \text{rank}(A_\varsigma C - CA_\pi), \tag{1.2}$$

is larger than 1, but it is still much less than the size of C. Such matrices are referred to as *Cauchy-like* matrices.

Similar observations can be made for all other patterns of structure discussed above. Simply for different kinds of structured matrices we need to use different auxiliary matrices $\{A_\pi, A_\varsigma\}$. Table 3 contains definitions for basic classes of structured matrices. Here

$$Z = \begin{bmatrix} 0 & \cdots & 0 & 0 \\ 1 & 0 & \cdots & 0 \\ \vdots & \ddots & \ddots & \vdots \\ 0 & \cdots & 1 & 0 \end{bmatrix}, \qquad D_x = \text{diag}(x_1, \cdots, x_n).$$

Matrices in Table 3 are called *matrices with displacement structure*, the number α in (1.2) is called the *displacement rank*. The name *"displacement"* originates in signal processing literature [**14, 5, 15**] where Toeplitz and Hankel matrices are of special interest. For these matrices the auxiliray matrices $\{A_\varsigma, A_\pi\}$ are shift (or displacement matrices) Z. The name *displacement structure* is now used also in connection to other classes of structured matrices in Table 3, though this terminology is not uniform (For example, in interpolation literature [**2**] they are called *null-pole coupling* matrices, in [**3**] they are referred to as matrices with low scaling rank).

1.3 A generator and superfast multiplication algorithms. It is now well-understood [**10**], [**3**], [**13**], [**17**] that a useful approach to design fast matrix algorithms is in avoiding operations on n^2 entries, and in operating instead on what is called a *generator* of a structured matrix.

If the displacement rank (1.2) of a structured matrix R is α, one can factor (non-uniquely)

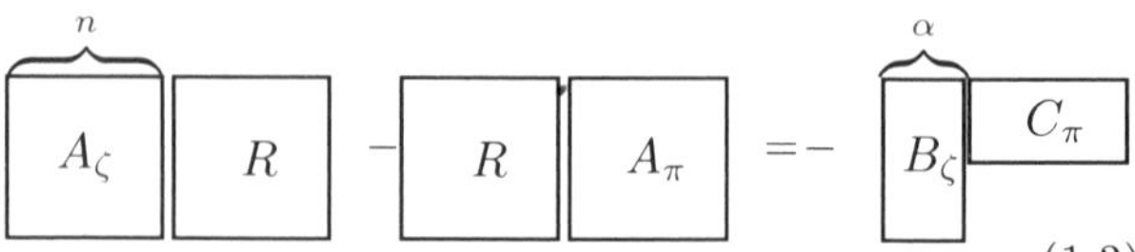

$$\underbrace{A_\varsigma}_{n} \; R - R \; A_\pi = - \underbrace{B_\varsigma}_{\alpha} \; C_\pi \tag{1.3}$$

where the two *rectangular* $\alpha \times n$ and $n \times \alpha$ matrices $\{C_\pi, B_\varsigma\}$ are called a *generator* of R.

Toeplitz-like	$\alpha M(n)$
Hankel-like	$\alpha M(n)$
Vandermonde-like	$\alpha M(n) \log n$
Cauchy-like	$\alpha M(n) \log n$
inverses of Toeplitz-like	$\alpha M(n) \log n$
inverses of Hankel-like	$\alpha M(n) \log n$
inverses of Vandermonde-like	$\alpha M(n) \log^2 n$
inverses of Cauchy-like	$\alpha M(n) \log^2 n$

TABLE 2. Complexities of multiplication by a vector for matrices with displacement structure

Let the two auxiliary matrices $\{A_\pi, A_\zeta\}$ be fixed. If the displacement equation (1.3) has the unique solution R, then the entire information on n^2 entries of R is conveniently compressed into only $2\alpha n$ entries of its generator $\{C_\pi, B_\zeta\}$.[1]

Avoiding operations on matrix entries and operating directly on a generator allows us to design fast and superfast algorithms. In particular, superfast algorithms to multiply by vectors matrices in Table 3 can be found in [**9**]. Table 4 lists the corresponding complexity bounds.

Notice that the problems of multiplying with the inverse is equivalent to solving the corresponding linear system of equations.

1.4. First problem: matrix-vector product for confluent Cauchy-like matrices. Notice that the auxiliary matrices $\{A_\zeta, A_\pi\}$ in Table 3 are all either shift or diagonal matrices, i.e., they all are special cases of the Jordan canonical form. Therefore it is natural to consider the more general class of structured matrices R defined by using the displacement equation $\text{rank}(A_\zeta R - RA_\pi) = B_\zeta C_\pi$ of the form (1.3), where

$$A_\zeta = \text{diag}\{J_{m_1}(x_1) \oplus \ldots \oplus J_{m_s}(x_s)\}^T,$$

$$A_\pi = \text{diag}\{J_{k_1}(y_1) \oplus \ldots \oplus J_{k_t}(y_t)\}, \tag{1.4}$$

are the general Jordan form matrices. We suggest to call such matrices R *confluent Cauchy-like* matrices. Notice that the class of confluent Cauchy-like matrices is the most general class of structured matrices, containing all other classes listed in Table 3 as special cases[2]. Therefore it is of interest to design a **uniform** superfast algorithm to multiply a confluent Cauchy-like matrix by a vector: such an algorithm would then contain all algorithms in Tables 2 and 4 (e.g., convolution, FFT, rational and polynomial multipoint evaluation and interpolation) as special cases.

Though such a generalized problem is of interest by itself, we were motivated to study it by a rather general tangential interpolation problem formulated in the next section.

---

[1]Because of page limitations we do not provide details on what happens if there are multiple solutions R to the displacement equation. We only mention that our algorithm admits a modification to handle this situation, and that all our complexity bounds fully apply to this more involved case. We would like to mention an interesting connection of this case to the rational interpolation problem discussed in Sec. 1.5 below. Specifically, this degenerate case corresponds to the case of *boundary* interpolation treated in [**12**].

[2]It also contains new classes of matrices not studied earlier, e.g., *scaled confluent Cauchy* matrices defined in Sec. 2 below.

1.5. Second problem: Confluent rational matrix interpolation problem. Rational functions appear as transfer functions of linear time-invariant systems, and in the MIMO (Multi-Input Multi-Output) case the corresponding function is a rational *matrix* function (i.e., an $N \times M$ matrix $F(z)$ whose entries are rational functions $\frac{p_{ij}(z)}{q_{ij}(z)}$). It is often important to identify the transfer function $F(x)$ via certain interpolation conditions, one such rather general problem is formulated below. There is an extensive mathematical and electrical engineering literature on such problems, some of the pointers can be found in [**18**].

Tangential confluent rational interpolation problem.

Given: r distinct points $\{z_k\}$ in the open right-half-plane Π^+, with their multiplicities $\{m_k\}$.
r nonzero chains of $N \times 1$ vectors $\{x_{k,1}, \ldots, x_{k,m_k}\}$,
r nonzero chains of $M \times 1$ vectors $\{y_{k,1}, \ldots, y_{k,m_k}\}$.

Construct: a rational $N \times M$ matrix function $F(x)$ such that

1. $F(z)$ is *analytic* inside the right half plane (i.e., all the poles are in the left half plane).
2. $F(z)$ is *passive*, which by definition means that

$$\sup_{z \in \Pi^+ \cup i\mathbb{R}} \|F(z)\| \leq 1. \tag{1.5}$$

3. $F(z)$ meets the tangential *confluent* interpolation conditions ($k = 1, 2, \ldots, r$):

$$\begin{bmatrix} x_{k1} & \cdots & x_{k,m_k} \end{bmatrix} \begin{bmatrix} F(z_k) & F'(z_k) & \cdots & \frac{F^{(m_k-1)}(z_k)}{(m_k-1)!} \\ 0 & F(z_k) & \ddots & \vdots \\ \vdots & & \ddots & F'(z_k) \\ 0 & \cdots & \cdots & F(z_k) \end{bmatrix} =$$

$$\begin{bmatrix} y_{k,1} & \cdots & y_{k,m_k} \end{bmatrix}. \tag{1.6}$$

We next briefly clarify the terminology.

- The *passivity* condition (1.5) is naturally imposed by the conservation of energy. Indeed, it means that if $F(z)$ is seen as a *transfer function* of a certain linear time-invariant system then the energy of the output
$$y(z) = u(z)F(z)$$
does not exceed the energy of input $u(z)$.
- The term *tangential* was suggested by Mark Grogorievich Krein. It means that it is not the full matrix value $F(z_k)$ that is given here, rather the action of $F(z_k)$ for certain directions $\{x_{k,1}\}$ is prescribed.
- The term *confluent* emphasizes the condition $m_k > 1$, i.e., not only $F(z)$ involved in the interpolation conditions, but also its derivatives $F'(z), \ldots, F^{(m_k-1)}(z)$.

We next consider several clarifying special cases to show the fundamental nature of the above interpolation problem.

EXAMPLE 1.1. **The tangential Nevanlinna-Pick problem.** *In the case of simple multiplicities $m_k = 1$ the interpolation condition (1.6) reduces to the usual tangential (i.e., x's and y's are vectors) interpolation condition*

$$x_k \cdot F(x_k) = y_k$$

which in the case of the scalar $F(z)$ further reduces to the familiar interpolation condition of the form

$$F(x_k) = \frac{y_k}{x_k}.$$

EXAMPLE 1.2. **The Caratheodory-Fejer problem.** *Let the vectors $\{x_{k,j}\}$ are just the following scalars:*

$$\begin{bmatrix} x_{k,1} & \cdots & x_{k,m_k} \end{bmatrix} = \begin{bmatrix} 1 & 0 & \cdots & 0 \end{bmatrix}.$$

Clearly, in this case (1.6) this is just a Hermite-type rational passive interpolation problem

$$F(z_k) = y_{k,1}, \quad F'(z_k) = y_{k,2}, \quad \cdots \quad \frac{F^{(m_k-1)}(z_k)}{(m_k-1)!} = y_{k,m_k}.$$

EXAMPLE 1.3. **Linear matrix pencils.** *Let $F(z) = A - zI$, then $F'(z) = -I$, $F''(z) = 0$. If the condition (1.6) has the form*

$$\begin{bmatrix} u_{k1} & u_{k2} & u_{k3} \end{bmatrix} \begin{bmatrix} (A - z_k I) & -I & 0 \\ 0 & (A - z_k I) & -I \\ 0 & 0 & (A - z_k I) \end{bmatrix} =$$

$$\begin{bmatrix} 0 & 0 & 0 \end{bmatrix}$$

then u_{k1} is the (left)eigenvector: $u_{k1}(A - z_k I) = 0$, whereas u_{k2} is the first generalized eigenvector: $u_{k2}(A - z_k I) = u_{k1}$, etc. In this simplest case the confluent interpolation problem reduces to recovering a matrix from its eigenvalues and eigenvectors.

To sum up, the confluent tangential rational interpolation problem is a rather general inverse eigenvalue problem that captures several classical interpolation problems as its special cases.

1.6. Main result. In this paper we describe a superfast algorithm to solve the confluent tangential rational interpolation problem. The running time of our algorithm is

$$Compl(n) = O\left(M(n) \log n \cdot \left[1 + \sum_{k=1}^{r} \frac{m_k}{n} \log \frac{n}{m_k} \right] \right) \tag{1.7}$$

where the multipliciteies $\{m_k\}$ are defined in Sec 1.4, $n = \sum m_k$, and $M(n)$ is defined in (1.1).

To understand the bound (1.7) it helps to consider two extreme cases.

- First, if all $m_k = 1$ (The Nevanlinna-Pick case: n points with simple multiplicities) then

$$Compl(n) = M(n) \log^2(n),$$

(or $n \log^3(n)$ if K supports FFT).

- Secondly, if $m_1 = n$ (The Caratheodory-Fejer case: one point with full multiplicity) then

$$Compl(n) = M(n)\log n,$$

$$(\text{or } n\log^2 n \text{ if } K \text{ supports FFT}).$$

The algorithm is based on a reduction of the above analytical problem to a structured linear algebra problem, namely to the problem of multiplication by a vector of a confluent Cauchy-like matrix. The corresponding running time is shown to be

$$O(M(n)\cdot[1+\sum_{k=1}^{r}\frac{m_k}{n}\log\frac{n}{m_k}+\sum_{k=1}^{s}\frac{t_k}{n}\log\frac{n}{t_k}]),$$

where $\{m_k\}$ are the sizes of the Jordan blocks of A_ζ and $\{t_k\}$ are the sizes of the Jordan blocks of A_π, see, e.g., (1.4).

1.7. Derivation of the algorithm and the structure of the paper. The overall algorithm is derived in several steps, using quite different techniques. Interestingly, sometimes purely matrix methods are advantageous, and other times analytic arguments help.

1. First, the confluent tangential rational interpolation problem is reduced to the problem of multiplying by a vector a confluent Cauchy matrix R whose generator is composed from the interpolation data. Namely, R is defined via

$$A_\zeta R + RA_\zeta^* = B_\zeta J B_\zeta^*,$$

where

$$A_\zeta = \begin{bmatrix} \boxed{J_{m_1}(z_1)^T} & & & \\ & \boxed{J_{m_2}(z_2)^T} & & \\ & & \ddots & \\ & & & \boxed{J_{m_r}(z_r)^T} \end{bmatrix},$$

$$B_\zeta = \begin{bmatrix} x_{11} & -y_{11} \\ \vdots & \vdots \\ x_{1,m_1} & -y_{1,m_1} \\ \hline \vdots & \vdots \\ \hline x_{r1} & -y_{r1} \\ \vdots & \vdots \\ x_{r,m_n} & -y_{r,m_n} \end{bmatrix}, J = \begin{bmatrix} I_M & 0 \\ 0 & -I_N \end{bmatrix}$$

2. Secondly, the problem for the confluent Cauchy-like R above is reduced to the analogous problem for the *scaled confluent Cauchy* matrix (i.e., not *-like*) defined in section 2.
3. Then the problem for the scaled confluent Cauchy matrix is further reduced in Sec. 3 to the following two problems. One is to multiply a confluent Vandermonde matrix by a vector, and the second is to multiply the inverse of a confluent Vandermonde matrix by a vector. The solution for

$$C_{i,j} =$$
$$\begin{bmatrix} B(x_i,y_j) & \partial_y^1 B(x_i,y_j) & \partial_y^2 B(x_i,y_j) & \cdots & \partial_y^{k_j-1} B(x_i,y_j) \\ \partial_x^1 B(x_i,y_j) & \partial_x^1 \partial_y^1 B(x_i,y_j) & \partial_x^1 \partial_y^2 B(x_i,y_j) & \cdots & \partial_x^1 \partial_y^{k_j-1} B(x_i,y_j) \\ \partial_x^2 B(x_i,y_j) & \partial_x^2 \partial_y^1 B(x_i,y_j) & \partial_x^2 \partial_y^2 B(x_i,y_j) & \cdots & \partial_x^2 \partial_y^{k_j-1} B(x_i,y_j) \\ \vdots & \vdots & \vdots & & \vdots \\ \partial_x^{m_i-1} B(x_i,y_j) & \partial_x^{m_i-1} \partial_y^1 B(x_i,y_j) & \partial_x^{m_i-1} \partial_y^2 B(x_i,y_j) & \cdots & \partial_x^{m_i-1} \partial_y^{k_j-1} B(x_i,y_j) \end{bmatrix},$$

FIGURE 3. The structure of C_{ij} in (2.9)

the second problem is available in the literature (Hermite-type polynomial interpolation).

4. The solution for the remaining problems (multiplication of a confluent Vandermonde matrix by a vector) is equivalent to the problem of multipoint Hermite-type evaluation, and the algorithm for it is described in the last section 4.

2. Scaled confluent Cauchy matrices

2.1. Definition. Suppose we have two sets of n nodes each

$$\{\overbrace{\underbrace{x_1,\ldots,x_1}_{m_1},\ldots,\underbrace{x_s,\ldots,x_s}_{m_s}}^{n}\}$$

$$\{\overbrace{\underbrace{y_1,\ldots,y_1}_{k_1},\ldots,\underbrace{y_t,\ldots,y_t}_{k_t}}^{n}\},$$

so that $n = m_1 + m_2 + \ldots + m_s$, and $n = k_1 + k_2 + \ldots + k_t$. We do not assume that the $s+t$ nodes $x_1,\ldots,x_s,y_1,\ldots,y_t$ are pairwise distinct. For a scalar bivariate function

$$B(x,y) = \frac{b(x)-b(y)}{x-y}, \tag{2.8}$$

where

$$b(x) = (x-y_1)^{k_1}\cdot(x-y_2)^{k_2}\cdot\ldots\cdot(x-y_s)^{k_t}$$

we define the block matrix

$$C = \left[\ C_{i,j}\ \right]_{1\le i\le s, 1\le j\le t}, \tag{2.9}$$

where the $m_i \times k_j$ block $C_{i,j}$ has the form shown in Table 5, where we denote

$$\partial_z^k := \frac{1}{k!}\frac{\partial^k}{\partial z^k}. \tag{2.10}$$

Before giving a special name to C let us notice that we do not assume that $\{x_1,\ldots,x_s,y_1,\ldots,y_n\}$ are pairwise distinct, so in the case $x_i = y_j$ the denominator in (2.8) is zero, hence we need to clarify what we mean by $B(x_i,x_i)$ and its partial derivatives. Using the following combinatorial identity

$$\partial_x^p\ \partial_y^q B(z,z) = \partial_z^{p+q+1} b(z).$$

we see that in the case $x_i = y_j$ the definition (2.8) implies that the block C_{ij} is a Hankel matrix of the following form

$$\begin{bmatrix} 0 & \cdots & 0 & \times \\ \vdots & \iddots & \iddots & \vdots \\ 0 & \iddots & & \vdots \\ \times & & & \vdots \\ \vdots & & & \vdots \\ \times & \cdots & \cdots & \times \end{bmatrix}$$

if $m_i > k_j$ or

$$\begin{bmatrix} 0 & \cdots & \cdots & \cdots & 0 & \times \\ \vdots & & & \iddots & \iddots & \vdots \\ 0 & \cdots & 0 & \times & \cdots & \times \end{bmatrix}$$

if $m_i < k_j$.

For example, if $m_i > k_j$ then $C_{i,j}$ has the following Hankel structure:

$$C_{ij} = \begin{bmatrix} 0 & \cdots & 0 & \partial^{k_j} b(x_i) \\ \vdots & \iddots & \iddots & \partial_x^{k_j+1} b(x_i) \\ 0 & \partial_x^{k_j} b(x_i) & \iddots & \vdots \\ \partial_x^{k_j} b(x_i) & \partial_x^{k_j+1} b(x_i) & \cdots & \partial_x^{2k_j-1} b(x_i) \\ \partial_x^{k_j+1} b(x_i) & \vdots & & \vdots \\ \vdots & \vdots & & \vdots \\ \partial_x^{m_i} b(x_i) & \partial_x^{m_i+1} b(x_i) & \cdots & \partial_x^{m_i+k_j-1} b(x_i) \end{bmatrix}.$$

We shall refer to C in (2.9) as a *scaled confluent Cauchy matrix*, and to explain the name we consider next a simple example.

EXAMPLE 2.1. *Let $m_i = k_j = 1$ (so that $s = t = n$), and $\{x_1, \dots, x_n, y_1, \dots, y_n\}$ are pairwise distinct. Since $b(y_k) = 0$ by (2.8), we have:*

$$C = \operatorname{diag}\{b(x_1), b(x_2), \dots, b(x_n)\} \cdot \begin{bmatrix} \frac{1}{x_1-y_1} & \cdots & \frac{1}{x_1-y_n} \\ \vdots & & \vdots \\ \frac{1}{x_n-y_1} & \cdots & \frac{1}{x_n-y_n} \end{bmatrix}.$$

The above example presents the case of simple multiplicities; in the general situation of higher multiplicities $\{m_i\}$ and $\{k_j\}$ we call C in (2.9) a *confluent* scaled Cauchy matrix.

2.2. Analytic interpretation. We have mentioned that multiplication of structured matrices by a vector often has an analytic interpretation, see, e.g., Table 2. Confluent Cauchy matrices are not an exception, and multiplying C in (2.9) by a vector

$$\left[\begin{array}{ccc|c|ccc} a_{0,1} & \cdots & a_{k_1-1,1} & \cdots & a_{0,t} & \cdots & a_{k_t-1,t} \end{array}\right]^T$$

$$A_\varsigma C - CA_\pi =$$

$$\left[\begin{array}{cccc|c|cccc}
b(x_1) & 0 & \cdots & 0 & & b(x_1) & 0 & \cdots & 0 \\
\partial_x^1 b(x_1) & 0 & \cdots & 0 & \cdots \quad \cdots & \partial_x^1 b(x_1) & 0 & \cdots & 0 \\
\vdots & \vdots & & \vdots & & \vdots & \vdots & & \vdots \\
\partial_x^{m_1-1} b(x_1) & 0 & \cdots & 0 & & \partial_x^{m_1-1} b(x_1) & 0 & \cdots & 0 \\
\hline
 & \vdots & & & & & \vdots & & \\
 & \vdots & & & & & \vdots & & \\
\hline
b(x_s) & 0 & \cdots & 0 & & b(x_s) & 0 & \cdots & 0 \\
\partial_x^1 b(x_s) & 0 & \cdots & 0 & \cdots \quad \cdots & \partial_x^1 b(x_s) & 0 & \cdots & 0 \\
\vdots & \vdots & & \vdots & & \vdots & \vdots & & \vdots \\
\partial_x^{m_s-1} b(x_s) & 0 & \cdots & 0 & & \partial_x^{m_s-1} b(x_s) & 0 & \cdots & 0
\end{array}\right]$$

FIGURE 4. Displacement of a scaled confluent Cauchy matrix C

is equivalent to multipoint evaluation of a rational function (with fixed poles $\{y_j\}$ of the orders $\{t_j\}$)

$$r(x) = \sum_{j=1}^{t} \sum_{l=0}^{k_j-1} a_{l,j} \partial_y^l B(x, y_j)$$

at points $\{x_i\}$ as well as of its derivatives up to the orders $\{m_i\}$.

2.3. Displacement structure. Let the auxiliary matrices $\{A_\varsigma, A_\pi\}$ be defined as in (1.4), and define the class of *confluent Cauchy-like matrices* as those having a low displacement rank (1.2) with these $\{A_\varsigma, A_\pi\}$. The following example justifies the latter name.

EXAMPLE 2.2. *Our aim here is to show that for the scaled confluent Cauchy matrix R in (2.9) we have*

$$\operatorname{rank}(A_\varsigma R - RA_\pi) = 1 \tag{2.11}$$

where $\{A_\varsigma, A_\pi\}$ are as in (1.4).

Indeed, applying $\partial_x^p \partial_y^q$ (recursively for $p, q = 0, 1, 2, \ldots$) to the both sides of

$$xB(x, y) - yB(x, y) = b(x) - b(y)$$

we obtain

$$x\partial_x^p \partial_y^q R(x, y) + \partial_x^{p-1} \partial_y^q R(x, y) - y\partial_x^p \partial_y^q R(x, y) -$$

$$-\partial_x^p \partial_y^{q-1} R(x, y) = \begin{cases} 0 & \text{if } p > 0 \text{ and } q > 0 \\ b^i(x) & \text{if } q = 0 \\ b^j(y) & \text{if } p = 0 \end{cases}$$

Now arranging the latter equation in a matrix form we obtain

$$J_{m_i}(x_i)^T C_{ij} - C_{ij} J_{k_j}(y_j) = \begin{bmatrix} b(x_i) & 0 & \cdots & 0 \\ \partial_x^1 b(x_i) & 0 & \cdots & 0 \\ \vdots & \vdots & & \vdots \\ \partial_x^{m_i-1} b(x_i) & 0 & \cdots & 0 \end{bmatrix}$$

Combining such equations for each block R_{ij} together we finally obtain the formula shown in Table 6, which yields (2.11).

Thus, the displacement rank of scaled confluent Cauchy matrices is 1, so we coin the name *confluent Cauchy-like* with the matrices whose displacement rank, though higher than 1, still much smaller than the size n.

3. Factorization of confluent Cauchy matrices. Confluent Vandermonde matrices

We shall need a superfast algorithm to multiply a confluent Cauchy matrix by a vector, this algorithm will be based on the formula (3.12) derived next.

THEOREM 3.1. *Let C be the confluent Cauchy-like matrix defined in (2.9), $b(x)$ is defined in (2.8) and ∂_x is defined in (2.10). Then*

$$C = V_P(x) V_P(y)^{-1} diag\{B_1, B_2, \dots B_n\}, \tag{3.12}$$

where

$$V_P(x) = col \begin{pmatrix} V_P(x_1) \\ \vdots \\ V_P(x_k) \end{pmatrix}$$

with $V_P(x_i) =$

$$\begin{bmatrix} P_0(x_i) & P_1(x_i) & \cdots & P_{n-1}(x_i) \\ \partial_x P_0(x_i) & \partial_x P_1) & \cdots & \partial_x P_{n-1}(x_i) \\ \partial_x^2 P_0(x_i) & \partial_x^2 P_1(x_i) & \cdots & \partial_x^2 P_{n-1}(x_i) \\ \vdots & \vdots & & \vdots \\ \partial_x^{m_i-1} P_0(x_i) & \partial_x^{m_i-1} P_1(x_i) & \cdots & \partial_x^{m_i-1} P_{n-1}(x_i) \end{bmatrix},$$

and

$$B_i = \begin{bmatrix} 0 & \cdots & 0 & \partial_x^{m_1} b(x_i) \\ \vdots & \cdot^{\cdot^{\cdot}} & \cdot^{\cdot^{\cdot}} & \partial_x^{m_1} b(x_i) \\ 0 & \partial_x^{m_1} b(x_i) & \cdot^{\cdot^{\cdot}} & \vdots \\ \partial_x^{m_1} b(x_i) & \partial_x^{(m_1+1)} b(x_i) & \cdots & \partial_x^{(2m_1-1)} b(x_i) \end{bmatrix}, \tag{3.13}$$

The proof is based on the following formula which is of interest by itself:

$$V_P(y)^{-1} = \tilde{I} V_{\hat{P}}^T(y) \cdot \text{ diag } (B_1^{-1}, B_2^{-1}, \dots, B_k^{-1}),$$

where $\tilde{I}$ stands for the flip permutation matrix, and $\{\widehat{P}\}$ denotes the *associated* system of polynomials defined in [**11**] (see also [**16**] for a connection with inversion of discrete transmission lines) .

We now turn to the computational aspects of the formula (3.12). Though the formula is given for an arbitrary polynomial system $\{P\}$ we shall use it here only for the usual power basis $P_k(x) = x^k$ (other choices, e.g., the Chebyshev polynomials are useful from the numerical point of view). The formula reduces the problem of multiplication of C by a vector to the same problem for the three matrices on the right hand side of (3.12). Each of the three problems is treated next (it is convenient to discuss these three problem in a reversed order, i.e., first step 3, then step 2, and finally step 1).

Step 3. Multipoint Hermite-type evaluation: The problem of multiplying $V(x)$ by a vector $[a_k]$ is equivalent to the problem of multipoint Hermite-type evaluation. Indeed, let us use the entries $[a_k]$ of to define the polynomial $a(x) = \sum_{k=0}^{n-1} a_k x^k$. Then it is clear that we need to evaluate the value of $a(x)$ at each x_i as well as the values of its first m_i (scaled) derivatives $\frac{a^{(s)}(x_i)}{s!}$ (notice that the rows of $V(x_i)$ are obtained by differentiation of their predecessors). The algorithm for this problem is offered in the next section.

Step 2. Hermite interpolation: The problem of multiplying $V_P(x)^{-1}$ by a vector is the problem that is inverse to the one discussed in the step 3 above. The superfast algorithm for this problem can be found in [**1**]. The complexity is the same as in the step 3.

Step 1. Convolution: The problem of multiplication of $diag\{B_1, B_2, \ldots, B_n\}$ by a vector is clearly reduced to the convolution, since all diagonal blocks (3.13) have a Hankel structure (constant along anti-diagonals). We also need to compute the entries of all these blocks B_k, i.e., for each point x_k we need to compute the first $2m_i$ Taylor coefficients $\partial_x^s b(x_k)$ $(s = 0, 1, \ldots 2m_i - 1)$. This is exactly the problem we treat in the step 3 above.

4. Multipoint Hermite-type evaluation

4.1. The problem. The problem we discuss here is that of computing the Taylor expansion of a univariate polynomial at different points. More precisely, for $f \in K[x]$ and $\xi \in K$ there are uniquely determined elements $f_{0,\xi}, f_{1,\xi}, \ldots$ such that

$$f = f_{0,\xi} + f_{1,\xi}(x - \xi) + f_{2,\xi}(x - \xi)^2 + \cdots .$$

The sequence $T(f, \xi, n) := (f_{0,\xi}, \ldots, f_{n-1,\xi})$ is called the sequence of Taylor coefficients of f up to order n. Let $\xi_1, \ldots, \xi_t$ be elements in K and $d_1, \ldots, d_t$ be positive integers, and suppose that f is a univariate polynomial of degree $< n := d_1 + \cdots + d_t$. We want to develop a fast algorithm for computing the sequences

$$T(f, \xi, d_1), \ldots, T(f, \xi, d_t).$$

Our algorithm will use as a subroutine a fast multiple evaluation algorithm. Its running time involves the entropy of the sequence $(d_1, \ldots, d_t)$ defined by

$$\mathcal{H}(d_1, \ldots, d_t) := -\sum_{i=1}^{t} \frac{d_i}{n} \log \frac{d_i}{n},$$

where $n = \sum_i d_i$ and log is the logarithm to the base 2.

4.2. Our Algorithm. The algorithm we present for the problem stated above consists of several steps.

ALGORITHM 4.1. *On input $\xi \in K$ and $d \geq 1$ the algorithm computes the polynomials $\Pi_{\ell,\xi} := (x-\xi)^{2^\ell}$ for $\ell = 0, \ldots, \lceil \log d \rceil$, as well as the polynomial $(x-\xi)^d$.*

(1) *Put $\Pi_{0,\xi} := (x-\xi)$.*
(2) *For $\ell := 1, \ldots, \lceil \log d_i \rceil$ put $\Pi_{\ell,\xi} := \Pi_{\ell-1,\xi}^2$.*
(3) *Let $d = 2^{\ell_1} + 2^{\ell_2} + \ldots + 2^{\ell_s}$ with $\ell_1 < \cdots < \ell_s$. Put $P := \Pi_{\ell_1,\xi}$.*
(4) *For $i = 2, \ldots, s$ set $P := P \cdot \Pi_{\ell_i,\xi}$. The final value of P equals $(x-\xi)^d$.*

LEMMA 4.2. *Algorithm 4.1 correctly computes its output with $O(M(d))$ operations.*

The proof of this and the following assertions are omitted and will be included in the final version of the paper.

The next step of our presentation is the solution to our problem in case $t = 1$.

ALGORITHM 4.3. *On input a positive integer d, and element $\xi \in K$, and a univariate polynomial $f \in K[x]$ of degree less than d, the algorithm computes $T(f, \xi, d)$. We assume that we have precomputed the polynomials $(x-\xi), (x-\xi)^2, \ldots, (x-\xi)^{2^\tau}$ where $\tau = \lceil \log d \rceil$.*

(1) *If $d = 1$, return f, else perform Step (2).*
(2) *Compute $f_0 := f \bmod (x-\xi)^{2^{\tau-1}}$ and $f_1 := (f - f_0)/(x-\xi)^{2^{\tau-1}}$ and run the algorithm recursively on f_0 and f_1. Output the concatenation of their outputs.*

LEMMA 4.4. *The above algorithm correcly computes its output in time $O(M(d))$.*

Now we are ready for our final algorithm.

ALGORITHM 4.5. *On input positive integers $d_1, \ldots, d_t$, elements $\xi_1, \ldots, \xi_t \in K$, and a polynomial $f \in K[x]$ of degree less than $n := \sum_{i=1}^t d_i$, the algorithm computes the sequences $T(f, \xi_i, d_i)$, $i = 1, \ldots, t$.*

(1) *We use Algorithm 4.1 to the inputs (ξ, d_i) for $i = 1, \ldots, t$. At this stage, we have in particular computed $(x-\xi_i)^{d_i}$ for $i = 1, \ldots, t$.*
(2) *Now we use the multiple evaluation algorithm given in the proof of* [**1**, Th. (3.19)] *to compute $f_i := f \bmod (x-\xi_i)^{d_i}$, $i = 1, \ldots, t$.*
(3) *Use Algorithm 4.3 on input (f, ξ_i, d_i) to compute $T(f, \xi_i, d_i)$ for $i = 1, \ldots, t$.*

THEOREM 4.6. *The above algorithm correctly computes its output in time $O(M(n)\mathcal{H})$ where $\mathcal{H}$ is the entropy of the sequence $(d_1, \ldots, d_t)$.*

We remark that the algorithm can obviously be customized to run in parallel time $O(\mathcal{H})$ on $O(n)$ processors.

References

[1] P. Bürgisser, M. Clausen and A. Shokrollahi, Algebraic Complexity Theory, series = Grundlehren der Mathematischen Wissenschaften, vol. 315, Springer Verlag, Heidelberg, 1996.

[2] J. Ball, I. Gohberg and L. Rodman, *Interpolation of rational matrix functions*, OT45, Birkhäuser Verlag, Basel, 1990.

[3] D. Bini and V. Pan, *Polynomial and Matrix Computations*, Volume 1, Birkhauser, Boston, 1994.

[4] Ph. Delsarte, Y. Genin and Y. Kamp, *On the role of the Nevanlinna-Pick problem in circuit and system theory*, Circuit Theory and Appl., **9** (1981), 177-187.

[5] B. Friedlander, M. Morf, T. Kailath and L. Ljung, "New inversion formulas for matrices classified in terms of their distance from Toeplitz matrices," Linear Algebra and Appl., **27**, 31-60, 1979.

[6] A. Gerasoulis, A fast algorithm for the multiplication of generalized Hilbert matrices with vectors, *Math. of Computation*, **50** (No. 181), 1987, 179 – 188.

[7] I. Gohberg and V. Olshevsky, "Fast state space algorithms for matrix Nehari and Nehari-Takagi interpolation problems," *Integral Equations and Operator Theory,* **20**, No. 1, pp. 44-83, 1994.

[8] I. Gohberg and V. Olshevsky, *Fast inversion of Chebyshev-Vandermonde matrices, Numerische Mathematik*, **67 (1)** (1994), 71-92.

[9] I. Gohberg and V. Olshevsky, *Complexity of multiplication with vectors for structured matrices, Linear Algebra and Its Applications*, **202** (1994), 163-192.

[10] G. Heinig and K. Rost, *Algebraic methods for Toeplitz-like matrices and operators*, Operator Theory, vol 13, Birkhauser, Basel, 1984.

[11] T. Kailath and V. Olshevsky, *Displacement Structure Approach to Polynomial Vandermonde and Related Matrices, Linear Algebra and Its Applications*, **261** (1997), 49-90.

[12] T. Kailath and V. Olshevsky, *Diagonal Pivoting for Partially Reconstructible Cauchy-like Matrices, With Applications to Toeplitz-like Linear Equations and to Boundary Rational Matrix Interpolation Problems, Linear Algebra and Its Applications*, **254** (1997), 251-302.

[13] T. Kailath and A.H. Sayed, *Displacement structure : Theory and Applications*, SIAM Review, **37** No.3 (1995), 297-386.

[14] M.Morf, "Fast algorithms for multivariable systems," Ph.D. thesis, Department of Electrical Engineering, Stanford University, 1974.

[15] M.Morf, "Doubling Algorithms for Toeplitz and Related Equations," *Proc. IEEE Internat. Conf. on ASSP*, pp. 954-959, IEEE Computer Society Press, 1980.

[16] V. Olshevsky, Eigenvector computation for almost unitary Hessenberg matrices and inversion of Szego-Vandermonde matrices via discrete transmission lines. Linear Algebra and Its Applications, (285)1-3 (1998) pp. 37-67

[17] V. Olshevsky, "Pivoting for structured matrices with applications," to appear in Linear Algebra and Its Applications, 2000;
available on http://www.cs.gsu.edu/~matvro

[18] V. Olshevsky and V. Pan, "A superfast state-space algorithm for tangential Nevanlinna-Pick interpolation problem," in *Proceedings of the 39th IEEE Symposium on Foundations of Computer Science*, pp. 192–201, 1998.

[19] V. Olshevsky and A. Shokrollahi, "A Displacement Approach to Efficient Decoding of Algebraic Codes," in *Proceedings of the 31st Annual ACM Symposium on Theory of Computing*, 235–244, 1999.

[20] V. Rokhlin, Rapid Solution of Integral Equations of Classical Potential Theory, *J. Comput. Physics*, **60**, 187–207, 1985.

Department of Mathematics and Statistics, Georgia State University, Atlanta, GA 30303
E-mail address: `volshevsky@gsu.edu`

Digital Fountain, 600 Alabama Street, San Francisco, CA 94110
E-mail address: `amin@digitalfountain.com`

Contemporary Mathematics
Volume **280**, 2001

The Maximal-Volume Concept in Approximation by Low-Rank Matrices

S. A. Goreinov and E. E. Tyrtyshnikov

ABSTRACT. It is shown how the maximal-volume concept from interpolation theory can be formulated for matrix approximation problems using low-rank matrices.

1. Introduction

Consider a compact domain $V \subset \mathbb{R}^p$, a function $f(x)$ continuous for $x \in V$, and some "simple" functions $u_1(x)$, …, $u_k(x)$ continuous for $x \in V$. We want to approximate $f(x)$ by a linear combination of these functions,

$$f(x) \approx \alpha_1 u_1(x) + \dots \alpha_k u_k(x),$$

so that the interpolation property

$$f(x_i) = \alpha_1 u_1(x_i) + \dots + \alpha_k u_k(x_i), \quad i = 1, \dots, k,$$

holds for some set of pairwise distinct nodes $x_i \in V$. If we were allowed to choose the nodes, what would be the best or near-to-best choice?

To formulate the above question in rigorous terms, bring in the norm

$$||u||_C \equiv \sup_{x \in V} |u(x)|.$$

For a given collection $x_1, \dots, x_k$ of nodes, the number

$$E_{int} \equiv ||f - \alpha_1^{int} u_1 - \dots - \alpha_k^{int} u_k||_C,$$

where the coefficients $\alpha_1^{int}, \dots, \alpha_k^{int}$ are given by the interpolation condition, measures the approximation quality. Let

$$E_{best} \equiv \inf_{\alpha_1, \dots, \alpha_k} ||f - \alpha_1 u_1 - \dots - \alpha_k u_k||_C.$$

Obviously, $E_{int} \geq E_{best}$. An intelligent choice of the nodes should guarantee that E_{int} is not too much larger than E_{best}.

In the unidimensional case, if V is a closed interval then the Chebyshev nodes are of good choice: one can prove (using Bernstein's results; cf [**12**]) that $E_{int} \leq c \log k \ E_{best}$, where $c > 0$ is a constant independent of k. That is not easy to

---

1991 *Mathematics Subject Classification.* 15A60.
This work was supported by the Russian Fund of Basic Research under Grant 99-01-00017.

generalize to the multidimensional case. However, there is an estimate for some other choice of the nodes which is a little bit weaker but quite universal (and much easier to prove). This choice is based on the following *maximal-volume concept*: find $x_1, \ldots, x_k \in V$ so that

$$|\det M| \equiv |\det M(x_1, \ldots, x_k)|$$

becomes maximal, where

$$M \equiv |\det M(x_1, \ldots, x_k)| \begin{bmatrix} u_1(x_1) & \ldots & u_1(x_k) \\ \ldots & \ldots & \ldots \\ u_k(x_1) & \ldots & u_k(x_k) \end{bmatrix}.$$

By the *volume* of a matrix, we mean the absolute value of its determinant (more generally, for an overdetermined matrix, it would mean the volume of the parallelepiped spanned by its columns). A consequence of this choice is the following classical result (see [**2**]).

THEOREM 1.1. *Let the interpolation nodes maximize the volume of the corresponding matrix M. Then*

$$E_{int} \leq (k+1)\, E_{best}.$$

This theorem has an elementary proof.

Denote by M_i the matrix coinciding with M everywhere except for the ith column, which is substituted by $[u_1(x), \ldots, u_k(x)]^T$. Then it can be verified straightforwardly that

$$u_{int} \equiv \alpha_1^{int} u_1 + \ldots \alpha_k^{int} u_k = f(x_1)\frac{\det M_1}{\det M} + \cdots + f(x_k)\frac{\det M_k}{\det M}.$$

Let $u_{best} = \alpha_1^{best} u_1 + \cdots + \alpha_k^{best} u_k$ be the function such that $E_{best} = ||f - u_{best}||_C$. Then

$$\begin{aligned} |f - u_{int}| &\leq |f - u_{best}| + |u_{best} - u_{int}| \\ &\leq E_{best} + \sum_{i=1}^{n} |f(x_i) - u_{best}(x_i)|\, |\frac{\det M_i}{\det M}|. \end{aligned}$$

By the maximal-volume assumption, $|\det M_i| \leq |\det M|$, and this completes the proof.

The purpose of this note is to present an algebraic version of this theorem for the approximation by low-rank matrices.

2. Main results

For any matrix A, denote by $||A||_C$ the maximal modulus of its entries. Let $\sigma_1(A) \geq \sigma_2(A) \ldots$ signify the singular values of A in the non-ascending order.

THEOREM 2.1. *Suppose that A is a block matrix of the form*

$$\begin{bmatrix} A_{11} & A_{12} \\ A_{21} & A_{22} \end{bmatrix}, \tag{2.1}$$

where A_{11} is nonsingular, $k \times k$, and of maximal volume among all $k \times k$ submatrices. Then

$$||A_{22} - A_{21}A_{11}^{-1}A_{12}||_C \leq (k+1)\,\sigma_{k+1}(A). \tag{2.2}$$

Proof. We have to show that the modulus of each entry of $E \equiv A_{22} - A_{21}A_{11}^{-1}A_{12}$ is not greater than $(k+1)\,\sigma_{k+1}$.

Consider any $(k+1)\times(k+1)$ submatrix $\hat{A}$ which is an augmentaion of A_{11} by one column and one row of A:

$$\hat{A} = \begin{bmatrix} A_{11} & b \\ c^T & a \end{bmatrix},$$

and set $\gamma = a - c^T A_{11}^{-1} b$. Obviously, $\gamma = 0$ if and only if $\hat{A}$ is singular. In the latter case, the corresponding entry of E is equal to zero.

It remains to consider the case $\gamma \neq 0$. We infer that $\sigma_{k+1}(A) > 0$, otherwise $E = 0$ and, hence, $\gamma = 0$. As is readily seen,

$$|\det \hat{A}|\;|\det A_{11}| = |\gamma^{-1}|,$$

and together with the maximal-volume property of A_{11} this implies that

$$|\gamma^{-1}| = ||\hat{A}^{-1}||_C.$$

Now, take into account that

$$\sigma_{k+1}^{-1}(A) \leq \sigma_{k+1}^{-1}(\hat{A}) \leq ||\hat{A}^{-1}||_C\;(k+1).$$

It immediately follows that

$$|\gamma| = ||\hat{A}^{-1}||_C^{-1} \leq (k+1)\;\sigma_{k+1}(A),$$

which completes the proof.

The maximal-volume assumption can be quite reasonably relaxed.

THEOREM 2.2. *Under the notation of Theorem 2.1, assume that A_{11} is nonsingular, $k\times k$, and its volume is not less than ν times the maximal volume among all $k\times k$ submatrices. Then*

$$||A_{22} - A_{21}A_{11}^{-1}A_{12}||_C \leq \nu^{-1}\;(k+1)\;\sigma_{k+1}(A). \tag{2.3}$$

Proof. Instead of $|\gamma^{-1}| = ||\hat{A}^{-1}||_C$, we obtain

$$|\gamma^{-1}| \geq \nu\;||\hat{A}^{-1}||_C.$$

Consequently,

$$|\gamma| \leq \nu^{-1}\;||\hat{A}^{-1}||_C^{-1} \leq \nu^{-1}\;(k+1)\;\sigma_{k+1}(A),$$

which will do the proof.

COROLLARY 2.3. *Let*

$$C = \begin{bmatrix} A_{11} \\ A_{21} \end{bmatrix}, \quad R = \begin{bmatrix} A_{11} & A_{12} \end{bmatrix}.$$

Then, under the assumptions of Theorem 2.2,

$$||A - CA_{11}^{-1}R||_C \leq \nu^{-1}\;(k+1)\;\sigma_{k+1}(A). \tag{2.4}$$

Frequently, A acquires the form (2.1) after some permutations of its columns and rows. To avoid these permutations in the formulation, we can denote by C and R the submatrices consisting of the corresponding columns and rows of A with the intersection block A_{11} (not necessarily located in the upper left corner of A). Then (2.4) remains valid.

If A_{11} is singular, then (2.4) can be transformed into the following inequality:

$$||A - CGR||_C \leq \nu^{-1}\,(k+1)\,\sigma_{k+1}(A), \tag{2.5}$$

where G is an appropriately chosen square matrix of order k. In [**5**], a matrix of the form CGR was called a pseudo-skeleton component of A.

3. Discussion

Recently, there have been a series of papers devoted to fast numerical algorithms for large, dense, and unstructured matrices coming from integral equations [**6, 7, 8, 13, 11**]. From the matrix algebra point of view, the techniques proposed in those papers suggest (implicitly) that a coefficient matrix is approximated by some structured matrices. In particular, it is intrumental to use matrices that could be split into blocks which are mostly of low rank. This algebraic point of view was put forth in [**9, 10**], some advanced estimates of the approximation accuracy were given in [**3**].

In [**4, 5**], it was discovered that a low-rank approximation of reasonable quality for an individual block can be computed through only a small part of the entries of this block. The entries in question belong to a cross comprised by the above-mentioned matrices C and R. A difficult problem is how could we select a proper cross. The choice discussed in [**4, 5**] was based in effect on the singular vectors for the block, which is not feasible. Now we have proved that a good choice results from pursueing the maximal-volume concept. Although we obtained now a short and elementary proof, the way to this result was not immediate (on the whole, it took us several years).

The maximal volume concept has certainly something to do with the pivoting ordering of the rows (and columns). For some structured matrices (Vandermonde-like and Cauchy-like) it can be done in advance and fast [**1**].

References

[1] T. Boros, T. Kailath, and V. Olshevsky, Fast Björk-Pereyra-type algorithm for parallel solution of Cauchy linear equations, *Linear Algebra Appl.* 302-303: 265–293, 1999.

[2] D. Gaier, *Vorlesungen über Approximation im Complexen*, Birkhäuser Verlag, Basel-Boston-Berlin, 1980.

[3] S. A. Goreinov, Mosaic-skeleton approximations of matrices generated by asymptotically smooth and oscillatory kernels, *Matrix Methods and Computations*, INM RAS, Moscow, pp. 41–76, 1999. (In Russian.)

[4] S. A. Goreinov, E. E. Tyrtyshnikov, N. L. Zamarashkin, Pseudo-Skeleton Approximations of Matrices, *Reports of the Russian Academy of Sciences*, 343(2): 151–152, 1995. (In Russian.)

[5] S. A. Goreinov, E. E. Tyrtyshnikov, N. L. Zamarashkin, A Theory of Pseudo-Skeleton Approximations, *Linear Algebra Appl.* 261: 1–21, 1997.

[6] L. Greengard and V. Rokhlin, A fast algorithm for particle simulations. *J. Comput. Physics* 73: 325–348, 1987.

[7] W. Hackbusch and Z. P. Novak, On the fast matrix multiplication in the boundary element method by panel clustering, *Numer. Math.* 54(4): 463–491, 1989.

[8] S. V. Myagchilov and E. E. Tyrtyshnikov, A fast matrix-vector multiplier in discrete vortex method, *Russian J. Numer. Anal. Math. Modelling* 7(4): 325–342, 1992.

[9] E. E. Tyrtyshnikov, Mosaic ranks and skeletons, in *L. Vulkov et al. (eds.), Lecture Notes in Computer Science 1196: Numerical Analysis and Its Applications. Proceedings of WNAA-96.* Springer-Verlag, 1996. pp. 505–516.

[10] E. E. Tyrtyshnikov, Mosaic–skeleton approximations, *Calcolo*, 33(1-2): 47–57, 1996.

[11] E. Tyrtyshnikov, Methods for fast multiplication and solution of equations, *Matrix Methods and Computations*, INM RAS, Moscow, pp. 4–40, 1999. (In Russian.)

[12] E. Tyrtyshnikov, *A Brief Introduction to Numerical Analysis*, Birkhauser, Boston, 1997.
[13] V. V. Voevodin, On an order-reduction method for matrices arising in the solution of integral equations, *Numerical Analysis on FORTRAN*, Moscow State University Press, 21–26, 1979. (In Russian.)

INSTITUTE OF NUMERICAL MATHEMATICS OF RUSSIAN ACADEMY OF SCIENCES, GUBKINA 8, MOSCOW 117333, RUSSIA
E-mail address: tee@inm.ras.ru

Contemporary Mathematics
Volume **280**, 2001

A Matrix Interpretation of the Extended Euclidean Algorithm

Martin H. Gutknecht

Abstract. The extended Euclidean algorithm for polynomials and formal power series that is used for the recursive computation of Padé approximants can be viewed in various ways as a sequence of successive matrix multiplications that are applied to a Sylvester matrix with the original data. Here we present this result in a general version that includes the treatment of the Cabay–Meleshko look-ahead algorithm, which generalizes the extended Euclidean algorithm and yields a weakly stable (forward stable) method for computing Padé fractions if it is combined with an appropriate rule for choosing the look-ahead step length. Moreover, we choose for the matrix interpretation a particularly appealing form where also the product of all the matrices that are applied has a meaning: this product yields at the end four Toeplitz blocks with the coefficients of the polynomials (which belong to Padé forms) that are generated by the extended Euclidean algorithm in addition to those resulting from the ordinary Euclidean algorithm.

1. Introduction and notation

The Euclidean algorithm for polynomials or power series is closely related to Padé approximation. It is, in fact, up to minor modifications, one of the standard algorithms to compute Padé approximants along a diagonal of the Padé table and to generate the corresponding continued fraction (called a P-fraction), although the name Euclid comes only up in part of the Padé literature. An excellent, detailed account of these connections is given by Bultheel and Van Barel [**BV97**]; some of the main aspects are also treated by Gragg and Gutknecht [**GG94**] and others.

Since the Euclidean algorithm (in the form discussed here) concerns linear combinations of polynomials and power series, it is no surprise that one can interpret it in terms of matrices. Recall that the ring of formal power series is isomorphic to the ring of infinite lower triangular Toeplitz matrices. Such matrix interpretations of the Euclidean algorithm for polynomials are not very common, but some sketches can be found in the literature, at least for the (generic) case where in each step the degrees of the polynomials reduce only by 1; see, for example, Householder and Stewart [**HS69**] — we do not attempt to give a complete set of references here. A

1991 *Mathematics Subject Classification.* Primary 41A21; Secondary 30E05.

This paper was first presented at the conference "Orthogonal Polynomials in Numerical Analysis" in Oberwolfach, Germany, March 23–27, 1998.

single step of the algorithm is described as a matrix multiplication, and the whole algorithm takes the form of a matrix product consisting of a Sylvester matrix with the original data multiplied by a sequence of matrices that contain the quotients from the Euclidean algorithm. However, there is some freedom in arranging the data and in defining the factors of the product. We have succeeded to do this in a way so that the product of the matrix factors becomes a 4×4 block matrix with Toeplitz blocks that contain the coefficients of the Padé forms that are implicitly constructed in the process. In the terminology of Padé approximation, the Euclidean algorithm produces residuals of Padé approximants, while the Padé approximants itself, or rather the Padé forms (which are pairs of polynomials that are the numerators and denominators of the Padé approximants, respectively) are constructed additionally in the so-called extended Euclidean algorithm [**BV97**]. The matrix identity that we derive links the Sylvester matrix containing the given data not only to the recurrence coefficients (quotients) of the Euclidean algorithm but also with the six different sets of polynomials that are generated by the extended version. Moreover, the well-known symmetric LDU and inverse symmetric LDU decompositions of the Hankel moment matrix associated with the Padé problem can be read off the matrix identity.

Our motivation for this work came from section 9 of [**GH95a**], where, building up on earlier work by Delosme and Ipsen [**DI89**], Gutknecht and Hochbruck derived an analogous matrix representation of the look-ahead Levinson and Schur(-Bareiss) algorithms introduced in that paper. There too the resulting matrix identities can be interpreted in many ways. In particular, they are linked to the LDU and inverse LDU decomposition of the Toeplitz matrix of the given trigonometric moments.

Before we can start with the derivation and description of our matrix form of the Euclidean algorithm we present in sections 2–6 the prerequisites. This partly well-known material is extracted from [**GG94**] and suitably adapted to the situation and needs of this paper.

Notation. We let $\mathcal{L}$ be the set of formal Laurent series

$$h_0(\zeta) = \sum_{k=-\infty}^{\infty} \mu_k \zeta^k$$

with complex coefficients. The following subsets of $\mathcal{L}$ will play a role:

$$\begin{aligned}
\mathcal{L}_{l:m} &:\equiv \{h_0 \in \mathcal{L};\ \mu_k = 0 \text{ if } k < l \text{ or } k > m\}, \\
\mathcal{L}_m^* &:\equiv \mathcal{L}_{-\infty:m} = \{h_0 \in \mathcal{L};\ \mu_k = 0 \text{ if } k > m\}, \\
\mathcal{L}_0^* &:\equiv \mathcal{L}_{-\infty:0} = \text{formal power series in } \zeta^{-1}, \\
\mathcal{P}_m &:\equiv \mathcal{L}_{0:m} = \text{polynomials in } \zeta \text{ of degree } m, \\
\mathcal{P} &:\equiv \bigcup \mathcal{P}_m = \text{polynomials in } \zeta.
\end{aligned}$$

For $h_0 \in \mathcal{L}_m^*$ or $h_0 \in \mathcal{P}$ we define the *exact degree* ∂h_0 by

$$\partial h_0 :\equiv \{n \in \mathbb{Z};\ \mu_n \neq 0, \mu_k = 0 \text{ for } k > n\}.$$

We also use the notation $h_0(\zeta) = \mathcal{O}_-(\zeta^m)$ to indicate that $h_0 \in \mathcal{L}_m^*$, while $h_0(\zeta) \equiv \mathcal{O}_-(\zeta^m)$ means that $h_0(\zeta) \in \mathcal{L}_m^*$, but $h_0(\zeta) \notin \mathcal{L}_{m-1}^*$; in other words, it means the same as $\partial h_0 = m$. Moreover, if $h_0 \in \mathcal{L}_0^*$ we define $h_0(\infty) :\equiv \mu_0$.

Finally, $\mathcal{R}_{m,n}$ denotes the set of rational functions of type (m, n). Such a function r that can be represented as the quotient of two not necessarily relatively prime polynomials, $r = p/q$, where the numerator p has degree at most m, and the denominator q has degree at most n and is not the zero polynomial.

While $:\equiv$ is used to denote definitions, $:=$ indicates assignments that may be used in an algorithm. (Sometimes either one could be applied.)

2. Padé approximants of a pair of formal power series

Given a pair of formal power series in ζ^{-1},

$$(2.1)\qquad \begin{array}{lll} f_0 \in \mathcal{L}^*_{-1}, & f_0 \neq 0, & f_0(\zeta) \equiv: \sum_{j=1}^{\infty} \phi_{j,0}\zeta^{-k}, \\ g_0 \in \mathcal{L}^*_0, & g_0(\infty) = 1, & g_0(\zeta) \equiv: \sum_{j=0}^{\infty} \gamma_{j,0}\zeta^{-k}, \end{array}$$

we consider, for fixed $n > 1$, pairs of pairs of polynomials,

$$(2.2)\qquad (\grave{p}_n, \grave{q}_n) \in \mathcal{P}_{n-2} \times \mathcal{P}_{n-1}, \qquad (p_n, q_n) \in \mathcal{P}_{n-1} \times \mathcal{P}_n \quad \text{with } q_n \text{ monic},$$

satisfying the following conditions: there are two new formal power series in ζ^{-1} of the form

$$(2.3)\qquad f_n \in \mathcal{L}^*_{-n-1}, \qquad g_n \in \mathcal{L}^*_{-n} \quad \text{with } g_n(\zeta) = \zeta^{-n} + \mathcal{O}_-(\zeta^{-n-1})$$

such that

$$(2.4a)\qquad \underbrace{g_0(\zeta)\grave{p}_n(\zeta)}_{\in \mathcal{L}^*_{n-2}} + \underbrace{f_0(\zeta)\grave{q}_n(\zeta)}_{\in \mathcal{L}^*_{n-2}} = \underbrace{g_n(\zeta)}_{\in \mathcal{L}^*_{-n}},$$

$$(2.4b)\qquad \underbrace{g_0(\zeta)p_n(\zeta)}_{\in \mathcal{L}^*_{n-1}} + \underbrace{f_0(\zeta)q_n(\zeta)}_{\in \mathcal{L}^*_{n-1}} = \underbrace{f_n(\zeta)}_{\in \mathcal{L}^*_{-n-1}}.$$

This means that (p_n, q_n) and $(\grave{p}_n, \grave{q}_n)$ are *Padé forms* of the pair (g_0, f_0) at $\zeta = \infty$. The Padé form (p_n, q_n) consists of the numerator p_n and the denominator q_n of the (n, n) *Padé approximant* $r_{n,n} :\equiv p_n/q_n$, and, likewise, $(\grave{p}_n, \grave{q}_n)$ contains the (differently normalized, not necessarily mutually prime) numerator and denominator of a $(n-1, n-1)$ Padé approximant, if we consider the fractions as functions of ζ^{-1}. Therefore, these Padé approximants lie on the main diagonal of the *Padé table* of (g_0, f_0). We call the formal power series g_n and f_n in ζ^{-1} *residuals* of (p_n, q_n) and $(\grave{p}_n, \grave{q}_n)$, respectively. (In the literature, often the shifted series $\zeta^n g_n(\zeta)$ and $\zeta^n f_n(\zeta)$ are referred to as the residuals; but here the formulas will be simpler when we let the leading n or $n+1$ coefficients be zero.)

Note that (2.4a) implies that the $2n-2$ coefficients of the terms $\zeta^{-n+1}, \ldots, \zeta^{n-2}$ on the left-hand side must be set to 0, and by (2.3) the one of ζ^{-n} is 1. These $2n-1$ conditions are matched by the same number of degrees of freedom in the coefficients of $\grave{p}_n$ and $\grave{q}_n$. Likewise, in (2.4b) the $2n$ coefficients of the terms $\zeta^{-n}, \ldots, \zeta^{n-1}$ are zero, and the polynomials p_n and q_n have exactly $2n$ free coefficients since q_n must be monic.

Normally, Padé approximation is defined for a single formal power series. We can return to this situation by noting that due to $g_0(\infty) = 1$ the quotient $h_0(\zeta) :\equiv -f_0(\zeta)/g_0(\zeta) \in \mathcal{L}^*_{-1}$ is well defined. If we then redefine in the above formulas

$$(2.5)\qquad g_0(\zeta) :\equiv 1, \qquad f_0(\zeta) :\equiv -h_0(\zeta),$$

it is easy to verify that (2.4a)–(2.4b) match with the usual Padé conditions. The series

$$h_0(\zeta) \equiv: \sum_{k=1}^{\infty} \frac{\mu_k}{\zeta^k} \quad \in \mathcal{L}_{-1}^*, \tag{2.6}$$

is the generating function of the *Markov parameters* or *moments* $\{\mu_k\}_{k=1}^{\infty}$ and can be viewed as the *symbol* of an infinite *Hankel matrix* M defined by $m_{k,l} :\equiv \mu_{k+l-1}$.

It is well known and easy to verify, see for example [**GG94**], that unique solutions $(\grave{p}_n, \grave{q}_n)$ and (p_n, q_n) to the conditions (2.2)–(2.4b) exist if and only if the nth principal leading submatrix

$$\mathsf{M}_n := \begin{bmatrix} \mu_1 & \mu_2 & \cdots & \mu_n \\ \mu_2 & & \cdot^{\cdot^{\cdot}} & \vdots \\ \vdots & \cdot^{\cdot^{\cdot}} & & \vdots \\ \mu_n & \cdots & \cdots & \mu_{2n-1} \end{bmatrix} \tag{2.7}$$

of M is nonsingular. In this case, we call q_n a *regular formal orthogonal polynomial* or *regular FOP*, the pair of denominators $(\grave{q}_n, q_n)$ a normalized nth *regular pair* and n a *regular index*. Heinig and Rost [**HR84**] refer to the pair as *fundamental solutions*, since it allows them to give an explicit formula for M^{-1}. Note that in the situation (2.5) it is trivial to determine the numerators $\grave{p}_n$ and p_n once the denominators $\grave{q}_n$ and q_n are known.

3. The recursive computation of Padé approximants by a generalized extended Euclidean algorithm

By using a 2×2 matrix with polynomial entries and row vectors with formal power series as entries we can write (2.4a)–(2.4b) as

$$\begin{bmatrix} g_0 & f_0 \end{bmatrix} \begin{bmatrix} \grave{p}_n & p_n \\ \grave{q}_n & q_n \end{bmatrix} = \begin{bmatrix} g_n & f_n \end{bmatrix}. \tag{3.1}$$

Note that if we extracted from the row vector on the right-hand side the common factor ζ^{-n}, the remainder would be of the same form as the row vector $[\ g_0 \quad f_0\ ]$. This indicates, but does not prove yet, that we can set up a recursive procedure. One still has to show that the corresponding 2×2 matrices fit together appropriately.

Generalizing earlier work of Gragg, Gustavson, Warner, and Yun [**GGWY82**] that was restricted to the generic case (where *all* leading principal submatrices of M are nonsingular) Cabay and Meleshko [**CM93**] established stable recurrences of this sort for computing well-conditioned Padé approximants along a diagonal of the Padé table. Here, well-conditioned means that the submatrix M_n, which could alternatively be used to determine these Padé approximants, is in a certain sense (left vague here) well conditioned. These Cabay-Meleshko recurrences are a special case of the following theorem, which could be used to compute *any* ordered sequence of regular pairs along the main diagonal of the table. The generalization to other diagonals is trivial. In the current notation this theorem is given and proved in [**GG94**].

THEOREM 3.1. *Let* $[\ \grave{q}_n \quad q_n\ ]$ *be a normalized* n*th regular pair for* $[\ g_0 \quad f_0\ ]$, *so that* (3.1) *holds with* (2.2)–(2.3). *Likewise, let* $[\ \grave{a} \quad a\ ]$ *be a normalized* k*th regular pair for* $\zeta^n[\ g_n \quad f_n\ ]$, *so that*

$$(\grave{b}, \grave{a}) \in \mathcal{P}_{k-2} \times \mathcal{P}_{k-1}, \qquad (b, a) \in \mathcal{P}_{k-1} \times \mathcal{P}_k \quad \textit{with } a \textit{ monic,} \tag{3.2}$$

and the residuals

$$[\ g_{n+k} \quad f_{n+k}\ ] := [\ g_n \quad f_n\ ] \begin{bmatrix} \grave{b} & b \\ \grave{a} & a \end{bmatrix} \tag{3.3}$$

satisfy

$$f_{n+k} \in \mathcal{L}^*_{-n-k-1}, \qquad g_{n+k} \in \mathcal{L}^*_{-n-k} \tag{3.4}$$

with

$$g_{n+k}(\zeta) = \zeta^{-n-k} + \mathcal{O}_-(\zeta^{-n-k-1}). \tag{3.5}$$

Then

$$\begin{bmatrix} \grave{p}_{n+k} & p_{n+k} \\ \grave{q}_{n+k} & q_{n+k} \end{bmatrix} := \begin{bmatrix} \grave{p}_n & p_n \\ \grave{q}_n & q_n \end{bmatrix} \begin{bmatrix} \grave{b} & b \\ \grave{a} & a \end{bmatrix} \tag{3.6}$$

yields a normalized $(n+k)$*th regular pair* $[\ \grave{q}_{n+k} \quad q_{n+k}\ ]$ *for* $[\ g_0 \quad f_0\ ]$ *as well as the corresponding numerators* $[\ \grave{p}_{n+k} \quad p_{n+k}\ ]$.

The corresponding pair of residuals is $[\ g_{n+k} \quad f_{n+k}\ ]$. *Hence,* (3.3) *is the recurrence for the residuals.*

To obtain stable recurrences one needs a rule for identifying well-conditioned normalized regular pairs; see Cabay and Meleshko [**CM93**] and, for further results and comments on this and related matter, Beckermann [**Bec96**] and Gragg and Gutknecht [**GG94**]. The basis of this rule is Heinig's inversion formula for M_n, an improved variation of the well-known Gohberg-Semencul formula; see Heinig and Rost [**HR84**] and Gutknecht and Hochbruck [**GH95b**].

The condition (3.3) for the "recurrence coefficients", that is, the Padé forms $(\grave{b}, \grave{a})$ and (b, a), translates easily into a pair of linear systems for the coefficients of these Padé forms. Let

$$g_n(\zeta) \equiv: \sum_{j=0}^{\infty} \gamma_{j,n} \zeta^{-n-j}, \qquad f_n(\zeta) \equiv: \sum_{j=1}^{\infty} \phi_{j,n} \zeta^{-n-j}, \tag{3.7}$$

with $\gamma_{0,n} = 1$, and consider the $2k \times 2k$ *Sylvester matrix*

$$\mathsf{S}_k(g_n, f_n) :\equiv \left[\begin{array}{cccc|cccc} 0 & \cdots & 0 & 1 & 0 & \cdots & \cdots & 0 \\ \vdots & \iddots & 1 & \gamma_{1,n} & \vdots & & 0 & \phi_{1,n} \\ 0 & \iddots & \iddots & \vdots & \vdots & \iddots & \iddots & \vdots \\ 1 & \iddots & & \gamma_{k-1,n} & 0 & \iddots & & \phi_{k-1,n} \\ \gamma_{1,n} & & \iddots & \gamma_{k,n} & \phi_{1,n} & & \iddots & \phi_{k,n} \\ \vdots & \iddots & \iddots & \vdots & \vdots & \iddots & \iddots & \vdots \\ \gamma_{k-1,n} & \iddots & & \vdots & \phi_{k-1,n} & \iddots & & \vdots \\ \gamma_{k,n} & \cdots & \cdots & \gamma_{2k-1,n} & \phi_{k,n} & \cdots & \cdots & \phi_{2k-1,n} \end{array}\right]. \tag{3.8}$$

Moreover, set

$$\mathsf{e}_{2k,2k} :\equiv [\ 0 \quad \cdots \quad 0 \quad 1\ ]^\top \in \mathbb{C}^{2k}, \qquad \mathsf{f}_{2k,n} :\equiv -[\ \phi_{1,n} \cdots \phi_{2k,n}\ ]^\top, \tag{3.9}$$

and let

$$\grave{\mathsf{b}}, \quad \grave{\mathsf{a}}, \quad \mathsf{b} \in \mathbb{C}^k, \qquad \mathsf{a} = \begin{bmatrix} \dot{\mathsf{a}} \\ 1 \end{bmatrix} \in \mathbb{C}^{k+1} \tag{3.10}$$

be the coefficient vectors of the polynomials $\grave{b}$, $\grave{a}$, b, and a, the first one being augmented by a zero component. Then (3.3) becomes, when the last component 1 of a is moved to the right-hand side,

$$\mathsf{S}_k(g_n, f_n) \begin{bmatrix} \grave{\mathsf{b}} & \mathsf{b} \\ \grave{\mathsf{a}} & \dot{\mathsf{a}} \end{bmatrix} = [\ \mathsf{e}_{2k,2k} \quad \mathsf{f}_{2k,n}\ ]. \tag{3.11}$$

When $k = 1$, which is the generic case, (3.11) reduces to

$$\begin{bmatrix} 1 & 0 \\ \gamma_{1,n} & \phi_{1,n} \end{bmatrix} \begin{bmatrix} \grave{\beta}_0 & \beta_0 \\ \grave{\alpha}_0 & \alpha_0 \end{bmatrix} = \begin{bmatrix} 0 & -\phi_{1,n} \\ 1 & -\phi_{2,n} \end{bmatrix}, \tag{3.12}$$

from which it follows that

$$\grave{\beta}_0 = 0, \qquad \grave{\alpha}_0 = \frac{1}{\phi_{1,n}}, \qquad \beta_0 = -\phi_{1,n}, \qquad \alpha_0 = \gamma_{1,n} - \frac{\phi_{2,n}}{\phi_{1,n}}. \tag{3.13}$$

4. J-fractions, P-fractions, and the extended Euclidean algorithm for power series

The recurrences of Theorem 3.1 are very general, and it is not so obvious what they have in common with the Euclidean algorithm. To explain this, we consider here two special cases, namely the one where $k = 1$ for all n, and thus all n are regular indices, and the one where $k > 1$ is allowed, but n and $n + k$ are two successive regular indices.

Therefore, for a moment assume that for all n, q_n is a regular FOP. Then the recursion (3.6) can be applied for all n with $k = 1$, where the coefficients are given by (3.13). Recall that this holds, if and only if $\phi_{1,n} \neq 0$ for all n. Then

$$(\grave{p}_{n+1}, \grave{q}_{n+1}) = (p_n, q_n)/\phi_{1,n}, \tag{4.1}$$

and hence

$$\begin{bmatrix} p_n & p_{n+1} \\ q_n & q_{n+1} \end{bmatrix} := \begin{bmatrix} p_{n-1} & p_n \\ q_{n-1} & q_n \end{bmatrix} \begin{bmatrix} 0 & -\beta_{n+1} \\ 1 & \zeta - \alpha_{n+1} \end{bmatrix}, \tag{4.2}$$

where

$$\beta_{n+1} := \frac{\phi_{1,n}}{\phi_{1,n-1}}, \qquad \alpha_{n+1} := \frac{\phi_{2,n}}{\phi_{1,n}} - \gamma_{1,n}, \tag{4.3}$$

with $\phi_{1,-1} := -1$, $\phi_{k,0} := \phi_k$, $\gamma_{k,0} := \gamma_k$.

Eliminating the trivial part of (4.2) we find the well-known generic diagonal recurrences for Padé forms and FOPs,

$$\begin{aligned} p_n(\zeta) &:= (\zeta - \alpha_n)p_{n-1}(\zeta) - \beta_n p_{n-2}(\zeta), \\ q_n(\zeta) &:= (\zeta - \alpha_n)q_{n-1}(\zeta) - \beta_n q_{n-2}(\zeta), \end{aligned} \tag{4.4}$$

with initial values

$$(p_{-1}(\zeta), q_{-1}(\zeta)) :\equiv (1, 0), \qquad (p_0(\zeta), q_0(\zeta)) :\equiv (0, 1). \tag{4.5}$$

They reveal that p_n and q_n are the nth numerator and denominator of the *Jacobi fraction (J-fraction)* of $h_0 = -f_0/g_0$,

$$h_0(\zeta) = -\frac{f_0(\zeta)}{g_0(\zeta)} = \frac{\beta_1 |}{| \zeta - \alpha_1} - \frac{\beta_2 |}{| \zeta - \alpha_2} - \cdots, \tag{4.6}$$

that is,

$$\frac{p_n(\zeta)}{q_n(\zeta)} = \frac{\beta_1 |}{| \zeta - \alpha_1} - \cdots - \frac{\beta_n |}{| \zeta - \alpha_n}. \tag{4.7}$$

This J-fraction does not exist if $\phi_{1,n} = 0$ for some n. It may have very large $|\alpha_{n+1}|$ and $|\beta_{n+2}|$ if $|\phi_{1,n}|$ is small.

In exact arithmetic the case $\phi_{1,n} = 0$ can be treated as follows: if

$$\phi_{0,n} = \cdots = \phi_{k-1,n} = 0, \qquad \phi_{k,n} \neq 0, \tag{4.8}$$

then $\mathsf{S}_1, \ldots, \mathsf{S}_{k-1}$ are singular, but S_k is nonsingular. The left system in (3.11) has then the solution

$$\grave{b}(\zeta) \equiv 0, \qquad \grave{a}(\zeta) \equiv \frac{1}{\phi_{k,n}}. \tag{4.9}$$

Consequently,

$$(\grave{p}_{n+k}, \grave{q}_{n+k}) = \frac{1}{\phi_{k,n}}(p_n, q_n), \qquad g_{n+k} = \frac{\zeta^k f_n}{\phi_{k,n}}. \tag{4.10}$$

In this way, we can proceed from any regular Padé form (p_n, q_n) and its upper-left neighbor $(\grave{p}_n, \grave{q}_n)$ to the next one, (p_{n+k}, q_{n+k}), and its upper-left neighbor $(\grave{p}_{n+k}, \grave{q}_{n+k})$, where k is determined by (4.8).

Let $\{n_j\}_{j=0}^{J}$ $(J \leq \infty)$ be the sequence of *all* regular indices, starting with $n_0 = 0$, and let

$$k_j :\equiv n_{j+1} - n_j \qquad (j \geq 1). \tag{4.11}$$

Additionally, we set $n_{-1} := -1$, $k_0 := 1$. Then, $\phi_{k_j, n_j} \neq 0$ $(\forall j)$. In analogy to (4.2), recurrence (3.6) becomes

$$\begin{bmatrix} p_{n_j} & p_{n_{j+1}} \\ q_{n_j} & q_{n_{j+1}} \end{bmatrix} := \begin{bmatrix} p_{n_{j-1}} & p_{n_j} \\ q_{n_{j-1}} & q_{n_j} \end{bmatrix} \begin{bmatrix} 0 & -\beta_{j+1} \\ 1 & a_{j+1}(\zeta) \end{bmatrix} \qquad (j \geq 0), \tag{4.12}$$

with

$$\beta_{j+1} := \frac{\phi_{k_j, n_j}}{\phi_{k_{j-1}, n_{j-1}}}, \qquad a_{j+1}(\zeta) := q_{k_j}^{(n_j)}(\zeta) \tag{4.13}$$

and initial conditions (4.5).

(4.12) is equivalent to the three-term recurrence of Magnus [**Mag62**] and Struble [**Str63**]:

$$\begin{aligned} p_{n_j}(\zeta) &:= a_j(\zeta) p_{n_{j-1}}(\zeta) - \beta_j\, p_{n_{j-2}}(\zeta), \\ q_{n_j}(\zeta) &:= a_j(\zeta) q_{n_{j-1}}(\zeta) - \beta_j\, q_{n_{j-2}}(\zeta), \end{aligned} \qquad (j \geq 1) \tag{4.14}$$

with the initial conditions (4.5); see also [**Gut92**] and further references given there.

They correspond to the Magnus *P-fraction*

$$h_0(\zeta) = -\frac{f_0(\zeta)}{g_0(\zeta)} = \frac{\beta_1 |}{| a_1(\zeta)} - \frac{\beta_2 |}{| a_2(\zeta)} - \cdots. \tag{4.15}$$

whose partial numerators and denominators are p_{n_j} and q_{n_j}, respectively.

In view of (4.10), the recurrence (3.3) for the residuals becomes

$$f_{n_j}(\zeta) = a_j(\zeta) f_{n_{j-1}}(\zeta) - \beta_j\, f_{n_{j-2}}(\zeta) \quad (j \geq 1) \tag{4.16}$$

with initial conditions

$$f_{n_{-1}}(\zeta) := f_{-1}(\zeta) := -g_0(\zeta), \qquad f_{n_0}(\zeta) := f_0(\zeta). \tag{4.17}$$

Recall that

$$\partial f_{n_{j-1}} = -n_j > \partial f_{n_j} = -n_{j+1} > \partial f_{n_{j+1}} \tag{4.18}$$

and

$$k_{j+1} = \partial a_{j+1} = \partial f_{n_{j-1}} - \partial f_{n_j}. \tag{4.19}$$

β_j is just used to make a_{j+1} monic.

This reveals that the P-fraction can be constructed by a simple generalization of the *Euclidean algorithm* for polynomials. In fact, the Euclidean algorithm for two polynomials f_{-1} and f_0 with $\partial f_{-1} > \partial f_0$, generates a finite sequence of polynomials f_j of decreasing degree according to

$$f_j(\zeta) := a_j(\zeta) f_{j-1}(\zeta) - \beta_j f_{j-2}(\zeta) \qquad (j \geq 1), \tag{4.20}$$

with $0 \neq \beta_j \in \mathbb{C}$ arbitrary and $a_j \in \mathcal{P}$. Note that this is the same formula as (4.16) except for a slightly different indexing scheme.

(4.14) and (4.16) define the *extended Euclidean algorithm for power series in* ζ^{-1}. The word "extended" refers to the fact that not only the residuals f_{n_j} are updated, which are the tails of the P-fraction (4.15), but also the corresponding partial numerators and denominators p_{n_j} and q_{n_j}; see Bultheel and Van Barel [**BV97**] for a detailed discussion of the extended Euclidean algorithm.

In the generic case, where $a_j(\zeta) = \zeta - \alpha_j$, the Euclidean algorithm for power series is also referred to as *Chebyshev algorithm*. An alternative generalization of the latter to the case where $k > 1$ for some n was proposed by Golub and Gutknecht [**GG90**].

Dividing (4.16) by $f_{n_{j-1}}(\zeta)$ shows that a_j is the *polynomial* or *principal part* of the formal Laurent series at ∞ of $\beta_j f_{n_{j-2}}/f_{n_{j-1}}$. This fact gave rise to the name P-fraction.

5. Inner polynomials

Let us return to the Padé approximation problems for (g_0, f_0) or, equivalently, for $h_0 :\equiv -f_0/g_0$ that we considered in section 2. If n is not regular, that is, if M_n of (2.7) is singular, there is no or not a unique polynomial $q_n \in \mathcal{P}_n$ satisfying the corresponding Padé condition (2.4b) with some $p_n \in \mathcal{P}_{n-1}$ and some $f_n \in \mathcal{L}^*_{-n-1}$.

However, there are still solutions to a *relaxed Padé condition*: if n is regular and $k > 0$, we can always find a Padé form

$$(p_{n+k}, q_{n+k}) \in \mathcal{P}_{n+k-1} \times \mathcal{P}_{n+k} \quad \text{with } q_{n+k} \text{ monic} \tag{5.1}$$

such that the corresponding residual f_{n+k} satisfies

$$f_{n+k} :\equiv g_0 p_{n+k} + f_0 q_{n+k} \in \mathcal{L}^*_{-n-1}. \tag{5.2}$$

The quotient $r_{n+k,n+k} :\equiv p_{n+k}/q_{n+k}$ has been called an *underdetermined Padé approximant* in [**Gut93**]. In conjunction with the algorithm of Theorem 3.1, which allows us to compute a sequence of regular indices and pairs, we will refer to those q_{n+k} that are not part of this sequence but still satisfy (5.1)–(5.2) as *inner polynomials*.

The following theorem, which has been proved in [**GG94**], establishes not only the well-known existence of such inner polynomials and underdetermined Padé approximants, but also how to compute them from an earlier regular pair.

THEOREM 5.1. *Let* $[\ \grave{q}_n \quad q_n\ ]$ *be a normalized* n*th regular pair for* $[\ g_0 \quad f_0\ ]$, *and let* k *be a positive integer. Consider any pair* (p_{n+k}, q_{n+k}) *constructed according to*

$$\begin{bmatrix} p_{n+k} \\ q_{n+k} \end{bmatrix} := \begin{bmatrix} \grave{p}_n & p_n \\ \grave{q}_n & q_n \end{bmatrix} \begin{bmatrix} b \\ a \end{bmatrix} \tag{5.3}$$

from a pair

$$(b, a) \in \mathcal{P}_{k-1} \times \mathcal{P}_k \quad \text{with a monic} \tag{5.4}$$

that satisfies

$$f_{n+k} :\equiv g_n b + f_n a \in \mathcal{L}^*_{-n-1}. \tag{5.5}$$

Then (5.1) *and* (5.2) *hold.*

Condition (5.5) *is fulfilled when*

$$b(\zeta) := -\frac{\zeta^k f_n(\zeta)}{g_n(\zeta)} + \mathcal{O}_-(\zeta^{-1}), \qquad a(\zeta) := \zeta^k\,; \tag{5.6}$$

that is,

$$b(\zeta) := \mu_1^{(n)} \zeta^{k-1} + \mu_2^{(n)} \zeta^{k-2} + \cdots + \mu_k^{(n)} \tag{5.7}$$

if

$$h_n(\zeta) :\equiv -\frac{f_n(\zeta)}{g_n(\zeta)} \equiv: \mu_1^{(n)} \zeta^{-1} + \mu_2^{(n)} \zeta^{-2} + \cdots. \tag{5.8}$$

The computation of $b(\zeta)$ in (5.6) requires only the solution of a $k \times k$ triangular Toeplitz system, and as k is growing, these systems are nested, so that the solution of the last one contains the solution for all.

6. Matrix representations of Padé conditions and orthogonality

The most relevant aspect of the Padé condition (2.4b) for the Padé form (p_n, q_n) is that the terms in $\zeta^{-n}, \ldots, \zeta^{n-1}$ on the left-hand side cancel. If $g_0(\zeta) \equiv 1$ as in (2.5), $p_n \in \mathcal{P}_{n-1}$ is chosen so that the positive powers cancel, while q_n must be chosen so that the $n+1$ terms in $\zeta^{-n}, \ldots, \zeta^{-1}, \zeta^0$ cancel. If we choose some fixed $N > n$ and assume that all $n < N$ are regular indices, this condition can be formulated in terms of the $N \times N$ Hankel moment matrix M_N for $h_0 :\equiv -f_0/g_0$ as

$$\mathsf{M}_N \, \mathsf{R}_N = \mathsf{F}_N, \tag{6.1}$$

where

$$\mathsf{R}_N :\equiv \begin{bmatrix} 1 & \rho_{0,1} & \cdots & \cdots & \rho_{0,N-1} \\ & 1 & \cdots & \cdots & \rho_{1,N-1} \\ & & \ddots & & \vdots \\ & & & \ddots & \rho_{N-2,N-1} \\ & & & & 1 \end{bmatrix}, \tag{6.2}$$

is a unit upper triangular matrix containing in its columns the coefficients of the polynomials $q_0, q_1, \dots, q_{N-1}$, while F_N is a lower triangular matrix

$$\mathsf{F}_N :\equiv \begin{bmatrix} \phi_{1,0} & & & \\ \phi_{2,0} & \phi_{1,1} & & \\ \vdots & \vdots & \ddots & \\ \phi_{N,0} & \phi_{N-1,1} & \cdots & \phi_{1,N-1} \end{bmatrix} \tag{6.3}$$

whose columns contain the coefficients, down to those of the term in ζ^{-N}, of $f_0, f_1, \dots, f_{N-1}$.

If we multiply (6.1) from the left by $\mathsf{R}_N^\top$, we obtain

$$\mathsf{R}_N^\top \mathsf{M}_N \mathsf{R}_N = \mathsf{R}_N^\top \mathsf{F}_N,$$

where the left-hand side is clearly symmetric and the right-hand side is clearly lower triangular, hence both sides must equal a diagonal matrix D_N: on the left,

$$\mathsf{R}_N^\top \mathsf{M}_N \mathsf{R}_N = \mathsf{D}_N :\equiv \operatorname{diag}[\delta_0 \cdots \delta_{N-1}], \tag{6.4}$$

where

$$\delta_n :\equiv \mathsf{r}_n^\top \mathsf{M}_{n+1} \mathsf{r}_n, \qquad \mathsf{r}_n :\equiv [\ \rho_{0,n} \quad \dots \rho_{n,n}\ ]^\top. \tag{6.5}$$

Eq. (6.4) is often referred to as an *inverse symmetric LDU decomposition* of M_N. On the right, $\mathsf{R}_N^\top \mathsf{F}_N = \mathsf{D}_N$, or, $(\mathsf{R}_N^\top)^{-1} = \mathsf{F}_N \mathsf{D}_N^{-1}$, and, therefore, substituting R_N in (6.4) yields a *symmetric LDU decomposition* of M_N:

$$\mathsf{M}_N = \mathsf{F}_N \mathsf{D}_N^{-1} \mathsf{F}_N^\top. \tag{6.6}$$

The decompositions (6.4) and (6.6) are well known and are fully analogous to decomposition related to the Levinson and Schur algorithms for Toeplitz matrices, except that the latter are not necessarily symmetric; see, for example, **[GH95a]** for the decompositions of Toeplitz matrices. In particular, (6.4) is just a matrix representation of the mutual formal orthogonality of the FOPs.

If q_n is not regular for some n, the decompositions (6.4) and (6.6) do not exist in this form. But by applying Theorem 5.1 we can redefine all those q_n and f_n so that (6.4) and (6.6) hold as a *inverse symmetric block LDU decomposition* and a *symmetric block LDU decomposition*, respectively. Then, D_N is *block diagonal*, R_N is still *unit upper triangular*, and F_n is *lower block triangular*.

We may make the blocks bigger than theoretically needed and only keep those regular q_n that are well-conditioned. Then the columns of R_n and F_n with well-conditioned regular index n still contain the coefficients of the FOPs and the corresponding residuals. They are the first columns of a block, while the other columns contain the coefficients of the inner polynomials and the corresponding residuals, respectively.

If we want to aim at an analogous treatment of the general Padé conditions (2.4a)–(2.4b) for pairs of Padé forms $(\grave{p}_n, \grave{q}_n)$, (p_n, q_n) that interpolate pairs of power series (g_0, f_0), then we need to include the powers $\zeta^0, \dots, \zeta^{n-1}$. Using our notation (3.8) for Sylvester matrices we can write (with e_1 and e_N the first and the last standard unit vector of $\mathbb{C}^N$)

$$\mathsf{S}_N(g_0, f_0) \left[\begin{array}{c|c} \grave{\mathsf{P}}_N & \mathsf{P}_N \\ \hline \grave{\mathsf{R}}_N & \mathsf{R}_N \end{array}\right] = \left[\begin{array}{c|c} \mathsf{e}_N^\top \mathsf{e}_1 & \mathsf{O} \\ \hline \mathsf{G}_N & \mathsf{F}_N \end{array}\right], \tag{6.7}$$

where R_N is defined in (6.2) and $\grave{\mathsf{R}}_N$, P_N, and $\grave{\mathsf{P}}_N$ are analogous upper triangular matrices with the coefficients of the polynomials $\grave{q}_n$, p_n, and $\grave{p}_n$ $(0 \leq n < N)$, respectively, while G_N is a $N \times N$ lower Hessenberg matrix that contains in its columns the coefficients (down to those of ζ_{-N}) of the residuals g_n (except that $\gamma_{0,0} = 1$ is missing at the top of the first column), and F_N is a $N \times N$ lower triangular matrix with the coefficients of the residuals f_n as before.

However, compared to these residuals, those whose coefficients are in F_N of (6.3) are devided by $-g_0$, because this is how the general case of (2.4a)–(2.4b) is related to the special case of (2.5). Therefore, to return from (6.7) to the special case (6.1), we multiply (6.7) from the left by a $2N \times 2N$ lower triangular Toeplitz matrix that contains the first $2N$ coefficients of the formal power series of $-1/g_0$. Then, after redefining F_N, the $(2,2)$ block of the resulting block matrix identity yields exactly (6.1).

7. Matrix form of the extended Euclidean algorithm

Now, we can finally aim at the intended reformulation of the extended Euclidean algorithm as a matrix product applied to the Sylvester matrix with the given data. First, we discuss a single step and return to (3.3),

$$\begin{bmatrix} g_n & f_n \end{bmatrix} \begin{bmatrix} \grave{b} & b \\ \grave{a} & a \end{bmatrix} = \begin{bmatrix} g_{n+k} & f_{n+k} \end{bmatrix},$$

and its matrix form (3.11), rewritten with two additional columns and two additional rows as

$$\mathsf{S}_{k+1}(g_n, f_n) \left[\begin{array}{c|c} \grave{\mathsf{b}} & \mathsf{b} \\ 0 & 0 \\ \hline \grave{\mathsf{a}} & \mathsf{a} \\ 0 & 1 \end{array}\right] = \begin{bmatrix} \mathsf{e}_{2k+1,2k+1} & \mathsf{o}_{2k+1} \\ \gamma_{1,n+k} & \phi_{1,n+k} \end{bmatrix} \tag{7.1}$$

(where o_{2k+1} is the zero vector in $\mathbb{C}^{2k+1}$). This corresponds to the terms from ζ^{-n+k} downto ζ^{-n-k-1} of (3.3). We can take into account as many leading terms as we want by adding additional rows to the Sylvester matrix and to the right-hand side. In particular, if $g_n \in \mathcal{L}^*_{-n}$ and $f_n \in \mathcal{L}^*_{-n-1}$ are polynomials in ζ^{-1} of degree at most $2N$, we can process all their coefficients if the Sylvester matrix is of size $(2N - n + 2) \times (2k + 2)$.

Starting from a pair

$$f_0 \in \mathcal{L}_{-2N:-1}, \qquad g_0 \in \mathcal{L}_{-2N:0}$$

with $f_0 \neq 0$ and $g_0(\infty) :\equiv \gamma_{0,0} = 1$, we will apply (3.3) recursively until we reach some $f_n \in \mathcal{L}^*_{-N-1}$. This will happen for some regular $n \equiv: n_J \leq N$. We let $\{n_j\}_{j=0}^J$ be the corresponding sequence of regular indices, starting with $n_0 = 0$, and set

$$k_j := n_{j+1} - n_j \qquad (j \geq 1).$$

Then, $\phi_{k_j,n_j} \neq 0$ $(\forall j)$. Additionally, we set $n_{-1} := -1$, $k_0 := 1$. The sequence $\{n_j\}_{j=0}^J$ need not contain *all* regular indices, but only the chosen well-conditioned ones.

Adding further terms of order $\mathcal{O}_-(-2N-1)$ to f_0 and g_0 will not change the resulting pair (p_{n_J}, q_{n_J}).

If, in the step starting at $n = n_j$, we write $\grave{a} \equiv: \grave{a}_j$, $\grave{b} \equiv: \grave{b}_j$, $\grave{\mathsf{a}} \equiv: \grave{\mathsf{a}}_j$, $\grave{\mathsf{b}} \equiv: \grave{\mathsf{b}}_j$, etc., then (3.3) and (3.6) yield

$$\begin{bmatrix} g_{n_{j+1}} & f_{n_{j+1}} \end{bmatrix} := \begin{bmatrix} g_{n_j} & f_{n_j} \end{bmatrix} \begin{bmatrix} \grave{b}_j & b_j \\ \grave{a}_j & a_j \end{bmatrix} \qquad (j \geq 0), \tag{7.2}$$

and

$$\begin{bmatrix} \grave{p}_{n_{j+1}} & p_{n_{j+1}} \\ \grave{q}_{n_{j+1}} & q_{n_{j+1}} \end{bmatrix} := \begin{bmatrix} \grave{p}_{n_j} & p_{n_j} \\ \grave{q}_{n_j} & q_{n_j} \end{bmatrix} \begin{bmatrix} \grave{b}_j & b_j \\ \grave{a}_j & a_j \end{bmatrix} \qquad (j \geq 0). \tag{7.3}$$

Consequently,

$$\begin{bmatrix} g_{n_J} & f_{n_J} \end{bmatrix} := \begin{bmatrix} g_{n_0} & f_{n_0} \end{bmatrix} \prod_{j=0}^{J-1} \begin{bmatrix} \grave{b}_j & b_j \\ \grave{a}_j & a_j \end{bmatrix} \tag{7.4}$$

and

$$\begin{bmatrix} \grave{p}_{n_J} & p_{n_J} \\ \grave{q}_{n_J} & q_{n_J} \end{bmatrix} := \begin{bmatrix} \grave{p}_{n_0} & p_{n_0} \\ \grave{q}_{n_0} & q_{n_0} \end{bmatrix} \prod_{j=1}^{J-1} \begin{bmatrix} \grave{b}_j & b_j \\ \grave{a}_j & a_j \end{bmatrix} \tag{7.5}$$

with $f_{n_J} \in \mathcal{L}^*_{-N-1}$. Obviously, we could combine the two equations (7.4) and (7.5) into one.

Let us again look first at the *generic case*, where $k = 1$ for all n, and thus $n_j = n$. Here, (7.1) reduces to

$$\left[\begin{array}{cc|cc} 0 & 1 & 0 & 0 \\ 1 & \gamma_{1,n} & 0 & \phi_{1,n} \\ \gamma_{1,n} & \gamma_{2,n} & \phi_{1,n} & \phi_{2,n} \\ \gamma_{2,n} & \gamma_{3,n} & \phi_{2,n} & \phi_{3,n} \end{array}\right] \left[\begin{array}{c|c} 0 & \beta_0^{(j)} \\ 0 & 0 \\ \hline \grave{\alpha}_0^{(j)} & \alpha_0^{(j)} \\ 0 & 1 \end{array}\right] = \left[\begin{array}{c|c} 0 & 0 \\ 0 & 0 \\ 1 & 0 \\ \gamma_{1,n+1} & \phi_{1,n+1} \end{array}\right]. \tag{7.6}$$

This implies trivially that

$$\left[\begin{array}{cc|cc} 0 & 1 & 0 & 0 \\ 1 & \gamma_{1,n} & 0 & \phi_{1,n} \\ \gamma_{1,n} & \gamma_{2,n} & \phi_{1,n} & \phi_{2,n} \\ \gamma_{2,n} & \gamma_{3,n} & \phi_{2,n} & \phi_{3,n} \end{array}\right] \left[\begin{array}{cc|cc} 1 & 0 & 0 & \beta_0^{(j)} \\ 0 & 0 & 0 & 0 \\ \hline 0 & \grave{\alpha}_0^{(j)} & 1 & \alpha_0^{(j)} \\ 0 & 0 & 0 & 1 \end{array}\right] =$$

$$\left[\begin{array}{cc|cc} 0 & 0 & 0 & 0 \\ 1 & 0 & 0 & 0 \\ \gamma_{1,n} & 1 & \phi_{1,n} & 0 \\ \gamma_{2,n} & \gamma_{1,n+1} & \phi_{2,n} & \phi_{1,n+1} \end{array}\right].$$

Note that the two additional columns in the second matrix allow us to "store" the coefficients of the residuals in the resulting matrix on the right-hand side.

To obtain a representation that can accommodate all N steps we need to "imbed" this relationship into $2N \times 2N$ matrices. For example, in the first step, where $n = 0$, we get

$$\mathsf{S}_N(g_0, f_0)\mathsf{C}_1 = \mathsf{V}_1$$

with

$$\mathsf{S}_N(g_0, f_0) :\equiv \left[\begin{array}{cccc|cccc}
0 & \cdots & 0 & 1 & 0 & \cdots & \cdots & 0 \\
\vdots & \iddots & 1 & \gamma_{1,0} & \vdots & & 0 & \phi_{1,0} \\
0 & \iddots & \iddots & \vdots & \vdots & \iddots & \iddots & \vdots \\
1 & \iddots & & \vdots & 0 & \iddots & & \vdots \\
\gamma_{1,0} & & & \gamma_{N,0} & \phi_{1,0} & & & \phi_{N,0} \\
\vdots & & \iddots & \vdots & \vdots & & \iddots & \vdots \\
\vdots & \iddots & & \vdots & \vdots & \iddots & & \vdots \\
\gamma_{N,0} & \cdots & \cdots & \gamma_{2N-1,0} & \phi_{N,0} & \cdots & \cdots & \phi_{2N-1,0}
\end{array}\right],$$

$$\mathsf{C}_1 :\equiv \left[\begin{array}{cccc|cccc}
1 & 0 & & & 0 & \beta_0^{(0)} & & \\
 & 0 & \ddots & & & 0 & \ddots & \\
 & & \ddots & 0 & & & \ddots & \beta_0^{(0)} \\
 & & & 0 & & & & 0 \\
\hline
0 & \grave{\alpha}_0^{(0)} & & & 1 & \alpha_0^{(0)} & & \\
 & 0 & \ddots & & & 1 & \ddots & \\
 & & \ddots & \grave{\alpha}_0^{(0)} & & & \ddots & \alpha_0^{(0)} \\
 & & & 0 & & & & 1
\end{array}\right],$$

$$\mathsf{V}_1 :\equiv \left[\begin{array}{c|ccc|c|ccc}
0 & 0 & \cdots & 0 & 0 & 0 & \cdots & 0 \\
\vdots & \vdots & \iddots & 0 & \vdots & \vdots & \iddots & 0 \\
0 & 0 & \iddots & 1 & \vdots & 0 & \iddots & 0 \\
1 & 0 & \iddots & \gamma_{1,1} & 0 & 0 & \iddots & \phi_{1,1} \\
\gamma_{1,0} & 1 & \iddots & \vdots & \phi_{1,0} & 0 & \iddots & \vdots \\
\vdots & \gamma_{1,1} & & \vdots & \vdots & \phi_{1,1} & & \vdots \\
\vdots & \vdots & & \vdots & \vdots & \vdots & & \vdots \\
\gamma_{N,0} & \gamma_{N-1,1} & \cdots & \gamma_{2N-3,1} & \phi_{N,0} & \phi_{N-1,1} & \cdots & \phi_{2N-3,1}
\end{array}\right].$$

In the second step,

$$\mathsf{V}_1 \mathsf{C}_2 = \mathsf{V}_2,$$

C_2 will have two 2×2 unit matrices where C_1 had two 1's, and therefore the columns with indices $1, 2, N+1, N+2$ of V_1 will remain untouched in the transition to V_2:

$$
\mathsf{V}_2 = \left[\begin{array}{cc|cccc|cc|cccc}
0 & 0 & 0 & \cdots & & 0 & 0 & 0 & 0 & \cdots & & 0 \\
\vdots & \vdots & \vdots & \iddots & & \vdots & \vdots & \vdots & \vdots & \iddots & & \vdots \\
0 & 0 & 0 & & & 0 & \vdots & 0 & 0 & & & \vdots \\
1 & 0 & \vdots & \iddots & & 1 & 0 & \vdots & \vdots & & & 0 \\
\gamma_{1,0} & 1 & 0 & \iddots & & \gamma_{1,2} & \phi_{1,0} & 0 & \vdots & \iddots & & \phi_{1,2} \\
\vdots & \gamma_{1,1} & 1 & \iddots & & \vdots & \vdots & \phi_{1,1} & 0 & \iddots & & \vdots \\
\vdots & \vdots & \gamma_{1,2} & & & \vdots & \vdots & \vdots & \phi_{1,2} & & & \vdots \\
\vdots & \vdots & \vdots & & & \vdots & \vdots & \vdots & \vdots & & & \vdots \\
\gamma_{N,0} & \gamma_{N-1,1} & \gamma_{N-2,2} & \cdots & & \gamma_{2N-5,2} & \phi_{N,0} & \phi_{N-1,1} & \phi_{N-2,2} & \cdots & & \phi_{2N-5,2}
\end{array}\right]
$$

while all the Toeplitz blocks in C_2 and V_2 are by one column and one or two rows smaller than the corresponding ones in C_1 and V_1.

By now it should be clear, how, in the generic case, the matrices V_n change from step to step until at the end,

$$
\mathsf{V}_N = \left[\begin{array}{c|c}
\mathsf{e}_N^\top \mathsf{e}_1 & \mathsf{O} \\
\hline
\mathsf{G}_N & \mathsf{F}_N
\end{array}\right]
$$

is the matrix on the right-hand side of (6.7).

Now, we turn to the *look-ahead case* where $n = n_j$ and $k = k_j > 1$, either due to the situation (4.8) or because of a general look-ahead step that also leaves out ill-conditioned regular pairs. We append $N - n_j - k_j - 1 = N - n_{j+1} - 1$ rows at the bottom of the $(2k_j+2) \times (2k_j+2)$ matrix $\mathsf{S}_{k_j+1}(g_{n_j}, f_{n_j})$ to get the $(N - n_j + k_j + 1) \times (2k_j + 2)$ matrix

$$
\underline{\mathsf{S}}_{k_j+1}(g_{n_j}, f_{n_j}) :\equiv
$$

$$
\left[\begin{array}{cccc|cccc}
0 & \cdots & 0 & 1 & 0 & \cdots & \cdots & 0 \\
\vdots & \iddots & 1 & \gamma_{1,n_j} & \vdots & & 0 & \phi_{1,n_j} \\
0 & \iddots & \iddots & \vdots & \vdots & \iddots & \iddots & \vdots \\
1 & \iddots & & \vdots & 0 & \iddots & & \vdots \\
\gamma_{1,n_j} & & & \gamma_{k+1,n_j} & \phi_{1,n_j} & & & \phi_{k+1,n_j} \\
\vdots & & \iddots & \gamma_{k+2,n_j} & \vdots & & \iddots & \phi_{k+2,n_j} \\
\vdots & \iddots & \iddots & \vdots & \vdots & \iddots & \iddots & \vdots \\
\gamma_{k+1,n_j} & \iddots & & \gamma_{2k+1,n_j} & \phi_{k+1,n_j} & \iddots & & \phi_{2k+1,n_j} \\
\hline
\gamma_{k+2,n_j} & \cdots & \cdots & \gamma_{2k+2,n_j} & \phi_{k+2,n_j} & \cdots & \cdots & \phi_{2k+2,n_j} \\
\vdots & & & \vdots & \vdots & & & \vdots \\
\gamma_{N-n_j,n_j} & \cdots & \cdots & \gamma_{N-n_{j+1},n_j} & \phi_{N-n_j,n_j} & \cdots & \cdots & \phi_{N-n_{j+1},n_j}
\end{array}\right]
$$

(for simplicity, we write k instead of k_j in the indices). Now, (7.1) reads

$$\underline{\mathsf{S}}_{k_j+1}(g_{n_j}, f_{n_j}) \left[\begin{array}{c|c} 0 & \beta_0^{(j)} \\ \grave{\beta}_1^{(j)} & \beta_1^{(j)} \\ \vdots & \vdots \\ \grave{\beta}_{k-1}^{(j)} & \beta_{k-1}^{(j)} \\ 0 & 0 \\ \hline \grave{\alpha}_0^{(j)} & \alpha_0^{(j)} \\ \grave{\alpha}_1^{(j)} & \alpha_1^{(j)} \\ \vdots & \vdots \\ \grave{\alpha}_{k-1}^{(j)} & \alpha_{k-1}^{(j)} \\ 0 & 1 \end{array}\right] = \left[\begin{array}{c|c} 0 & 0 \\ 0 & 0 \\ \vdots & \vdots \\ \vdots & \vdots \\ 0 & 0 \\ 1 & 0 \\ \gamma_{1,n_{j+1}} & \phi_{1,n_{j+1}} \\ \hline \gamma_{2,n_{j+1}} & \phi_{2,n_{j+1}} \\ \vdots & \vdots \\ \gamma_{N-n_j-k,n_{j+1}} & \phi_{N-n_j-k,n_{j+1}} \end{array}\right]. \tag{7.7}$$

This system contains $N - n_{j+1} + 1$ equations and reflects the terms from $\zeta^{-n_{j+1}-1}$ downto ζ^{-N+1} of (3.3). We could add $N + n_j - k_j - 1$ zero rows at the top to get a total of $2N$ rows and to cover the terms upto ζ^N.

Next we want to include the inner polynomials. To this end, we multiply the $(N - n_j + k_j + 1) \times (2k_j + 2)$ matrix $\underline{\mathsf{S}}_{k_j+1}(g_{n_j}, f_{n_j})$ by the $(2k_j + 2) \times (2k_j + 2)$ matrix

$$\mathsf{C}_j^\circ :\equiv \left[\begin{array}{c|c} \grave{\mathsf{B}}_j^\circ & \mathsf{B}_j^\circ \\ \hline \grave{\mathsf{A}}_j^\circ & \mathsf{A}_j^\circ \end{array}\right] :\equiv \left[\begin{array}{ccccc|ccccc} 1 & 1 & \cdots & 1 & 0 & 0 & \mu_1^{(j)} & \cdots & \mu_k^{(j)} & \beta_0^{(j)} \\ 0 & 0 & \cdots & 0 & \grave{\beta}_1^{(j)} & & 0 & \ddots & \vdots & \beta_1^{(j)} \\ \vdots & & & \vdots & \vdots & & & \ddots & \mu_1^{(j)} & \vdots \\ \vdots & & & \vdots & \grave{\beta}_{k-1}^{(j)} & & & & 0 & \beta_{k-1}^{(j)} \\ 0 & 0 & \cdots & 0 & 0 & & & & & 0 \\ \hline 0 & 0 & \cdots & 0 & \grave{\alpha}_0^{(j)} & 1 & 0 & \cdots & 0 & \alpha_0^{(j)} \\ \vdots & & & \vdots & \grave{\alpha}_1^{(j)} & & 1 & \ddots & \vdots & \alpha_1^{(j)} \\ \vdots & & & \vdots & \vdots & & & \ddots & 0 & \vdots \\ \vdots & & & \vdots & \grave{\alpha}_{k-1}^{(j)} & & & & 1 & \alpha_{k-1}^{(j)} \\ 0 & 0 & \cdots & 0 & 0 & & & & & 1 \end{array}\right].$$

This yields

$$\underline{\mathsf{S}}_{k_j+1}(g_{n_j}, f_{n_j})\, \mathsf{C}_j^\circ = \begin{bmatrix} S_{11} & S_{12} \\ S_{21} & S_{22} \end{bmatrix}$$

with

$$S_{11} = \begin{bmatrix} 0 & \cdots & 0 & 0 \\ \vdots & & \vdots & \vdots \\ 0 & \cdots & 0 & \vdots \\ 1 & \cdots & 1 & \vdots \\ \gamma_{1,n_j} & \cdots & \gamma_{1,n_j} & \vdots \\ \vdots & & \vdots & 0 \\ \vdots & & \vdots & 1 \\ \gamma_{k+1,n_j} & \cdots & \gamma_{k+1,n_j} & \gamma_{1,n_{j+1}} \end{bmatrix},$$

$$S_{12} = \begin{bmatrix} 0 & \cdots & 0 & 0 \\ \vdots & & \vdots & \vdots \\ \vdots & & \vdots & \vdots \\ 0 & \cdots & 0 & \vdots \\ \phi_{1,n_j} & \cdots & \phi_{1,n_{j+1}-1} & \vdots \\ \vdots & & \vdots & 0 \\ \vdots & & \vdots & 0 \\ \phi_{k+1,n_j} & \cdots & \phi_{k+1,n_{j+1}-1} & \phi_{1,n_{j+1}} \end{bmatrix},$$

$$S_{21} = \begin{bmatrix} \gamma_{k+2,n_j} & \cdots & \gamma_{k+2,n_j} & \gamma_{2,n_{j+1}} \\ \vdots & & \vdots & \vdots \\ \gamma_{N-n_j,n_j} & \cdots & \gamma_{N-n_j,n_j} & \gamma_{N-n_j-k,n_{j+1}} \end{bmatrix},$$

$$S_{22} = \begin{bmatrix} \phi_{k+2,n_j} & \cdots & \phi_{k+2,n_{j+1}-1} & \phi_{2,n_{j+1}} \\ \vdots & & \vdots & \vdots \\ \cdots & \phi_{N-n_j,n_{j+1}-1} & \phi_{N-n_j-k,n_{j+1}} & \end{bmatrix}.$$

We still need to imbed these matrices into even bigger ones of size $2N \times 2N$. First, as mentioned, $\underline{\mathsf{S}}_{k_j+1}(g_{n_j}, f_{n_j})$ is extended on top by $N + n_j - k_j - 1$ zero rows to make it a $2N \times (2k_j + 2)$ matrix. Then each half of it is extended on the right by $N - n_j - k_j - 1$ columns, so that each half has size $2N \times (N - n_j)$. Finally, at the left of each half, n_j columns with the previously computed coefficients of the residuals g_n, $n = 0, \ldots n_j - 1$, and f_n, $n = 0, \ldots n_j - 1$, are appended. This yields a $2N \times 2N$ matrix that we call again V_{j-1}. At the beginning and at the end, we have, respectively,

$$\mathsf{V}_0 = \mathsf{S}_N(g_0, f_0), \qquad \mathsf{V}_J = \left[\begin{array}{c|c} \mathsf{e}_N^\top \mathsf{e}_1 & \mathsf{O} \\ \hline \mathsf{G}_N & \mathsf{F}_N \end{array}\right], \tag{7.8}$$

where the last matrix is the same one that appears in (6.7).

To imbed C_j°, each of the four square blocks of size $k_j + 1$ of C_j° is first extended to a square block of size $N - n_j + 1$ by appending $N - n_j - k_j$ rows and columns, so that the last $N - n_j - k_j + 1$ columns have Toeplitz structure. This yields the blocks $\grave{\mathsf{B}}_j$, B_j, $\grave{\mathsf{A}}_j$, and A_j, which are imbedded into a $2N \times 2N$ matrix C_j as follows:

$$\mathsf{C}_j :\equiv \left[\begin{array}{cc|cc} \mathsf{I}_{n_j-1} & \mathsf{O} & \mathsf{O}_{n_j-1} & \mathsf{O} \\ \mathsf{O} & \grave{\mathsf{B}}_j & \mathsf{O} & \mathsf{B}_j \\ \hline \mathsf{O}_{n_j-1} & \mathsf{O} & \mathsf{I}_{n_j-1} & \mathsf{O} \\ \mathsf{O} & \grave{\mathsf{A}}_j & \mathsf{O} & \mathsf{A}_j \end{array}\right]. \tag{7.9}$$

Note that in view of the 1's in the upper left corners of $\grave{\mathsf{B}}_j$ and A_j the blocks I_{n_j-1} are part of unit matrices of size n_j, whose boundaries do not match those of the blocks $\grave{\mathsf{B}}_j$, $\grave{\mathsf{A}}_j$, B_j, and A_j, however. Now we have

$$\mathsf{V}_{j-1}\mathsf{C}_j = \mathsf{V}_j \tag{7.10}$$

and, consequently,

$$\mathsf{S}_N(g_0, f_0) \prod_{j=1}^{J} \mathsf{C}_j = \left[\begin{array}{c|c} \mathsf{e}_N^{\mathsf{T}}\mathsf{e}_1 & \mathsf{O} \\ \hline \mathsf{G}_N & \mathsf{F}_N \end{array}\right], \tag{7.11}$$

where G_N is now lower block Hessenberg, and F_N is lower block triangular. So, finally, the comparison with (6.7) reveals that

$$\prod_{j=1}^{J} \mathsf{C}_j = \left[\begin{array}{c|c} \grave{\mathsf{P}}_N & \mathsf{P}_N \\ \hline \grave{\mathsf{R}}_N & \mathsf{R}_N \end{array}\right]. \tag{7.12}$$

Eqs. (7.8)–(7.12) are the matrix relations we have been looking for. Eq. (7.10) describes a single step of the Euclidean algorithm as a matrix multiplication, and (7.11) summarizes the whole process as a matrix product with the factors C_j. Moreover, (7.11) shows how the coefficients of the Padé forms constructed are also obtained by multiplying all these matrices C_j. Finally, as we have seen at the end of section 6, it is an easy matter to link (7.8) to the block LDU decomposition and to the inverse block LDU decomposition of the Hankel matrix M_n.

Acknowledgement. The author would like to thank Paul van Dooren for pointing out some references to earlier related work such as [**HS69**]. He also mentioned that his diploma student Paul Bailey wrote a related thesis, which, however, was not available to the author.

References

[Bec96] B. Beckermann, *The stable computation of formal orthogonal polynomials*, Numerical Algorithms **11** (1996), 1–23.

[BV97] A. Bultheel and M. Van Barel, *Linear algebra, rational approximation and orthogonal polynomials*, Elsevier, Amsterdam, 1997.

[CM93] S. Cabay and R. Meleshko, *A weakly stable algorithm for Padé approximants and the inversion of Hankel matrices*, SIAM J. Matrix Anal. Appl. **14** (1993), 735–765.

[DI89] J.-M. Delosme and I. C. F. Ipsen, *From Bareiss' algorithm to the stable computation of partial correlations*, J. Comput. Appl. Math. **27** (1989), 53–91.

[GG90] G. H. Golub and M. H. Gutknecht, *Modified moments for indefinite weight functions*, Numer. Math. **57** (1990), 607–624.

[GG94] W. B. Gragg and M. H. Gutknecht, *Stable look-ahead versions of the Euclidean and Chebyshev algorithms*, Approximation and Computation (R. V. M. Zahar, ed.), ISNM, vol. 119, Birkhäuser Verlag, Basel-Boston-Berlin, 1994, pp. 231–260.

[GGWY82] W. B. Gragg, F. G. Gustavson, D. D. Warner, and D. Y. Y. Yun, *On fast computation of superdiagonal Padé fractions*, Math. Programming Stud. **18** (1982), 39–42.

[GH95a] M. H. Gutknecht and M. Hochbruck, *Look-ahead Levinson and Schur algorithms for non-Hermitian Toeplitz systems*, Numer. Math. **70** (1995), 181–227.

[GH95b] M. H. Gutknecht and M. Hochbruck, *The stability of inversion formulas for Toeplitz matrices*, Linear Algebra Appl. **223/224** (1995), 307–324.

[Gut92] M. H. Gutknecht, *A completed theory of the unsymmetric Lanczos process and related algorithms, Part I*, SIAM J. Matrix Anal. Appl. **13** (1992), 594–639.

[Gut93] M. H. Gutknecht, *Stable row recurrences in the Padé table and generically superfast lookahead solvers for non-Hermitian Toeplitz systems*, Linear Algebra Appl. **188/189** (1993), 351–421.

[HR84] G. Heinig and K. Rost, *Algebraic methods for Toeplitz-like matrices and operators*, Akademie-Verlag, Berlin, and Birkhäuser, Basel/Stuttgart, 1984.

[HS69] A. S. Householder and G. W. Stewart, *Bigradients, Hankel determinants, and the Padé table*, Constructive aspects of the fundamental theorem of algebra (B. Dejon and P. Henrici, eds.), Wiley-Interscience, 1969, pp. 131–150.

[Mag62] A. Magnus, *Certain continued fractions associated with the Padé table*, Math. Z. **78** (1962), 361–374.

[Str63] G. W. Struble, *Orthogonal polynomials: variable-signed weight functions*, Numer. Math **5** (1963), 88–94.

Seminar for Applied Mathematics, ETH-Zurich, ETH-Zentrum HG, CH-8092 Zurich
E-mail address: `mhg@sam.math.ethz.ch`

Contemporary Mathematics
Volume **280**, 2001

The essential polynomial approach to convergence of matrix Padé Approximants

Victor M. Adukov

ABSTRACT. The method of essential polynomials (see, e.g., [**1, 2**]) is applied to a problem of convergence of Padé approximants for meromorphic matrix functions. The approach is based on the stability of indices of a sequence. The first case of the stability is carefully studied. As a result, we obtain a matrix analog of the Montessus theorem.

1. Introduction

The notion of Padé approximants can be generalized to matrix power series. In this case there are several different types of matrix Padé approximants (see, e.g., [**3, 4, 5**]). For goals of the present article it is sufficient to use the classical definition [**6**] .

Let $a(z) = \sum_{i=0}^{\infty} a_i z^i$, $a_i \in \mathbb{C}^{p\times q}$, be a matrix (formal) power series. The *classical right Padé approximant* of types (n, m) for $a(z)$ is a rational matrix function $\pi^R_{n,m}(z) = P^R_{n,m}(z)Q^R_{n,m}(z)^{-1}$ such that the $p \times q$ matrix polynomial $P^R_{n,m}(z)$ and the $q \times q$ matrix polynomial $Q^R_{n,m}(z)$ satisfy the following conditions:

(1) $\deg P^R_{n,m}(z) \leq n$, $\deg Q^R_{n,m}(z) \leq m$,
(2) $\det Q^R_{n,m}(z) \not\equiv 0$,
(3) $a(z)Q^R_{n,m}(z) - P^R_{n,m}(z) = O(z^{n+m+1})$, $z \to 0$.

If, in addition, the Baker condition

4. $Q^R_{n,m}(0)$ is invertible

is fulfilled, then the condition 3) can be rewritten in the equivalent form

3B. $a(z) - \pi^R_{n,m}(z) = O(z^{n+m+1})$, $z \to 0$.

We can define a left Padé approximant $\pi^L_{n,m}(z)$ in a similar way.

Unfortunately, in contrast to the scalar case, the classical matrix Padé approximants for $p \neq q$ may not exist. In order to obtain conditions for the existence of Padé approximants, we must study in detail the kernel structure for block Toeplitz matrices. To do this, we use in Section 2 the notion of essential polynomials which was first introduced in [**7**] (see also [**1, 2**]).

1991 *Mathematics Subject Classification.* Primary 41A21; Secondary 15A57.

In this work we consider the Padé approximants for meromorphic matrix functions. Let $a(z)$ be a matrix function which is analytic at the origin $z = 0$, meromorphic in the disk $|z| < R$, and has finitely many poles in this disk. Without loss of generality we will suppose that $R > 1$ and all poles lie in the unit disk $|z| < 1$. Let us represent $a(z)$ in the following form:

$$a(z) = b(z) + r(z).$$

Here the matrix function $b(z)$ is analytic in $|z| < R$ and $r(z)$ is a strictly proper rational matrix function which is the sum of principal parts of the Laurent series of $a(z)$ in neighborhoods of the poles of $a(z)$. Obviously, $r(z)$ is analytic at $z = 0$.

In the work [**8**] the problem was formulated of how the passage from the analytic function $b(z)$ to the meromorphic function $b(z) + r(z)$ influences the convergence of Padé approximants. In [**8, 9**] the convergence of the diagonal scalar Padé approximants was studied for the case when $b(z)$ is the Markov function for a measure with a compact support on $\mathbb{R}$. Essential in the considerations of those works was the fact that the asymptotic behavior of denominators of Padé approximants for $b(z)$ was well-studied and $r(z)$ was considered as a kind of perturbation.

In contrast with [**8, 9**] we will study the convergence of the row matrix Padé approximants of $a(z)$ as the result of a small perturbation of the Padé approximants of the rational matrix function $r(z)$ by $b(z)$. To do this, we will use the technique that is different from the method of above-mentioned works. First, we find the indices and the essential polynomials of a sequence consisting of the Taylor coefficients of $r(z)$ (Section 3). We obtain the description of the indices and essential polynomials in terms of coprime fractional factorizations of $r(z)$ and minimal solutions of the Bezout equations. Using this description, we find four stability cases for this sequence.

In Section 4 we study the first stability case which leads to a matrix analog of the Montessus theorem. The basic tool for analysis of this case is the perturbation theory for Fredholm operators in Banach spaces [**10**]. We prove that in the first stability case there exist the right (left) Padé approximants of type (n, m) for $a(z)$ if $p \leq q$ $(p \geq q)$, $n \to \infty$, and $m = \frac{\lambda+\rho}{2}$ (Corollary 4.1). Here we suppose that the denominators $D_R(z)$ and $D_L(z)$ of the right and left coprime fractional factorizations of $r(z)$ have invertible leading coefficients and λ, ρ are the degrees of $D_R(z)$, $D_L(z)$, respectively. The next step is to prove that the denominators $Q^R_{n,m}(z)$ $(Q^L_{n,m}(z))$ of the Padé approximants converge to $D_R(z)$ $(D_L(z))$ as $n \to \infty$ (Theorem 4.3). Now the uniform convergence of $\pi^{R,L}_{n,m}(z)$ to $a(z)$ on compact sets which do not contain the poles of $a(z)$ is proved almost as in the scalar case (Theorem 4.4). If $p = q = 1$ we get the Montessus theorem.

2. A necessary condition for the existence of classical Padé approximants

For a matrix A we denote its jth row (column) by $[A]_j$ $([A]^j)$. It is easily seen from the definition of the classical Padé approximants that the coefficients of the vector polynomials $[Q^R_{n,m}(z)]^j$, $j = 1, \ldots, q$, satisfy a linear homogenious system with the coefficient matrix

$$T_{n+1} = \begin{pmatrix} a_{n+1} & a_n & \dots & a_{n-m+1} \\ a_{n+2} & a_{n+1} & \dots & a_{n-m+2} \\ \vdots & \vdots & & \vdots \\ a_{n+m} & a_{n+m-1} & \dots & a_n \end{pmatrix},$$

i.e. the vector of the coefficients belongs to the right kernel of the block Toeplitz matrix T_{n+1}. Similarly, if we write the coefficients of the vector polynomials $[Q^L_{n,m}(z)]_j$ in the reverse order, we get the vector from the left kernel of the block Toeplitz matrix

$$T_n = \begin{pmatrix} a_n & a_{n-1} & \dots & a_{n-m+1} \\ a_{n+1} & a_n & \dots & a_{n-m+2} \\ \vdots & \vdots & & \vdots \\ a_{n+m} & a_{n+m-1} & \dots & a_{n+1} \end{pmatrix}.$$

Thus, we arrive at the problem of description of the kernel structure for block Toeptitz matrices. This problem was first studied in [**7**], where the kernel structure for block Toeplitz matrices was described in terms of indices and essential polynomials. The description was also repeated in [**11**] in connection with an explicit construction of a Wiener-Hopf factorization for meromorphic matrix functions (see [**1**] for detailed proofs). Later the same description for block Hankel matrices was obtained in [**12**] whose authors were probably not aware of the above-mentioned works.

Recall the basic definitions and results on the kernel structure. The extensive exposition can be found in our summarizing work [**2**].

Let $\mathbb{C}^{p\times q}$ be the set of complex $p \times q$ matrices. For a matrix A we will denote by $\ker_R A$ its right kernel and by $\ker_L A$ its left kernel:

$$\ker_R A = \{x \mid Ax = 0\}, \ \ker_L A = \{y \mid yA = 0\}.$$

Let A be a block matrix with blocks in $\mathbb{C}^{p\times q}$ and let A has the block size $(n+1) \times (m+1)$. We partition the column $R \in \ker_R A$ into $m+1$ blocks (the size of blocks is $q \times 1$):

$$R = \begin{pmatrix} r_0 \\ r_1 \\ \vdots \\ r_m \end{pmatrix}$$

and for R we define its generating vector polynomial in the variable z to be the polynomial

$$R(z) = r_0 + r_1 z + \ldots + r_m z^m.$$

Similarly, for a row in $\ker_L A$ we define the generating vector polynomial in z^{-1}.

Let $a_M^N = (a_M, a_{M+1}, \dots, a_N)$, $M < N$, be a finite sequence of complex $p \times q$ matrices. By $a_M^N(z)$ we denote its generating function. Let us form the family of block Toeplitz matrices $T_k(a_M^N) = \|a_{i-j}\|_{\substack{i = k, k+1, \dots, N \\ j = 0, 1, \dots, k-M}}$, $k = M, M+1, \dots, N$, and describe the structure of right and left kernels of $T_k(a_M^N)$. For brevity, if there will be no possibility of misinterpretation, then we will use the notation T_k instead of $T_k(a_M^N)$.

Since it is more convenient to deal not with vectors but with generating vector polynomials, we pass from the spaces $\ker_R T_k$ and $\ker_L T_k$ to the isomorphic spaces of generating vector polynomials in z or in z^{-1}. By $\mathcal{N}_k^R$ $(M \le k \le N)$ we denote the space of generating polynomials of vectors in $\ker_R T_k$. For convenience, we put $\mathcal{N}_{M-1}^R = 0$ and denote by $\mathcal{N}_{N+1}^R$ the $(N-M+2)q$-dimensional space of all vector polynomials in z of formal degree $N-M+1$.

Similarly, the space $\ker_L T_k$ is naturally isomorphic to the space $\mathcal{N}_k^L$ of generating vector polynomials in z^{-1}. We put $\mathcal{N}_{N+1}^L = 0$ and denote by $\mathcal{N}_{M-1}^L$ the $(N-M+2)p$-dimensional space of all vector polynomials in z^{-1} of formal degree $N-M+1$.

Let $\alpha = \dim \mathcal{N}_M^R$ and $\omega = \dim \mathcal{N}_N^L$. The sequence $a_M, a_{M+1}, \dots, a_N$ is called *left regular* (*right regular*) if $\alpha = 0$ $(\omega = 0)$. The sequence is said to be *regular*, if $\alpha = \omega = 0$. The integer α (ω) will be called the *left (right) defect* of the sequence.

By d_k^R (d_k^L) we denote the dimension of the space $\mathcal{N}_k^R$ $(\mathcal{N}_k^L)$. Let $\Delta_k^R = d_k^R - d_{k-1}^R$ $(M \le k \le N+1)$, $\Delta_k^L = d_k^L - d_{k+1}^L$ $(M-1 \le k \le N)$. Using the Grassmann formula, it is not difficult to prove that the following inequalities hold

$$\alpha = \Delta_M^R \le \Delta_{M+1}^R \le \dots \le \Delta_N^R \le \Delta_{N+1}^R = p+q-\omega,$$

$$p+q-\alpha = \Delta_{M-1}^L \ge \Delta_M^L \ge \dots \ge \Delta_{N-1}^L \ge \Delta_N^L = \omega.$$

It follows from these inequalities that there exist $p+q-\alpha-\omega$ integers $\mu_{\alpha+1} \le \dots \le \mu_{p+q-\omega}$ such that

$$\begin{array}{rcccccl} \Delta_M^R & = & \dots & = & \Delta_{\mu_{\alpha+1}}^R & = & \alpha, \\ & & \dots & & & & \\ \Delta_{\mu_i+1}^R & = & \dots & = & \Delta_{\mu_{i+1}}^R & = & i, \\ & & \dots & & & & \\ \Delta_{\mu_{p+q-\omega}+1}^R & = & \dots & = & \Delta_{N+1}^R & = & p+q-\omega. \end{array} \tag{2.1}$$

If the ith row in these relations is absent, then we assume that $\mu_i = \mu_{i+1}$. By definition, we put $\mu_1 = \dots = \mu_\alpha = -m-1$ if $\alpha \ne 0$ and $\mu_{p+q-\omega+1} = \dots = \mu_{p+q} = 0$ if $\omega \ne 0$.

The integers $\mu_1, \dots, \mu_{p+q}$ defined in (2.1) will be called the *indices* of the sequence $a_M, a_{M+1}, \dots, a_N$.

Since $\Delta_k^L = p+q-\Delta_{k+1}^R$, we can obtain for Δ_k^L the relations that are similar to (2.1).

It is easily seen (see [**2**]) that $\mathcal{N}_k^R + z\mathcal{N}_k^R$ is a subspace of $\mathcal{N}_{k+1}^R$ and the dimension h_{k+1}^R of a complement $\mathcal{H}_{k+1}^R$ of this subspace is equal to $\Delta_{k+1}^R - \Delta_k^R$. It follows from (2.1) that $h_{k+1}^R \ne 0$ iff $k = \mu_j$ $(j = \alpha+1, \dots, p+q-\omega)$. In this case h_{k+1}^R coincides with the multiplicity k_j of the index μ_j. Hence for $k \ne \mu_j$

$$\mathcal{N}_{k+1}^R = \mathcal{N}_k^R + z\mathcal{N}_k^R \tag{2.2}$$

and for $k = \mu_j$

$$\mathcal{N}_{k+1}^R = (\mathcal{N}_k^R + z\mathcal{N}_k^R) \dot{+} \mathcal{H}_{k+1}^R. \tag{2.3}$$

DEFINITION 2.1. *If $\alpha \ne 0$, then any polynomials $R_1(z), \dots, R_\alpha(z)$ that form a basis for the space $\mathcal{N}_M^R$ will be called right essential polynomials of the sequence a_M, $a_{M+1}, \dots, a_N$ corresponding to the index $\mu_1 = \dots = \mu_\alpha$.*

Any polynomials $R_j(z), \dots, R_{j+k_j-1}(z)$ that form a basis for the complement $\mathcal{H}_{\mu_j+1}^R$ will be called right essential polynomials of the sequence corresponding to the index μ_j, $\alpha+1 \le j \le p+q-\omega$.

Similarly, for $k \neq \mu_j$

$$\mathcal{N}^L_{k-1} = \mathcal{N}^L_k + z^{-1}\mathcal{N}^L_k$$

and for $k = \mu_j$

$$\mathcal{N}^L_{k-1} = (\mathcal{N}^L_k + z^{-1}\mathcal{N}^L_k)\dot{+}\mathcal{H}^L_{k-1}.$$

Choosing bases for the spaces $\mathcal{N}^L_N$ (if $\omega \neq 0$) and $\mathcal{H}^L_{\mu_j-1}(\alpha+1 \leq j \leq p+q-\omega)$, we obtain a sequence of vector polynomials $L_{\alpha+1}(z), \dots, L_{p+q}(z)$ that will be called *left essential polynomials* of the sequence $a_M, a_{M+1}, \dots, a_N$.

Therefore for any sequence $a_M, a_{M+1}, \dots, a_N$ there are $p+q$ indices, $p+q-\omega$ right essential polynomials and $p+q-\alpha$ left essential polynomials.

It turns out that we can always supplement the system $R_1(z), \dots, R_{p+q-\omega}(z)$ and the system $L_{\alpha+1}(z), \dots, L_{p+q}(z)$ to full systems consisting of $p+q$ polynomials. The supplement procedure was described in [**2**]

Now we can obtain a necessary condition for the existence of Padé approximants.

PROPOSITION 2.1. *If for $a(z)$ the right (left) classical Padé approximants of type (n,m) exist, then for the index μ_q (μ_{q+1}) of the sequence $a_{n-m+1}, a_{n-m+1}, \dots, a_{n+m}$ the following inequality holds:*

$$n \geq \mu_q \quad (n+1 \leq \mu_{q+1}).$$

Proof. Let $n \in [\mu_i; \mu_{i+1})$ for some i, $1 \leq i \leq p+q$. Suppose that $n < \mu_q$. Then $i < q$ and a basis of $\mathcal{N}^R_{n+1}$ is

$$\left\{R_j(z), zR_j(z), \dots, z^{n-\mu_j}R_j(z)\right\}_{j=1}^{i}$$

Hence

$$\left[Q^R_{n,m}(z)\right]^k = q_{1k}(z)R_1(z) + \dots + q_{ik}(z)R_i(z),$$

$k = 1, \dots, q$. Here $q_{jk}(z)$ is a scalar polynomial in z and $\deg q_{jk}(z) \leq n - \mu_j$. Thus

$$Q^R_{n,m}(z) = (R_1(z) \dots R_i(z))\, Q(z).$$

The first factor has sizes $q \times i$ and $Q(z)$ is $i \times q$ matrix. Since $i < q$, by the Cauchy–Binet formula we have $\det Q^R_{n,m}(z) \equiv 0$. Therefore, if $n < \mu_q$, then the right classical Padé approximant of type (n,m) does not exist.

The second inequality is proved similarly. ∎

REMARK 2.1. *¿From the proof we see that if $n \geq \mu_q$ and $\det(R_1(z) \dots R_q(z)) \not\equiv 0$, then the matrix polynomial $(R_1(z) \dots R_q(z))$ can be choosen as a denominator of the right Padé approximants of type (n,m) . Here $R_1(z), \dots, R_q(z)$ are the first q right essential polynomials of the sequence $a_{n-m+1}, a_{n-m+2}, \dots, a_{n+m}$. A similar statement holds for the left Padé approximants.*

For necessary and sufficient conditions of the existence and the uniqueness of the matrix Padé approximants in the sence of Baker we refer to [**13, 14, 15**].

3. Indices and essential polynomials for a sequence of Taylor coefficiens of $r(z)$

As we have seen in Section 2 for the construction of Padé approximants indices and essential polynomials of a corresponding sequence are required. In this section we find indices and essential polynomials for a sequence of the Taylor coefficients $r_M, r_{M+1}, \dots, r_N$ of the strictly proper rational matrix function $r(z)$ that is analytic at $z = 0$. For the sequence of Laurent coefficients of $r(z)$ a similar problem was studied in [**16**]. For this reason we omit the proofs of those statements that have analogs in the above-mentioned work.

The indices and essential polynomials are constructed in terms of coprime fractional factorizations of $r(z)$. Recall the definitions. Let $r(z)$ be a strictly proper rational $p \times q$ matrix function. The following matrix fractional factorization

$$r(z) = D_L^{-1}(z)N_L(z) \tag{3.1}$$

plays an important role in the system theory (see, e.g., [**17**]). Here $D_L(z), N_L(z)$ are left coprime matrix polynomials in z and $D_L(z)$ is nonsingular, that is, $\det D_L(z) \not\equiv 0$. The left coprimeness condition is equivalent to the solvability of the following Bezout equation [**17**]:

$$D_L(z)U_L(z) + N_L(z)V_L(z) = I_p.$$

Here $U_L(z)$ and $V_L(z)$ are matrix polynomials in z. As is well known, an arbitrary strictly proper rational matrix function $r(z)$ admits the representation (3.1) that is called the *left coprime fractional factorization* of $r(z)$.

For any nonsingular $p \times p$ matrix polynomials $\bar{L}(t)$ there exists a unimodular matrix polynomial $S(t)$ such that $L(t) = S(t)\bar{L}(t)$ is row proper. This means that the constant matrix L^{row} consisting of the coefficients of the highest degrees in each row of $L(t)$ is invertible. Hence we can choose the denominator in (3.1) in row proper form. Moreover, we can always assume that the denominator $D_L(z)$ is both row proper and column proper, that is, $D_L(z)$ is in Popov form (or in polynomial-echelon form). An algorithm of a transformation of a polynomial matrix to the polynomial-echelon form can be found in [**17**, section 6.7.2]. In the system theory it is shown that the *row degrees* $\lambda_1, \ldots, \lambda_p$ of the row proper denominator are uniquely determinated by $r(z)$. They will be minimal (or Kroneker) indices of $r(z)$ [**17, 18**].

The sum δ of the indices coincides with the degree of $\det D_L(z)$ and is called *McMillan degree* of $r(z)$. We can suppose that $\lambda_1 \geq \lambda_2 \geq \ldots \geq \lambda_p$.

We define the *right coprime fractional factorization* of $r(z)$ in an analogous manner:

$$r(z) = N_R(z)D_R^{-1}(z). \tag{3.2}$$

Here $N_R(z), D_R(z)$ are right coprime matrix polynomials. This is equivalent to the solvability of the Bezout equation

$$U_R(z)D_R(z) + V_R(z)N_R(z) = I_q,$$

where $U_R(z)$ and $V_R(z)$ are matrix polynomials.

We may suppose that $D_R(z)$ is a column and row proper $p \times p$ matrix polynomial, and the *column degrees* ρ_j are in increasing order. The sum of the column degrees coincides with the degree of $\det D_R(z)$ and is equal to δ. By D_R^{col} we denote the invertible matrix consisting of the coefficients of the highest degrees in each column of $D_R(z)$. It is easily seen that all zeros of $D_R(z)$ $(D_L(z))$ are poles of $r(z)$. (We note that the multiplicity of a pole can be less than the multiplicity of a zero.) Hence, if $r(z)$ is analytic at $z = 0$, then $D_R(0)$ and $D_L(0)$ are invertible.

Besides the Bezout equations, we will use the following Diophantine equations:

$$U_R^k(z)D_R(z) + V_R^k(z)N_R(z) = z^k I_q, \tag{3.3}$$

$$D_L(z)U_L^k(z) + N_L(z)V_L^k(z) = z^k I_p, \tag{3.4}$$

for $k = 0, 1 \dots$. These equations will be called the *Bezout equations of order k*. Obviously, the equations are solvable. Moreover, it is known (see, e.g., [**19**]) that there exist the unique solutions such that

$$\deg[V_R^k(z)]^j \ < \ \deg[D_L(z)]^j \equiv \lambda_j,$$

if $\lambda_j > 0$ and $[V_R^k(z)]^j \equiv 0$, if $\lambda_j = 0$, $j = 1, \dots, p$;

$$\deg[V_L^k(z)]_j \ < \ \deg[D_R(z)]_j \equiv \rho_j,$$

if $\rho_j > 0$ and $[V_L^k(z)]_j \equiv 0$, if $\rho_j = 0$, $j = 1, \dots, q$.

These solutions will be called *minimal solutions* of the equations.

PROPOSITION 3.1. *Let $\rho \equiv \rho_q$ and $\lambda \equiv \lambda_1$ be the degrees of the matrix polynomials $D_R(z)$, $D_L(z)$, respectively. Let*

$$D_R(z) = d_0 + d_1 z + \dots + d_\rho z^\rho, \quad D_L(z) = \delta_0 + \delta_1 z + \dots + \delta_\lambda z^\lambda.$$

Then for any $k \geq \rho$ the minimal solutions of the equations (3.3) are solutions of the following difference equations:

$$d_\rho V_R^k(z) + d_{\rho-1} V_R^{k-1}(z) + \dots + d_0 V_R^{k-\rho}(z) = 0, \tag{3.5}$$

$$d_\rho U_R^k(z) + d_{\rho-1} U_R^{k-1}(z) + \dots + d_0 U_R^{k-\rho}(z) = z^{k-\rho} I_q.$$

Similarly, for any $k \geq \lambda$ the minimal solutions of (3.4) satisfy the equations

$$V_R^k(z)\delta_\lambda + V_R^{k-1}(z)\delta_{\lambda-1} + \dots + V_R^{k-\lambda}(z)\delta_0 = 0, \tag{3.6}$$

$$U_R^k(z)\delta_\lambda + U_R^{k-1}(z)\delta_{\lambda-1} + \dots + U_R^{k-\lambda}(z)\delta_0 = z^{k-\lambda} I_p.$$

Proof. Let $\left(U_R^i(z), V_R^i(z)\right)$ be the minimal solution of the Bezout equation of order i:

$$U_R^i(z)D_R(z) + V_R^i(z)N_R(z) = z^i I_q.$$

Multiplying this equation on the right by $d_{\rho-k+i}$, $k - \rho \leq i \leq k$, and summing, we get

$$\sum_{i=k-\rho}^{k} d_{\rho-k+i}\, U_R^i(z)\, D_R(z) + \sum_{i=k-\rho}^{k} d_{\rho-k+i}\, V_R^i(z)\, N_R(z) = z^{k-\rho} D_R(z).$$

This means that the matrix polynomials $X_R(z) = \sum_{i=k-\rho}^{k} d_{\rho-k+i}\, U_R^i(z) - z^{k-\rho} I_q$ and $Y_R(z) = \sum_{i=k-\rho}^{k} d_{\rho-k+i}\, V_R^i(z)$ satisfy the equation

$$X_R(z)D_R(z) + Y_R(z)N_R(z) = 0.$$

Since $\left(U_R^i(z), V_R^i(z)\right)$ is the minimal solution, we have $\deg[Y_R(z)]^j < \lambda_j$. It is easily seen from the uniqueness of the minimal solution that $X_R(z) \equiv 0$, $Y_R(z) \equiv 0$.

The second part of the proposition is proved similarly. ∎

PROPOSITION 3.2. *Suppose that d_ρ and δ_λ are invertible and all poles of $r(z)$ lie in the disk $|z| < 1$. Denote by $v_k^{ij} \in \mathbb{C}^{q\times 1}$ the coefficient in z^i of the vector polynomial $[V_R^k(z)]^j$, $j = 1, \dots, p$; $i = 0, \dots, \lambda_j - 1$. Then the sequence $\{v_k^{ij}\}_{k=0}^{\infty}$ belongs to $l_{q\times 1}^1$.*

Similarly, if $\mathrm{v}_k^{ij} \in \mathbb{C}^{1\times p}$ is the coefficient in z^i of the vector polynomial $[V_L^k(z)]_j$, $j = 1, \dots, q$; $i = 0, \dots, \rho_j - 1$, then the sequence $\{\mathrm{v}_k^{ij}\}_{k=0}^{\infty} \in l_{1\times p}^1$.

Proof. For brevity, we will use the designation $v_k = v_k^{ij}$. It follows from (3.5) that the sequence $\{v_k\}_{k=0}^{\infty}$ satisfies the following infinite system of equations:

$$\begin{pmatrix} d_0 & d_1 & \dots & d_\rho & 0 & 0 & \dots \\ 0 & d_0 & \dots & d_{\rho-1} & d_\rho & 0 & \dots \\ \vdots & \vdots & \ddots & & \ddots & \ddots & \end{pmatrix} \begin{pmatrix} v_0 \\ v_1 \\ \vdots \end{pmatrix} = \begin{pmatrix} 0 \\ 0 \\ \vdots \end{pmatrix}. \tag{3.7}$$

Since d_ρ is invertible, any solution of this system is uniquely determined by initial conditions

$$v_0 = \alpha_0, \dots, v_{\rho-1} = \alpha_{\rho-1}.$$

To prove the statement of the proposition, we will use the theory of Toeplitz operators (see, e.g., [**20**]). Denote $c(z) = D_R(z^{-1})$ and consider the Toeplitz operator $\mathbb{T}_c$ with the symbol $c(z)$. Since all poles of $r(z)$ lie in the disk $|z| < 1$, the symbol $c(z)$ is invertible and $c(z) = z^{-\rho}c_+(z)$, where $c_+(z) = d_\rho + d_{\rho-1}z + \dots + d_0 z^\rho$, is a Wiener-Hopf factorization of the symbol $c(z)$. Hence the kernel of $\mathbb{T}_c$ in the space $l^1_{q\times 1}$ is found by the formula

$$\ker \mathbb{T}_c = \mathbb{T}_{c_+^{-1}} \ker \mathbb{T}_{z^{-\rho}I_q}.$$

It follows from this formula that $\ker \mathbb{T}_c$ consists of vectors of the form

$$\begin{pmatrix} \gamma_0 & 0 & 0 & \dots \\ \gamma_1 & \gamma_0 & 0 & \dots \\ \gamma_2 & \gamma_1 & \gamma_0 & \dots \\ \vdots & \vdots & \vdots & \ddots \end{pmatrix} \begin{pmatrix} \beta_0 \\ \vdots \\ \beta_{\rho-1} \\ 0 \\ \vdots \end{pmatrix},$$

where $c_+^{-1}(z) = \gamma_0 + \gamma_1 z + \dots$, and $\beta_0, \dots, \beta_{\rho-1}$ are arbitrary vectors in $\mathbb{C}^{q\times 1}$. Since $\gamma_0 = d_\rho^{-1}$ is invertible, for arbitrary vectors $\alpha_0, \dots, \alpha_{\rho-1}$ we can find $\beta_0, \dots, \beta_{\rho-1}$ such that

$$\begin{array}{lllllll} \gamma_0\beta_0 & & & & & = & \alpha_0, \\ \gamma_1\beta_0 & + & \gamma_0\beta_1 & & & = & \alpha_1, \\ & & & \dots & & & \\ \gamma_{\rho-1}\beta_0 & + & \gamma_{\rho-2}\beta_1 & + \;\dots\; + & \gamma_0\beta_{\rho-1} & = & \alpha_{\rho-1}. \end{array}$$

Hence there exists a vector of $\ker \mathbb{T}_c$ such that its first ρ coordinates coincide with $\alpha_0, \dots, \alpha_{\rho-1}$. Taking into account that a solution of the system (3.7) is uniquely determined by $\alpha_0, \dots, \alpha_{\rho-1}$, we see that the vector $\{v_k\}_{k=0}^{\infty}$ belongs to $\ker \mathbb{T}_c \subset l^1_{q\times 1}$.

The second part of the proposition is proved similarly. ∎

Now we formulate the basic result of this section. The proof of this result is similar to the proof of Theorem 7 in [**16**].

THEOREM 3.1. *Let $r(z)$ be a strictly proper rational matrix function that is analytic at $z = 0$. Let*

$$r(z) = N_R(z)D_R^{-1}(z), \quad r(z) = D_L^{-1}(z)N_L(z)$$

be the right and left coprime fractional factorizations of $r(z)$, respectively. Suppose that the denominators $D_R(z)$, $D_L(z)$ have both row proper and column proper form and

$$\rho_1 \le \dots \le \rho_q, \quad \lambda_1 \ge \dots \ge \lambda_p$$

are the column degrees of $D_R(z)$ *and the row degrees of* $D_L(z)$, *respectively. Let* $(U_R^k(z),\ V_R^k(z))$ *and* $(U_L^k(z),\ V_L^k(z))$ *be the minimal solution of the Bezout equations (3.3), (3.4) of order* k.

Then for any sequence $r_M, r_{M+1}, \ldots, r_N$ *of the Taylor coefficients of* $r(z)$ *such that* $M \geq 0$, $N - M \geq \lambda_1 + \rho_q - 2$, *the left (right) defect of the sequence coincides with the number of row (column) degrees of* $D_R(z)$ $(D_L(z))$ *that are equal to zero, the integers*

$$M + \rho_1 - 1, \ldots, M + \rho_q - 1, N - \lambda_1 + 1, \ldots, N - \lambda_p + 1 \tag{3.8}$$

are the indices; the vector polynomials

$$[D_R(z)]^1,\ \ldots,\ [D_R(z)]^q,\ [V_L^{N+1}(z)]^1, \ldots, [V_L^{N+1}(z)]^p \tag{3.9}$$

form a full system of right essential polynomials; and the vector polynomials

$$\begin{gathered} z^{-(N-M-\rho_1+2)}[V_R^{N+1}(z)]_1,\ \ldots,\ z^{-(N-M-\rho_q+2)}[V_R^{N+1}(z)]_q, \\ z^{-\lambda_1}[D_L(z)]_1,\ \ldots,\ z^{-\lambda_p}[D_L(z)]_p \end{gathered} \tag{3.10}$$

form a full system of left essential polynomials.

■

Now we can find under what conditions the indices of the sequence are stable under small perturbations. It is easily seen that if $k \leq \mu_1$, then T_k is left invertible and if $k \geq \mu_{p+q}$, then T_k is right invertible. Since the set of one-sided invertible operators is open, the indices $\bar{\mu}_1, \ldots, \bar{\mu}_{p+q}$ of a perturbation sequence satisfy the inequality

$$\mu_1 \leq \bar{\mu}_1 \leq \ldots \leq \bar{\mu}_{p+q} \leq \mu_{p+q}.$$

Hence, if $\mu_{p+q} - \mu_1 \leq 1$, then the indices $\mu_1, \ldots, \mu_{p+q}$ are stable under small perturbations.

Now from the previous theorem we obtain the following four stability cases for the sequence $r_M, r_{M+1}, \ldots, r_N$.

COROLLARY 3.1. *The sequence* $r_M, r_{M+1}, \ldots, r_N$ *for* $M \geq 0$ *and* $N - M \geq \lambda_1 + \rho_q - 2$, *has the stable indices if one of the following conditions is fulfilled:*

(1) $\lambda_1 - \lambda_p = 0, \quad \rho_q - \rho_1 = 0, \quad N - M = \lambda_1 + \rho_q - 1,$
(2) $\lambda_1 - \lambda_p = 0, \quad \rho_q - \rho_1 = 0, \quad N - M = \lambda_1 + \rho_q - 2,$
(3) $\lambda_1 - \lambda_p = 0, \quad \rho_q - \rho_1 = 1, \quad N - M = \lambda_1 + \rho_q - 2,$
(4) $\lambda_1 - \lambda_p = 1, \quad \rho_q - \rho_1 = 0, \quad N - M = \lambda_1 + \rho_q - 2.$

■

4. The Montessus theorem for matrix Padé approximants

Let $a(z)$ be a $p \times q$ matrix function which is analytic at the origin $z = 0$, meromorphic in the disk $|z| < R$, and has finitely many poles in this disk. Without loss of generality we will suppose that $R > 1$ and all poles lie in the unit disk $|z| < 1$. Let $r(z)$ be the sum of principal parts of the Laurent series of $a(z)$ in neighborhoods of the poles of $a(z)$. Obviously, $r(z)$ is a strictly proper rational matrix function that is analytic at $z = 0$. Let us represent $a(z)$ in the form:

$$a(z) = b(z) + r(z),$$

where $p \times q$ matrix function $b(z)$ is analytic in $|z| < R$. In order to construct the Padé approximants of type (n, m) for $a(z)$ we need the sequence $a_{n-m+1}, a_{n-m+2},$

$\ldots, a_{n+m}$ consisting of the Taylor coefficient of $a(z)$. We will consider the sequence $a_{n-m+1}, a_{n-m+2}, \ldots, a_{n+m}$ as a perturbation of the sequence $r_{n-m+1}, r_{n-m+2}, \ldots, r_{n+m}$ by the sequence $b_{n-m+1}, \ldots, b_{n+m}$ of the Taylor coefficients of $b(z)$ that is infinitesimal as $n \to \infty$. By this reason, the indices of $r_{n-m+1}, r_{n-m+2}, \ldots, r_{n+m}$ must be stable. In the present work we will study the first stability case only. In this case for the matrix function $a(z)$ the following conditions are fulfilled:

(1) the denominators $D_R(z)$, $D_L(z)$ have the invertible leading coefficients, that is $\rho_1 = \ldots = \rho_q \equiv \rho$ and $\lambda_1 = \ldots = \lambda_p \equiv \lambda$,
(2) $\lambda + \rho$ is even.

Here $\rho = \deg D_R(z)$, $\lambda = \deg D_L(z)$.

THEOREM 4.1. *If the conditions 1)–2) are fulfilled,* $m = \frac{\lambda+\rho}{2}$, *and* $n \geq m-1$, *then the sequence* $r_{n-m+1}, r_{n-m+2}, \ldots, r_{n+m}$ *has the stable indices*

$$\underbrace{n-\nu, \ldots, n-\nu}_{q}, \underbrace{n-\nu+1, \ldots, n-\nu+1}_{p},$$

where $\nu = \frac{\lambda-\rho}{2}$. *Moreover, if* $D_R(z) = d_0 + d_1 z + \ldots + d_\rho z^\rho$, $\quad D_L(z) = \delta_0 + \delta_1 z + \ldots + \delta_\lambda z^\lambda$, *then the matrices*

$$\mathcal{P}^R_{n-\nu+1}(r^{n+m}_{n-m+1}) = \begin{pmatrix} I_q & 0 & \ldots & 0 \\ d_1 d_0^{-1} & 0 & \ldots & 0 \\ \vdots & \vdots & & \vdots \\ d_\rho d_0^{-1} & 0 & \ldots & 0 \end{pmatrix},$$

$$\mathcal{P}^L_{n-\nu}(r^{n+m}_{n-m+1}) = \begin{pmatrix} 0 & \ldots & 0 & 0 \\ \vdots & & \vdots & \vdots \\ 0 & \ldots & 0 & 0 \\ \delta_0^{-1}\delta_\lambda & \ldots & \delta_0^{-1}\delta_1 & I_p \end{pmatrix}$$

are the projectors on the spaces $\ker_R T_{n-\nu+1}(r^{n+m}_{n-m+1})$ *and* $\ker_L T_{n-\nu}(r^{n+m}_{n-m+1})$, *respectively. Here we consider* $\mathcal{P}^L_{n-\nu}(r^{n+m}_{n-m+1})$ *as an operator of right multiplication by the matrix.*

Proof. The formulas for the indices follow at once from Theorem 3.1. In order to obtain the formula for the projector $\mathcal{P}^R_{n-\nu+1}(r^{n+m}_{n-m+1})$ we consider the truncated sequence $r_{n-m+1}, \ldots, r_{n+m-1}$. It is easily seen from Theorem 3.1 that the indices of this sequence coincide with the integers $\underbrace{n-\nu, \ldots, n-\nu}_{p+q}$. Hence the matrix $T_{n-\nu}(r^{n+m-1}_{n-m+1})$ is invertible.

Since the matrix $T_{n-\nu+1}(r^{n+m}_{n-m+1})$ has the form

$$T_{n-\nu+1}(r^{n+m}_{n-m+1}) = \begin{pmatrix} R & T_{n-\nu}(r^{n+m-1}_{n-m+1}) \end{pmatrix},$$

where $R = \begin{pmatrix} r_{n-\nu+1} \\ \vdots \\ r_{n+m} \end{pmatrix}$, it is easily seen that the matrix

$$T^\dagger_{n-\nu+1}(r^{n+m}_{n-m+1}) = \begin{pmatrix} 0 \\ T^{-1}_{n-\nu}(r^{n+m-1}_{n-m+1}) \end{pmatrix}$$

is the right inverse of $T_{n-\nu+1}(r_{n-m+1}^{n+m})$. Then

$$\hat{\mathcal{P}}_{n-\nu+1}^{R}(r_{n-m+1}^{n+m}) = I_{(\rho+1)q} - T_{n-\nu+1}^{\dagger}(r_{n-m+1}^{n+m})T_{n-\nu+1}(r_{n-m+1}^{n+m}) =$$

$$\begin{pmatrix} I_q & 0 \\ -T_{n-\nu+1}^{-1}(r_{n-m+1}^{n+m-1})\,R & 0 \end{pmatrix}$$

is the projector on $\ker_R T_{n-\nu+1}(r_{n-m+1}^{n+m})$.

On the other hand, by Theorem 3.1, the space $\mathcal{N}_{n-m+1}^{R}(r_{n-m+1}^{n+m})$ of generating polynomials of vectors in $\ker_R T_{n-\nu+1}(r_{n-m+1}^{n+m})$ coincides with the span of the vector polynomials $[D_R(z)]^1, \dots, [D_R(z)]^q$. It easily follows from this that $\mathcal{P}_{n-\nu+1}^{R}(r_{n-m+1}^{n+m})$ is also a projector on $\ker_R T_{n-\nu+1}(r_{n-m+1}^{n+m})$. Hence

$$\operatorname{Im} \mathcal{P}_{n-\nu+1}^{R}(r_{n-m+1}^{n+m}) = \operatorname{Im} \hat{\mathcal{P}}_{n-\nu+1}^{R}(r_{n-m+1}^{n+m}).$$

Taking into account the form of these projectors, we can easily verify that they coincide. Thus the projector $\mathcal{P}_{n-\nu+1}^{R}(r_{n-m+1}^{n+m})$ is generated by the right inverse $T_{n-\nu+1}^{\dagger}(r_{n-m+1}^{n+m})$.

The second part of the theorem is proved similarly. ■

Now we can prove that the indices of the sequence $a_{n-m+1}, a_{n-m+2}, \dots, a_{n+m}$ coincide with the indices of $r_{n-m+1}, r_{n-m+2}, \dots, r_{n+m}$ for $m = \frac{\lambda+\rho}{2}$ and for n sufficiently large. The proof of this fact is based on the stability of the invertibility property for linear bounded operators in Banach spaces.

For matrices in $\mathbb{C}^{l\times k}$ we will use the following norms:

$$\|A\|_1 = \max_{1\le j\le l} \sum_{i=1}^{k} |A_{ij}|$$

and

$$\|A\|_\infty = \max_{1\le i\le k} \sum_{j=1}^{l} |A_{ij}|.$$

For the sequence $a_M^N = (a_M, a_{M+1}, \dots, a_N)$ we will also apply two norms:

$$\|a_M^N\|_1 = \sum_{i=1}^{N-M+1} \|a_{M+i}\|_1,$$

$$\|a_M^N\|_\infty = \sum_{i=1}^{N-M+1} \|a_{M+i}\|_\infty,$$

The same norms we will use for the generating polynomial $a_M^N(z) = a_M z^M + a_{M+1}z^{M+1} + \dots + a_N z^N$ of this sequence. It is easily seen that

$$\|T_k(a_M^N)\|_1 \le \|T_M(a_M^N)\|_1 \le \|a_M^N\|_1,$$

$$\|T_k(a_M^N)\|_\infty \le \|T_N(a_M^N)\|_\infty \le \|a_M^N\|_\infty$$

for $M \le k \le N$.

Now we can prove an analog of Theorem 4.1 for the perturbed sequence $a_{n-m+1}, \dots, a_{n+m}$.

THEOREM 4.2. *If the conditions 1)–2) are fulfilled, $m = \frac{\lambda+\rho}{2}$, and n is sufficiently large, then the sequence $a_{n-m+1}, a_{n-m+2}, \dots, a_{n+m}$ has the stable indices*

$$\underbrace{n-\nu, \dots, n-\nu}_{q}, \underbrace{n-\nu+1, \dots, n-\nu+1}_{p},$$

where $\nu = \frac{\lambda-\rho}{2}$. *Moreover, there exist right essential polynomials* $R_1^n(z), \ldots, R_q^n(z)$ *of degree* ρ *and left essential polynomials* $L_{q+1}^n(z), \ldots, L_{p+q}^n(z)$ *of degree* λ *such that* $Q_n^R(0) = I_q$, $Q_n^L(0) = I_p$. *Here*

$$Q_n^R(z) = \begin{pmatrix} R_1^n(z) \ldots R_q^n(z) \end{pmatrix}, \quad Q_n^L(z) = z^\lambda \begin{pmatrix} L_{q+1}^n(z) \\ \vdots \\ L_{p+q}^n(z) \end{pmatrix}.$$

If $Q_n^R(z) = I_q + d_1^n z + \ldots + d_\rho^n z^\rho$, $\quad Q_n^L(z) = I_p + \delta_1^n z + \ldots + \delta_\lambda^n z^\lambda$, *then the matrices*

$$\mathcal{P}_{n-\nu+1}^R(a_{n-m+1}^{n+m}) = \begin{pmatrix} I_q & 0 & \ldots & 0 \\ d_1^n & 0 & \ldots & 0 \\ \vdots & \vdots & & \vdots \\ d_\rho^n & 0 & \ldots & 0 \end{pmatrix},$$

$$\mathcal{P}_{n-\nu}^L(a_{n-m+1}^{n+m}) = \begin{pmatrix} 0 & \ldots & 0 & 0 \\ \vdots & & \vdots & \vdots \\ 0 & \ldots & 0 & 0 \\ \delta_\lambda^n & \ldots & \delta_1^n & I_p \end{pmatrix}$$

are the projectors on the spaces $\ker_R T_{n-\nu+1}(a_{n-m+1}^{n+m})$ *and* $\ker_L T_{n-\nu}(a_{n-m+1}^{n+m})$, *respectively.*

Proof. It is easily seen that

$$\|T_{n-\nu}(a_{n-m+1}^{n+m-1}) - T_{n-\nu}(r_{n-m+1}^{n+m-1})\|_1 \le \|b_{n-m+1}^{n+m-1}\|_1 = \|\sum_{i=n-m+1}^{n+m-1} b_i z^i\|_1.$$

Since the series $\sum_{i=0}^{\infty} b_i z^i$ is absolutely convergent at $z = 1$, for any $\varepsilon > 0$ there exists the number $\bar{n}_0$ such that for all $n \ge \bar{n}_0$ we have

$$\|T_{n-\nu}(a_{n-m+1}^{n+m-1}) - T_{n-\nu}(r_{n-m+1}^{n+m-1})\|_1 < \varepsilon.$$

In Theorem 4.1 we prove that the matrix $T_{n-\nu}(r_{n-m+1}^{n+m-1})$ is invertible. Let us prove that the matrix $T_{n-\nu}(a_{n-m+1}^{n+m-1})$ is also invertible for n sufficiently large. To do this, we must estimate the norm of the inverse of $T_{n-\nu}(r_{n-m+1}^{n+m-1})$. The problem of the inversion of block Toeplitz and Hankel matrices was a subject of numerous investigations. Some relevant references are [**21**] – [**24**].

We will use an explicit formula for the inverse of a block Toeplitz matrix from the work [**2**, eq. 5.13]. This formula contains the matrices $\mathcal{R}(z)$, $\mathcal{L}(z)$ of right and left essential polynomials of the truncated sequence $r_{n-m+1}, \ldots, r_{n+m-1}$. By Theorem 3.1, we have

$$\mathcal{R}(z) = \begin{pmatrix} D_R(z) \; V_L^{n+m}(z) \end{pmatrix}, \quad \mathcal{L}(z) = z^{-\lambda} \begin{pmatrix} V_R^{n+m}(z) \\ D_L(z) \end{pmatrix},$$

where $\deg \mathcal{R}(z) = \rho$, $\deg \mathcal{L}(z) = \lambda$. Let

$$\mathcal{R}(z) = \mathcal{R}_0 + \mathcal{R}_1 z + \ldots + \mathcal{R}_\rho z^\rho, \quad \mathcal{L}(z) = \mathcal{L}_0 + \mathcal{L}_{-1} z^{-1} + \ldots + \mathcal{L}_{-\lambda} z^{-\lambda},$$

$\mathcal{R}_i \in \mathbb{C}^{q\times(p+q)}$, $\mathcal{L}_{-i} \in \mathbb{C}^{(p+q)\times p}$. Then, by the above-mentioned formula, we have

$$T_{n-\nu}^{-1}(r_{n-m+1}^{n+m-1}) = \begin{pmatrix} \mathcal{R}_0 & 0 & \dots & 0 \\ \mathcal{R}_1 & \mathcal{R}_0 & \dots & 0 \\ \vdots & \vdots & \ddots & \vdots \\ \mathcal{R}_{\rho-1} & \mathcal{R}_{\rho-2} & \dots & \mathcal{R}_0 \end{pmatrix} \Pi \begin{pmatrix} \mathcal{L}_0 & \mathcal{L}_{-1} & \dots & \mathcal{L}_{-\lambda+1} \\ 0 & \mathcal{L}_0 & \dots & \mathcal{L}_{-\lambda+2} \\ \vdots & \vdots & \ddots & \vdots \\ 0 & 0 & \dots & \mathcal{L}_0 \end{pmatrix}.$$

Here

$$\Pi = \begin{pmatrix} I_{p+q} & \dots & 0 & \dots & 0 \\ \vdots & \ddots & \vdots & & \vdots \\ 0 & \dots & I_{p+q} & \dots & 0 \end{pmatrix}$$

if $\rho \leq \lambda$, or

$$\Pi = \begin{pmatrix} I_{p+q} & \dots & 0 \\ \vdots & \ddots & \vdots \\ 0 & \dots & I_{p+q} \\ \vdots & & \vdots \\ 0 & \dots & 0 \end{pmatrix}$$

if $\rho \geq \lambda$.

Obviously, $\|\Pi\|_1 = 1$ and

$$\|T_{n-\nu}^{-1}(r_{n-m+1}^{n+m-1})\|_1 \leq \lambda(p+q)\ \|\mathcal{R}(z)\|_1\ \|\mathcal{L}(z)\|_\infty.$$

Here we take into account that $\|\cdot\|_1 \leq \lambda(p+q)\ \|\cdot\|_\infty$. By Proposition 3.2, the coefficients of the matrix polynomials $V_L^k(z)$, $V_R^k(z)$ converge to 0 as $k \to \infty$. Hence there esists the number $\tilde{n}_0$ such that

$$\|V_L^{n+m}(z)\|_1 < \|D_R(z)\|_1, \quad \|V_R^{n+m}(z)\|_\infty < \|D_L(z)\|_\infty$$

for all $n \geq \tilde{n}_0$. Thus

$$\|\mathcal{R}(z)\|_1 \leq \|D_R(z)\|_1, \quad \|\mathcal{L}(z)\|_\infty \leq \|D_L(z)\|_\infty.$$

Now, if we take $\varepsilon = \frac{1}{\lambda(p+q)\ \|D_R(z)\|_1\ \|D_L(z)\|_\infty}$, then we see that there exist the number n_0 such that

$$\|T_{n-\nu}(a_{n-m+1}^{n+m-1}) - T_{n-\nu}(r_{n-m+1}^{n+m-1})\|_1 < \frac{1}{\|T_{n-\nu}^{-1}(r_{n-m+1}^{n+m-1})\|}$$

for $n \geq n_0$. This means that $T_{n-\nu}(a_{n-m+1}^{n+m-1})$ is invertible for n sufficiently large. Hence the matrix

$$T_{n-\nu+1}(a_{n-m+1}^{n+m}) = \begin{pmatrix} A & T_{n-\nu}(a_{n-m+1}^{n+m-1}) \end{pmatrix},$$

where $A = \begin{pmatrix} a_{n-\nu+1} \\ \vdots \\ a_{n+m} \end{pmatrix}$, is right invertible, and the matrix

$$Q_n^R \equiv \begin{pmatrix} I_q \\ d_1^n \\ \vdots \\ d_\rho^n \end{pmatrix} = \begin{pmatrix} I_q \\ -T_{n-\nu}^{-1}(a_{n-m+1}^{n+m-1})A \end{pmatrix}$$

satisfy the equation $T_{n-\nu+1}(a_{n-m+1}^{n+m})Q_n^R = 0$. It follows from this that the columns of Q_n^R form a basis of the space $\ker_{\mathrm{R}} T_{n-\nu+1}(a_{n-m+1}^{n+m})$ and the matrix polynomial $Q_n^R(z) = I_q + d_1^n z + \ldots + d_\rho^n z^\rho$ is the matrix of right essential polynomials $R_1^n(z), \ldots, R_q^n(z)$. It is evident that

$$T_{n-\nu+1}^{\dagger}(a_{n-m+1}^{n+m}) = \begin{pmatrix} 0 \\ T_{n-\nu}^{-1}(a_{n-m+1}^{n+m-1}) \end{pmatrix}$$

is the right inverse of $T_{n-\nu+1}(a_{n-m+1}^{n+m})$ and

$$I_{(\rho+1)q} - T_{n-\nu+1}^{\dagger}(a_{n-m+1}^{n+m}) T_{n-\nu+1}(a_{n-m+1}^{n+m}) = \mathcal{P}_{n-\nu+1}^R(a_{n-m+1}^{n+m}).$$

The second part of the theorem is proved similarly. ■

By Remark 2.1, $Q_n^R(z)$ $(Q_n^L(z))$ is the denominator of the right (left) Padé approximants of type (n, m) for $a(z)$. Since $T_{n-\nu}(a_{n-m+1}^{n+m-1})$ is invertible, the denominator $Q_n^R(z)$ $(Q_n^L(z))$ is unique if we normalize it by the condition $Q_n^R(0) = I_q$ $(Q_n^L(0) = I_p)$. This leads to the following statement.

COROLLARY 4.1. *If the conditions 1)–2) are fulfilled,* $m = \frac{\lambda+\rho}{2}$, *and* n *is sufficiently large, then for* $a(z)$ *there exists the right (left) Padé approximant of type* (n, m) *in the sense of Baker iff* $p \leq q$ $(p \geq q)$. *The denominator* $Q_n^R(z)$ $(Q_n^L(z))$ *of this approximant is unique if we normalize it by the condition* $Q_n^R(0) = I_q$ $(Q_n^L(0) = I_p)$. ■

Now we ready for the proof of the main part of the Montessus theorem. We will need the continuity of projectors on the kernels of Fredholm operators. We reformulate this well-known result (see, e.g., [**10**], Theorem 11.3) in the form that is suitable for us.

LEMMA 4.1. *Let* A *be a right invertible operator from a Banach space* E_1 *into a Banach space* E_2 *such that* $\ker A$ *is finite-dimensional. Let* $A^\dagger$ *be any right inverse of* A *and* $P_A = I - A^\dagger A$ *is the projector on* $\ker A$. *Then for any operator* B *such that* $\|A - B\| < \frac{1}{2\|A^\dagger\|}$ *it is fulfilled:*

(1) B *is the right invertible and* $B^\dagger = C^{-1}A^\dagger$ *is the right inverse of* B. *Here* $C = I - A^\dagger(A - B)$, $C^{-1} = \sum_{i=0}^{\infty} \left(A^\dagger(A - B)\right)^i$.
(2) *For the projector* $P_B = I - B^\dagger B = C^{-1}P_A C$ *on* $\ker B$ *the following inequality holds*

$$\|P_A - P_B\| < 4\,\|P_A\|\,\|A^\dagger\|\,\|A - B\|.$$

■

THEOREM 4.3. *Suppose the conditions 1)–2) are fulfilled and* $m = \frac{\lambda+\rho}{2}$. *Let* $Q_n^R(z)$ $(Q_n^L(z))$ *be the (unique) denominator of the Padé approximant of type* (n, m) *for* $a(z)$ *and* $Q_n^R(0) = I_q$ $(Q_n^L(0) = I_p)$. *Then*

$$\lim_{n\to\infty} Q_n^R(z) = D_R^0(z), \quad \lim_{n\to\infty} Q_n^L(z) = D_L^0(z),$$

where $D_R^0(z) = D_R(z)D_R^{-1}(0)$, $D_L^0(z) = D_L^{-1}(0)D_L(z)$.

Proof. In Lemma 4.1 we put $A = T_{n-\nu+1}(r_{n-m+1}^{n+m})$ and $B = T_{n-\nu+1}(a_{n-m+1}^{n+m})$. Let

$$A^\dagger = \begin{pmatrix} 0 \\ T_{n-\nu}^{-1}(r_{n-m+1}^{n+m-1}) \end{pmatrix}.$$

Then $P_A = \mathcal{P}^R_{n-\nu+1}(r^{n+m}_{n-m+1})$. Now we will find the projector $P_B = C^{-1}P_AC$. Since

$$A^\dagger(A-B) = -\begin{pmatrix} 0 \\ T^{-1}_{n-\nu}(r^{n+m-1}_{n-m+1}) \end{pmatrix} T_{n-\nu+1}(b^{n+m}_{n-m+1}),$$

the matrix $A^\dagger(A-B)$ has the following structure:

$$A^\dagger(A-B) = \begin{pmatrix} 0 & \dots & 0 \\ * & \dots & * \\ \vdots & & \vdots \\ * & \dots & * \end{pmatrix}.$$

Here 0 is the $q \times q$ zero matrix and by $*$ we denote unessential elements of the matrix. It follows from this that the matrices C and C^{-1} have the form:

$$\begin{pmatrix} I_q & 0 & \dots & 0 \\ * & * & \dots & * \\ \vdots & \vdots & & \vdots \\ * & * & \dots & * \end{pmatrix}.$$

Hence

$$P_B = \begin{pmatrix} I_q & 0 & \dots & 0 \\ * & 0 & \dots & 0 \\ \vdots & \vdots & & \vdots \\ * & 0 & \dots & 0 \end{pmatrix}.$$

Taking into account the form of the projectors P_B and $\mathcal{P}^R_{n-\nu+1}(a^{n+m}_{n-m+1})$ and the equality $\operatorname{Im} P_B = \operatorname{Im} \mathcal{P}^R_{n-\nu+1}(a^{n+m}_{n-m+1})$ one may now conclude that $P_B = \mathcal{P}^R_{n-\nu+1}(a^{n+m}_{n-m+1})$. Since $\|P_A\|_1 = \|D^0_R(z)\|_1$ and

$$\|A^\dagger\|_1 \le \lambda(p+q)\,\|D_R(z)\|_1\,\|D_L(z)\|_\infty,$$

we obtain from Lemma 4.1

$$\|\mathcal{P}^R_{n-\nu+1}(r^{n+m}_{n-m+1}) - \mathcal{P}^R_{n-\nu+1}(a^{n+m}_{n-m+1})\|_1 \le \text{const}\,\|b^{n+m}_{n-m+1}\|_1.$$

But the series $\sum_{i=1}^\infty b_i z^i$ is absolutely converge at $z = 1$, hence

$$\mathcal{P}^R_{n-\nu+1}(a^{n+m}_{n-m+1}) \to \mathcal{P}^R_{n-\nu+1}(r^{n+m}_{n-m+1})$$

as $n \to \infty$. By virtue of the form of the projectors $\mathcal{P}^R_{n-\nu+1}(r^{n+m}_{n-m+1})$ and $\mathcal{P}^R_{n-\nu+1}(a^{n+m}_{n-m+1})$, we obtain

$$Q^R_n(z) \to D^0_R(z)$$

as $n \to \infty$.

The second part of the theorem is proved similarly. ■

Now we can finish the proof of a matrix analog of the Montessus theorem.

THEOREM 4.4. *Suppose the conditions 1)–2) are fulfilled and* $m = \frac{\lambda+\rho}{2}$. *Let* K *be any compact subset of the disk* $|z| < R$ *and also* $a(z)$ *has no poles in* K. *Let* $\pi^R_n(z)$ $(\pi^L_n(z))$ *be the right (left) Padé approximant of type* (n, m) *for* $a(z)$. *Then*

$$\lim_{n\to\infty} \|a(z) - \pi^R_n(z)\|_{C(K)} = 0, \qquad \lim_{n\to\infty} \|a(z) - \pi^L_n(z)\|_{C(K)} = 0.$$

Proof. From the equations $a(z) = b(z) + r(z)$ and $r(z) = D_L^{-1}(z)N_L(z)$ we have

$$a(z) = D_L^{-1}(z)C_L(z),$$

where the matrix function $C_L(z) = D_L(z)b(z) + N_L(z)$ is analytic in $|z| < R$. Hence

$$a(z) - \pi_n^R(z) = D_L^{-1}(z)\left[C_L(z)Q_n^R(z) - D_L(z)P_n^R(z)\right]Q_n^R(z)^{-1}.$$

Since $\det D_R^0(z)$ is bounded away from zero on K and $\det Q_n^R(z) \to \det D_R^0(z)$, we obtain

$$\|Q_n^R(z)^{-1} - D_R^0(z)^{-1}\|_{C(K)} \to 0$$

as $n \to \infty$. Thus

$$\|a(z) - \pi_n^R(z)\|_{C(K)} \le \text{const}\|C_L(z)Q_n^R(z) - D_L(z)P_n^R(z)\|_{C(K)}.$$

Now we prove that $\|C_L(z)Q_n^R(z) - D_L(z)P_n^R(z)\|_{C(K)} \to 0$ as $n \to \infty$. We can do it as in the scalar case (see [**6**], Theorem 6.2.1). Since $Q_n^R(z) \in \mathcal{N}_{n-\nu+1}^R(a_{n-m+1}^{n+m})$, we have $\deg P_n^R(z) \le n - \nu$ and $\deg D_L(z)P_n^R(z) \le n + m$. It follows from the definition of Padé approximants of type (n, m) that

$$C_L(z)Q_n^R(z) = D_L(z)P_n^R + O(z^{n+m+1}), \quad z \to 0.$$

Hence

$$C_L(z)Q_n^R(z) - D_L(z)P_n^R(z) = \sum_{j=0}^{\rho}\left(\sum_{i=n+m+1}^{\infty} c_{i-j}z^i\right)d_j^n.$$

Here c_i are the Taylor coefficients of $C_L(z)$ and d_j^n are the coefficients of the matrix polynomial $Q_n^R(z)$. Let K is a subset of the disk $|z| \le R_0 < R$. Then

$$\|C_L(z)Q_n^R(z) - D_L(z)P_n^R(z)\|_{C(K)} \le \text{const}\sum_{j=0}^{\rho}\sum_{i=n+m+1}^{\infty}\|c_{i-j}\|R_0^i.$$

Here $\|\cdot\|$ is any matrix norm on $\mathbb{C}^{p\times q}$.

Since the matrix power series $\sum_{i=0}^{\infty} c_i z^i$ is absolutely converge on $|z| = R_0$, we get

$$\|C_L(z)Q_n^R(z) - D_L(z)P_n^R(z)\|_{C(K)} \to 0$$

as $n \to \infty$.

The second part of the theorem is proved similarly. ■

We note that we can study the second stability case in the same manner.

If $p = q = 1$, then only the first stability case is realized and in this case we get the Montessus theorem.

The author would like to thank the referee for pointing out some references and helpful comments.

References

[1] V. M. Adukov, Wiener-Hopf factorization of meromorphic matrix-valued functions (in Russian) *Algebra i Analiz* **4** (1992), No. 1, pp. 54–74; English transl., *St. Petersburg Math. J.* **4** (1993), No. 1, pp. 51–69.

[2] V. M. Adukov, Generalized inversion of block Toeplitz matrices, *Linear Algebra Appl.* **274** (1998), pp. 85-124.

[3] A. Bultheel and M. Van Barel, A matrix euclidean algorithm and the matrix minimal Padé approximation problem, in *Continued Fractions and Padé approximants*, (C. Brezinski, ed.) pp. 11–51, 1990.

[4] G. Labahn and S. Cabay, Matrix Padé fraction and their computation, *SIAM J. Comput.* **18** (1989), No. 4, pp 639–657.

[5] B. Beckermann and G. Labahn, A uniform approach for the fast, reliable computation of Matrix-type Padé approximants, *SIAM J. Matrix Anal. Appl.*, **15** (1994), pp. 804–823.
[6] G. A. Baker Jr. and P. Graves-Morris, *Padé Approximants*. Addison-Wesley, MA, 1981.
[7] V. M. Adukov, The structure of the kernel and inversion of block Toeplitz matrices (in Russian), Manuscript 9030-B85, deposited at VINITI, 1985.
[8] A. A. Gončar, On convergence of Padé approximants for some classes of meromorphic functions (in Russian), *Matem. Sbornik*, **97** (1975), No. 4, pp. 607–629.
[9] E. A. Rakhmanov, On convergence of Padé approximants of meromorphic functions (in Russian), *Matem. Sbornik*, **104** (1977), No. 2, pp. 271–291.
[10] I. C. Gohberg and N. Ya. Krupnik, *Einführung in die Theorie der Eindimensionalen Singulären Integraloperatoren*. Birkhäuser, Boston, 1979.
[11] V. M. Adukov, On Wiener-Hopf factorization of meromorphic matrix functions, *Integral Equations and Operator Theory*, **14** (1991), pp. 767–774.
[12] G. Heinig and P. Jankowski, Kernel structure of block Hankel and Toeplitz matrices and partial realization, *Linear Algebra Appl.*, **175** (1992), pp. 1–30.
[13] Xu Guo-liang, Existence and uniqueness of matrix Padé approximants, *J. Compt. Math.*, **8** (1990), No. 1, pp.65–74.
[14] Xu Guo-liang and Li Jiakai, Generalized matrix Padé approximants, *Approx. Theory Appl.*, **5** (1989), No. 4, pp.47–60.
[15] Xu Guo-liang and A. Bultheel, Matrix Padé approximation: Definition and properties, *Linear Algebra Appl.*, **137/138** (1990), pp.67–137.
[16] V. M. Adukov, The essential polynomial approach to coprime factorizations. Submitted.
[17] T. Kailath, *Linear Systems*. Prentice-Hall, Englewood Cliffs, N.-Y., 1980.
[18] G. D. Forney, Minimal bases of rational vector spaces, with applications to multivariable linear systems, *SIAM J. Control*, **13** (1975), no. 3, pp.493–520.
[19] W. A. Wolovich and P. A. Antsaklis, The canonical Diophantine equations with applications, *SIAM J. Control and Optimization*, **22** (1984), No. 5, pp. 777–787.
[20] I. C. Gohberg and I. A. Feldman, *Convolution Equations and Projection Method for Their Solution*, Amer. Math. Soc. Transl. Math. Monographs **41**, Providence, 1974.
[21] L. Lerer and M. Tismenetsky, Generalized Bezoutians and the inversion problem for block matrices, *Integral Equations and Operator Theory*, **9** (1986), pp. 790–819.
[22] I. C. Gohberg and T. Shalom, On inversion of square matrices partitioned into non-square blocks, *Integral Equations and Operator Theory*, **12** (1989), pp. 539–566.
[23] I. C. Gohberg and V. Olshevsky, Circulants, displacements and decompositions of matrices, *Integral Equations and Operator Theory*, **15** (1992), pp. 731–743.
[24] G. Labahn, B. Beckermann and S. Cabay, Inversion of mosaic Hankel matrices via matrix polynomial system, *Linear Algebra Appl.*, **221** (1995), pp. 253–279.

Department of Differential Equations and Dynamical Systems, South Ural State University, Chelyabinsk, Russian Federation
E-mail address: adukov@math.tu-chel.ac.ru

PART II. System Theory, Signal and Image Processing

Contemporary Mathematics
Volume **280**, 2001

Systems of low Hankel rank: a survey

P. Dewilde

ABSTRACT. During the last ten years a new class of structured systems has appeared and many of its properties have been thoroughly investigated. It is the class of matrices whose entries can be described by a low order time-varying system, recently also called 'quasi-separable systems'. This class is a subclass of the class of 'locally finite time-varying systems'. We give a survey of the main properties, to wit: system representation, factorization theorems, inversion theory and approximation theory. It turns out that most calculations can be done 'locally' on minimal algebraic representations yielding algorithms of very low complexity, determined by the number of free parameters. In particular, we give a algorithms for matrix inversion based on URV factorization and for matrix approximation which are linear in the size of the matrix.

1. Introduction

The basic object under study is an operator T which represents a linear map mapping a (possibly infinite) sequence of vectors $u = [u_i]_{i=-\infty}^{\infty}$ to a (possibly infinite) sequence of vectors $y = [y_k]_{k=-\infty}^{\infty}$:

$$y = uT.$$

Each u_i (y_k) is a vector belonging to a finite dimensional vector space of dimension m_i (respect. n_k). These dimensions may vary and even vanish, in which case the entry simply disappears, it is replaced by a 'placeholder' ('$\cdot$' - with the convention that the product of a matrix of dimensions $m \times 0$ with one of dimensions $0 \times n$ is a zero matrix of dimensions $m \times n$). This formalism has the advantage that it contains finite vectors and matrices, but also regular time invariant systems as special cases. T has a matrix representation $T = [T_{i,k}]$ with $T_{i,k}$ a matrix of dimensions $m_i \times n_k$. In the sequel we shall assume that T is bounded as an operator on ℓ_2 sequences, i.e. that

$$\|T\| = \sup_{\|u\|_2=1} \|uT\|_2 < \infty,$$

for $\|u\|_2 = \sqrt{\sum_{i=-\infty}^{\infty} \|u_i\|_2^2}$ and $\|u_i\|_2$ the regular Euclidean norm.

We say that T is *locally finite* if it possesses a time varying system representation or realization to be defined as follows: if T is block-upper triangular (i.e. if $T_{i,k} =$

1991 *Mathematics Subject Classification.* Primary 93B25, 93B28; Secondary 93B40.

0 for $i > k$), there exist matrices $\{A_k, B_k, C_k, D_k\}$ for each integer k (possibly vanishing!) such that

$$\begin{array}{rcl} T_{k,k} & = & D_k \\ (\text{for } i < k)\ T_{i,k} & = & B_i A_{i+1} \cdots A_{k-1} C_k. \end{array}$$

Alternatively, one can say that there is a time-varying 'system realization' that produces the operator T via the 'local' state space equations:

$$\left\{ \begin{array}{rcl} x_{k+1} & = & x_k A_k + u_k B_k \\ y_k & = & x_k C_k + u_k D_k. \end{array} \right.$$

The representation is said to be of low Hankel rank if the dimensions b_k of the state vectors x_k at each time point k is small or alternatively, if the dimensions of the matrices A_k is small. The dimension of x_k is called the local degree of T at that point. The dimensions of all the system realization matrices must of course be compatible. The dimensions of A_k, B_k, C_k and D_k are respectively $b_k \times b_{k+1}$, $m_k \times b_k$, $b_k \times n_k$ and $m_k \times n_k$. We shall also generally assume that the realization for T is *uniformly exponentially stable - ues* in the classical sense for time-varying systems, i.e. that there are uniform bounds on the matrices A_k, B_k, C_k and D_k, and that there exists a real number ρ with $0 \le \rho < 1$ such that, uniformly over k,

$$\lim\ \sup_{\ell \to \infty} \|A_{k+1} \cdots A_\ell\| \le \rho^{\ell - k}.$$

These conditions make T automatically a bounded operator on ℓ_2 sequences. (In the case of finite matrices the ues condition is trivially satisfied.)

In the case that T has both upper and lower parts, we shall assume that also its lower part has a ues realization, i.e. there exist matrices $\{A'_k, B'_k, C'_k\}$ such that for $i > k$,

$$T_{i,k} = B'_i A'_{i-1} \cdots A'_{k+1} C'_k,$$

corresponding to the backward realization

$$\left\{ \begin{array}{rcl} x'_{k-1} & = & x_k A'_k + u_k B'_k \\ y'_k & = & x_k C'_k \end{array} \right.$$

and satisfying ues conditions for the realization with primes.

It is notationally convenient and often computationally attractive to collect the local realization operators in global 'diagonal' operators. Connected to the series $[A_k]$ we define the operator A as

$$A = \begin{bmatrix} \ddots & & & & \\ & A_{-1} & & & \\ & & \boxed{A_0} & & \\ & & & A_1 & \\ & & & & \cdots \end{bmatrix},$$

in which we distinguish the (0,0)'th element by boxing it. Likewise for B, C and D. If, moreover, we introduce the 'causal shift' Z by the rule

$$[\cdots, u_{-1}, \boxed{u_0}, u_1, \cdots] Z = [\cdots, \boxed{u_{-1}}, u_0, u_1, \cdots]$$

and by Z^* its inverse, then T will have the representation

$$T = B'Z^*(I - A'Z^*)^{-1}C' + D + BZ(I - AZ)^{-1}C \tag{1}$$

in terms of its realization. The property of ues assures the existence of the inverse of $(I - AZ)$ as an upper operator and of $(I - A'Z^*)$ as a lower one. A representation need not be given originally, it can be obtained either via 'system identification' or 'system approximation' - see further sections.

Equivalence. Two upper realizations $\{A_k, B_k, C_k, D_k\}$ and $\{\hat{A}_k, \hat{B}_k, \hat{C}_k, \hat{D}_k\}$ are *strictly (or Lyapunov) equivalent* if there exists a uniformily bounded sequence of invertible square matrices $\{R_k\}$ such that the $\{R_k^{-k}\}$ are uniformly bounded as well, and such that $\hat{A}_k = R_k^{-1} A_k R_{k+1}$, $\hat{B}_k = B_k R_{k+1}$, $\hat{C}_k = R_k^{-1} C_k$ and $\hat{D}_k = D_k$ (the collection of $\{R_k\}'s$ is called a Lyapunov state transformation). An equivalent realization is then given by the quadruple:

$$\begin{bmatrix} R_k^{-1} A_k R_{k+1} & R_k^{-1} C_k \\ B_k R_{k+1} & D_k \end{bmatrix}. \tag{2}$$

A Lyapunov-equivalence preserves the degree and the ues property, in fact with the same spectral bound ρ. A realization is minimal if the dimension of all A_k's is as small as possible. Minimal realizations can be found by the procedure described in the section on realizations. A realization is in *input normal form* if all the pairs $\begin{bmatrix} A_k \\ B_k \end{bmatrix}$ are co-isometric (i.e. $A_k^* A_k + B_k^* B_k = I$) and in *output normal form* if all the pairs $[A_k, C_k]$ are isometric. It is an interesting question whether a given realization can be brought in input, respect. output normal form through a strict equivalency. If one starts from a ues realization $\{A, B, C, D\}$ then the answer is as follows. Let $\{M_k\}$ be the bounded solution of the collection of Lyapunov-Stein equations

$$M_{k+1} = A_k^* M_k A_k + B_k^* B_k. \tag{3}$$

The classical Lyapunov-Stein theorem states that these equations will indeed have a bounded solution (if the system is ues) given by the series expansion

$$M_k = \sum_{i=1}^{\infty} A_{k-1}^* \cdots A_{k-i+1}^* B_{k-i}^* B_{k-i} A_{k-i+1} \cdots A_{k-1}. \tag{4}$$

This bounded solution will of course be positive semidefinite. If the solution is actually strictly positive definite, i.e. if there exists an ϵ such that for all k, $M_k > \epsilon I$, then we call the system *strictly reachable*, and an adequate set of state transformation matrices is found from $M_k = R_k^{-*} R_k^{-1}$. The corresponding strictly equivalent realization $\{\hat{A}$ etc.$\}$ will be both ues and in input normal form. The M_k obtained in the procedure have an important physical interpretation, they form the Gramians of the reachability operator at each time point of the system under consideration. The case for the output normal form is dual, the observability Gramians are given by the backward equation

$$N_k = A_k N_{k+1} A_k^* + C_k C_k^*.$$

Extended input/output spaces. Let $[\mathcal{M}_i]$ be the sequence of input spaces, each u_i belongs to $\mathcal{M}_i$, then by $\ell_2^{\mathcal{M}}$ we denote the space of quadratically summable input series. We call

$$\mathcal{M} \triangleq \oplus_{i=-\infty}^{\infty} \mathcal{M}_i$$

the base space of $\ell_2^{\mathcal{M}}$. T is by definition a bounded map from $\ell_2^{\mathcal{M}}$ to $\ell_2^{\mathcal{N}}$. For the purposes of representation theory, it is advantageous to augment these spaces as follows. Let, for each fixed integer k, $U_{k,i}$ be a time series in i belonging to $\ell_2^{\mathcal{M}}$, in such a way that $\|U\| \triangleq \sqrt{\sum_{i,k=-\infty}^{\infty} \|U_{k,i}\|^2} < \infty$, then we say that U belongs to the Hilbert-Schmidt space $\mathcal{X}_2^{\mathcal{M}}$. The action of T on U is obtained by 'stacking', i.e. $Y_{(k,\cdot)} = U_{(k,\cdot)}T$ for each k. (This stacking principle is of course dimensionally compatible with matrix calculus.) It is easy to see that $\|T\|$ is the same whether T is interpreted as an operator between $\mathcal{X}_2$ spaces or ℓ_2 spaces.

The present paper gives a short, hopefully coherent summary of the theory and is intended as a didactical presentation of the main ideas. An extensive treatment is to be found in the book [**6**] which appeared last year. There were some forerunners of the theory, we mention in particular the work of Koltracht and co-authors [**10**] on 'semi-separable' matrices, which form a special case where the local state dimension is just one, the work of Dym and Gohberg [**9**] on extension theory of matrices and of Deprettere [**3**] for the extension of the Schur algorithm to the matrix case.

2. Kronecker or realization theory

The question whether a given upper operator T can be realized using a finite state space model is answered by considering the Hankel operator connected to T. The time varying realization theory parallels in this the classical Kronecker degree theory, or its generalization to multiport systems, the Ho-Kalman theory. At each time point k a Hankel operator H_k can be defined which maps an input time series with support on the $\mathcal{M}_\ell$ with $\ell < k$ to the restriction of the response on the semi-infinite interval $\mathcal{N}_\ell$ for $\ell \geq k$. If we express H_k in terms of any realization, we find

$$H_k = \begin{bmatrix} \vdots \\ B_{k-2}A_{k-1} \\ B_{k-1} \end{bmatrix} [C_k \; A_kC_{k+1} \; A_kA_{k+1}C_{k+2} \; \cdots].$$

Hence, the rank of H_k is at most the dimension b_k of the state x_k. Conversely, suppose that the dimension of each H_k is finite, then each H_k has a minimal factorization

$$H_k = \mathcal{R}_k\mathcal{O}_k$$

and we can define

$$\begin{aligned} C_k &= \text{first entry of } \mathcal{O}_k \\ B_k &= \text{bottom entry of } \mathcal{R}_{k+1} \\ A_k &= \mathcal{O}_k^{\leftarrow}\mathcal{O}_{k+1}^{\dagger} \end{aligned}$$

in which $\mathcal{O}_k^{\leftarrow}$ indicates the remainder of $\mathcal{O}_k$ when the first entry has been chopped off, and the (right) pseudo-inverse $\mathcal{O}_{k+1}^{\dagger}$ exists since minimal factorizations have been used. The generalized Kronecker theorem then states that this is a genuine minimal realization for T [**15**]. The operator $\mathcal{R}_k$ is called the kth reachability operator of T, while $\mathcal{O}_k$ is the kth observability operator.

The realization will be in output normal form if in the factorization of each H_k the observability operator is isometric (i.e. $\mathcal{O}_k\mathcal{O}_k^* = I$), and it will be in input normal form if all reachability operators are co-isometric, meaning that for all k, $\mathcal{R}_k^*\mathcal{R}_k = I$. We shall say that a system is strictly observable if it has an output normal form which is ues, while it will be strictly reachable if it has an input normal form which is ues.

3. Factorization theory and the generalized Beurling-Lax theorem

We consider two types of factorizations:

[A] External factorizations.

We try to represent T as

$$T = \Delta_\ell^* V_\ell = V_r \Delta_r^*$$

in which Δ_ℓ, Δ_r are upper and V_ℓ and V_r are *inner*, i.e. upper unitary. This is the LTV generalization of the classical 'coprime factorization' for a causal rational transfer function, for example: $T = \frac{z-\alpha^*}{1-\beta z} = \frac{z-\alpha^*}{z-\beta^*} \cdot \frac{z-\beta^*}{1-\beta z}$. The inner factors V_ℓ and V_r capture the dynamics of the system.

B Outer-Inner factorizations.

Here we try to factor T as

$$T = T_{o\ell} V$$

in which $T_{o\ell}$ is in some sense invertible and V is a candidate for inner embedding.

It turns out that external factorization is considerably simpler than outer-inner factorization, but both types have important applications. The key theorem leading to these factorizations is a generalization of the Beurling-Lax theorem to the time-varying setting, see also the generalization given by Arveson to related 'nest algebras' [**1**].

THEOREM 1 (Generalized Beurling-Lax [**13, 6**]). *Suppose that $\mathcal{K}$ is a closed subspace of $\mathcal{U}_2^{\mathcal{M}}$ which is both invariant for left multiplication with arbitrary conformal diagonal operators and for left multiplication with the shift Z (i.e. $D\mathcal{K} \subset \mathcal{K}$ for arbitrary D and $Z\mathcal{K} \subset \mathcal{K}$), then there exist a base space $\mathcal{N}_0$ and an isometric and upper operator V such that*

$$\mathcal{K} = \mathcal{U}_2^{\mathcal{N}_0} V.$$

The proof of the theorem is pretty simple, and goes via construction of a so called sliced orthonormal basis for the wandering subspace $\mathcal{L} \stackrel{\Delta}{=} \mathcal{K} \ominus Z\mathcal{K}$. This basis actually produces V.

Interestingly, the operator appearing in the generalized Beurling-Lax theorem is not inner, it is just isometric. It is important to characterize it further. We have:

THEOREM 2 (Characterization of upper inner operator [**6**]). *Suppose that V is locally finite and isometric, and let $\mathcal{H}_O$ be the observability space of V. Then $\mathcal{H}_O$ is closed and*

$$\mathcal{U}_2^{\mathcal{N}} = \mathcal{H}_O \oplus \mathcal{U}_2^{\mathcal{N}_0} \oplus \ker(\cdot V^*|_{\mathcal{U}_2^{\mathcal{M}}}).$$

V will be inner iff $\ker(\cdot V^*|_{\mathcal{X}_2^{\mathcal{M}}}) = 0$.

The distinction between $\ker(\cdot V^*|_{\mathcal{U}_2})$ and $\ker(\cdot V^*|_{\mathcal{X}_2})$ should be noted. The first is an invariant subspace of $\mathcal{U}_2$. Thanks to the Beurling-Lax theorem, it can be represented as $\mathcal{U}_2^{\mathcal{N}_1} U$ for some causal isometric U and an appropriate base space $\mathcal{N}_1$. It turns out that $\mathcal{K}_0^{''} \stackrel{\Delta}{=} \ker(\cdot U^*|_{\mathcal{X}_2}) = \ker(\cdot V^*|_{\mathcal{X}_2})$, the two operators have the same *defect space*. The operator

$$W = \begin{bmatrix} V \\ U \end{bmatrix}$$

is again isometric, possesses a unitary realization, but is inner iff $\mathcal{K}_0^{''} = 0$. These considerations allow us to solve the two factorization problems mentioned at the beginning of the section. We shall say that an invariant subspace $\mathcal{K} = V\mathcal{U}_2 \subset \mathcal{U}_2$ has *full range* when V is inner, i.e. when the defect space of V is empty.

External Factorization T will possess a left external factorization $T = \Delta_\ell^* V_\ell$ with Δ_ℓ causal iff the observability nullspace $\mathcal{K}_0$ of T has full range. A sufficient condition for this to happen is that T possesses a ues output normal form.

The external factorization is easy to produce. Let $\{A, B, C, D\}$ be a realization of T in output normal form, then $[A, C]$ is isometric, and a realization for V_ℓ is found by unitary completion: $[A, B_u, C, D_u]$, while the realization for Δ_ℓ is given by

$$\begin{bmatrix} A & AB^* + CD^* \\ B_u & B_u B^* + D_u D^* \end{bmatrix}$$

(assuming output normal form of T!). Hence, an external factorization is obtained numerically simply by solving a Lyapunov-Stein equation recursively. This can be done in a numerically stable way, 'in square root form' and reduces to a classical QR-type factorization on a small matrix at each time step [**6**].

Outer-inner factorization In this case we write

$$\mathcal{K} \stackrel{\Delta}{=} \overline{\mathcal{U}_2^{\mathcal{M}} T}$$

then by the generalized Beurling-Lax theorem we have

$$\mathcal{K} = \mathcal{U}_2^{\mathcal{N}_0} V$$

for some isometric V and a base space $\mathcal{N}_0$. It then follows that $T = T_{o\ell} V$ in which $T_{o\ell}$ is left-outer, i.e. $\overline{\mathcal{U}_2^{\mathcal{M}} T_{o\ell}} = \mathcal{U}_2^{\mathcal{N}_o}$, hence it is causally invertible, be it that the inverse has dense domain but is not uniformily bounded. $T_{o\ell}$ can be further factorized as $T_{o\ell} = UT_o$ in which U is co-isometric ($U^*U = I$) and T_o is outer. We have obtained a representation for T as

$$T = UT_o V.$$

The Moore-Penrose pseudoinverse of T now follows as

$$T^\dagger = V^* T_o^{-1} U^*$$

in which T_o^{-1} exists as a causal operator on a dense domain.

This inversion procedure can easily be generalized to mixed causal-anticausal operators T with a locally finite representation for their causal and anti-causal parts - we skip the details. We terminate this section by giving the recursive algorithm capable of computing the outer-inner factorization first presented in [**14**]. It should be remarked that all operations (in which U, T_0, V are computed) are local, hence of low complexity, and linear in the dimension of T - see the discussion at the end of the paper.)

Recursive algorithm for outer-inner factorization. The recursive algorithm for the left outer-inner factorization $T = T_{o\ell} V$ assumes knowledge of a minimal state space representation $\{A, B, C, D\}$ for T, and knowledge of an intermediate matrix Y_{k+1}. It then computes isometric realizations $\{A_{Vk}, B_{Vk}, C_{Vk}, D_{Vk}\}$ and $\{A_{Vk}, B_{Uk}, C_{Vk}, D_{Uk}\}$ for V and the orthonormal complement U of V respectively, a realization of $\{A_k, B_k, C_{ok}, D_{ok}\}$ of $T_{o\ell}$ all at time point k and an update Y_k of the connecting intermediate matrix, via the recursive RQ factorization:

$$\begin{bmatrix} A_k Y_{k+1} & C_k \\ B_k Y_{k+1} & D_k \end{bmatrix} = \begin{bmatrix} 0 & Y_k & C_{ok} \\ 0 & 0 & D_{ok} \end{bmatrix} W_k \tag{5}$$

where W_k is the a unitary matrix which decomposes conformally as

$$W_k = \begin{bmatrix} B_{U_k} & C_{U_k} \\ A_{V_k} & C_{V_k} \\ B_{V_k} & D_{V_k} \end{bmatrix}.$$

This computational scheme consists of an RQ factorization applied to the left hand side, so that the resulting upper 'R' matrix is in upper column echelon form. This guarantees that $\ker(Y_k\cdot) = 0$ and $\ker(D_{ok}\cdot) = 0$ as required from the extended Beurling-Lax theorem. The solution of (5) is dependent on the existence of a starting value Y_{k+1} for some large k. In the case of finite matrices the start of the recursion is trivial, one just takes $Y_{k+1} = [\cdot]$ - an empty matrix - for the first value (typically k is then the dimension of the matrix). In the more general case, more information on the structure of T for large k is needed. For example, if it is LTI, the LTI value can be taken. The recursion is based on orthogonal transformations and is numerically stable in a broad sense, namely in the sense that numerical errors are not amplified in subsequent steps. However, QR factorizations do not offer an absolute guarantee for correct determination of the rank of a matrix and more sophisticated reductions in the style of 'rank revealing QR' or SVD may be needed - this topic goes beyond the purposes of the present paper and needs further investigations.

Interestingly, the recursive algorithm amounts to a generalization of the Kalman square root algorithm [**12**], in which the matrices C_k and D_k have a special form. Conversely, one can state that the Kalman filter recursion is equivalent to an inner-outer factorization of a special kind, and also to the Cholesky factorization of the covariance matrix build on the observation data.

4. Low Hankel rank operator approximation theory

Another important result of the theory of locally finite systems (of which low Hankel rank matrices form a special case) is the possibility to approximate a locally finite operator of 'large' Hankel rank with one of low Hankel rank, in such a way that the approximation minimizes the complexity as measured by the degree of the realization, given a tolerance on the approximation. Suppose for example that we know a diagonal expansion of the operator, which we break off at a certain high level k:

$$T = T_1 Z + T_2 Z^{[2]} + T_3 Z^{[3]} + \cdots T_k Z^{[k]}$$

in which we indicate by $Z^{[k]}$ a product of k shifts with conformal dimensions. Such a series expansion can be represented trivially by a realization

$$A = \begin{bmatrix} 0 & & & 0 \\ I & \ddots & & \\ & \ddots & \ddots & \\ 0 & & I & 0 \end{bmatrix}, \quad C = \begin{bmatrix} I \\ 0 \\ \vdots \\ 0 \end{bmatrix}.$$

$$B = \begin{bmatrix} T_1 & \cdots & T_k \end{bmatrix}, \quad D = [\,0\,].$$

This realization is not necessarily minimal but it is in output normal form.

Let Γ be a diagonal matrix which represents the desired degree of approximation (it could be an 'epsilon', but we allow time-varying approximation.) Our first goal will be to find an operator T' of possibly mixed type ($T' \in \mathcal{X}$) such that

$$\|\Gamma^{-1}(T - T')\| < 1.$$

If T_a is taken as the strictly upper part of T', then it turns that T_a approximates T in 'Hankel norm', i.e.

$$\|T - T'\| = \|H_{T-T_a}\| \triangleq \|T - T_a\|_H$$

in which H_T indicates the Hankel operator associated to the strictly upper operator T. This property is known as 'Nehari's theorem'. The Hankel norm is a pretty strong norm on $\mathcal{U}Z$.

The solution of the problem, first derived in [**8**] leans heavily on the properties of causal, ues J-unitary matrices, which we now introduce briefly. We work with a splitting of the bases spaces in two orthogonal complements at each time point: $\mathcal{M} \triangleq \mathcal{M}_1 \oplus \mathcal{M}_2$, $\mathcal{N} = \mathcal{N}_1 \oplus \mathcal{N}_2$. We also allow splittings of the series of state spaces, which we call $\mathcal{B} = \mathcal{B}_+ \oplus \mathcal{B}_-$ here. To these split spaces we attach a mixed positive-negative metric, represented by J-unitary matrices:

$$J_1 = \begin{bmatrix} I_{\mathcal{M}_1} & \\ & -I_{\mathcal{M}_2} \end{bmatrix}, \quad J_2 = \begin{bmatrix} I_{\mathcal{N}_1} & \\ & -I_{\mathcal{N}_2} \end{bmatrix}, \quad J_{\mathcal{B}} = \begin{bmatrix} I_{\mathcal{B}_+} & \\ & -I_{\mathcal{B}_-} \end{bmatrix}$$

The characterization of upper J-unitary operators that we need is given by the following theorem:

THEOREM 3. *Let*

$$\theta \triangleq \left[\begin{array}{c|cc} A & C_1 & C_2 \\ \hline B_1 & D_{11} & D_{12} \\ B_2 & D_{21} & D_{22} \end{array}\right]$$

be a realization of a locally finite ues system, and assume further that

$$\theta^* \begin{bmatrix} J_{\mathcal{B}} & \\ & J_1 \end{bmatrix} \theta = \begin{bmatrix} J_{\mathcal{B}}^{(-1)} & \\ & J_2 \end{bmatrix}$$

(i.e. each θ_i is J-unitary), then the corresponding transfer map Θ, which is causal and upper, satisfies

$$\Theta^* J_1 \Theta = J_2, \quad \Theta J_2 \Theta^* = J_1.$$

Starting out from a realization for T in output normal form, the method to find T' now proceeds as follows:
(1) compute $T = \Delta_\ell^* U_\ell$ a left coprime external factorization for T;
(2) find a minimal, causal and J-unitary operator Θ such that

$$[U_\ell^* \;\; T^*\Gamma^{-1}]\Theta = [A' \;\; B'] \in \mathcal{U}. \tag{6}$$

If Θ can be found (see further for the algorithm), then the solution exists and is given by:

$$T' = \Gamma\Theta_{22}^{-*}B'^*$$

and

$$T_a = \text{stricly upper part of } T'.$$

That the latter is the case, if Θ exists, is pretty easy to see. From (6) and $S_{\text{in}} \triangleq -\Theta_{12}\Theta_{22}^{-1}$ we find

$$\Gamma^{-1}(T - T') = S_{\text{in}}^* U_\ell.$$

The J-unitarity of Θ guarantees the existence of Θ_{22} and the contractivity of S_{in}. The norm inequality then follows easily.

Remains to find Θ. It turns out that that is a fairly simple problem of the interpolation type. Without giving a lengthy motivation, we simply state the result. Let

$$\begin{bmatrix} A & C \\ B_u & D_u \end{bmatrix} \tag{7}$$

be a unitary realization for U_r. There will be a solution for Θ if the (block diagonal) operator

$$\begin{bmatrix} A \\ \hline B_u \\ \Gamma^{-1}B \end{bmatrix}$$

is such that we can find a state transformation R and a state space splitting $\mathcal{B} = \mathcal{B}_+ \oplus \mathcal{B}_-$ such that (using the notation $T^{(-1)} \triangleq Z^*TZ$)

$$A^*(R^*J_{\mathcal{B}}R)A + B_u^*B_u - B^*\Gamma^{-2}B = (R^*J_{\mathcal{B}}R)^{(-1)}.$$

(i.e. such that there exists a state transformation which makes the reachability operator J-unitary). Let M be the solution of the Lyapunov-Stein system of equations

$$A^*MA + B_u^*B_u - B^*\Gamma^{-2}B = M^{(-1)} \tag{8}$$

then the condition reduces to requiring that the solution M is non-singular, in which case it can be written as $M = R^*J_{\mathcal{B}}R$ via local determination of its inertia at each index k. The fact that the inertia has no zeros can be formulated as 'the columns of (7) form a Krein space'. Whence the theorem

THEOREM 4 (Hankel norm complexity reduction theorem [**8, 16**]). *Let be given a locally finite operator T with a realization $\{A, B, C, D\}$ in output normal form and let Γ be a given positive definite diagonal matrix. Let $\{A, B_u, C, D_u\}$ be a realization of U_ℓ in the corresponding left external factor $T = \Delta_\ell^* U_\ell$, and suppose that (7) forms a Krein space, i.e. is such that the recursive Lyapunov-Krein equation (8) yields a non-singular inertia. Then there exists a causal (J_1, J_2)-unitary operator Θ such that*
(1) (7) is a reachability pair for Θ;
(2) $B' \triangleq U_\ell^*\Theta_{12} + T^*\Gamma^{-1}\Theta_{22}$ is upper;
(3) $T_a \triangleq$ strictly upper part of $\Gamma\Theta_{22}^{-} * B'^*$ is a locally finite causal operator with the lowest possible local degree for each time point satisfying

$$\|T - T_a\|_H < \Gamma.$$

The local degree of T_a - solution to the problem - is determined by the Hankel rank of the strictly upper part of Θ^{-*}, namely the number of eigenvalues of M that are less than zero (i.e. the local dimension of $\mathcal{B}_-$). A state realization for T_a can be derived in a straightforward way, the details go beyond the present paper. In case M is not definite at some time point, then a slight modification of Γ will yield the desired result. Many more results are available in this area, in particular:
- all solutions can be expressed in terms of the entries of Θ,
- there are especially attractive forms for the case of matrix approximations.

The theory amounts to a generalization of the classical model reduction theory in the sense of Schur-Takagi and Adamyan-Arov-Krein. It solves a central problem of linear algebra: the approximation of a general matrix by one of minimal computational complexity.

5. Further results and discussion

Some additional results are:
- interpolation theory for matrices and time-varying systems, both of the Schur kind [**4**] and the Nevanlinna-Pick type [**7, 5, 2**],
- spectral factorization theory [**6**],
- embedding theory [**6**],
- time-varying versions of the bounded-real and positive-real lemma's [**13**],
- the combination of the low displacement rank theory with the low Hankel rank property [**6**],
- the solution of minimal sensitivity problems and robust control problems in a time-varying setting [**11, 17**] including the four block problem.

A further interesting point is the complexity of computations obtained by exploiting the low Hankel rank structure of the operators and/or matrices involved. A realization theory using a minimal number of algebraic parameters is available [**6**]. Assuming equal local degree b everywhere and size n of the original operator, the number of parameter needed to represent the original is roughly nb. The computation of the inverse, whether directly (if possible) or via the URV method explained above, would lay between nb^2 and nb depending on the sophistication of the local computations. In ideal circumstances, that is if minimal algebraic representations and a minimal number of computations are used, nb is feasible, since the inverse of system can be found by simple graph manipulations on the signal flow graph and the minimality of the representation is preserved.

References

[1] W. Arveson. Interpolation problems in nest algebras. *J. Functional Anal.*, 20:208–233, 1975.
[2] J.A. Ball, I. Gohberg, and M.A. Kaashoek. Nevanlinna-Pick interpolation for time-varying input-output maps: the discrete case. In I. Gohberg, editor, *Time-Variant Systems and Interpolation*, volume 56 of *Operator Theory: Advances and Applications*, pages 1–51. Birkhäuser Verlag, 1992.
[3] E. Deprettere. Mixed-form time-variant lattice recursions. In *Outils et Modèles Mathématiques pour l'Automatique, l'Analyse de Systèmes et le Traitement du Signal*, Paris, 1981. CNRS.
[4] P. Dewilde and Ed.F. Deprettere. The generalized schur algorithm: approximation and hierarchy. *Operator Theory: Advances and Applications*, 16:437–503, 1988.
[5] P. Dewilde and H. Dym. Interpolation for upper triangular operators. *Operator Theory and Applications*, pages 153–260, 1992.
[6] P. Dewilde and A.-J. van der Veen. *Time-varying Systems and Computations*. Kluwer, 1998.
[7] P.M. Dewilde. A course on the algebraic Schur and Nevanlinna-Pick interpolation problems. In Ed. F. Deprettere and A.J. van der Veen, editors, *Algorithms and Parallel VLSI Architectures*, volume A, pages 13–69. Elsevier, 1991.
[8] P.M. Dewilde and A.J. van der Veen. On the Hankel-norm approximation of upper-triangular operators and matrices. *Integral Eq. Operator Th.*, 17(1):1–45, 1993.
[9] H. Dym and I. Gohberg. Extensions of band matrices with band inverses. *Lin. Alg. Appl.*, 36:1–24, 1981.
[10] I. Gohberg, T. Kailath, and I. Koltracht. Linear complexity algorithms for semiseparable matrices. *Integral Equations and Operator Theory*, 8:780–804, 1985.
[11] A. Halanay and V. Ionescu. *Time-Varying Discrete Linear Systems*, volume 68 of *Operator Theory: Advances and Applications*. Birkhäuser, 1994.
[12] M. Morf and T. Kailath. Square-root algorithms for Least-Squares Estimation. *IEEE Trans. Automat. Control*, 20(4):487–497, 1975.

[13] A.J. van der Veen. *Time-Varying System Theory and Computational Modeling: Realization, Approximation, and Factorization.* PhD thesis, Delft University of Technology, Delft, The Netherlands, June 1993.

[14] A.J. van der Veen. Time-varying lossless systems and the inversion of large structured matrices. *Archiv f. Elektronik u. Übertragungstechnik*, 49(5/6):372–382, September 1995.

[15] A.J. van der Veen and P.M. Dewilde. Time-varying system theory for computational networks. In P. Quinton and Y. Robert, editors, *Algorithms and Parallel VLSI Architectures, II*, pages 103–127. Elsevier, 1991.

[16] A.J. van der Veen and P.M. Dewilde. On low-complexity approximation of matrices. *Linear Algebra and its Applications*, 205/206:1145–1201, July 1994.

[17] Xiaode Yu. *Time-varying System Identification, J-lossless Factorization, and H_∞ Control.* PhD thesis, Delft Univ. of Technology, Delft, The Netherlands, May 1996.

DIMES, DELFT UNIVERSITY OF TECHNOLOGY POB 5031, 2600GA DELFT, THE NETHERLANDS.

E-mail address: Dewilde@DIMES.tudelft.nl

Contemporary Mathematics
Volume **280**, 2001

Tensor Approximation and Signal Processing Applications

Eleftherios Kofidis and Phillip A. Regalia

Abstract. Recent advances in application fields such as blind deconvolution and independent component analysis have revived the interest in higher-order statistics (HOS), showing them to be an indispensable tool for providing piercing solutions to these problems. Multilinear algebra and in particular tensor analysis provide a natural mathematical framework for studying HOS in view of the inherent multiindexing structure and the multilinearity of cumulants of linear random processes. The aim of this paper is twofold. First, to provide a semitutorial exposition of recent developments in tensor analysis and blind HOS-based signal processing. Second, to expose the strong links connecting recent results on HOS blind deconvolution with parallel, seemingly unrelated, advances in tensor approximation. In this context, we provide a detailed convergence analysis of a recently introduced tensor-equivalent of the power method for determining a rank-1 approximant to a matrix and we develop a novel version adapted to the symmetric case. A new effective initialization is also proposed which has been shown to frequently outperform that induced by the Tensor Singular-Value Decomposition (TSVD) in terms of its closeness to the globally optimal solution. In light of the equivalence of the symmetric high-order power method with the so-called superexponential algorithm for blind deconvolution, also revealed here, the effectiveness of the novel initialization shows great promise in yielding a clear solution to the local-extrema problem, ubiquitous in HOS-based blind deconvolution approaches.

1. Introduction

The development and understanding of *blind* methods for system identification/deconvolution has been an active and challenging topic in modern signal processing research [**1**]. *Blindness* refers to an algorithm's ability to operate with no knowledge of the input signal(s) other than that of some general statistical characteristics, and is imposed either by the very nature of the problem, as in noncooperative communications, or by practical limitations such as bandwidth constraints in mobile communications. Although blind methods using only second-order statistics (SOS) of the signals involved exhibit very good performance in terms of estimation accuracy and convergence speed, they can sometimes be overly sensitive to mismodeling effects [**43**]. Methods relying on HOS criteria, on the other hand,

1991 *Mathematics Subject Classification.* 93B30.

E. Kofidis was supported by a Training and Mobility of Researchers (TMR) grant of the European Commission, contract no. ERBFMBICT982959.

are observed to be more robust in this respect and are therefore still a matter of interest, despite the long data lengths required to estimate higher order statistics [**50**].

The most widely used and studied family of higher-order blind deconvolution criteria is based on seeking maxima of normalized cumulants of the deconvolved signal [**5**]. These so-called "Donoho" or "minimum-entropy" criteria (after [**22**]) have been shown to possess strong information-theoretic foundations [**22, 15**] and have resulted in efficient algorithmic schemes for blind equalization [**56, 57, 52**] and blind source separation [**11, 28**]. Furthermore, as shown in [**51**], they are equivalent with the popular Godard criterion [**29, 30**], being at the same time directly applicable to both sub-Gaussian and super-Gaussian sources. A well-known efficient algorithm resulting from a gradient optimization of the normalized cumulant is the *Super-Exponential Algorithm* (SEA), so called because its convergence rate may be shown faster than exponential [**57**] in specific circumstances. It was recently rederived as a natural procedure stemming from a characterization of the Donoho cost stationary points [**52**] and proved to result from a particular choice of the step size in a more general gradient search scheme [**46**]. Like many blind algorithms, however, it usually exhibits multiple convergent points, underscoring the need for an effective initialization strategy.

Several blind source separation (BSS) approaches that exploit multilinearity by reducing the problem to that of cumulant tensor[1] decomposition have been reported and proven successful both for the fewer sources than sensors case [**15**] as well as the more challenging setting of non-full column rank mixing matrix [**9, 18**]. Criteria such as the maximization of the sum of squares of autocumulants of the deconvolved signal vector are shown to be equivalent to the diagonalization of the corresponding tensor, performed either directly with the aid of linear algebra tools ([**7**]–[**10**], [**12**]–[**16**]) or by exploiting the isomorphism of symmetric tensors with homogeneous multivariate polynomials to take advantage of classic polynomial decomposition theorems [**19, 18**]. For the particular, yet important, case of a single deconvolution output, the above criterion reduces to the maximization of the square of the normalized cumulant, which may in turn be seen to correspond to the best least squares (LS) rank-1 approximation of the source cumulant tensor.

An appealing method for numerically solving the latter problem was developed in [**35**] as a generalization of the power method to higher-order arrays, and demonstrated through simulation results to converge to the globally-optimum rank-1 approximant when initialized with the aid of a Tensor extension of the Singular Value Decomposition (TSVD). However, this *Tensor Power Method (TPM)*, as presented in [**35**], appears inefficient for the symmetric tensor case, as encountered in blind equalization problems, and moreover no clear convergence proof is given. The tensor power method is revisited in this paper and adapted to the symmetric problem. Convergence proofs for both the general and the symmetric version are devised that help to better understand the underlying mechanism. Furthermore, the equivalence of the symmetric tensor power method and the super-exponential algorithm is brought out. An alternative initialization method is likewise proposed, which is observed in simulations to compare favourably with that of TSVD in terms of the closeness of the initial point to the global optimum.

[1]By the term *tensor* we mean a multi-indexed array (see Section 2).

This paper is organized as follows: Section 2 contains definitions of tensor-related quantities as well as a review of the tensor decompositions proposed in the literature. The powerful and elegant notation of Tucker product along with its properties is also introduced as it plays a major role in formulating the developments in later sections. A brief presentation of the blind source separation and deconvolution problems is given in Section 3, emphasizing the tensor representation of the relevant criteria. A recently introduced formulation of the blind noisy multisource deconvolution problem is employed here, which avoids assumptions on the source cumulants and takes into account the various undermodeling sources in a unified manner [**53**]. This section serves mainly as a motivation for the study of the low-rank tensor approximation problem in the sequel of the paper. In Section 4 the latter problem is treated both in its general and its symmetric version. It is shown to be equivalent to a maximization problem which may be viewed as a generalization of the normalized cumulant extremization problems. The algorithm of [**35**] along with the initialization suggested therein is presented and proved to be convergent. The stationary points of the optimization functional and those among them that correspond to stable equilibria are completely characterized. A version of the algorithm, adapted to the symmetric case, is developed and also shown to be convergent under convexity (or concavity) conditions on an induced polynomial functional. Expressing the tensor approximation problem as an equivalent matrix one, we are led to a novel initialization method, which appears more effective than that based on the TSVD. Moreover, bounds on the initial value of the associated criterion are also derived via the above analysis. The fact that the super-exponential algorithm for blind deconvolution is simply the tensor power method applied to a symmetric tensor is demonstrated and commented upon in Section 5, where representative simulation results are provided as well to demonstrate the observed superiority of the new initialization method over that of [**35**]. Conclusions are drawn in Section 6.

Notation. Vectors will be denoted by bold lowercase letters (e.g., $\boldsymbol{s}$) while bold uppercase letters (e.g., $\boldsymbol{A}$) will denote tensors of 2nd- (i.e., matrices) or higher-order. High-order tensors will be usually denoted by bold, calligraphic uppercase letters (e.g., $\boldsymbol{\mathcal{T}}$). The symbols $\boldsymbol{I}$ and $\mathbf{0}$ designate the identity and zero matrices, respectively, where the dimensions will be understood from the context. The superscripts ^{T} and $^\#$ are employed for transposition and (pseudo)inversion, respectively. The $(i, j, k, \ldots, l)$ element of a tensor $\boldsymbol{\mathcal{T}}$ is denoted by $\mathcal{T}_{i,j,k,\ldots,l}$. All indices are assumed to start from 1. We shall also use the symbol $\otimes$ to denote the (right) Kronecker product. In order to convey the main ideas of this paper in the simplest possible way, relieved from any unnecessary notational complications, we restrict our exposition to the real case.

2. Higher-Order Tensors: Definitions, Properties, and Decompositions

By *tensor* we will mean any *multiway* array. An array of n ways (dimensions) will be called a tensor of order n. For example, a matrix is a 2nd-order tensor, a vector is a 1st-order tensor, and a scalar is a zeroth-order tensor. Although the tensor admits a more rigorous definition in terms of a tensor product induced by a multilinear mapping [**25**], the definition adopted here is more suitable for the purposes of this presentation and is commonly encountered in the relevant literature [**16**].

A class of tensors appearing often in signal processing applications is that of *(super)symmetric* tensors:

DEFINITION 1 (Symmetric tensor). *A tensor is called supersymmetric or more simply symmetric if its entries are invariant under any permutation of their indices.*

The above definition extends the familiar definition of a symmetric matrix, where the entries are "mirrored" across its main diagonal. In a symmetric higher-order tensor this "mirroring" extends to all of its main diagonals.

The notions of scalar product and norm, common for one-dimensional vectors, can be generalized for higher-order tensors [**36**]:

DEFINITION 2 (Tensor scalar product). *The scalar product of two tensors $\boldsymbol{\mathcal{A}}$ and $\boldsymbol{\mathcal{B}}$, of the same order, n, and same dimensions, is given by*

$$\langle \boldsymbol{\mathcal{A}}, \boldsymbol{\mathcal{B}} \rangle = \sum_{i_1, i_2, \ldots, i_n} \mathcal{A}_{i_1, i_2, \ldots, i_n} \mathcal{B}_{i_1, i_2, \ldots, i_n}$$

DEFINITION 3 (Frobenius norm). *The Frobenius norm of a tensor $\boldsymbol{\mathcal{A}}$ of order n is defined as*

$$\|\boldsymbol{\mathcal{A}}\|_F = \sqrt{\langle \boldsymbol{\mathcal{A}}, \boldsymbol{\mathcal{A}} \rangle} = \left(\sum_{i_1, i_2, \ldots, i_n} \mathcal{A}^2_{i_1, i_2, \ldots, i_n} \right)^{1/2}$$

The process of obtaining a matrix by an outer-product of two vectors can be generalized to a matrix outer-product yielding higher-order tensors with the aid of the *Tucker product* notation [**26**]:

DEFINITION 4 (Tucker product). *The Tucker product of n matrices $\{\boldsymbol{A}_i\}_{i=1}^n$ of dimension $M_i \times L$ each, yields an nth-order tensor $\boldsymbol{\mathcal{A}}$ of dimensions $M_1 \times M_2 \times \cdots \times M_n$ as:*

$$\mathcal{A}_{i_1, i_2, \ldots, i_n} = \sum_{l=1}^{L} (\boldsymbol{A}_1)_{i_1, l} (\boldsymbol{A}_2)_{i_2, l} \cdots (\boldsymbol{A}_n)_{i_n, l}$$

and is denoted by

$$\boldsymbol{\mathcal{A}} = \boldsymbol{A}_1 \star \boldsymbol{A}_2 \star \cdots \star \boldsymbol{A}_n.$$

We will find it also useful to define the *weighted Tucker product*:

DEFINITION 5 (Weighted Tucker product). *The weighted Tucker product (or $\boldsymbol{\mathcal{T}}$-product), with kernel an $L_1 \times L_2 \times \cdots \times L_n$ tensor $\boldsymbol{\mathcal{T}}$, of n matrices $\{\boldsymbol{A}_i\}_{i=1}^n$ of dimensions $M_i \times L_i$ yields an nth-order $M_1 \times M_2 \times \cdots \times M_n$ tensor as:*

$$\mathcal{A}_{i_1, i_2, \ldots, i_n} = \sum_{l_1=1}^{L_1} \sum_{l_2=1}^{L_2} \cdots \sum_{l_n=1}^{L_n} \mathcal{T}_{l_1, l_2, \ldots, l_n} (\boldsymbol{A}_1)_{i_1, l_1} (\boldsymbol{A}_2)_{i_2, l_2} \cdots (\boldsymbol{A}_n)_{i_n, l_n}$$

and is denoted by

$$\boldsymbol{\mathcal{A}} = \boldsymbol{A}_1 \overset{\mathcal{T}}{\star} \boldsymbol{A}_2 \overset{\mathcal{T}}{\star} \cdots \overset{\mathcal{T}}{\star} \boldsymbol{A}_n.$$

Clearly, the $\boldsymbol{\mathcal{T}}$-product reduces to the standard Tucker product if $\boldsymbol{\mathcal{T}}$ is the identity tensor, i.e., $\mathcal{T}_{i_1, i_2, \ldots, i_n} = \delta(i_1, i_2, \ldots, i_n)$ where $\delta(\cdots)$ is the n-variate Kronecker delta.

The $\boldsymbol{\mathcal{T}}$-product, which will prove to be of great help to our subsequent developments, may also be written using the Kronecker product and the high-order extension of the vec-operator [**2, 27**]. If we define $\mathrm{vec}(\boldsymbol{\mathcal{S}})$ as the vector resulting from arranging in a column the entries of the tensor $\boldsymbol{\mathcal{S}}$ in a lexicographic order:

$$\mathrm{vec}(\boldsymbol{\mathcal{S}}) = \left[\begin{array}{cccccc} \mathcal{S}_{11\ldots1} & \mathcal{S}_{21\ldots1} & \cdots & \mathcal{S}_{12\ldots1} & \mathcal{S}_{22\ldots1} & \cdots \end{array}\right]^T$$

then it can be shown that [**26**]:

FACT 1 (Matrix-vector Tucker product). *The $\boldsymbol{\mathcal{T}}$-product of Definition 5 can be expressed as:*

$$\mathrm{vec}(\boldsymbol{\mathcal{A}}) = (\boldsymbol{A}_1 \otimes \boldsymbol{A}_2 \otimes \cdots \otimes \boldsymbol{A}_n)\,\mathrm{vec}(\boldsymbol{\mathcal{T}}). \tag{1}$$

A particular higher-order outer product, that will be of use in the sequel, is the following:

DEFINITION 6 (Matrix outer product). *The outer product of n matrices $\{\boldsymbol{A}_i\}_{i=1}^n$ of dimensions $M_i \times L_i$ yields a $2n$th-order tensor $\boldsymbol{\mathcal{A}}$ of dimensions $M_1 \times L_1 \times M_2 \times L_2 \times \cdots \times M_n \times L_n$ as:*

$$\mathcal{A}_{i_1,i_2,i_3,i_4,\ldots,i_{2n-1},i_{2n}} = (A_1)_{i_1,i_2}(A_2)_{i_3,i_4}\cdots(A_n)_{i_{2n-1},i_{2n}}$$

and is denoted by

$$\boldsymbol{\mathcal{A}} = \boldsymbol{A}_1 \circ \boldsymbol{A}_2 \circ \cdots \circ \boldsymbol{A}_n.$$

Extending the notion of rank from matrices to higher-order tensors is more difficult. As is well known [**24**], the rank of a matrix, $\boldsymbol{A}$, equals the minimum number of terms in a representation of $\boldsymbol{A}$ as a sum of rank-1 matrices (vector outer products). This number, $r = \mathrm{rank}(\boldsymbol{A})$, coincides also with the number of independent rows of the matrix and the number of independent columns, and therefore is always upper-bounded by the lesser of the two dimensions of $\boldsymbol{A}$. It can be seen that, although the above definitions of rank can be also extended to a tensor $\boldsymbol{\mathcal{A}}$ of order $n \geq 3$, they generally do not yield the same number [**3**]. To go a bit further into this, let us generalize the notion of row- and column-vectors of a matrix for the case of a high-order tensor [**36**]:

DEFINITION 7 (Mode-k vectors). *The mode-k vectors of an $M_1 \times M_2 \times \cdots \times M_n$ tensor $\boldsymbol{\mathcal{A}}$ with entries $\mathcal{A}_{i_1,i_2,\ldots,i_n}$ are the M_k-dimensional vectors obtained from $\boldsymbol{\mathcal{A}}$ by varying the index i_k and keeping the other indices fixed. We will also call the space spanned by these vectors the mode-k space of $\boldsymbol{\mathcal{A}}$.*

It is easy to see that the columns and rows of a matrix correspond to its mode-1 and mode-2 vectors, respectively. Hence, based on the definition of $\mathrm{rank}(\boldsymbol{A})$ as the dimension of the column- and row-spaces of $\boldsymbol{A}$, the following "modal" ranks for a tensor come up naturally [**36**]:

DEFINITION 8 (Mode-k rank). *The mode-k rank, r_k, of a tensor $\boldsymbol{\mathcal{A}}$ is the dimension of its mode-k space and we write $r_k = \mathrm{rank}_k(\boldsymbol{\mathcal{A}})$.*[2]

Contrary to what holds for 2nd-order tensors, the mode ranks of a high-order tensor $\boldsymbol{\mathcal{A}}$ are generally not equal to each other, unless, of course, $\boldsymbol{\mathcal{A}}$ is symmetric.

A single "rank" for a tensor can be defined by following the definition of the matrix rank associated to its decomposition in rank-1 terms [**36**]:

[2]These quantities also appear in [**32, 33**] where they are denoted as $\dim_k(\boldsymbol{\mathcal{A}})$.

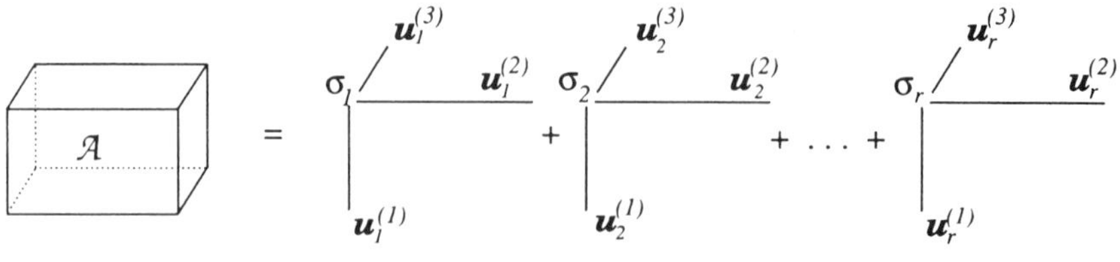

FIGURE 1. PARAFAC of a 3rd-order tensor.

DEFINITION 9 (Tensor rank). *The rank, r, of an $M_1 \times M_2 \times \cdots \times M_n$ tensor $\mathcal{A}$ is the minimal number of terms in a finite decomposition of $\mathcal{A}$ of the form:*

$$\mathcal{A} = \sum_{\rho=1}^{r} \underbrace{\boldsymbol{v}_\rho^{(1)} \star \boldsymbol{v}_\rho^{(2)} \star \cdots \star \boldsymbol{v}_\rho^{(n)}}_{\mathcal{A}_\rho} \tag{2}$$

where $\boldsymbol{v}_\rho^{(i)}$ are M_i-dimensional column vectors, and is denoted by $\operatorname{rank}(\mathcal{A})$. *Each term $\mathcal{A}_\rho$ is a rank-1 tensor with elements:*

$$(\mathcal{A}_\rho)_{i_1,i_2,\ldots,i_n} = (v_\rho^{(1)})_{i_1}(v_\rho^{(2)})_{i_2}\cdots(v_\rho^{(n)})_{i_n}$$

A (symmetric) tensor that can be written as in (2) with all the $\mathcal{A}_\rho$ being symmetric will be said to be *Tucker factorable.* It can be shown that $r_k \leq r$ for all $k \in \{1, 2, \ldots, n\}$ [**32**]. The upper bounds on the rank of a matrix induced by its dimensions do not carry over to higher-order tensors. There are examples of tensors where r is strictly greater than any of the tensor's dimensions [**3**]. It is for this ambiguity in defining and analyzing dimensional quantities in multiway arrays that more than one decomposition aiming at extending SVD to higher orders have been proposed and applied in multiway data analysis [**20, 36**]. That among them which is more like the matrix SVD is the so-called *CANonical DECOMPosition (CANDECOMP)* [**36**] or otherwise called *PARAllel FACtor analysis (PARAFAC)* [**3**], described in (2) and defined in more detail below:

DEFINITION 10 (PARAFAC). *A Parallel Factor (PARAFAC) analysis of an nth-order $M_1 \times M_2 \times \cdots \times M_n$ tensor $\mathcal{A}$ is a decomposition of $\mathcal{A}$ in a sum of $r = \operatorname{rank}(\mathcal{A})$ rank-1 terms:*

$$\mathcal{A} = \sum_{\rho=1}^{r} \sigma_\rho \boldsymbol{u}_\rho^{(1)} \star \boldsymbol{u}_\rho^{(2)} \star \cdots \star \boldsymbol{u}_\rho^{(n)} \tag{3}$$

with σ_ρ being scalars and $\boldsymbol{u}_\rho^{(i)}$ being M_i-dimensional unit-norm vectors.

Fig. 1 visualizes the PARAFAC for a tensor of order 3. This decomposition was originally conceived and analyzed by researchers in mathematical psychology and later applied in chemometrics [**3**]. Note that for $n = 2$, eq. (3) takes the form of the SVD. To see this more clearly, we can employ the Tucker product notation introduced above and write the PARAFAC for $\mathcal{A}$ as follows:

$$\begin{aligned} \mathcal{A} &= \begin{bmatrix} \boldsymbol{u}_1^{(1)} & \boldsymbol{u}_2^{(1)} & \cdots & \boldsymbol{u}_r^{(1)} \end{bmatrix} \overset{\mathcal{S}}{\star} \begin{bmatrix} \boldsymbol{u}_1^{(2)} & \boldsymbol{u}_2^{(2)} & \cdots & \boldsymbol{u}_r^{(2)} \end{bmatrix} \overset{\mathcal{S}}{\star} \\ &\qquad \cdots \overset{\mathcal{S}}{\star} \begin{bmatrix} \boldsymbol{u}_1^{(n)} & \boldsymbol{u}_2^{(n)} & \cdots & \boldsymbol{u}_r^{(n)} \end{bmatrix} \\ &= \boldsymbol{U}^{(1)} \overset{\mathcal{S}}{\star} \boldsymbol{U}^{(2)} \overset{\mathcal{S}}{\star} \cdots \overset{\mathcal{S}}{\star} \boldsymbol{U}^{(n)} \end{aligned} \tag{4}$$

where the nth-order (core) tensor $\boldsymbol{\mathcal{S}}$ is diagonal, with $\mathcal{S}_{\rho,\rho,\ldots,\rho} = \sigma_\rho$. It suffices then to observe that for $n = 2$ the above becomes

$$\boldsymbol{\mathcal{A}} = \boldsymbol{U}^{(1)} \overset{\mathcal{S}}{\star} \boldsymbol{U}^{(2)} = \boldsymbol{U}^{(1)} \boldsymbol{\mathcal{S}} (\boldsymbol{U}^{(2)})^T.$$

Despite its apparent similarity with matrix SVD, the properties of the decomposition (3) are in general rather different from those of its 2nd-order counterpart. Thus, contrary to matrix SVD, the matrices $\boldsymbol{U}^{(i)}$ need not have orthogonal columns to ensure uniqueness of the decomposition. PARAFAC uniqueness conditions have been derived for both 3-way [**32, 41**] and multi-way [**33, 58**] arrays, and are based on generalized notions of linear independence of the columns of $\boldsymbol{U}^{(i)}$'s. A PARAFAC decomposition with orthogonal matrices $\boldsymbol{U}^{(i)}$ need not exist [**31, 21**].

An SVD-like tensor decomposition that trades diagonality of the kernel (core) tensor for orthogonality of the factor matrices was proposed in [**37**] and successfully applied in independent component analysis applications among others [**36, 38, 39**]. This *Tensor Singular Value Decomposition (TSVD)* [**37**] (or as otherwise called *Higher-Order SVD (HOSVD)* [**38, 36**]) generalizes the so-called *Tucker3* model, originally proposed for 3-way arrays [**31, 4**], to higher orders and is defined below:

DEFINITION 11 (TSVD). *The TSVD of an nth-order $M_1 \times M_2 \times \cdots \times M_n$ tensor $\boldsymbol{\mathcal{A}}$ is given by*

$$\boldsymbol{\mathcal{A}} = \boldsymbol{U}^{(1)} \overset{\mathcal{S}}{\star} \boldsymbol{U}^{(2)} \overset{\mathcal{S}}{\star} \cdots \overset{\mathcal{S}}{\star} \boldsymbol{U}^{(n)} \tag{5}$$

where:

- $\boldsymbol{U}^{(i)}$, *$i = 1, 2, \ldots, n$, are orthogonal $M_i \times M_i$ matrices, and*
- *the core tensor $\boldsymbol{\mathcal{S}}$ is of the same size as $\boldsymbol{\mathcal{A}}$ and its subtensors $\boldsymbol{\mathcal{S}}_{i_k=\alpha}$, obtained by fixing the kth index to α, have the properties of:*
  - *all-orthogonality: two subtensors $\boldsymbol{\mathcal{S}}_{i_k=\alpha}$ and $\boldsymbol{\mathcal{S}}_{i_k=\beta}$ are orthogonal for any possible values of k and $\alpha \neq \beta$, in the sense that:*

$$\langle \boldsymbol{\mathcal{S}}_{i_k=\alpha}, \boldsymbol{\mathcal{S}}_{i_k=\beta} \rangle = 0$$

  - *ordering: For all k,*

$$\|\boldsymbol{\mathcal{S}}_{i_k=1}\|_F \geq \|\boldsymbol{\mathcal{S}}_{i_k=2}\|_F \geq \cdots \geq \|\boldsymbol{\mathcal{S}}_{i_k=r_k}\|_F > \|\boldsymbol{\mathcal{S}}_{i_k=r_k+1}\|_F = \cdots = \|\boldsymbol{\mathcal{S}}_{i_k=M_k}\|_F = 0$$

It is evident that the SVD of a matrix satisfies the conditions of the above definition. The uniqueness of the TSVD follows from its definition and relies on the orthogonality of the factors $\boldsymbol{U}^{(i)}$. This decomposition is directly computable via a series of matrix SVD's applied to properly "unfolded" forms of $\boldsymbol{\mathcal{A}}$. Thus, the matrix $\boldsymbol{U}^{(i)}$ can be found as the matrix of the left singular vectors of the $M_i \times M_1 M_2 \cdots M_{i-1} M_{i+1} \cdots M_n$ matrix built with the mode-i vectors of $\boldsymbol{\mathcal{A}}$. In other words, the first r_i columns of $\boldsymbol{U}^{(i)}$ constitute an orthonormal basis for the mode-i space of $\boldsymbol{\mathcal{A}}$. After having computed the orthogonal factors in (5), $\boldsymbol{\mathcal{S}}$ is determined by

$$\boldsymbol{\mathcal{S}} = (\boldsymbol{U}^{(1)})^T \overset{\mathcal{A}}{\star} (\boldsymbol{U}^{(2)})^T \overset{\mathcal{A}}{\star} \cdots \overset{\mathcal{A}}{\star} (\boldsymbol{U}^{(n)})^T \tag{6}$$

An alternative way of approaching the diagonality of the core tensor is by requiring that the sum of the squares of its diagonal entries is maximum over all tensors generated by (6) when the $\boldsymbol{U}^{(i)}$'s range over all orthogonal $M_i \times M_i$ matrices. This criterion of *maximal LS diagonality* [**39**] can be viewed as a direct extension of the Jacobi criterion for symmetric matrix diagonalization [**24**] to multiway arrays.

It is thus of no surprise that iterative Jacobi algorithms have been developed for calculating this type of SVD for symmetric tensors [**12, 13, 14, 15**].

Before concluding this section, let us make a brief reference to a few more recent works on tensor decomposition. By viewing a symmetric tensor as the array of the coefficients of an homogeneous multivariate polynomial, it is possible to express the problem of computing the PARAFAC of a tensor as one of expanding a multivariate polynomial as a sum of powers of linear forms [**19**]. This formulation of the problem permits the use of powerful theorems from the algebra of multilinear forms in deriving decomposability conditions as well as estimates on the *generic* rank of symmetric tensors as a function of their dimensions [**19**]. Numerical algorithms have also resulted from such considerations, though still restricted to small problem sizes [**18**]. A conceptually simple approach for computing a factorization for a higher-order tensor, resembling to a TSVD, was originally reported in [**55**] and recently reappeared in [**45**] where it was applied to deriving approximate factorizations for multivariate polynomials. The algorithm (multiway Principal Component Analysis) consists of a series of successive matrix SVD steps, each computed for the folded right singular vectors of the previous one. The decomposition starts with the SVD of the $M_1 \times M_2 M_3 \cdots M_n$ mode-1 matrix of the tensor. In spite of its simplicity, the practical value of this method is somewhat limited, as it tends to produce many more terms than necessary in the final expansion. A number of other decomposition approaches may be found in the relevant literature (e.g., [**20**]).

3. Blind Source Separation: A Tensor Decomposition Problem

The blind recovery of the component signals of a noisy linear mixture, called Blind Source Separation (BSS) or otherwise Independent Component Analysis (ICA) [**11, 28**], is a topic that has attracted much interest in the last decade, in view of its numerous applications, including random vector modeling, blind channel identification/equalization and image restoration, voice-control, circuit analysis and semiconductor manufacturing [**59**]. It is well-known that higher-order statistics are necessary for unambiguously identifying a multi-input multi-output (MIMO) system in a blind way [**59, 44**]. Furthermore, even in situations where methods using only second-order statistics are theoretically applicable, approaches that take advantage of the additional information contained in higher order statistics have been proven more successful (e.g., [**8**]). In this section we formulate the blind separation and deconvolution problems using tensor formalism, with the aim of bringing out their strong tensor-decomposition-flavour.

3.1. The Cumulant Tensor: Definition and Properties.

DEFINITION 12 (Cumulants). *The (nth-order) cumulant of a set of random variables $\{x_i\}_{i=1}^n$ (not necessarily distinct), denoted as* $\mathrm{cum}(x_1, x_2, \ldots, x_n)$, *is defined as the coefficient of the monomial $\omega_1\omega_2\cdots\omega_n$ in the Taylor series expansion (around zero) of the second characteristic function of the random vector $\boldsymbol{x} = [x_1, x_2, \ldots, x_n]^T$:*

$$\Psi_x(\boldsymbol{\omega}) = \ln E\{\exp(j\boldsymbol{\omega}^T\boldsymbol{x})\}$$

with $\boldsymbol{\omega} = [\omega_1, \omega_2, \ldots, \omega_n]^T$ and $E\{\cdot\}$ denoting the expectation operator.

The above definition applies to the cumulants of any order and shows that their ensemble completely specifies the statistics, i.e., the joint probability density function (pdf), of the random vector $\boldsymbol{x}$.[3] Moreover, it follows from their definition, that the cumulants are related to the high-order moments of the random variables in question. For the first few orders, these relations assume rather simple forms and play sometimes the role of a definition [**48**]:

FACT 2 (Cumulants vs moments). *If $x_i, x_j, \dots$ are zero-mean random variables, their cumulants of order one to four are given by:*

$$\begin{aligned}
\mathrm{cum}(x_i) &= E\{x_i\} \\
\mathrm{cum}(x_i, x_j) &= E\{x_i x_j\} \\
\mathrm{cum}(x_i, x_j, x_k) &= E\{x_i x_j x_k\} \\
\mathrm{cum}(x_i, x_j, x_k, x_l) &= E\{x_i x_j x_k x_l\} \\
&\quad - E\{x_i x_j\}E\{x_k x_l\} - E\{x_i x_k\}E\{x_j x_l\} - E\{x_i x_l\}E\{x_j x_k\}
\end{aligned}$$

The following properties of the cumulants explain their popularity over the moments in signal processing [**48**]:

FACT 3 (Properties of cumulants). *For any sets of random variables $\{x_i\}$, $\{y_i\}$ the following hold true:*

(1) *The value of $\mathrm{cum}(x_{i_1}, x_{i_2}, \dots, x_{i_n})$ is invariant to a permutation of the x_i's.*

(2) *For any scalars λ_i,*

$$\mathrm{cum}(\lambda_1 x_1, \lambda_2 x_2, \dots, \lambda_n x_n) = \left(\prod_{i=1}^{n} \lambda_i\right) \mathrm{cum}(x_1, x_2, \dots, x_n)$$

(3) *The cumulant is additive in any of its arguments:*

$$\mathrm{cum}(x_1 + y_1, x_2, x_3, \dots, x_n) = \mathrm{cum}(x_1, x_2, x_3, \dots, x_n) + \mathrm{cum}(y_1, x_2, x_3, \dots, x_n)$$

(4) *If the random variables $\{x_i\}$ are independent of $\{y_i\}$,*

$$\mathrm{cum}(x_1 + y_1, x_2 + y_2, \dots, x_n + y_n) = \mathrm{cum}(x_1, x_2, \dots, x_n) + \mathrm{cum}(y_1, y_2, \dots, y_n)$$

(5) *If the set $\{x_1, x_2, \dots, x_n\}$ can be partitioned in two subsets of variables that are independent one of the other, then*

$$\mathrm{cum}(x_1, x_2, \dots, x_n) = 0.$$

(6) *If the random variables x_i, $i = 1, 2, \dots, n$ are jointly Gaussian and $n \geq 3$,*

$$\mathrm{cum}(x_1, x_2, \dots, x_n) = 0.$$

Properties 1–3 above imply that $\mathrm{cum}(\cdot)$ is a multilinear function [**47**] and justify the following definition:

DEFINITION 13 (Cumulant tensor). *Given a random vector $\boldsymbol{x} = [x_1, x_2, \dots, x_M]^T$ and a positive integer n we define its nth-order cumulant tensor, $\boldsymbol{\mathcal{C}}_{x,n}$, as the nth-way $M \times M \times \cdots \times M$ array having the value $\mathrm{cum}(x_{i_1}, x_{i_2}, \dots, x_{i_n})$ as its $(i_1, i_2, \dots, i_n)$-element.*

It is then readily seen that Fact 3 implies the following properties for the cumulant tensor:

[3]Sometimes, the terms cumulants and HOS are used interchangeably in the literature [**48**].

FACT 4 (Properties of the cumulant tensor). *Let $\boldsymbol{x}, \boldsymbol{y}$ and $\boldsymbol{z}$ be random vectors of dimensions M_x, M_y and M_z, respectively, and n be a positive integer. Then the following hold true:*

(1) $\mathcal{C}_{x,n}$ *is symmetric.*

(2) *If $\boldsymbol{y} = \boldsymbol{A}\boldsymbol{x}$ for some $M_y \times M_x$ matrix $\boldsymbol{A}$, then*

$$\mathcal{C}_{y,n} = \underbrace{\boldsymbol{A} \overset{\mathcal{C}_{x,n}}{\star} \boldsymbol{A} \overset{\mathcal{C}_{x,n}}{\star} \cdots \overset{\mathcal{C}_{x,n}}{\star} \boldsymbol{A}}_{n \ \ times} \triangleq \boldsymbol{A}^{\overset{\mathcal{C}_{x,n}}{\star} n}.$$

(3) *If $\boldsymbol{x}$ and $\boldsymbol{y}$ are independent and $\boldsymbol{z} = \boldsymbol{x} + \boldsymbol{y}$, then*

$$\mathcal{C}_{z,n} = \mathcal{C}_{x,n} + \mathcal{C}_{y,n}.$$

(4) *If the components of $\boldsymbol{x}$ are independent of each other, then*

$$(\mathcal{C}_{x,n})_{i_1,i_2,\ldots,i_n} = \operatorname{cum}(x_{i_1}, x_{i_1}, \ldots, x_{i_1})\delta(i_1, i_2, \ldots, i_n)$$

that is, the associated cumulant tensor is diagonal.

(5) *If $\boldsymbol{z} = \boldsymbol{x} + \boldsymbol{y}$ where $\boldsymbol{x}$ and $\boldsymbol{y}$ are independent and $\boldsymbol{y}$ is Gaussian, then*

$$\mathcal{C}_{z,n} = \mathcal{C}_{x,n} \text{ for } n \geq 3.$$

3.2. BSS via Normalized Cumulant Extremization. The model employed in the source separation problems is of the form

$$\boldsymbol{u} = \boldsymbol{H}\boldsymbol{a} + \boldsymbol{v} \tag{7}$$

where $\boldsymbol{u}$ is a $P \times 1$ *observation vector*, $\boldsymbol{a}$ is a $K \times 1$ *source vector*, and $\boldsymbol{v}$ is a $P \times 1$ *noise vector*. All vectors are considered as being random. The $P \times K$ matrix $\boldsymbol{H}$ "mixes" the sources $\boldsymbol{a}$ and is referred to as the *mixing matrix*. The goal of BSS is to identify the matrix $\boldsymbol{H}$ and/or recover the sources $\boldsymbol{a}$ with the aid of realizations of the observation vector only, with no knowledge of the source or noise vectors apart from general assumptions on their statistical characteristics. Numerous signal processing problems fall into this framework. In particular, blind deconvolution comes as a special case of the above separation problem, with the matrix $\boldsymbol{H}$ constrained to have a Toeplitz structure [**15**]. For the time being we are concerned with the static (memoryless) mixture case. The blind deconvolution setting will be considered in more detail in the next subsection, where we shall also extend the model description (7) so as to explicitly accommodate convolutive mixtures and noise dynamics.[4]

Commonly made assumptions for the source and noise statistics, usually met in practice, include:

- All random vectors have zero mean.
- The components of $\boldsymbol{a}$ are independent and have unit variance.
- $\boldsymbol{v}$ is independent of $\boldsymbol{a}$ and has unit-variance components.

Moreover, if all sources are to be recovered at once, at most one source signal is assumed to be Gaussian distributed [**6**]. Note that, since, as shown in [**59**], $\boldsymbol{H}$ can only be identified up to a scaling and a permutation, the unit-variance assumptions above entail no loss of generality, as the real variances can be incorporated in the sought-for parameters. Due to lack of information on the noise structure, the noise is often "neglected" in the study of separation algorithms or assumed Gaussian in methods using HOS only, since it is annihilated when considering higher-order

[4]Nevertheless, it can be seen that a dynamic BSS problem with independent source signals can always be expressed as in (7) if the latter are also assumed to be i.i.d.

cumulants (see Fact 4.5). For the purposes of this discussion we shall also omit noise from our considerations. In the next subsection, we will see how this assumption can be relaxed.

The independence assumption for $\boldsymbol{a}$ can then be shown to reduce the problem to that of rendering the components of the observation vector independent as well, via a linear transformation [**15**]:

$$\boldsymbol{y} = \boldsymbol{G}\boldsymbol{u} = \boldsymbol{G}\boldsymbol{H}\boldsymbol{a}.$$

In view of Fact 4.4, this implies that $\boldsymbol{G}$ must be chosen such that the tensors $\boldsymbol{\mathcal{C}}_{y,n}$ are diagonal for all n. Obviously, if $\boldsymbol{H}$ is identified and has full column rank (hence $P \geq K$), the sought-for linear transformation is simply $\boldsymbol{G} = \boldsymbol{H}^{\#}$, i.e., the pseudoinverse of $\boldsymbol{H}$, and $\boldsymbol{a}$ is perfectly recovered (the mixing is undone) in $\boldsymbol{y}$.[5] In such a case (i.e., more sensors than sources) $\boldsymbol{H}$ can be determined as a basis, $\boldsymbol{T}$, for the range space of the autocorrelation matrix of $\boldsymbol{u}$, but only up to an orthogonal factor, $\boldsymbol{Q}$. This indeterminacy can only be removed by resorting to higher order statistics of the *standardized* (whitened) observations $\tilde{\boldsymbol{u}} = \boldsymbol{T}\boldsymbol{u}$. Then, our goal is to determine $\boldsymbol{Q}$ such that the vector $\boldsymbol{y} = \boldsymbol{Q}\tilde{\boldsymbol{u}}$ have all its cumulant tensors diagonal.

Fortunately, not all *standardized* cumulant tensors have to be considered. It can be shown that only the 4th-order crosscumulants need to be nulled [**15**].[6] In fact, any order greater than 2 will do [**17**].[7] The criterion

$$\max_{QQ^T=I}\left(\psi(\boldsymbol{Q}) = \sum_i \operatorname{cum}(y_i, y_i, y_i, y_i)^2\right) \tag{8}$$

can be shown to possess the desired factor $\boldsymbol{Q}$ as its global solution [**15**]. Notice that, since $\tilde{\boldsymbol{u}}$ has the identity as autocorrelation matrix and $\boldsymbol{Q}$ is orthogonal, the vector $\boldsymbol{y}$ is also standardized, i.e., $\boldsymbol{\mathcal{C}}_{y,2} = \boldsymbol{I}$. Hence, the cumulants involved in the maximization (8) are in fact *normalized*, so that the cost function $\psi(\cdot)$ can also be expressed as

$$\psi(\boldsymbol{Q}) = \sum_i \left[\frac{\operatorname{cum}(y_i, y_i, y_i, y_i)}{(\operatorname{cum}(y_i, y_i))^2}\right]^2 \tag{9}$$

The 4th-order cumulant, normalized by the variance squared,

$$k_y(4,2) = \frac{\operatorname{cum}(y, y, y, y)}{(\operatorname{cum}(y, y))^2} \tag{10}$$

is commonly known as the *kurtosis* of y [**5**]. According to Fact 3.6, $k_y(4,2) = 0$ if y is Gaussian. Depending on whether $k_y(4,2)$ is positive or negative, y is referred to as *super-Gaussian (leptokurtic)* or *sub-Gaussian (platykurtic)*, respectively [**51**].

Problem (8) can be viewed as the tensor analog of Jacobi's diagonalization criterion for symmetric matrices [**24**]. The idea is that, due to the orthogonality of $\boldsymbol{Q}$, $\|\boldsymbol{\mathcal{C}}_{y,4}\|_F$ is constant and hence (8) is equivalent to minimizing the "energy" in the off-diagonal entries (crosscumulants). Similar, yet less strict, criteria based on 4th-order cumulants have also been proposed in [**13, 39**]. Iterative algorithms,

---

[5] Subject, of course, to some scaling and permutation, which are implicit in the problem statement.

[6] Assuming that at most one source signal has zero 4th-order cumulant.

[7] One has often to employ 4th-order cumulants, even when 3rd-order ones are theoretically sufficient. This is because odd-order cumulants vanish for signals with symmetric pdf, frequently met in practical applications [**48**].

composed of Jacobi iterations, have been developed for implementing these criteria and shown to perform well [**13, 15, 39**], though their convergence behavior is still to be analyzed. Similar algorithms, formulated as joint diagonalizations of sets of output-data induced matrices, have also been reported in the context of array processing [**60**].

A second class of blind separation algorithms, based exclusively on higher-order statistics, avoid the data standardization step, thereby gaining in robustness against insufficient information on the noise structure and allowing mixing matrix identification in the more challenging case of more sources than sensors (i.e., $P < K$). The fundamental idea underlying these methods (e.g., [**7**]–[**9**]) is that, subject to a Gaussianity assumption for the noise, the 4th-order cumulant tensor of the observations satisfies the polynomial relation (see Fact 4.2)

$$\mathcal{C}_{u,4} = \boldsymbol{H} \overset{\mathcal{C}_{a,4}}{\star} \boldsymbol{H} \overset{\mathcal{C}_{a,4}}{\star} \boldsymbol{H} \overset{\mathcal{C}_{a,4}}{\star} \boldsymbol{H} \tag{11}$$

which, in view of the fact that $\mathcal{C}_{a,4}$ is diagonal (Fact 4.4), represents a decomposition of $\mathcal{C}_{u,4}$ in additive rank-1 terms. Since all factors involved are equal (due to the supersymmetry of $\mathcal{C}_{u,4}$), (11) is referred to as a *tetradic decomposition* [**10**]. Furthermore, one can view $\mathcal{C}_{u,4}$ as a symmetric linear operator over the space of real matrices [**23**], through the so-called *contraction* operation, $\boldsymbol{N} = \mathcal{C}_{u,4}(\boldsymbol{M})$, where $\boldsymbol{N}$ and $\boldsymbol{M}$ are $P \times P$ matrices and $N_{i,j} = \sum_{k,l}(\mathcal{C}_{u,4})_{i,j,k,l} M_{l,k}$. Hence $\mathcal{C}_{u,4}$ admits an eigenvalue decomposition, in the form:

$$\mathcal{C}_{4,u} = \sum_{\rho} \lambda_\rho \boldsymbol{M}_\rho \circ \boldsymbol{M}_\rho \tag{12}$$

where the *eigenmatrices* $\boldsymbol{M}_\rho$ are orthogonal to each other in the sense that $\langle \boldsymbol{M}_\rho, \boldsymbol{M}_\mu \rangle = 0$ for $\rho \neq \mu$. The above eigen-decomposition can be computed by solving the equivalent matrix eigen-decomposition problem that results from unfolding the $P \times P \times P \times P$ tensor in a $P^2 \times P^2$ matrix [**7**]. In the case that the projectors on the spaces spanned by the columns of $\boldsymbol{H}$ are independent of each other, it can be shown that the decomposition (11) (a PARAFAC decomposition in that case) is unique and can be determined via the eigen-decomposition of (12) [**9**]. Note that the above independence condition is trivially satisfied when $\boldsymbol{H}$ has full column rank (hence $P \geq K$), in which case the expansion (12) for the standardized cumulant tensor provides us directly with the columns of $\boldsymbol{H}$ [**7, 8**]. Direct solutions to the cumulant tensor decomposition problem, able to cope with the $P < K$ case, have also been developed through the expansion in powers of linear forms of the corresponding multilinear form [**19, 18**].

3.3. Blind Deconvolution as a Nonlinear Eigenproblem. Although the problem of blind deconvolution (equalization) can be viewed as a special case of that of blind separation problem as described above, its dynamical nature justifies a separate treatment [**17**]. Moreover, further specializing our study to the case of a single source recovery helps clarifying the close connection of the problem with a low-rank tensor approximation one and revealing the super-exponential algorithm as a naturally emerging solution approach.

Consider the dynamic analog of the model (7), described in the $\mathcal{Z}$-transform domain as:

$$\boldsymbol{u}(z) = \boldsymbol{H}(z)\boldsymbol{a}(z) + \boldsymbol{v}(z) \tag{13}$$

where the involved quantities are of the same dimensions as in (7). The same statistical assumptions are made here, i.e., $\boldsymbol{a}(z)$ and $\boldsymbol{v}(z)$ are independent of each other and the sources $a_i(z)$ are independent identically distributed (i.i.d.) signals and independent of each other. Moreover, all vectors have zero mean and the components of $\boldsymbol{a}(z)$ and $\boldsymbol{v}(z)$ are of unit variance. The noise is not considered as negligible nor do we make any assuption on its temporal and/or spatial structure. Our only assumption about noise is that it is Gaussian. In order to simplify the appearance of the affine model (13) and reduce it to a linear form we incorporate the dynamics of the noise component into those of the system $\boldsymbol{H}(z)$. To this end, call $\boldsymbol{L}(z)$ the $P \times P$ innovations filter for the process $\boldsymbol{v}(z)$, i.e.,

$$\boldsymbol{v}(z) = \boldsymbol{L}(z)\boldsymbol{w}(z)$$

where $\boldsymbol{w}(z)$ is now white and Gaussian. Since for a Gaussian process uncorrelatedness implies independence, $\boldsymbol{w}(z)$ has also i.i.d. components, independent of each other. Then the model (13) can be rewritten as [**53**]:

$$\boldsymbol{u}(z) = \underbrace{\begin{bmatrix} \boldsymbol{H}(z) & \boldsymbol{L}(z) \end{bmatrix}}_{\boldsymbol{F}^T(z)} \underbrace{\begin{bmatrix} \boldsymbol{a}(z) \\ \boldsymbol{w}(z) \end{bmatrix}}_{\boldsymbol{\alpha}(z)} = \boldsymbol{F}^T(z)\boldsymbol{\alpha}(z) \tag{14}$$

Our aim is to blindly recover a source from $\boldsymbol{\alpha}(z)$ with the aid of a linear equalizer (deconvolution filter) described by a $P \times 1$ vector $\boldsymbol{g}(z)$:

$$y(z) = \boldsymbol{g}^T(z)\boldsymbol{u}(z) = \underbrace{\boldsymbol{g}^T(z)\boldsymbol{F}^T(z)}_{\boldsymbol{s}^T(z)}\boldsymbol{\alpha}(z) = \boldsymbol{s}^T(z)\boldsymbol{\alpha}(z) \tag{15}$$

where the $(K+P) \times 1$ vector $\boldsymbol{s}(z)$ is the cascade of the system $\boldsymbol{F}$ with the equalizer $\boldsymbol{g}$ and will be called *combined response*. Note that in the present setting we are not interested in recovering a particular source; any component of $\boldsymbol{\alpha}(z)$ will do. Additional information is normally required for letting a given source pass through (e.g., CDMA communications [**44**]). We would thus like to have

$$\boldsymbol{s}(z) = \begin{bmatrix} \boldsymbol{0} & cz^{-d} & \boldsymbol{0} \end{bmatrix}^T \tag{16}$$

for some scalar $c \neq 0$ and a nonnegative integer d. The position of the monomial cz^{-d} in the above vector is not crucial to our study.

It will also be useful to express the blind deconvolution problem in the time domain:

$$y_n = \boldsymbol{g}^T\boldsymbol{u}_n = \underbrace{\boldsymbol{g}^T\boldsymbol{F}^T}_{\boldsymbol{s}^T}\boldsymbol{\alpha}_n = \boldsymbol{s}^T\boldsymbol{\alpha}_n \tag{17}$$

where the vectors involved are built by a juxtaposition of the coefficients of the corresponding vector polynomials in (15) and $\boldsymbol{F}$ is a block-Toeplitz matrix summarizing the convolution operation in (15). Then, the perfect equalization condition (16) takes the form

$$\begin{aligned} \boldsymbol{s} &= c\begin{bmatrix} 0 & 0 & \cdots & 0 & 1 & 0 & \cdots & 0 & 0 \end{bmatrix}^T \\ &= c\boldsymbol{e}_m^T \end{aligned} \tag{18}$$

with $\boldsymbol{e}_m$ denoting the vector of all zeros with a 1 at the mth position. The uncertainty about the values of c and m is the form it takes in the present single source recovery problem the scaling and permutation indeterminacy, inherent in BSS.

It can be shown that the maximization of the square of the kurtosis of y, $|k_y(4,2)|^2$, has the perfect equalizer, that is that $\boldsymbol{g}$ leading to a combined response of the form of (18), as its global solution [**5, 53**]. This should be expected, as the kurtosis squared of y is one of the terms in the sum to be maximized for the more general case of all sources recovery problem (see (8), (9)). In fact, we can go even further and consider higher-order normalized cumulants [**5**]:

$$k_y(2p,2q) = \frac{\text{cum}\,\overbrace{(y,y,\ldots,y)}^{2p \text{ times}}}{(\text{cum}\,\underbrace{(y,y,\ldots,y)}_{2q \text{ times}})^{p/q}} \triangleq \frac{\text{cum}_{2p}(y)}{(\text{cum}_{2q}(y))^{p/q}} \qquad p,q \text{ positive integers}, p>q$$

Here we set, as it is usually the case in blind equalization studies, $q=1$. In view of our assumption for unit variance α_i's, $k_{\alpha_i}(2p,2)$ can thus be replaced by $\text{cum}_{2p}(\alpha_i)$. Moreover, to simplify notation, we shall henceforth omit the subscripts from the source cumulant tensor of order $2p$, $\boldsymbol{\mathcal{C}}_{\alpha,2p}$, and denote it simply as $\boldsymbol{\mathcal{C}}$. To appreciate the applicability of the criteria $\max_s |k_y(2p,2)|^2$ to the problem in question, let us express the above functions in terms of the combined response $\boldsymbol{s}$ with the aid of the multilinearity property of the cumulant tensor (Fact 4.2):

FACT 5 (Input-output relation for normalized cumulants). *If y and $\boldsymbol{\alpha}$ are related as in (17), their normalized cumulants satisfy the relation:*

$$k_y(2p,2) = \frac{\overbrace{\boldsymbol{s}^T \overset{\mathcal{C}}{\star} \boldsymbol{s}^T \overset{\mathcal{C}}{\star} \cdots \overset{\mathcal{C}}{\star} \boldsymbol{s}^T}^{2p \text{ times}}}{(\boldsymbol{s}^T \star \boldsymbol{s}^T)^p} \triangleq \frac{(\boldsymbol{s}^T)^{\overset{\mathcal{C}}{\star} 2p}}{\|\boldsymbol{s}\|_2^{2p}} \tag{19}$$

Note that, in view of the independence assumptions for the source signals, the tensor $\boldsymbol{\mathcal{C}}$ is diagonal. One can then see that[8] [**5, 61, 52, 53**]:

FACT 6 (Normalized cumulant extremization). *The solutions of the maximization problem*[9]

$$\max_{s \in \ell_2} |k_y(2p,2)|^2, \quad p \geq 2$$

are given by the combined responses (18).

We shall denote by $J_{2p}(\boldsymbol{s})$ the above cost as a function of $\boldsymbol{s}$:

$$J_{2p}(\boldsymbol{s}) = \frac{(\boldsymbol{s}^T)^{\overset{\mathcal{C}}{\star} 2p}}{\|\boldsymbol{s}\|_2^{2p}} \tag{20}$$

It must be emphasized that the statement of Fact 6 relies on the whole space ℓ_2 being the support of the maximization. However, by its very definition,

$$\boldsymbol{s} = \boldsymbol{F}\boldsymbol{g},$$

[8] This result holds also true for normalized cumulants of odd order, $k_y(2p+1,2)$, $p>0$, provided that all source cumulants $k_{\alpha_i}(2p+1,2)$ are equal in absolute value. An important special case of this is the blind equalization of single-input single-output systems [**5**].

[9] We consider the space ℓ_2 of square summable sequences as the general parameter space. For BIBO stable system and deconvolution filters, $\boldsymbol{s}$ belongs to ℓ_1, and therefore to ℓ_2 as well.

$\boldsymbol{s}$ is constrained to lie in the column space of the convolution matrix $\boldsymbol{F}$, which we will call the *space of attainable responses* and denote by $\mathcal{S}_A$:

$$\mathcal{S}_A \triangleq \{\boldsymbol{s} : \boldsymbol{s} = \boldsymbol{F}\boldsymbol{g} \text{ for some equalizer } \boldsymbol{g}\} \subset \ell_2 \tag{21}$$

Hence, our adopted criterion should be written as:

$$\max_{s \in \mathcal{S}_A} |J_{2p}(\boldsymbol{s})|^2 \tag{22}$$

The stationary points of $J_{2p}(\cdot)$ will thus be given by making the gradient $\nabla J_{2p}(\cdot)$ orthogonal to $\mathcal{S}_A$:

$$\boldsymbol{P}_A \nabla J_{2p}(\boldsymbol{s}) = \boldsymbol{0}, \tag{23}$$

where $\boldsymbol{P}_A$ denotes the orthogonal projector onto the space of attainable responses:

$$\boldsymbol{P}_A \triangleq \boldsymbol{F}(\boldsymbol{F}^T\boldsymbol{F})^{\#}\boldsymbol{F}^T. \tag{24}$$

If $\mathcal{S}_A \neq \ell_2$, that is, $\boldsymbol{P}_A \neq \boldsymbol{I}$, not all sequences $\boldsymbol{s}$ can be attained by varying the equalizer vector $\boldsymbol{g}$. In particular, perfect equalization, corresponding to a combined response as in (18), might then be impossible. In this case, we shall say that we have *undermodeling*. On the other hand, a setting with $\boldsymbol{P}_A = \boldsymbol{I}$ will be said to correspond to a *sufficient order* case. Notice that, when noise is nonnegligible in our model (14), the matrix $\boldsymbol{F}$ is always "tall", that is, it has more rows than columns. Hence undermodeling results regardless of the equalizer order. This means that in practical applications undermodeling is rather unavoidable and the role played by $\boldsymbol{P}_A$ in our analysis should not be overlooked. To make the fact that in general $\mathcal{S}_A \neq \ell_2$ explicit in the expression for $J_{2p}(\boldsymbol{s})$, we write (20) in the form

$$J_{2p}(\boldsymbol{s}) = \frac{((\boldsymbol{P}_A\boldsymbol{s})^T)^{\overset{C}{\star}2p}}{\|\boldsymbol{s}\|_2^{2p}} \tag{25}$$

Let us now derive a characterization of the stationary points of $J_{2p}(\cdot)$. Making use of the scale invariance of $J_{2p}(\cdot)$ we can always assume that $\|\boldsymbol{s}\|_2 = 1$. Differentiating (20) and substituting in (23) we arrive at the following [**53**]:

THEOREM 1 (Characterization of stationary points). *A candidate $\boldsymbol{s} \in \mathcal{S}_A$ (scaled to unit 2-norm) is a stationary point of $J_{2p}(\boldsymbol{s})$ over $\mathcal{S}_A$ if and only if*

$$\boldsymbol{P}_A(\underbrace{\boldsymbol{s}^T \overset{C}{\star} \boldsymbol{s}^T \overset{C}{\star} \cdots \overset{C}{\star} \boldsymbol{s}^T}_{2p-1 \text{ times}} \overset{C}{\star} \boldsymbol{I})^T = J_{2p}(\boldsymbol{s})\boldsymbol{s}. \tag{26}$$

Defining the polynomial transformation $\mathcal{T}(\boldsymbol{s}) = \boldsymbol{P}_A((\boldsymbol{s}^T)^{\overset{C}{\star}(2p-1)} \overset{C}{\star} \boldsymbol{I})^T$, eq. (26) above takes the form:

$$\mathcal{T}(\boldsymbol{s}) = J_{2p}(\boldsymbol{s})\boldsymbol{s}$$

which points to an interpretation of the stationary points of $J_{2p}(\boldsymbol{s})$ as the unit 2-norm "eigenvectors" of the transformation $\mathcal{T}$. The "eigenvalue" for an "eigenvector" $\boldsymbol{s}$ is found by calculating $J_{2p}(\cdot)$ at $\boldsymbol{s}$. The maximization (22) will then be equivalent to determining the "eigenvector(s)" of $\mathcal{T}$ corresponding to the "eigenvalue(s)" of maximal absolute value. It is well-known that the power method provides a solution to the analogous problem for linear transformations [**24**].

Further insight into the above interpretation can be gained by expressing the function $J_{2p}(\boldsymbol{s})$ in a scalar product form, which could be viewed as a higher-order analog of the Rayleigh quotient [**24**]. To this end, let us define the symmetric $2p$th-order tensor $\mathcal{T}$ by:

$$\mathcal{T} = \boldsymbol{P}_A^{\overset{\mathcal{C}}{\star}2p} \tag{27}$$

or equivalently,

$$\mathcal{T}_{i_1,i_2,\ldots,i_{2p}} = \sum_j \mathcal{C}_{j,j,\ldots,j}(\boldsymbol{P}_A)_{i_1,j}(\boldsymbol{P}_A)_{i_2,j}\cdots(\boldsymbol{P}_A)_{i_{2p},j} \tag{28}$$

where the fact that $\mathcal{C}$ is diagonal was taken into account. We can then express $J_{2p}(\cdot)$ from (25) as $J_{2p}(s) = \frac{(s^T)^{\overset{\mathcal{T}}{\star}2p}}{\|s\|_2^{2p}}$ or equivalently, using the tensor scalar product notation:

$$J_{2p}(\boldsymbol{s}) = \frac{\langle \mathcal{T}, \boldsymbol{s}^{\star 2p}\rangle}{\langle \boldsymbol{s}, \boldsymbol{s}\rangle^p} \tag{29}$$

4. The Tensor Power Method

It is well known that maximizing the absolute value of the Rayleigh quotient of a symmetric matrix amounts to determining its maximal (in absolute value) eigenvalue-eigenvector pair [**24**]. Equivalently, this yields the maximal singular value and corresponding singular vector, which in turn provide us with the best (in the LS sense) rank-1 approximation to the given matrix [**24**, Theorem 2.5.2]. This equivalence between the LS rank-1 approximation to a matrix, provided by its maximal singular triple, and the maximization of the absolute value of the associated bilinear form holds also true for general, nonsymmetric matrices [**24**]. Perhaps the most common way of determining the maximal eigenpair of a matrix $\boldsymbol{T}$ is the power method, described as:

Algorithm 1: Power Method

Initialization: $\boldsymbol{s}_{(0)}$ = a unit 2-norm vector

Iteration:

for $k = 1, 2, \ldots$

$$\tilde{\boldsymbol{s}}_{(k)} = \boldsymbol{T}\boldsymbol{s}_{(k-1)}$$

$$\boldsymbol{s}_{(k)} = \frac{\tilde{\boldsymbol{s}}_{(k)}}{\|\tilde{\boldsymbol{s}}_{(k)}\|_2}$$

$$\sigma_{(k)} = \langle \boldsymbol{s}_{(k)}, \boldsymbol{T}\boldsymbol{s}_{(k)}\rangle$$

end

It is a fixed-point algorithm, which is proven to converge to the absolutely maximal eigenvalue $\lambda_{\max}$ (in $\sigma_{(k)}$) and the corresponding unit-norm eigenvector $\boldsymbol{s}_{\max}$ (in $\boldsymbol{s}_{(k)}$), provided that $\lambda_{\max}$ is simple and the initialization $\boldsymbol{s}_{(0)}$ is not orthogonal to $\boldsymbol{s}_{\max}$. The speed of convergence depends on how much $|\lambda_{\max}|$ is greater than the next absolutely largest eigenvalue [**24**]. Applying the power method to a symmetrized variant of the rectangular matrix $\boldsymbol{T}$, for example $\begin{bmatrix} \boldsymbol{0} & \boldsymbol{T}^T \\ \boldsymbol{T} & \boldsymbol{0} \end{bmatrix}$, one computes its maximal singular value along with the corresponding left and right singular vectors [**24**].

In this section we discuss a higher-order extension of the above algorithm, the Tensor Power Method (TPM) (or as otherwise called, Higher-Order Power Method), developed in [**35**] for providing a LS rank-1 approximation to *nonsymmetric* tensors. We devise an alternative proof of the characterization of the stationary points, which moreover provides us a means of discriminating the stable equilibria from the saddle points. A clear proof of convergence is also presented. Contrary to the doubts expressed in [**35**] concerning the existence of a working constrained version of the method for symmetric tensors, such an algorithm is shown here to be convergent subject to some convexity (concavity) conditions, justified in several practical applications. For the symmetric tensor case, a new initialization method is developed, which has been seen, via extensive simulations, to provide an initial point closer to the global optimum than the TSVD-based initialization suggested in [**35**]. As a byproduct of the above analysis, we arrive at lower and upper bounds for the initially obtained cost value, whose tightness has been experimentally verified.

4.1. Equivalence of Tensor Rank-1 Approximation and Multilinear Form Maximization. We now clarify the connection of the problem of maximizing the absolute value of a multilinear form with that of obtaining the LS rank-1 approximant to its associated tensor [**34**]; this extends to higher-order arrays a well-known property for matrices [**24**].

THEOREM 2 (Tensor low-rank approximation). *Given an nth-order tensor $\boldsymbol{\mathcal{T}}$, consider the problem of determining a scalar σ and n vectors $\{\boldsymbol{s}^{(i)}\}_{i=1}^n$ minimizing the Frobenius norm squared*

$$\|\boldsymbol{\mathcal{T}} - \sigma[\boldsymbol{s}^{(1)} \star \boldsymbol{s}^{(2)} \star \cdots \star \boldsymbol{s}^{(n)}]\|_F^2 = \sum_{i_1,i_2,\ldots,i_n} \left(\mathcal{T}_{i_1,i_2,\ldots,i_n} - \sigma s_{i_1}^{(1)} s_{i_2}^{(2)} \cdots s_{i_n}^{(n)}\right)^2 \tag{30}$$

subject to the constraint that $\|\boldsymbol{s}^{(i)}\|_2 = 1$ for $i = 1, 2, \ldots, n$. Then the vectors $\{\boldsymbol{s}^{(i)}\}_{i=1}^n$ correspond to a local minimum of (30) if and only if they yield a local maximum of $|f(\boldsymbol{s}^{(1)}, \boldsymbol{s}^{(2)}, \ldots, \boldsymbol{s}^{(n)})|$, with

$$f(\boldsymbol{s}^{(1)}, \boldsymbol{s}^{(2)}, \ldots, \boldsymbol{s}^{(n)}) = \sum_{i_1,i_2,\ldots,i_n} \mathcal{T}_{i_1,i_2,\ldots,i_n} s_{i_1}^{(1)} s_{i_2}^{(2)} \cdots s_{i_n}^{(n)} = \langle \boldsymbol{\mathcal{T}}, \boldsymbol{s}^{(1)} \star \cdots \star \boldsymbol{s}^{(n)} \rangle. \tag{31}$$

The corresponding value of σ is $\sigma = f(\boldsymbol{s}^{(1)}, \boldsymbol{s}^{(2)}, \ldots, \boldsymbol{s}^{(n)})$.

Proof: Fixing the $\boldsymbol{s}^{(i)}$'s, (30) takes the form of a LS problem in the scalar σ, whose solution is readily found as:

$$\sigma_{\text{opt}} = \frac{\langle \boldsymbol{\mathcal{T}}, \boldsymbol{s}^{(1)} \star \boldsymbol{s}^{(2)} \star \cdots \star \boldsymbol{s}^{(n)} \rangle}{\langle \boldsymbol{s}^{(1)} \star \boldsymbol{s}^{(2)} \star \cdots \star \boldsymbol{s}^{(n)}, \boldsymbol{s}^{(1)} \star \boldsymbol{s}^{(2)} \star \cdots \star \boldsymbol{s}^{(n)} \rangle}$$

From the property (1) of the Tucker product it follows that $\|\boldsymbol{s}^{(1)} \star \boldsymbol{s}^{(2)} \star \cdots \star \boldsymbol{s}^{(n)}\|_F^2 = \|\boldsymbol{s}^{(1)} \otimes \boldsymbol{s}^{(2)} \otimes \cdots \otimes \boldsymbol{s}^{(n)}\|_2^2$, whereby, using well-known properties of the Kronecker product [**2**], we obtain

$$\|\boldsymbol{s}^{(1)} \star \boldsymbol{s}^{(2)} \star \cdots \star \boldsymbol{s}^{(n)}\|_F^2 = \prod_{i=1}^n \|\boldsymbol{s}^{(i)}\|_2^2 = 1.$$

Hence

$$\begin{aligned} \sigma_{\text{opt}} &= \langle \boldsymbol{\mathcal{T}}, \boldsymbol{s}^{(1)} \star \boldsymbol{s}^{(2)} \star \cdots \star \boldsymbol{s}^{(n)} \rangle \\ &= f(\boldsymbol{s}^{(1)}, \boldsymbol{s}^{(2)}, \ldots, \boldsymbol{s}^{(n)}) \end{aligned}$$

and the LS squared error becomes

$$\begin{aligned} \|\mathcal{T} - \sigma[\boldsymbol{s}^{(1)} \star \boldsymbol{s}^{(2)} \star \cdots \star \boldsymbol{s}^{(n)}]\|_F^2 &= \|\mathcal{T}\|_F^2 - \sigma_{\mathrm{opt}}^2 \\ &= \|\mathcal{T}\|_F^2 - |f(\boldsymbol{s}^{(1)}, \boldsymbol{s}^{(2)}, \ldots, \boldsymbol{s}^{(n)})|^2. \end{aligned} \tag{32}$$

Since $\|\mathcal{T}\|_F$ is constant, the latter relation completes the proof. ◇

4.2. Characterization of Stationary Points. In view of Theorem 2 the study of the extremization of multilinear forms (31) parallels that of higher-order tensor approximation. The characterization of the stationary points of (31), developed in this subsection, further enlightens this equivalence.

First of all, let us point out that the constrained problem

$$\max_{\|\boldsymbol{s}^{(i)}\|_2 = 1} |f(\boldsymbol{s}^{(1)}, \boldsymbol{s}^{(2)}, \ldots, \boldsymbol{s}^{(n)})|^2 \tag{33}$$

may equivalently be viewed as the unconstrained maximization of the functional

$$h(\boldsymbol{s}^{(1)}, \boldsymbol{s}^{(2)}, \ldots, \boldsymbol{s}^{(n)}) = \frac{|f(\boldsymbol{s}^{(1)}, \boldsymbol{s}^{(2)}, \ldots, \boldsymbol{s}^{(n)})|^2}{\|\boldsymbol{s}^{(1)}\|_2^2 \|\boldsymbol{s}^{(2)}\|_2^2 \cdots \|\boldsymbol{s}^{(n)}\|_2^2} = \frac{\langle \mathcal{T}, \boldsymbol{s}^{(1)} \star \cdots \star \boldsymbol{s}^{(n)} \rangle^2}{\langle \boldsymbol{s}_{(1)}, \boldsymbol{s}_{(1)} \rangle \cdots \langle \boldsymbol{s}_{(n)}, \boldsymbol{s}_{(n)} \rangle} \tag{34}$$

It is readily seen that $h(\cdot)$ is scale invariant, that is,

$$h(\beta_1 \boldsymbol{s}^{(1)}, \beta_2 \boldsymbol{s}^{(2)}, \cdots, \beta_n \boldsymbol{s}^{(n)}) = h(\boldsymbol{s}^{(1)}, \boldsymbol{s}^{(2)}, \ldots, \boldsymbol{s}^{(n)})$$

for any nonzero scalars β_i.

Before plunging into the analysis of the stationary points of h let us briefly recall the notion of directional derivative as it will be extensively employed in the sequel.

Definition 14 (Directional derivative). **[54]** *Let $h(\boldsymbol{s}^{(1)}, \boldsymbol{s}^{(2)}, \ldots, \boldsymbol{s}^{(n)})$ be a function from ℓ_2^n to $\mathbb{R}$. The following limit*

$$\begin{aligned} & h'(\boldsymbol{s}^{(1)}, \boldsymbol{s}^{(2)}, \ldots, \boldsymbol{s}^{(n)}; \boldsymbol{r}^{(1)}, \boldsymbol{r}^{(2)}, \ldots, \boldsymbol{r}^{(n)}) \\ & \quad \triangleq \lim_{t \to 0} \frac{h(\boldsymbol{s}^{(1)} + t\boldsymbol{r}^{(1)}, \ldots, \boldsymbol{s}^{(n)} + t\boldsymbol{r}^{(n)}) - h(\boldsymbol{s}^{(1)}, \ldots, \boldsymbol{s}^{(n)})}{t}, \end{aligned}$$

if it exists, is called the (first) directional derivative (or Gateaux differential) of h at the point $(\boldsymbol{s}^{(1)}, \boldsymbol{s}^{(2)}, \ldots, \boldsymbol{s}^{(n)})$ with respect to the directional vectors $(\boldsymbol{r}^{(1)}, \boldsymbol{r}^{(2)}, \ldots, \boldsymbol{r}^{(n)})$.

Using standard notation, we may write

$$h'(\boldsymbol{s}^{(1)}, \boldsymbol{s}^{(2)}, \ldots, \boldsymbol{s}^{(n)}; \boldsymbol{r}^{(1)}, \boldsymbol{r}^{(2)}, \ldots, \boldsymbol{r}^{(n)}) = \left. \frac{dh(\boldsymbol{s}^{(1)} + t\boldsymbol{r}^{(1)}, \ldots, \boldsymbol{s}^{(n)} + t\boldsymbol{r}^{(n)})}{dt} \right|_{t=0}$$

The second directional derivative is defined analogously:

$$h''(\boldsymbol{s}^{(1)}, \boldsymbol{s}^{(2)}, \ldots, \boldsymbol{s}^{(n)}; \boldsymbol{r}^{(1)}, \boldsymbol{r}^{(2)}, \ldots, \boldsymbol{r}^{(n)}) = \left. \frac{d^2 h(\boldsymbol{s}^{(1)} + t\boldsymbol{r}^{(1)}, \ldots, \boldsymbol{s}^{(n)} + t\boldsymbol{r}^{(n)})}{dt^2} \right|_{t=0}$$

For a function $h(\boldsymbol{s})$ with continuous partial derivatives, its first directional derivative is given by the projection of its gradient onto the directional vector, i.e.,

$$h'(\boldsymbol{s}; \boldsymbol{r}) = \langle \nabla h(\boldsymbol{s}), \boldsymbol{r} \rangle \tag{35}$$

and provides the rate of variation at $\boldsymbol{s}$ in the direction given by $\boldsymbol{r}$. If h is twice continuously differentiable, one can see that

$$h''(\boldsymbol{s}; \boldsymbol{r}) = \langle \boldsymbol{r}, \boldsymbol{H}(\boldsymbol{s}) \boldsymbol{r} \rangle \tag{36}$$

where $\boldsymbol{H}(\boldsymbol{s}) = \frac{\partial^2 h(s)}{\partial s^2}$ is the Hessian matrix of h at $\boldsymbol{s}$.

Call $d(\boldsymbol{s}^{(1)}, \ldots, \boldsymbol{s}^{(n)})$ the denominator function in (34). For convenience we shall denote by $g(t)$ the function $g(\boldsymbol{s}^{(1)} + t\boldsymbol{r}^{(1)}, \ldots, \boldsymbol{s}^{(n)} + t\boldsymbol{r}^{(n)})$ of t and by g', g'' its first and second derivatives with respect to t. Then $g'(0)$ and $g''(0)$ will equal the first and second directional derivatives of g, respectively. In view of the scale invariance of h, we may assume without loss of generality that the stationary points $(\boldsymbol{s}^{(1)}, \ldots, \boldsymbol{s}^{(n)})$ are normalized to unit norm, $\|\boldsymbol{s}^{(i)}\|_2 = 1$. Let the vectors $\boldsymbol{r}^{(i)}$ be decomposed as $\boldsymbol{r}^{(i)} = \alpha_i \boldsymbol{s}^{(i)} + \boldsymbol{v}^{(i)}$ with $\langle \boldsymbol{v}^{(i)}, \boldsymbol{s}^{(i)} \rangle = 0$. The stationary points of h are found as solutions to the equation:

$$0 = h'(0) = \left. \frac{2dff' - f^2 d'}{d^2} \right|_{t=0} \tag{37}$$

One can see that $d'(0) = 0$. Taking also into account the equality $d(0) = 1$, (37) becomes

$$0 = h'(0) = 2f(0)f'(0). \tag{38}$$

Clearly the case $f(0) = 0$ corresponds to a global minimum of h ($h \geq 0$). Thus consider the case that $f(0) \neq 0$, whereby $f'(0) = 0$. Calculating f' we obtain

$$f'(0) = \langle \boldsymbol{v}^{(1)}, \boldsymbol{x}^{(1)} \rangle + \langle \boldsymbol{v}^{(2)}, \boldsymbol{x}^{(2)} \rangle + \cdots + \langle \boldsymbol{v}^{(n)}, \boldsymbol{x}^{(n)} \rangle \tag{39}$$

where the vector $\boldsymbol{x}^{(i)}$ is obtained by excluding $\boldsymbol{s}^{(i)}$ from the sum (31):

$$\begin{aligned} x_{i_1}^{(1)} &= \sum_{i_2,\ldots,i_n} \mathcal{T}_{i_1,i_2,\ldots,i_n} s_{i_2}^{(2)} \cdots s_{i_n}^{(n)} \\ &\vdots \\ x_{i_n}^{(n)} &= \sum_{i_1,\ldots,i_{n-1}} \mathcal{T}_{i_1,i_2,\ldots,i_n} s_{i_1}^{(1)} \cdots s_{i_{n-1}}^{(n-1)} \end{aligned} \tag{40}$$

We should emphasize here that (39) holds for all vectors $\boldsymbol{v}^{(i)}$ orthogonal to $\boldsymbol{s}^{(i)}$. Hence one can set all but one, say the jth, equal to zero. Then the stationary point condition $f'(0) = 0$ implies that

$$\langle \boldsymbol{v}^{(j)}, \boldsymbol{x}^{(j)} \rangle = 0, \quad \text{for all } \boldsymbol{v}^{(j)} \text{ orthogonal to } \boldsymbol{s}^{(j)}.$$

This in turn implies that $\boldsymbol{x}^{(j)}$ is colinear with $\boldsymbol{s}^{(j)}$; that is,

$$\boldsymbol{x}^{(j)} = \sigma \boldsymbol{s}^{(j)}$$

with the scalar σ given by

$$\sigma = \langle \boldsymbol{x}^{(j)}, \boldsymbol{s}^{(j)} \rangle = \langle \boldsymbol{\mathcal{T}}, \boldsymbol{s}^{(1)} \star \boldsymbol{s}^{(2)} \star \cdots \boldsymbol{s}^{(n)} \rangle = f(\boldsymbol{s}^{(1)}, \ldots, \boldsymbol{s}^{(n)}). \tag{41}$$

Conversely, if $\boldsymbol{x}^{(j)} = f(0)\boldsymbol{s}^{(j)}$ for all j, then it follows directly from (39) that $f'(0) = 0$, that is the $\boldsymbol{s}^{(i)}$'s are a stationary point of h. We have thus shown the following:

THEOREM 3 (Characterization of stationary points). *The unit 2-norm vectors $\{\boldsymbol{s}^{(i)}\}_{i=1}^n$ are a stationary point of the functional h in (34), if and only if*

$$\begin{aligned}\sum_{i_2,\ldots,i_n} \mathcal{T}_{i_1,i_2,\ldots,i_n} s_{i_2}^{(2)} s_{i_3}^{(3)} \ldots s_{i_n}^{(n)} &= \sigma s_{i_1}^{(1)} \text{ for all } i_1 \\ &\vdots \\ \sum_{i_1,\ldots,i_{n-1}} \mathcal{T}_{i_1,\ldots,i_{n-1},i_n} s_{i_1}^{(1)} \ldots s_{i_{n-2}}^{(n-2)} s_{i_{n-1}}^{(n-1)} &= \sigma s_{i_n}^{(n)} \text{ for all } i_n\end{aligned}$$

with $\sigma = f(\boldsymbol{s}^{(1)}, \boldsymbol{s}^{(2)}, \ldots, \boldsymbol{s}^{(n)})$.

The above characterization was also derived in [**34**], with the aid of the Lagrangian function for the constrained problem (33). Nonetheless, the above analysis has the advantage of allowing a study of the stability of the stationary points as well. To this end, we consider the second directional derivative of h at a stationary point. A straightforward calculation, taking into account that $f'(0) = 0$, yields:

$$h''(0) = 2f(0)\left(f''(0) - \frac{1}{2}f(0)d''(0)\right). \tag{42}$$

The point in question is a local maximum (minimum) of h if $h''(0) \leq 0$ ($h''(0) \geq 0$) for all possible directions. In view of (36), this is to say that the Hessian matrix is seminegative- (semipositive-) definite [**54**]. Assume $f(0) > 0$. The case $f(0) < 0$ can be treated in an analogous way. In that case, the condition for a local maximum takes the form:

$$f''(0) - \frac{1}{2}f(0)d''(0) \leq 0 \tag{43}$$

It can be verified that

$$d''(0) = 2\sum_{i=1}^{n} \|\boldsymbol{v}^{(i)}\|_2^2 \tag{44}$$

$$f''(0) = 2\sum_{i=1}^{n}\sum_{\substack{j=1\\ j>i}}^{n} \langle \boldsymbol{v}^{(i)}, \boldsymbol{M}^{(i,j)}\boldsymbol{v}^{(j)}\rangle \tag{45}$$

where $\boldsymbol{M}^{(i,j)}$ is the matrix that results from excluding $\boldsymbol{s}^{(i)}$ and $\boldsymbol{s}^{(j)}$ from the sum (31). For example,

$$M_{i_1,i_2}^{(1,2)} = \sum_{i_3,\ldots,i_n} \mathcal{T}_{i_1,i_2,i_3,\ldots,i_n} s_{i_3}^{(3)} \cdots s_{i_n}^{(n)} \tag{46}$$

It is of interest to point out that *all the matrices $\boldsymbol{M}^{(i,j)}$ share a common singular value, namely* $f(0)$; indeed,

$$\boldsymbol{M}^{(i,j)}\boldsymbol{s}^{(j)} = \boldsymbol{x}^{(i)} = f(0)\boldsymbol{s}^{(i)}$$

and

$$(\boldsymbol{M}^{(i,j)})^T\boldsymbol{s}^{(i)} = \boldsymbol{x}^{(j)} = f(0)\boldsymbol{s}^{(j)}$$

where the characterization in Theorem 3 was used.

Condition (43) now reads as:

THEOREM 4 (Characterization of stable equilibria). *The stationary point* $(\boldsymbol{s}^{(1)}, \ldots, \boldsymbol{s}^{(n)})$ *of* h *in (34), with* $\|\boldsymbol{s}^{(i)}\|_2 = 1$, *corresponds to a local maximum of the function if and only if*

$$(47) \quad 2\sum_{i=1}^{n}\sum_{\substack{j=1\\j>i}}^{n}\langle \boldsymbol{v}^{(i)}, \boldsymbol{M}^{(i,j)}\boldsymbol{v}^{(j)}\rangle \leq f(\boldsymbol{s}^{(1)}, \ldots, \boldsymbol{s}^{(n)})\sum_{i=1}^{n}\|\boldsymbol{v}^{(i)}\|_2^2, \qquad f(\cdot) > 0$$

$$(48) \quad 2\sum_{i=1}^{n}\sum_{\substack{j=1\\j>i}}^{n}\langle \boldsymbol{v}^{(i)}, \boldsymbol{M}^{(i,j)}\boldsymbol{v}^{(j)}\rangle \geq f(\boldsymbol{s}^{(1)}, \ldots, \boldsymbol{s}^{(n)})\sum_{i=1}^{n}\|\boldsymbol{v}^{(i)}\|_2^2, \qquad f(\cdot) < 0$$

for all vectors $\boldsymbol{v}^{(i)}$ *orthogonal to* $\boldsymbol{s}^{(i)}$, *with the matrices* $\boldsymbol{M}^{(i,j)}$ *defined as in (46). For a local minimum, (47), (48) apply with their first inequalities being reversed.*

An interesting byproduct of the above results is summarized as:

COROLLARY 1 (Characterization of the functional extremal values). *The square root of the value assumed by* h *at a local maximum (minimum)* $(\boldsymbol{s}^{(1)}, \ldots, \boldsymbol{s}^{(n)})$ *with* $\|\boldsymbol{s}^{(i)}\|_2 = 1$, *is the largest (smallest) singular value of each of the matrices* $\boldsymbol{M}^{(i,j)}$.

Proof: Consider the case of a maximum where f is positive. The other cases follow similarly. Set in (47) all vectors $\boldsymbol{v}^{(i)}$ equal to zero, except for two, say $\boldsymbol{v}^{(j)}$ and $\boldsymbol{v}^{(k)}$, $j < k$, with $\|\boldsymbol{v}^{(j)}\|_2 = \|\boldsymbol{v}^{(k)}\|_2 = 1$. This yields

$$\langle \boldsymbol{v}^{(j)}, \boldsymbol{M}^{(j,k)}\boldsymbol{v}^{(k)}\rangle \leq f(0)$$

for all unit-norm $\boldsymbol{v}^{(j)}$ and $\boldsymbol{v}^{(k)}$ orthogonal to $\boldsymbol{s}^{(j)}$ and $\boldsymbol{s}^{(k)}$, respectively. The result then follows by considering the fact that $(f(0), \boldsymbol{s}^{(j)}, \boldsymbol{s}^{(k)})$ is a singular triple of $\boldsymbol{M}^{(j,k)}$. ◇

Theorem 4 takes a particularly elegant form in the case that $\mathcal{T}$ is symmetric. Note that in this case all $\frac{n(n-1)}{2}$ matrices $\boldsymbol{M}^{(i,j)}$ are symmetric and equal, say to $\boldsymbol{M}$, as well as the vectors $\boldsymbol{s}^{(1)} = \boldsymbol{s}^{(2)} = \cdots = \boldsymbol{s}^{(n)} = \boldsymbol{s}$ and $\boldsymbol{v}^{(1)} = \boldsymbol{v}^{(2)} = \cdots = \boldsymbol{v}^{(n)} = \boldsymbol{v}$. Then (47) (for $f > 0$) becomes

$$(49) \qquad (n-1)\frac{\langle \boldsymbol{v}, \boldsymbol{M}\boldsymbol{v}\rangle}{\langle \boldsymbol{v}, \boldsymbol{v}\rangle} \leq f(\boldsymbol{s})$$

which, in view of the fact that $f(\boldsymbol{s}) = \frac{\langle \mathcal{T}, \boldsymbol{s}^{\star n}\rangle}{\|\boldsymbol{s}\|_2^{2n}}$ at a maximum is the largest eigenvalue of $\boldsymbol{M}$, shows that it is the most positive eigenvalue by a factor of $n-1$.

4.3. TPM and TSVD-Initialization (Nonsymmetric Case). An iterative algorithm, extending Algorithm 1 to higher-order tensors, was developed in [**35**] based on the equalities satisfied by the stationary points of h in Theorem 3. The vectors $\boldsymbol{s}^{(i)}$ are each optimized alternately using the corresponding equality and considering the remaining vectors fixed to their current values. This algorithm is interpreted in [**34**] as a gradient ascent scheme for the maximization of h.[10] As the function h is multimodal (for $n > 2$), the initial values for the vectors $\boldsymbol{s}^{(i)}$ in this gradient scheme are critical as to whether it will converge to a local maximum or to its global one. Inspired from the role that plays in the 2nd-order problem the maximal singular triple of $\mathcal{T}$, the authors of [**35**] have proposed an initialization

[10] In fact, the TPM is nothing but the so-called *Alternating Least Squares (ALS)* procedure employed in nonlinear least-squares problems [**4**].

for TPM on the basis of the corresponding TSVD. Namely, $\boldsymbol{s}^{(i)}$ is initialized as the first column of the matrix $\boldsymbol{U}^{(i)}$ in the decomposition (5) of $\boldsymbol{\mathcal{T}}$. Although no formal proof is given, it is conjectured in [**35**], on the basis of simulation results, that this initialization is almost always found in a basin of attraction of the globally optimum point. The TPM is described below:

Algorithm 2: Tensor Power Method (Nonsymmetric case)

Initialization: $\boldsymbol{s}^{(i)}(0) =$ a unit 2-norm vector, $1 \leq i \leq n$

Iteration: for $k = 1, 2, \dots$

$$\tilde{s}^{(1)}_{i_1}(k) = \sum_{i_2, i_3, \dots, i_n} \mathcal{T}_{i_1, i_2, \dots, i_n} s^{(2)}_{i_2}(k-1) s^{(3)}_{i_3}(k-1) \cdots s^{(n)}_{i_n}(k-1)$$

$$\boldsymbol{s}^{(1)}(k) = \varepsilon \frac{\tilde{\boldsymbol{s}}^{(1)}(k)}{\|\tilde{\boldsymbol{s}}^{(1)}(k)\|_2}$$

$$\tilde{s}^{(2)}_{i_2}(k) = \sum_{i_1, i_3, \dots, i_n} \mathcal{T}_{i_1, i_2, \dots, i_n} s^{(1)}_{i_1}(k) s^{(3)}_{i_3}(k-1) \cdots s^{(n)}_{i_n}(k-1)$$

$$\boldsymbol{s}^{(2)}(k) = \varepsilon \frac{\tilde{\boldsymbol{s}}^{(2)}(k)}{\|\tilde{\boldsymbol{s}}^{(2)}(k)\|_2}$$

$$\vdots$$

$$\tilde{s}^{(n)}_{i_n}(k) = \sum_{i_1, i_2, \dots, i_{n-1}} \mathcal{T}_{i_1, i_2, \dots, i_n} s^{(1)}_{i_1}(k) s^{(2)}_{i_2}(k) \cdots s^{(n-1)}_{i_{n-1}}(k-1)$$

$$\boldsymbol{s}^{(n)}(k) = \varepsilon \frac{\tilde{\boldsymbol{s}}^{(n)}(k)}{\|\tilde{\boldsymbol{s}}^{(n)}(k)\|_2}$$

end

The factor ε is set to 1 if f is to be maximized and to -1 if it is to be minimized. Henceforth we shall only be concerned with the *maximization* of h. Then ε ideally equals the sign of the scalar σ in Theorem 3.

Although the convergence of TPM is implicit in [**35**], no clear proof is given for it. We provide one in the following:

THEOREM 5 (Convergence of TPM). *If $\varepsilon = 1$ ($\varepsilon = -1$), Algorithm 2 converges to a local maximum (minimum) of f for any initialization, except for saddle points or crest lines leading to such saddle points.*

Proof: Begin by writing f as (see (41))

$$f(\boldsymbol{s}^{(1)}(k-1), \boldsymbol{s}^{(2)}(k-1), \cdots, \boldsymbol{s}^{(n)}(k-1)) = \langle \boldsymbol{s}^{(1)}(k-1), \boldsymbol{x}^{(1)}(k-1) \rangle$$

where the vector $\boldsymbol{x}^{(1)}(k-1)$ is defined as in (40), i.e.,

$$x^{(1)}_{i_1}(k-1) = \sum_{i_2, \dots, i_n} \mathcal{T}_{i_1, i_2, \dots, i_n} s^{(2)}_{i_2}(k-1) \cdots s^{(n)}_{i_n}(k-1)$$

Applying the Cauchy-Schwarz inequality to the scalar product $\langle \boldsymbol{s}^{(1)}(k-1), \boldsymbol{x}^{(1)}(k-1) \rangle$ yields (recall that $\|\boldsymbol{s}^{(1)}(k-1)\|_2 = 1$)

$$-\|\boldsymbol{x}^{(1)}(k-1)\|_2 \leq \langle \boldsymbol{s}^{(1)}(k-1), \boldsymbol{x}^{(1)}(k-1) \rangle \leq \|\boldsymbol{x}^{(1)}(k-1)\|_2. \tag{50}$$

The upper (lower) bound in (50) is obtained if and only if $s^{(1)}(k-1) = \frac{x^{(1)}(k-1)}{\|x^{(1)}(k-1)\|_2}$ ($s^{(1)}(k-1) = -\frac{x^{(1)}(k-1)}{\|x^{(1)}(k-1)\|_2}$) which coincides with the first step in the iteration of Algorithm 2 for $s^{(1)}(k)$ when $\varepsilon = 1$ ($\varepsilon = -1$). Consider the case $\varepsilon = 1$ as the other case can be addressed in a dual manner. Then, if we are not at a stationary point, the above argument implies that $\langle s^{(1)}(k), x^{(1)}(k-1)\rangle - \langle s^{(1)}(k-1), x^{(1)}(k-1)\rangle > 0$, or equivalently,

$$f(s^{(1)}(k), s^{(2)}(k-1), \cdots, s^{(n)}(k-1)) > f(s^{(1)}(k-1), s^{(2)}(k-1), \cdots, s^{(n)}(k-1)).$$

By repeating this argument for the remaining vectors we obtain the chain of inequalities:

$$\begin{aligned} f(s^{(1)}(k), s^{(2)}(k), \cdots, s^{(n)}(k-1)) &> f(s^{(1)}(k), s^{(2)}(k-1), \cdots, s^{(n)}(k-1)) \\ &\vdots \\ f(s^{(1)}(k), s^{(2)}(k), \cdots, s^{(n)}(k)) &> f(s^{(1)}(k), s^{(2)}(k), \cdots, s^{(n)}(k-1)) \end{aligned}$$

showing f to be increasing. From (32) it is seen that $|f|$ is upper bounded by $\|\mathcal{T}\|_F$. We thus conclude that f is convergent to a (local) maximum. $\diamond$

4.4. Symmetric TPM. The TPM, as described above, when applied to a symmetric tensor $\mathcal{T}$, will necessarily converge to a point where all the vectors $s^{(i)}$ are equal. However, as pointed out in [**35**], the intermediate rank-1 approximants are generally nonsymmetric. Forcing the equality $s^{(1)} = s^{(2)} = \ldots = s^{(n)} = s$ all from the beginning in Algorithm 2 is criticized in [**35**] as being unreliable with respect to preserving the monotonicity of f. In this subsection we study a symmetric version of TPM and show it to be convergent for even n, subject to a condition of convexity (concavity) for f. As it will be seen in the next section, such conditions often occur in common blind deconvolution setups. Moreover, although the convergence proof provided here relies on this assumption on f, convergence has also been observed in all of our simulations conducted in the absence of this condition. The symmetric TPM results from Algorithm 2 above by equating all the vectors $s^{(i)}$:

Algorithm 3: Tensor Power Method for Symmetric Tensors

Initialization: $s(0)$ = a unit 2-norm vector (e.g., computed as in §4.5)

Iteration: for $k = 1, 2, \ldots$

$$\tilde{s}_{i_1}(k) = \sum_{i_2, i_3, \ldots, i_n} \mathcal{T}_{i_1, i_2, \ldots, i_n} s_{i_2}(k-1) s_{i_3}(k-1) \cdots s_{i_n}(k-1)$$

$$s(k) = \varepsilon \frac{\tilde{s}(k)}{\|\tilde{s}(k)\|_2}$$

end

As in Algorithm 2, the sign factor ε assumes the value 1 when f is to be maximized (convex case) and -1 when it is to be minimized (concave case).

THEOREM 6 (Convergence of symmetric TPM). *Let $\mathcal{T}$ be a symmetric $M \times M \times \cdots \times M$ tensor of even order n. If the function $f(s) = \langle \mathcal{T}, s^{\star n}\rangle$ is convex (concave) over $\mathbb{R}^M$, then Algorithm 3 with $\varepsilon = 1$ ($\varepsilon = -1$) will converge to a maximum (minimum) of the function $\frac{f(s)}{\langle s, s\rangle^{n/2}}$.*

We should emphasize that even when $f(\mathbf{s})$ is convex, the scaled function $\frac{f(s)}{\langle s,s\rangle^{n/2}}$ will not be.

Proof: Consider the case that f is convex, as the case of f being concave can be addressed similarly. The major implication of this assuption is that f satisfies the (sub-)gradient inequality [**54**]

$$f(\boldsymbol{w}^{(2)}) - f(\boldsymbol{w}^{(1)}) \geq \langle \boldsymbol{w}^{(2)} - \boldsymbol{w}^{(1)}, \nabla f(\boldsymbol{w}^{(1)})\rangle$$

for any vectors $\boldsymbol{w}^{(1)}, \boldsymbol{w}^{(2)} \in \mathbb{R}^M$ (regardless of how distant they may be). Recall from (39) that (see (35)):

$$[\nabla f(\boldsymbol{w})]_{i_1} = n \sum_{i_2,\ldots,i_n} \mathcal{T}_{i_1,i_2,\ldots,i_n} w_{i_2}\cdots w_{i_n}.$$

Now set $\boldsymbol{w}^{(2)} = \boldsymbol{s}(k)$ and $\boldsymbol{w}^{(1)} = \boldsymbol{s}(k-1)$ in the above inequality, to obtain

$$f(\boldsymbol{s}(k)) - f(\boldsymbol{s}(k-1)) \geq \langle \boldsymbol{s}(k), \nabla f(\boldsymbol{s}(k-1))\rangle - \langle \boldsymbol{s}(k-1), \nabla f(\boldsymbol{s}(k-1))\rangle \tag{51}$$

To show that $f(\boldsymbol{s}(k))$ is increasing it suffices then to prove that the right-hand side above is positive whenever $\boldsymbol{s}(k) \neq \boldsymbol{s}(k-1)$ (i.e., not a stationary point). Note that, for any unit-norm vector $\boldsymbol{s}$, the Cauchy-Schwarz inequality yields:

$$\langle \boldsymbol{s}, \nabla f(\boldsymbol{s}(k-1))\rangle \leq \|\nabla f(\boldsymbol{s}(k-1))\|_2 \tag{52}$$

with equality if and only if $\boldsymbol{s} = \frac{\nabla f(s(k-1))}{\|\nabla f(s(k-1))\|_2}$. But this is precisely the formula for $\boldsymbol{s}(k)$ in the Algorithm 3 for $\varepsilon = 1$. Hence

$$\langle \boldsymbol{s}(k), \nabla f(\boldsymbol{s}(k-1))\rangle - \langle \boldsymbol{s}(k-1), \nabla f(\boldsymbol{s}(k-1))\rangle > 0$$

which, in view of (51), implies that $f(\boldsymbol{s}(k))$ is increasing with k. The convergence to a (local) maximum follows again by considering the fact that $|f|$ is bounded from above. ◇

Note that the left-hand side of (52) equals the directional derivative of f with respect to the direction $\boldsymbol{s}$. Hence Algorithm 3 makes for $\boldsymbol{s}(k)$ the choice which leads to the greatest possible increase in f. We shall come back to this point in Section 5 when discussing the SEA.

4.5. A Novel Initialization Method and Some Performance Bounds. Although our simulations have verified the closeness of the TSVD-based initialization of [**35**] to the global extremum, an alternative initialization presented here proves to be more effective in approaching the globally optimum point. For the sake of simplicity our presentation will be confined to the case of a symmetric 4th-order tensor, which corresponds to the most common scenario in signal processing applications.

The fundamental idea stems from the fact that the functional $f(\boldsymbol{s})$ of Theorem 6 can be written in the form

$$f(\boldsymbol{s}) = \sum_{i_1,i_2,i_3,i_4} \mathcal{T}_{i_1,i_2,i_3,i_4} s_{i_1} s_{i_2} s_{i_3} s_{i_4} = \langle \boldsymbol{s}\otimes\boldsymbol{s}, \boldsymbol{T}(\boldsymbol{s}\otimes\boldsymbol{s})\rangle \tag{53}$$

where the $M^2 \times M^2$ matrix $\boldsymbol{T}$ is built out from the entries of the $M\times M\times M\times M$ tensor $\boldsymbol{\mathcal{T}}$ by grouping their first two indices in a row index and the last two in a column index. That is,

$$T_{i,j} = \mathcal{T}_{i_1,i_2,i_3,i_4}$$

where

$$i = M(i_1 - 1) + i_2, \quad 1 \leq i_1, i_2 \leq M$$
$$j = M(i_3 - 1) + i_4, \quad 1 \leq i_3, i_4 \leq M.$$

Due to the *super*symmetry of $\boldsymbol{T}$ it is readily verified that $\boldsymbol{T}$ is not only symmetric (i.e., $\boldsymbol{T}^T = \boldsymbol{T}$) but it also satisfies the equality $\boldsymbol{PTP} = \boldsymbol{T}$ where $\boldsymbol{P}$ denotes the $M^2 \times M^2$ *vec-permutation matrix* [**27**]. Letting $\boldsymbol{E}_{i,j}$ denote the matrix $\boldsymbol{e}_i\boldsymbol{e}_j^T$ of all zeros with a 1 at the position (i,j), $\boldsymbol{P}$ can be expressed as [**2, 27**]:

$$\boldsymbol{P} = \sum_{i=1}^{M}\sum_{j=1}^{M}(\boldsymbol{E}_{i,j} \otimes \boldsymbol{E}_{j,i})$$

whereby it is seen to be symmetric.

Consider the SVD of $\boldsymbol{T}$:

(54)

$$\boldsymbol{T} = \underbrace{\begin{bmatrix} \boldsymbol{\vartheta}_1 & \boldsymbol{\vartheta}_2 & \cdots & \boldsymbol{\vartheta}_{M^2} \end{bmatrix}}_{\boldsymbol{\Theta}} \underbrace{\begin{bmatrix} \sigma_1 & & & \bigcirc \\ & \sigma_2 & & \\ & & \ddots & \\ \bigcirc & & & \sigma_{M^2} \end{bmatrix}}_{\boldsymbol{\Sigma}} \underbrace{\begin{bmatrix} \boldsymbol{\xi}_1 & \boldsymbol{\xi}_2 & \cdots & \boldsymbol{\xi}_{M^2} \end{bmatrix}^T}_{\boldsymbol{\Xi}^T}$$

Since $\boldsymbol{T}$ is symmetric, its left and right singular vectors coincide modulo a sign ambiguity; that is, $\boldsymbol{\xi}_i = \pm\boldsymbol{\vartheta}_i$ for $i = 1, 2, \ldots, M^2$. From the identity $\boldsymbol{PTP} = \boldsymbol{T}$ it follows that $\boldsymbol{P\Xi} = \boldsymbol{\Xi}$ and hence $\boldsymbol{P\xi}_i = \boldsymbol{\xi}_i$.[11] If $\boldsymbol{s}$ has unit norm, then we have

$$|f(\boldsymbol{s})| = \left|\frac{\langle \boldsymbol{s} \otimes \boldsymbol{s}, \boldsymbol{T}(\boldsymbol{s} \otimes \boldsymbol{s})\rangle}{\langle \boldsymbol{s} \otimes \boldsymbol{s}, \boldsymbol{s} \otimes \boldsymbol{s}\rangle}\right| \leq \sigma_1 \tag{55}$$

with equality if and only if $\boldsymbol{s} \otimes \boldsymbol{s}$ coincides with a corresponding singular vector, $\boldsymbol{\xi}_1$ (or $\boldsymbol{\vartheta}_1 = \pm\boldsymbol{\xi}_1$). This remark would directly provide us with the global solution to the problem of maximizing $h(\boldsymbol{s}) = \frac{f^2(\boldsymbol{s})}{\langle \boldsymbol{s}, \boldsymbol{s}\rangle^n}$ if $\boldsymbol{\xi}_1$ could be written as a "Kronecker square" $\boldsymbol{s} \otimes \boldsymbol{s}$, which is in general not true.

Nonetheless, the above suggests an initialization for $\boldsymbol{s}$ as the best "Kronecker square root" of $\boldsymbol{\xi}_1$ in the LS sense. Since both these vectors have unit norm, this approximation problem amounts to that of maximizing the correlation $\langle \boldsymbol{\xi}_1, \boldsymbol{s} \otimes \boldsymbol{s}\rangle$. If we denote by $\text{unvec}(\boldsymbol{\xi}_1)$ the $M \times M$ matrix whose columns when put one above the other form $\boldsymbol{\xi}_1$, the above functional to be maximized takes the form

$$\langle \boldsymbol{\xi}_1, \boldsymbol{s} \otimes \boldsymbol{s}\rangle = \langle \boldsymbol{s}, \text{unvec}(\boldsymbol{\xi}_1)\boldsymbol{s}\rangle = \langle \text{unvec}(\boldsymbol{\xi}_1), \boldsymbol{s}^{\star 2}\rangle$$

and is thus seen to be equivalent to a rank-1 approximation to the matrix $\text{unvec}(\boldsymbol{\xi}_1)$.

LEMMA 1. *The symmetry relation $\boldsymbol{\xi}_1 = \boldsymbol{P\xi}_1$ is equivalent to the matrix* $\text{unvec}(\boldsymbol{\xi}_1)$ *being symmetric.*

---

[11]It is tacitly assumed here that the singular value σ_i is simple. Nevertheless, it incurs no loss of generality to assume that this relation holds true even when σ_i is associated to more than one left singular vectors, since one can always choose a vector from the space they span so as to satisfy the desired symmetry. An example might be the vector $\boldsymbol{\xi}_i' = \boldsymbol{P\xi}_i + \boldsymbol{\xi}_i$ normalized to unit 2-norm.

Proof: Let us denote by $\boldsymbol{a}_{(M)}$ the vector resulting from the $M^2 \times 1$ vector $\boldsymbol{a}$ by writing first every Mth element of $\boldsymbol{a}$ starting with the first one, then every Mth element starting with the second, etc. It can be shown that $\boldsymbol{a}_{(M)} = \boldsymbol{P}\boldsymbol{a}$ [**27**]. We thus have

$$\mathrm{vec}((\mathrm{unvec}(\boldsymbol{\xi}_1))^T) = (\boldsymbol{\xi}_1)_{(M)} = \boldsymbol{P}\boldsymbol{\xi}_1 = \boldsymbol{\xi}_1 = \mathrm{vec}(\mathrm{unvec}(\boldsymbol{\xi}_1)).$$

◇

Similarly with (55) we have now

$$|\langle \boldsymbol{\xi}_1, \boldsymbol{s} \otimes \boldsymbol{s}\rangle| = \left| \frac{\langle \boldsymbol{s}, \mathrm{unvec}(\boldsymbol{\xi}_1)\boldsymbol{s}\rangle}{\langle \boldsymbol{s}, \boldsymbol{s}\rangle} \right| \leq \varsigma_1 \tag{56}$$

where ς_1 is the largest singular value of the matrix $\mathrm{unvec}(\boldsymbol{\xi}_1)$. Notice that $\varsigma_1^2 \leq \|\mathrm{unvec}(\boldsymbol{\xi}_1)\|_F^2 = \|\boldsymbol{\xi}_1\|_2^2 = 1$. Taking $\boldsymbol{s}$ as the corresponding singular vector of $\mathrm{unvec}(\boldsymbol{\xi}_1)$ equality is achieved in (56) and a reasonable initialization for TPM is obtained. In summary, to obtain the initial vector $\boldsymbol{s}(0)$ for Algorithm 3, two matrix rank-1 approximation problems need to be solved:

New Initialization (Symmetric 4th-order tensor case)

1. Determine the maximal singular vector, $\boldsymbol{\xi}_1$, of the unfolded tensor $\boldsymbol{T}$.
2. Set $\boldsymbol{s}(0)$ equal to the maximal singular vector of the matrix $\mathrm{unvec}(\boldsymbol{\xi}_1)$

Having thus chosen $\boldsymbol{s}$, (56) becomes an equality and hence (53) along with (54) imply:

$$f(\boldsymbol{s}) = \langle \boldsymbol{s} \otimes \boldsymbol{s}, \boldsymbol{T}(\boldsymbol{s} \otimes \boldsymbol{s})\rangle = \varepsilon_1 \sigma_1 \varsigma_1^2 + \sum_{i=2}^{M^2} \varepsilon_i \sigma_i |\langle \boldsymbol{\xi}_i, \boldsymbol{s} \otimes \boldsymbol{s}\rangle|^2$$

where $\varepsilon_i = 1$ if $\boldsymbol{\vartheta}_i = \boldsymbol{\xi}_i$ and $\varepsilon_i = -1$ if $\boldsymbol{\vartheta}_i = -\boldsymbol{\xi}_i$. When $\boldsymbol{T}$ is sign-(semi)definite (a condition frequently met in practice), all sign factors ε_i are equal, and the latter relation implies:

$$|f(\boldsymbol{s})| \geq \sigma_1 \varsigma_1^2. \tag{57}$$

In conjuction with (55), this provides us with lower and upper bounds on the initial value of h:

THEOREM 7 (Bounds on initial value). *The value assumed by h by the suggested initialization is bounded as:*

$$\sigma_1^2 \varsigma_1^4 \leq h(\boldsymbol{s}(0)) \leq \sigma_1^2$$

where σ_1 and ς_1 are the largest singular values of the matrices $\boldsymbol{T}$ and $\mathrm{unvec}(\boldsymbol{\xi}_1)$, respectively. This initial value approaches to the global maximum as the vector $\boldsymbol{\xi}_1$ approaches Kronecker decomposability (i.e., as ς_1 approaches one).

5. The Super-Exponential Algorithm

In light of the developments in the previous section and, in particular, by comparing (29) with the cost function in Theorem 6, it becomes apparent that the blind equalization problem of Section 3.3 is not but a special case of the symmetric tensor approximation problem treated above. Applying then Algorithm 3, with $n = 2p$, to the tensor $\boldsymbol{T}$ of (27)–(28) results in the following algorithm, which is recognized as the SEA for multisource blind deconvolution [**53**]:

Algorithm 4: Super-Exponential Algorithm

Initialization: $\boldsymbol{s}(0)$ = a unit 2-norm vector in $\mathcal{S}_A$.

Iteration: for $k = 1, 2, \ldots$

$$\tilde{s}_i(k) = \sum_j (P_A)_{i,j}(\mathcal{C})_{j,j,\ldots,j} s_j^{2p-1}(k-1)$$

$$\boldsymbol{s}(k) = \text{sign}(J_{2p}(\boldsymbol{s}(k-1))) \frac{\tilde{\boldsymbol{s}}(k)}{\|\tilde{\boldsymbol{s}}(k)\|_2}$$

end

It is shown in [**53**] that the above recursion can be interpreted as a gradient ascent scheme where the step size has been so chosen as to attain the maximum increase of $|J_{2p}|$ at each step. This should be seen in conjuction with the pertinent remark we have done for the symmetric TPM above. Corollary 1 and (49) above verify the soundness of the "nonlinear eigenproblem" interpretation given in Section 3.3 to the problem of maximizing $|J_{2p}|^2$.

Suppose we are in the sufficient-order case ($\boldsymbol{P}_A = \boldsymbol{I}$) and all sources have nonpositive $2p$th-order cumulants. The latter requirement is met invariably in communications systems where all the common modulation formats imply negative 4th-order cumulants for the transmitted signals [**56**]. Then $\boldsymbol{\mathcal{T}}$ becomes equal to $\boldsymbol{\mathcal{C}}$ and

$$J_{2p}(\boldsymbol{s}) = \frac{\langle \boldsymbol{\mathcal{C}}, \boldsymbol{s}^{\star 2p} \rangle}{\langle \boldsymbol{s}, \boldsymbol{s} \rangle^p} = \frac{\sum_i k_{\alpha_i}(2p, 2) s_i^{2p}}{\|\boldsymbol{s}\|_2^{2p}}$$

turns out to have a concave numerator. Hence, for this scenario, Theorem 6 is applied and shows the SEA to be convergent to a local minimum of J_{2p}. In fact, it can be shown that, for this particular case, SEA converges always to the global minimum, given by the perfect sequences (18) [**53**] for any initialization (save, of course, for the exceptional initial conditions mentioned earlier). In this case, the bounds given by Theorem 7 are applied ($\boldsymbol{\mathcal{T}}$ is seminegative definite) and, in fact, they turn out to be tight, i.e., *the initial value suggested by our initialization yields directly the perfect solution.* Further insight into this can be gained by examining the equivalent minimization problem (cf. Theorem 2), namely that of minimizing the norm $\|\boldsymbol{\mathcal{C}} - \sigma \boldsymbol{s}^{\star 2p}\|_F$. Since $\boldsymbol{\mathcal{C}}$ is diagonal, one sees that the optimum $\boldsymbol{s}$ should be of the form (18) where the 1 is located at the position of the source having the largest (in absolute value) cumulant. In other words, the resulting deconvolution filter "passes" the least Gaussian source. The above discussion applies to the undermodeling case as well, except that only convergence to a local minimum is then guaranteed. Moreover, for $\boldsymbol{P}_A \neq \boldsymbol{I}$, the convergence of the SEA is no longer super-exponential [**52**].

Nevertheless, we should emphasize here that, although a convergence proof is still to be devised (for the case $\boldsymbol{P}_A \neq \boldsymbol{I}$), Algorithm 4 has been observed to converge to local extrema of J_{2p} even when the source cumulants are not all of the same sign. To give an illustrative example of the convergence of SEA and the superiority of the novel initialization over that of [**35**] we consider a static blind deconvolution problem modeled as in (14) with the mixing matrix $\boldsymbol{F}$ being of dimensions 10×3; that is, a scenario with many more sources than sensors is adopted. We consider the maximization of $|J_{2p}(\cdot)|^2$ with $p = 2$ and we are not making any assumptions on the signs of the source 4th-order cumulants. This problem corresponds evidently to

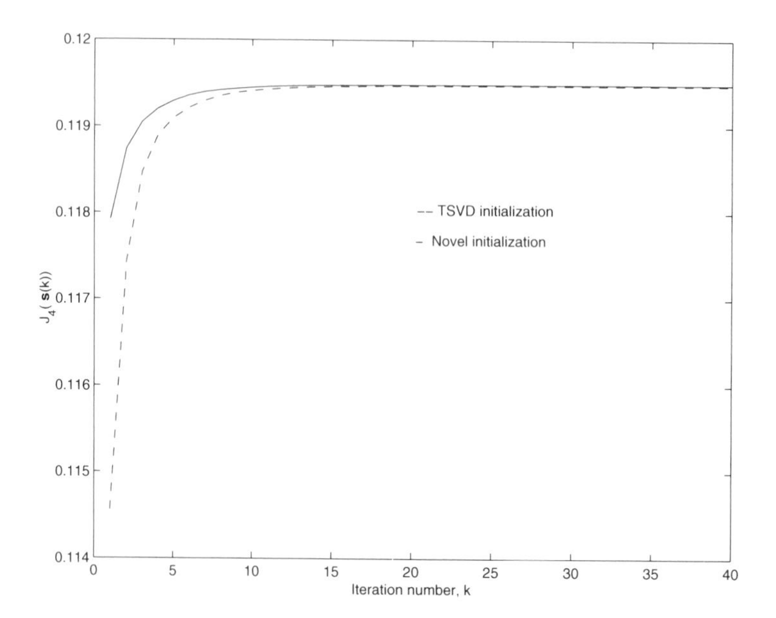

FIGURE 2. Average cost function evolution for the new and TSVD-based initializations.

an undermodeled case and a cost function J_{2p} whose numerator is neither convex nor concave. The evolution of J_4, averaged over 1000 independent Monte-Carlo runs where $\boldsymbol{F}$ and $k_{\alpha_i}(4,2)$ are randomly chosen, is plotted in Fig. 2. The TSVD-based initialization, though being quite close to the global optimum, is seen to be frequently outperformed by the new method.

6. Concluding Remarks

In this semitutorial paper we have attempted to provide a comprehensive exposition of recent advances in tensor decomposition/approximation and related approaches for blind source separation and linear system deconvolution. One of our main goals was to put into evidence the intimate relationship between the super-exponential blind deconvolution algorithm and a symmetric tensor power method developed here. We feel that the results presented may greatly enhance our understanding of the local convergence behavior of minimum-entropy blind estimation algorithms and hopefully pave the way for a fruitful interplay with the domain of tensor approximation.

We have seen that the convergence of the SEA is perfectly understood for sources with cumulants of the same sign. The case of mixed-sign source cumulants is also worthy of investigation as the observed ability of the algorithm to operate in such an environment enhances its applicability in applications such as multimedia communications, mixing e.g. speech (super-Gaussian) and data (sub-Gaussian), and renders the receiver robust against intentional interferences.

The method of initialization we developed here may prove to be a vehicle for an ultimate solution to the local extrema problem. Further theoretical analysis is needed though in order to provide a clear explanation of its observed superiority over the TSVD scheme as well as improvements aiming at reducing its rate of failure of ensuring global convergence (which has been observed to be at least as low as that of the TSVD-based method). Furthermore, fast implementations need to be developed as the computational burden grows quite fast with the problem size. A preliminary step towards this end in a blind deconvolution application can be the exploitation of the Tucker-product structure of the 4th-order cumulant tensor of the system output for reducing the decomposition problem to a 2nd-order one, with subsequent gains in computations and numerical accuracy.

Extensions of the TPM for approximating a tensor with one of given mode-ranks were reported in [**40, 34**] and can be viewed as higher-order analogs of the method of orthogonal iterations for matrices [**24**]. The results presented here could easily be generalized to this problem by replacing the vectors $\boldsymbol{s}^{(i)}$ by matrices with orthonormal columns [**34**]. However, the more interesting problem of approximating a *symmetric* tensor with one of a lower *tensor rank* remains open. The results of [**40, 34**] for a rank-$(r_1, r_2, \ldots, r_n)$ tensor do not generally provide an insight into this problem since there is no known relationship between these two rank notions. An approach which would successively subtract from a tensor its rank-1 approximant would not be successful in general, as pointed out in [**21, 18**] (unless, of course, the tensor is Tucker factorable).

As a final remark, the tensor formulation we have given to the BSS problem, assisted by an elegant and powerful notation, may prove to provide a means of coping with the more challenging problem of analysis of *nonindependent* components.

References

[1] Special issue on "Blind system identification and estimation," *Proc. IEEE,* vol. 86, no. 10, Oct. 1998.

[2] J. W. Brewer, "Kronecker products and matrix calculus in system theory," *IEEE Trans. Circuits and Systems,* vol. 25, no. 9, pp. 772–781, Sept. 1978.

[3] R. Bro, "PARAFAC: Tutorial and applications," *Chemometrics and Intelligent Laboratory Systems,* vol. 38, pp. 149–171, 1997.

[4] R. Bro, N. Sidiropoulos, and G. Giannakis, "A fast least squares algorithm for separating trilinear mixtures," *Proc. ICA'99,* Aussois, France, Jan. 11–15, pp. 289–294.

[5] J. A. Cadzow, "Blind deconvolution via cumulant extrema," *IEEE Signal Processing Magazine,* pp. 24–42, May 1996.

[6] X.-R. Cao and R. Liu, "General approach to blind source separation," *IEEE Trans. Signal Processing,* vol. 44, no. 3, pp. 562–571, March 1996.

[7] J.-F. Cardoso, "Eigen-structure of the fourth-order cumulant tensor with application to the blind source separation problem," *Proc. ICASSP'90,* pp. 2655–2658.

[8] J.-F. Cardoso, "Localisation et identification par la quadricovariance," *Traitement du Signal,* vol. 7, no. 5, pp. 397–406, Dec. 1990.

[9] J.-F. Cardoso, "Super-symmetric decomposition of the fourth-order cumulant tensor – Blind identification of more sources than sensors," *Proc. ICASSP'91,* Toronto, Canada, pp. 3109–3112.

[10] J.-F. Cardoso, "A tetradic decomposition of 4th-order tensors: Application to the source separation problem," pp. 375–382 in [**49**].

[11] J.-F. Cardoso, "Blind signal separation: Statistical principles," pp. 2009–2025 in [**1**].

[12] J.-F. Cardoso and P. Comon, "Independent component analysis: A survey of some algebraic methods," *Proc. ISCAS'96,* Atlanta, May 1996, pp. 93–96.

[13] J.-F. Cardoso and A. Souloumiac, "Blind beamforming for non-Gaussian signals," *Proc. IEE, Pt. F,* vol. 140, no. 6, pp. 362–370, Dec. 1993.

[14] P. Comon, "Remarques sur la diagonalization tensorielle par la méthode de Jacobi," *Proc. 14th Colloque GRETSI,* Juan les Pins, France, Sept. 13–16, 1993.

[15] P. Comon, "Independent component analysis, a new concept?," *Signal Processing,* vol. 36, pp. 287–314, 1994.

[16] P. Comon, "Tensor diagonalization: A useful tool in signal processing," *Proc. 10th IFAC Symposium on System Identification,* Copenhagen, Denmark, July 4–6, 1994, pp. 77–82.

[17] P. Comon, "Contrasts for multichannel blind deconvolution," *IEEE Signal Processing Letters,* vol. 3, no. 7, pp. 209–211, July 1996.

[18] P. Comon, "Blind channel identification and extraction of more sources than sensors," *Proc. SPIE Conf. Advanced Signal Processing VIII,* San Diego, pp. 2–13, July 22–24, 1998.

[19] P. Comon and B. Mourrain, "Decomposition of quantics in sums of powers of linear forms," *Signal Processing,* vol. 53, pp. 96–107, 1996.

[20] R. Coppi and S. Bolasco (Eds.), *Multiway Data Analysis,* Elsevier Science Publishers B.V. (North Holland), 1989.

[21] J.-B. Denis and T. Dhorne, "Orthogonal tensor decomposition of 3-way tables," pp. 31–37 in [**20**].

[22] D. Donoho, "On minimum entropy deconvolution," pp. 565–609 in D. F. Findley (ed.), *Applied Time Series Analysis II,* Academic, 1981.

[23] A. Franc, "Multiway arrays: Some algebraic remarks," pp. 19–29 in [**20**].

[24] G. H. Golub and C. F. Van Loan, *Matrix Computations,* Johns Hopkins University Press, 1989.

[25] W. H. Greub, *Multilinear Algebra,* Springer-Verlag, Berlin, 1967.

[26] V. S. Grigorascu and P. A. Regalia, "Tensor displacement structures and polyspectral matching," Chap. 9 in T. Kailath and A. H. Sayeed (eds.), *Fast, Reliable Algorithms for Matrices with Structure,* SIAM Publications, Philadelphia, PA, 1999.

[27] H. V. Henderson and S. R. Searle, "The vec-permutation matrix, the vec operator and Kronecker products: A review," *Linear and Multilinear Algebra,* vol. 9, pp. 271–288, 1981.

[28] A. Hyvärinen, "Survey on independent component analysis," *Neural Computing Surveys,* vol. 2, pp. 94–128, 1999.

[29] C. R. Johnson et al., "Blind equalization using the constant modulus criterion: A review," pp. 1927–1950 in [**1**].

[30] C. R. Johnson et al., "The core of FSE-CMA behavior theory," Chap. 2 in S. Haykin (ed.), *Unsupervised Adaptive Filtering, Vol. II,* John Wiley & Sons, 2000.

[31] P. M. Kroonenberg, "Singular value decompositions of interactions in three-way contingency tables," pp. 169–184 in [**20**].

[32] J. B. Kruskal, "Three-way arrays: Rank and uniqueness of trilinear decompositions, with application to arithmetic complexity and statistics," *Linear Algebra and Its Applications,* vol. 18, pp. 95–138, 1977.

[33] J. B. Kruskal, "Rank, decomposition, and uniqueness for 3-way and N-way arrays," pp. 7–18 in [**20**].

[34] L. De Lathauwer, *Signal Processing Based on Multilinear Algebra,* Ph.D. Dissertation, K.U.Leuven, Sept. 1997.

[35] L. De Lathauwer, P. Comon, B. De Moor, and J. Vandewalle, "Higher-order power method – Application in independent component analysis," *Proc. Int'l Symposium on Nonlinear Theory and its Applications (NOLTA'95),* Las Vegas, Dec. 10–14, 1995.

[36] L. De Lathauwer and B. De Moor, "From matrix to tensor: Multilinear algebra and signal processing," *Proc. 4th Int'l Conf. on Mathematics in Signal Processing,* Warwick, UK, Part I, pp. 1–11, Dec. 17–19, 1996.

[37] L. De Lathauwer, B. De Moor, and J. Vandewalle, "A singular value decomposition for higher-order tensors," *Proc. ATHOS Workshop on System Identification and Higher-Order Statistics,* Nice, France, Sept. 20–21, 1993.

[38] L. De Lathauwer, B. De Moor, and J. Vandewalle, "The application of higher order singular value decomposition to independent component analysis," pp. 383–390 in [**49**].

[39] L. De Lathauwer, B. De Moor, and J. Vandewalle, "Blind source separation by simultaneous third-order tensor diagonalization," *Proc. EUSIPCO'96,* Trieste, Italy, Sept. 1996, pp. 2089–2092.

[40] L. De Lathauwer, B. De Moor, and J. Vandewalle, "Dimensionality reduction in higher-order-only ICA," *Proc. IEEE Signal Processing Workshop on Higher-Order Statistics,* Banff, Alberta, Canada, July 21–23, 1997, pp. 316–320.

[41] S. E. Leurgans, R. T. Ross, and R. B. Abel, "A decomposition for three-way arrays," *SIAM Journal on Matrix Analysis and Applications,* vol. 14, no. 4, pp. 1064–1083, Oct. 1993.

[42] Y. Li and K. J. R. Liu, "Adaptive blind source separation and equalization for multiple-input/multiple-output systems," *IEEE Trans. Information Theory,* vol. 44, no. 7, pp. 2864–2876, Nov. 1998.

[43] A. P. Liavas, P. A. Regalia, and J.-P. Delmas, "Robustness of least-squares and subspace methods for blind channel identification/equalization with respect to channel undermodeling," *IEEE Trans. Signal Processing,* vol. 47, no. 6, pp. 1636–1645, June 1999.

[44] U. Madhow, "Blind adaptive interference suppression for direct-sequence CDMA," pp. 2049–2069 in [**1**].

[45] N. E. Mastorakis, "A method of approximate multidimensional factorization via the singular value decomposition," *Proc. DSP'97,* Santorini, Greece, July 1997, pp. 883–888.

[46] M. Mboup and P. A. Regalia, "On the equivalence between the super-exponential algorithm and a gradient search method," *Proc. ICASSP'99,* Phoenix, AZ, May 1999.

[47] P. McCullagh, *Tensor Methods in Statistics,* Chapman and Hall, NY, 1987.

[48] J. M. Mendel, "Tutorial on higher-order statistics (spectra) in signal processing and system theory: Theoretical results and some applications," *Proc. IEEE,* vol. 79, no. 3, pp. 278–305, March 1991.

[49] M. Moonen and B. De Moor (eds.), *SVD and Signal Processing, III: Algorithms, Architectures and Applications,* Elsevier Science B.V., 1995.

[50] C. L. Nikias and J. M. Mendel, "Signal processing with higher-order spectra," *IEEE Signal Processing Magazine,* pp. 10–37, July 1993.

[51] P. A. Regalia, "On the equivalence between the Godard and Shalvi-Weinstein schemes of blind equalization," *Signal Processing,* vol. 73, pp. 185–190, Feb. 1999.

[52] P. A. Regalia and M. Mboup, "Undermodeled equalization: A characterization of stationary points for a family of blind criteria," *IEEE Trans. Signal Processing,* vol. 47, no. 3, pp. 760–770, March 1999.

[53] P. A. Regalia and M. Mboup, "Undermodeled equalization of noisy multi-user channels," *Proc. IEEE Workshop on Nonlinear Signal and Image Processing,* Antalya, Turkey, June 1999.

[54] R. T. Rockafellar, *Convex Analysis,* Princeton University Press, 1970.

[55] M. Schmutz, "Optimal and suboptimal separable expansions for 3D-signal processing," *Pattern Recognition Letters,* vol. 8, pp. 217–220, Nov. 1988.

[56] O. Shalvi and E. Weinstein, "New criteria for blind deconvolution of nonminimum phase systems (channels)," *IEEE Trans. Information Theory,* vol. 36, pp. 312–321, March 1990.

[57] O. Shalvi and E. Weinstein, "Super-exponential methods for blind deconvolution," *IEEE Trans. Information Theory,* vol. 39, pp. 504–519, March 1993.

[58] N. D. Sidiropoulos and R. Bro, "On communication diversity for blind identifiability and the uniqueness of low-rank decompositions of N-way arrays," *Proc. ICASSP-2000,* Istanbul, Turkey, June 2000.

[59] L. Tong et al., "Indeterminacy and identifiability of blind identification," *IEEE Trans. Circuits and Systems,* vol. 38, no. 5, pp. 499–509, May 1991.

[60] A.-J. van der Veen, "Algebraic methods for deterministic blind beamforming," pp. 1987–2008 in [**1**].

[61] K. L. Yeung and S. F. Yau, "A cumulant-based super-exponential algorithm for blind deconvolution of multi-input multi-output systems," *Signal Processing,* vol. 67, pp. 141–162, 1998.

E-mail address: `kofidis@@sim.int-evry.fr`

DÉPT. SIGNAL & IMAGE, INSTITUT NATIONAL DES TÉLÉCOMMUNICATIONS, 9, RUE CHARLES FOURIER, F-91011 EVRY CEDEX FRANCE

E-mail address: `Phillip.Regalia@@int-evry.fr`

Contemporary Mathematics
Volume **280**, 2001

Exploiting Toeplitz-like structure in adaptive filtering algorithms using signal flow graphs

I.K. Proudler

ABSTRACT. In signal processing, adaptive filtering algorithms based on recursive least squares minimization are common. The solution to the problem involves explicitly or implicitly inverting a matrix (cf. the normal equations). When this matrix has a Toeplitz structure it is possible to reduce the amount of computation need to solve the problem by a factor proportional to the number of unknown parameters. If the matrix has some other Toeplitz-like structure such as Block-Toeplitz it is usually assumed that a suitably reduced complexity algorithm is also possible. Such algorithm derivations, were they exist, are usually based on linear algebra but rapidly become complicated and hence can be difficult to follow. In this paper we describe an alternative approach that is based on the manipulation of signal flow graphs that is visual yet mathematically rigorous. As an example, the derivations of the multi-channel least squares lattice and the fast transversal filters algorithms are discussed. Finally an attempt to find an efficient algorithm for a problem with a Toeplitz-block-Toeplitz structure is considered.

1. Introduction

In signal processing, one often comes across adaptive filtering problems that can be cast as least squares minimization problems. In many cases, the normal equations (equation (2.4)) that govern the solution to the problem contain a matrix that has a Toeplitz or Toeplitz-like structure. It has long been known[**ChK, H**] that if this matrix is Toeplitz then the amount of computation need to solve the problem can be reduced by a factor proportional to the number of unknown parameters. If the problem is solved once for a given 'block' of data, one has the Levinson[**LD**] or Schur algorithms[**Ka**]. If the problem is to be solved continuously as new data arrives, the recursive least squares (RLS) algorithms like the Fast Transversal Filters and Least Squares Lattices algorithms result[**H**].

1991 *Mathematics Subject Classification.* Primary 47B35, 65Y05, 93E24, 65F20; Secondary 60G35, 62M10, 62M20, 93E11, 62M15, 60G25.

This work was carried out as part of Technology Group TG10 of the MoD Corporate Research Programme.

In other situations, the matrix in the Normal equations could have a Toeplitz-like structure such as Block-Toeplitz or Toeplitz-Block-Toeplitz. It is usually assumed that this structure can also be used to obtain a reduced complexity algorithm. The task is then to find a way to use the structure to generate the algorithm. In the literature one usually finds algorithm derivations, were they exist, are based on linear algebra and their complexity is such that many signal processing engineers find it difficult to follow them. In this paper we describe an alternative to the matrix algebra approach for the RLS problem that is based on the manipulation of signal flow graphs(SFG)[**Ku**]. This leads to a very visual approach in which it is easy to see how the various different parts of the algorithm interact. In particular it is easy to see how the Toepltiz structure can be used to transform the SFG into one which has reduced complexity. Because all of the parts of the SFG are rigorously defined, any transformation of the SFG is well defined and the resulting algorithm is mathematically equivalent to the orginal. As an example, the derivations of the multi-channel least squares lattice[**HSMP**] and the fast transversal filters algorithms[**Pr**] are discussed. Finally an attempt to find an efficient algorithm for a problem with a Toeplitz-block-Toeplitz structure is considered.

2. Adaptive Filtering

Adaptive filtering problems are ubiquitous in signal processing: for instance channel equalization, signal modelling, radar and sonar processing, noise and echo cancellation. For the finite impulse response case with real signals, this is equivalent [**H**] to the problem of the least squares estimation of the scalar $y(n)$ by a linear combination of the p components of a vector $\underline{x}_p(n)$ containing entries based on the signal(s) to be filtered:

$$y(n) \approx \underline{x}_p^T(n)\underline{\omega}_p(n) \tag{2.1}$$

The (p dimensional) vector of optimum coefficients (or filter weights), at time n, $\underline{\omega}_p(n)$ is determined by

$$\underset{\underline{\omega}_p(n)}{Min}(\|\underline{y}(n) - X_p(n)\underline{\omega}_p(n)\|_2^2) \tag{2.2}$$

where $X_p(n)$ (the 'data' matrix) is a $n \times p$ matrix whose rows consist of the vectors $\underline{x}_p(m)$ $(1 \leq m \leq n)$, and $\underline{y}(n)$, the so-called 'desired signal' vector, is given by

$$\underline{y}(n) = \begin{bmatrix} y(1) & \ldots & y(n-1) & y(n) \end{bmatrix}^T \tag{2.3}$$

and may be fictitious in the sense that it, and $X_p(n)$, are derived from the actual data via a transformation. Alternatively it may consist of data that, for a good physical reason, is known to be linearly dependent on the data in $X_p(n)$.

The solution to equation (2.2) may be found by solving the normal equations:

$$X_p^T(n)X_p(n)\underline{\omega}_p(n) = X_p^T(n)\underline{y}(n) \tag{2.4}$$

assuming that $X_p(n)$ has full column rank. Inverting the 'covariance' matrix $X_p^T(n)X_p(n)$ requires $O(p^3)$ operations in general which is prohibitive if this has to be done every time instance. It is, however, possible to use the solution to the least squares problem at time $(n-1)$ to reduced the amount of computation needed to solve the problem at time n - hence 'recursive' least squares (RLS) minimization. There are many approaches to RLS minimization (see Haykin[**H**]) however it is convenient for our purposes to consider one based on the QR decomposition (QRD) of the data matrix. Let the QR decomposition of $X_p(n)$ be

$$X_p(n) = Q_p^T(n)\begin{bmatrix} R_p(n) \\ 0 \end{bmatrix} \tag{2.5}$$

where $Q_p(n)$ is a $n \times n$ orthogonal matrix and $R_p(n)$ is a $p \times p$ upper triangular matrix. Then the solution to equation (2.2) is given by

$$R_p(n)\underline{\omega}_p(n) = \underline{u}(n) \tag{2.6}$$

where $\underline{u}(n)$ consists of the upper p components of the vector $Q_p^T(n)\underline{y}(n)$ (see equation 4.1). Note that $R_p(n)$ is a Cholesky factor of the covariance matrix since

$$X_p^T(n)X_p(n) = R_p^T(n)Q_p(n)Q_p^T(n)R_p(n) = R_p^T(n)R_p(n) \tag{2.7}$$

The utility of this QR decomposition-based approach is that it leads to an efficient time recursive, parallel implementation[**McW**]. Consider equation (2.5) and note that

$$X_p(n) = \begin{bmatrix} X_p(n-1) \\ \underline{x}_p^T(n) \end{bmatrix} \tag{2.8}$$

so that we can write

$$X_p(n) = \hat{Q}_p^T(n)\begin{bmatrix} Q_p^T(n-1) & \underline{0} \\ \underline{0}^T & 1 \end{bmatrix}\begin{bmatrix} X_p(n-1) \\ \underline{x}_p^T(n) \end{bmatrix} = \hat{Q}_p^T(n)\begin{bmatrix} R_p(n-1) \\ \underline{x}_p^T(n) \end{bmatrix} \tag{2.9}$$

Updating $R_p(n-1)$ to $R_p(n)$ then only requires $O(p^2)$ operations per time instance. The determination of the least squares weights requires a back- substitution operation (equation (2.6)). However in signal processing problems the least squares residual is sometimes of more interest than the filter weights and it is possible to calculate it directly[**McW**] from the Cholesky factor update operation (equation (2.9)).

In the simple case of a single channel FIR filter, the rows of $X_p(n)$ consist of time delayed samples of a signal $x(n)$ (say):

$$X_p(n) = \begin{bmatrix} \cdots & \cdots & \cdots & \cdots \\ x(n-1) & x(n-2) & \cdots & x(n-p) \\ x(n) & x(n-1) & \cdots & x(n-p+1) \end{bmatrix} \tag{2.10}$$

and $X_p(n)$ is Toeplitz and $X_p^T(n)X_p(n)$ is (symmetric) Toeplitz. If we have a multi-channel filter, with possibly a different number of filter taps for each channel, then the rows of $X_p(n)$ consist of time delayed samples of several signals $x_i(n)$ $(0 \le i \le q-1)$. If there are a different number r_i of delays for each signal then:

(2.11)

$$X_p(n) = \begin{bmatrix} \cdots & \cdots & \cdots & \cdots & \cdots & \cdots & \cdots \\ x_0(n-1) & \cdots & x_0(n-r_0) & x_1(n-1) & \cdots & x_1(n-r_1) & \cdots \\ x_0(n) & \cdots & x_0(n-r_0+1) & x_1(n) & \cdots & x_1(n-r_1+1) & \cdots \end{bmatrix}$$

in which case $X_p(n)$ is 'column-block-Toeplitz' i.e.

$$X_p(n) = \begin{bmatrix} A_0 & A_1 & \ldots & A_{q-1} \end{bmatrix} \tag{2.12}$$

where A_i is a $n \times r_i$ $(0 \leq i \leq q-1)$ Toeplitz matrix. The covariance matrix $X_p^T(n)X_p(n)$ is then a (symmetric) matrix with rectangular Toeplitz blocks:

$$X_p^T(n)X_p(n) = \begin{bmatrix} A_0^T A_0 & A_0^T A_1 & \dots & A_0^T A_{q-1} \\ A_1^T A_0 & A_1^T A_1 & \dots & A_1^T A_{q-1} \\ \dots & \dots & \dots & \dots \\ A_{q-1}^T A_0 & A_{q-1}^T A_1 & \dots & A_{q-1}^T A_{q-1} \end{bmatrix}. \tag{2.13}$$

The matrix products $(A_i^T A_j)$ $(0 \leq i, j \leq q-1)$ are Toeplitz under the assumption that the signals $x_i(n)$ $(0 \leq i \leq q-1)$ are wide-sense satationary i.e. the (k, l) element of $(A_i^T A_j)$ is

$$[A_i^T A_j]_{k,l} = \sum_{m=1}^{n} x_i(m-k+1)x_j(m-l+1) = \rho_{i,j}(k-l) \tag{2.14}$$

where $\rho_{i,j}(\tau)$ is the cross-correlation function for $x_i(n)$ and $x_j(n)$.

In image processing, one can have a 2-D filtering problem (actually linear prediction) in which the data vector $\underline{x}_p(n)$ consists of samples from a rectangular region ($q \times t$ pixels say) from an image (of width d pixels). If the image is considered to be generated from a time series $x(n)$ by the usual raster scanning process i.e. the pixel at (i, j) is $x((i-1)d+j-1)$, the data matrix $X_p(n)$ is 'column-block-Toeplitz' but these column-blocks are related to each other by a shift of d rows i.e.

$$X_p(n) = \begin{bmatrix} A(0) & A(d) & \dots & A(q-1) \end{bmatrix} \tag{2.15}$$

where

$$A(m) = \begin{bmatrix} \dots & \dots & \dots \\ x(n-md-1) & \dots & x(n-md-t) \\ x(n-md) & \dots & x(n-md-t+1) \end{bmatrix} \tag{2.16}$$

The matrix $X_p^T(n)X_p(n)$ is then a (symmetric) block-Toeplitz matrix with Toeplitz blocks:

(2.17)

$$X_p^T(n)X_p(n) = \begin{bmatrix} A^T(0)A(0) & A^T(0)A(1) & \dots & A^T(0)A(q-1) \\ A^T(1)A(0) & A^T(1)A(1) & \dots & A^T(1)A(q-1) \\ \dots & \dots & \dots & \dots \\ A^T(q-1)A(0) & A^T(q-1)A(1) & \dots & A^T(q-1)A(q-1) \end{bmatrix}$$

where

$$[A^T(i)A(j)]_{k,l} = \sum_{m=1}^{n} x(m-id-k+1)x(m-jd-l+1) = \rho(k-l+(i-j)d) \tag{2.18}$$

where now $\rho(\tau)$ is the auto-correlation function for $x(n)$.

It is well known that a Toeplitz structure can be exploited to produce various $O(p)$ algorithms - the so called 'fast' RLS algorithms[**CMK, CiK, F, Li, PMS, Pr**]. In the case of the multi-channel problem, the special case of equal numbers of taps is well known and the block-Toeplitz structure leads[**Le, MP93, Y**] to an $O(rq^2)$ algorithm (assuming q blocks of size r). More recently[**GlK1, GlK2, HSMP**] this has been generalized to different numbers of taps in the different channels. The conventional approach to the derivation of such algorithms is via linear algebra but rapidly become very complex so that some signal processing engineers find then difficult to understand. Next we describe an alternative approach that is based on the manipulation of signal flow graphs[**Ku**].

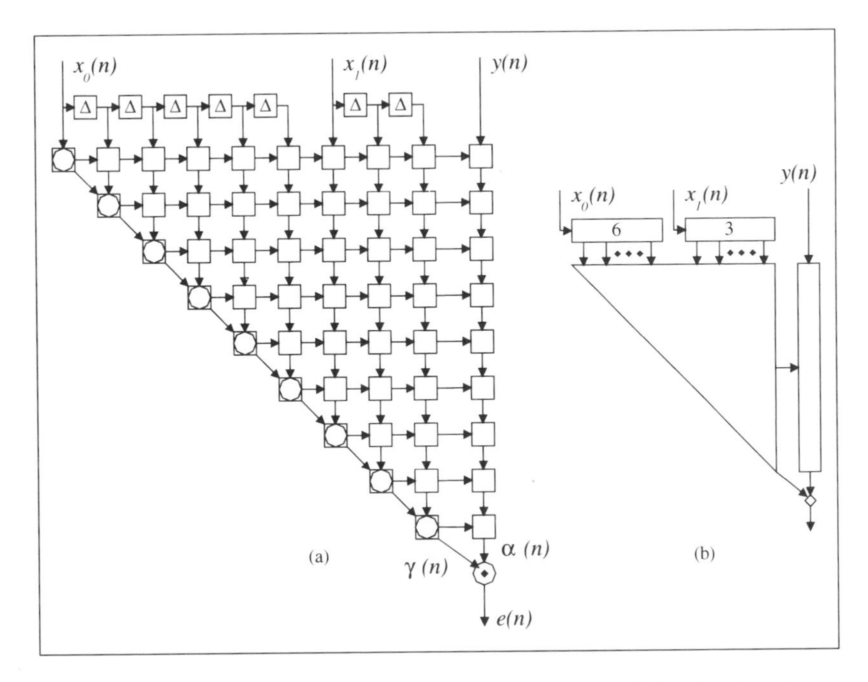

FIGURE 1. SFG for a two channel QRD-based RLS adaptive filtering algorithm

3. Signal Flow Graphs

A signal flow graph (SFG)[**Ku**] is a graphical representation of the operation of an algorithm. It consists of a set of operations, usually represented by geometrical symbols (squares, circles etc.) interconnected by lines that represent the data flow. Despite the similarity with engineering tools such as circuit diagrams, each operator block represents an instantaneous transformation of the inputs to the outputs - in essence a mathematical operator. Figure 1a shows a SFG for a two channel QRD-based RLS adaptive filtering algorithm (with 6 and 3 taps respectively). Each of the small 'cells' represent the transformations shown in figure 2. In practice, of course, the time required for each cell in figure 1 to perform its operations will not be negligible and it may be necessary to introduce some form of pipelining in order to be able to implement the circuit. However, the corresponding timing details can easily be inserted using the "cut" theorem at a later stage of the design process [**Ku**], and so they are omitted from the specification of basic building blocks. For example, the triangular systolic array proposed by Gentleman and Kung [**GeK**] may be derived very simply from the signal flow graph in figure 1a by introducing a pipeline cut between each diagonal row of processors. Where each cut crosses a data interconnection line, the circuit (a systolic array) simply requires a data storage or delay element. The underlying signal flow graph is thus sufficient to describe both the algorithm and architecture.

The SFG in figure 1a can also be considered to be a (more complex) transformation of inputs to outputs as shown in figure 1b. The tapped delay lines are represented in the obvious way by the rectangular block defined in figure 3 - see section 4 for more details. Hence the operations represented by a given block can

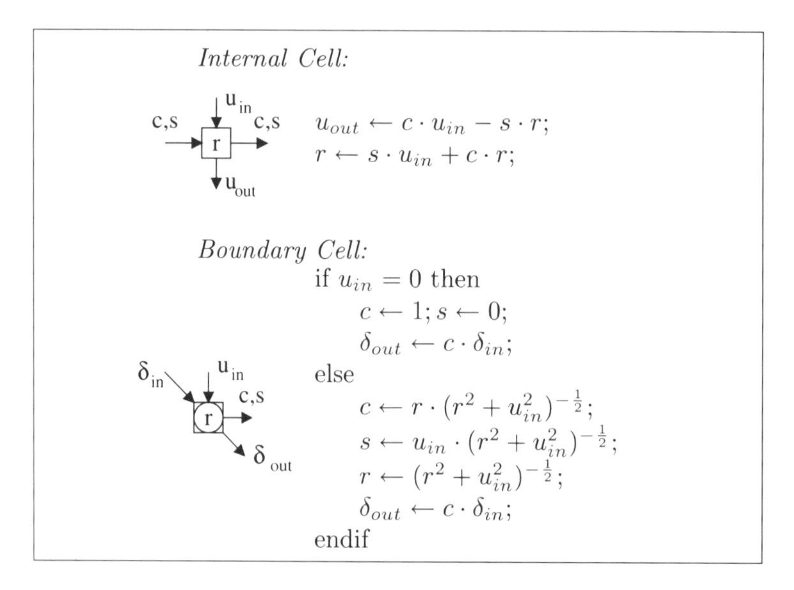

FIGURE 2. QRD-based RLS cells.

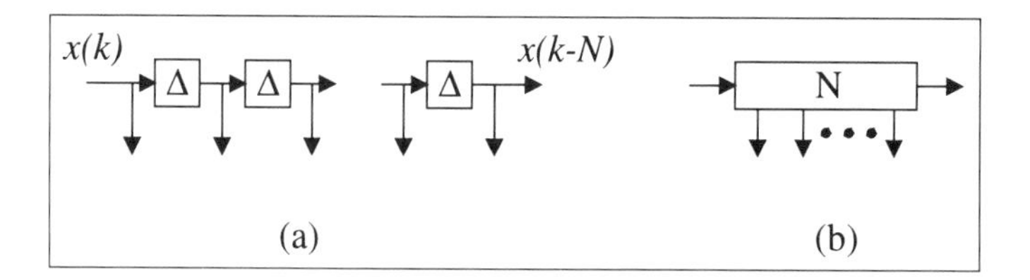

FIGURE 3. Representation of a Tapped Delay Line

be very complex and clearly must be properly defined if the SFG is to meaningful. By taking due account of these input-output transformations it is possible to manipulate the SFG in a mathematically rigorous manner. Hence for example it is clear that the two SFGs in figure 4 are identical since multiplication is distributive over addition. More significantly, it is possible to transform more complicated SFGs and in the process develop algorithms that have more desirable properties. In the following we outline the approach that has led to two fast RLS algorithms (for Toeplitz and block-Toeplitz structure respectively): the multi-channel RLS lattice[**HSMP**] and the QRD-based FTF algorithms[**Pr**]. The reader is referred to the appropriate reference for more details.

4. Multi-Channel RLS lattice[HSMP]

A RLS problem can be solved[**McW**] by recursively updating the QR factorization of the data matrix $X_p(n)$ (or the Cholesky factor of the covariance matrix $X_p^T(n)X_p(n)$). The SFG for such an algorithm is shown in figure 1. The triangular array of cells stores the Cholesky factor at time $(n-1)$: $R_p(n-1)$. The new data $\underline{x}_p^T(n)$ (see equation (2.8)) is input from the top and progressively eliminated by rotating it with each row of the stored triangular matrix $R_p(n-1)$ in turn. The appropriate rotation parameters are computed within each boundary cell and passed on to the internal cells in the same row to complete the rotation process

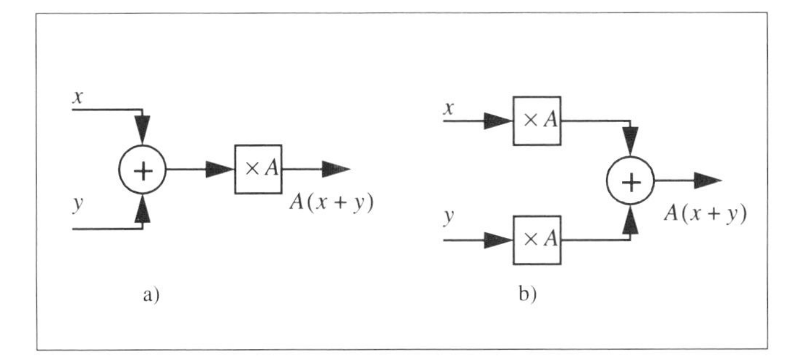

FIGURE 4. Simple SFG Transformation

(see figure 2). The updated triangular matrix $R_p(n)$ is computed in the course of eliminating the vector $\underline{x}_p^T(n)$ and stored within the array ready for the next input data cycle. In a similar manner, the right hand column of cells in the SFG evaluates and stores the p-element vector $\underline{u}(n)$ defined by

$$\hat{Q}^T(n)\begin{bmatrix}\underline{u}(n-1)\\ y(n)\end{bmatrix} = \begin{bmatrix}\underline{u}(n)\\ \alpha(n)\end{bmatrix} \tag{4.1}$$

The optimum weight vector is given by equation (2.6) but it is also well known[**McW**] that the scalar $\alpha(n)$ which emerges from the bottom cell in the right hand column of figure 1 is related to the least squares residuals as follows:

$$e(n) = y(n) + \underline{x}_p^T(n)\underline{\omega}(n) = \gamma(n)\alpha(n) \tag{4.2}$$

where $\gamma(n)$ is known as the likelihood variable and is given by

$$\gamma(n) = \hat{Q}^T(n)\begin{bmatrix}0 & \dots & 0 & 1\end{bmatrix}^T \tag{4.3}$$

and can be shown to be the product of the cosines of the elementary 2×2 rotations that make up $\hat{Q}(n)$ - see figure 2.

Figure 1 represents a multi-channel filtering algorithm and hence the input data vector $\underline{x}_p^T(n)$ is composed of r_i time delayed samples of several signals $x_i(n)$ $(0 \le i \le q-1)$ (equation (2.11)) - in figure 1 we assume $q = 2$, $r_0 = 6$ and $r_1 = 3$. The SGF representation of the generation of such a vector consists of q tapped delay lines of length r_i respectively. Clearly the operation count for figure 1 is $O((r_0+r_1)^2)$ per time instance since the triangular part of the SFG has dimension $(r_0+r_1)\times(r_0+r_1)$. Ideally we would like to use the Toeplitz-like structure to reduce this operation count in particular to remove the quadratic dependancy.

The block-Toeplitz structure of the data matrix $X_p(n)$ is a direct consequence of the fact that the data flowing into the RLS algorithm comes from a tapped delay line. Exploring the properties of the SFG in relation to the tapped delay line is thus the key to developing a more efficient algorithm. Figure 5 shows a QRD-RLS operator with a tapped delay line of length J at its input. Consider now a transformation of this operator that will prove useful in generating the fast algorithm. Recalling that the triangular 'block' is a RLS transformation (figure 1), it is easy to see that we can 'separate off' the right most column of cells. Furthermore since the least squares residual is invariant to a permutation of the inputs we can permute the inputs to the RLS operator without changing the output. Hence in figure 5b

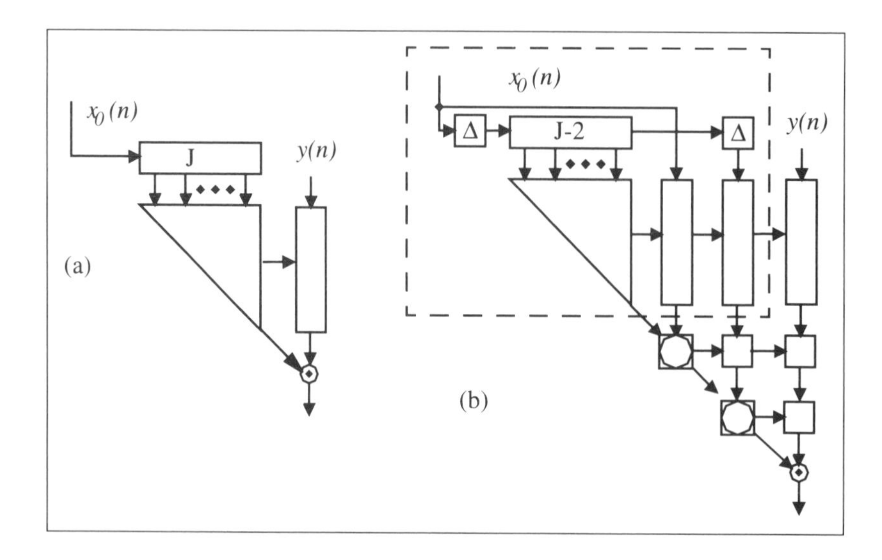

FIGURE 5. Transformation of the QRD-RLS Operator

we show the undelayed input sample appearing on the right of the other inputs. The tapped delay line is shown appropriately segmented.

That part of the SFG in figure 5b inside the dotted box can be seen to represent a $(J-2)$th order RLS linear prediction algorithm (cf. figure 5a). The right most column calculates the backward residual and the one to its left calculates the forward residual. This should not be unexpected since it is well known that forward and backward linear prediction are intimately involved with the development of fast RLS algorithms. Since there are now two prediction problems being solved in the SFG of figure 5b we could imagine separating them (by duplicating the $(J-2)$th order RLS operator). The backward problem would them look as shown in figure 6a. In this SFG note that the input signal passes thorough a delay element. Now it is easy to see that the output of a algorithm that is fed with a delayed input signal is the same as delaying the output from an algorithm whose input is not delayed. Hence one can transform the SFG by moving the delay from the input to the outputs. One has to take care here since there are stored parameters in the SFG (see figure 2). However if it is assumed that these parameters are initialised at zero then the transformation is still valid.

Using such elementary manipulations, one can transform the SFG of the multichannel least squares lattice algorithm (figure 1) can be transformed into that shown in figure 7. Here we see that the multi- channel filtering problem of order $(6,3)$ has been transformed into one of order $(5,2)$ plus some additional operations (below the line CD). Note that the small triangular blocks in figure 7 also represent QRD-RLS operators but of reduced order. We use the convention that the order of the RLS problem is determined from the number of inputs.

Having now reduced the $(6,3)$ problem to one of order $(5,2)$, this order reduction can be iterated until one of the input signal no longer has any taps left ($(3,0)$ in this case). At this point it is possible to transform the SFG in a similar manner to that discussed above so that the input signal with no taps is 'moved' from the input to triangular operator to a separate right hand column (cf. the signal $y(n)$

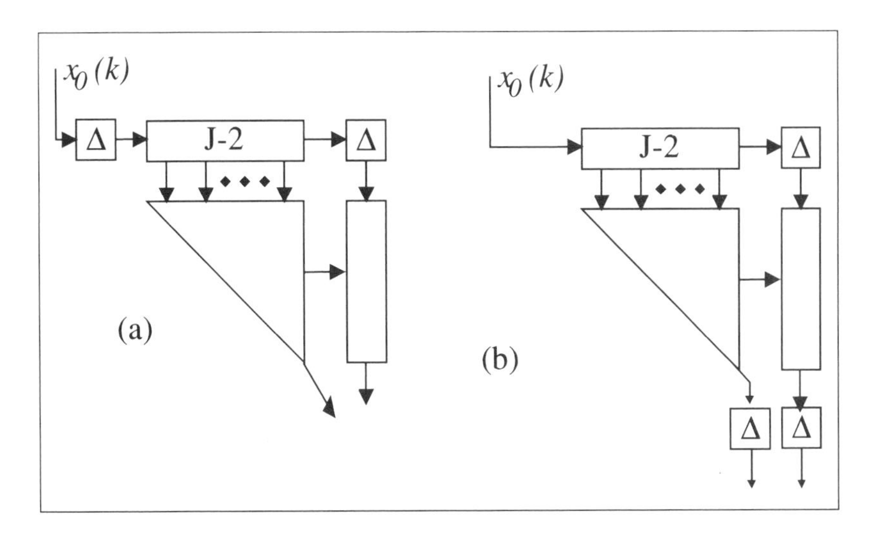

FIGURE 6. Transformation of the Backward Linear Prediction SFG

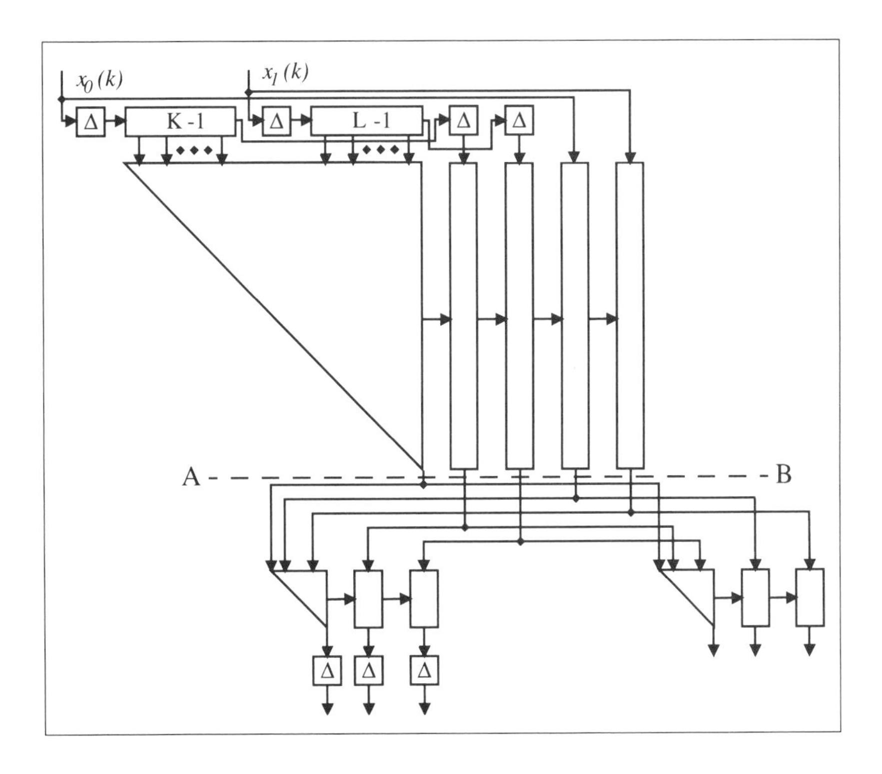

FIGURE 7. Partially Transformed Multi-Channel Latiice SFG

in figure 1). The order reduction on the remaining signals can proceed as before. Ultimately a reduced complexity SFG results (figure 8). The operation count for figure 1 is $O((r_0+r_1)^2)$ whereas that for figure 8 is clearly $O(k_0r_0+k_1r_1)$, for some constants k_1, k_2.

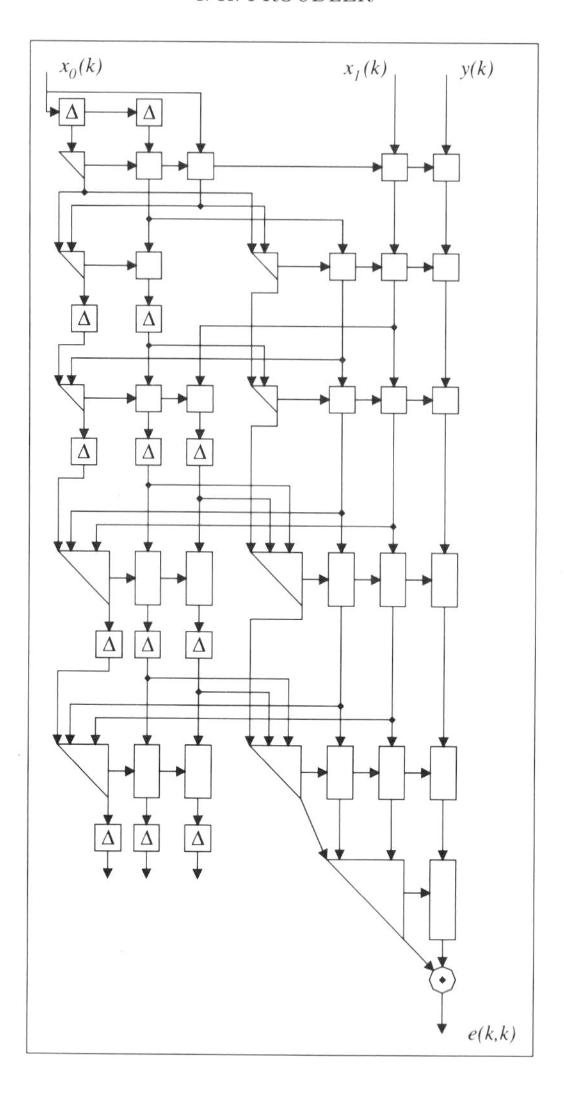

FIGURE 8. Multi-channel LSL Algorithm

5. QRD-Based FTF Algorithm[Pr]

Our second example is a single channel algorithm but this time it is based not on the QRD-RLS algorithm as above but on the so-called Inverse QRD-RLS algorithm[**A, N, PP**]. This algorithm is based on the update in time of the inverse of the Cholesky factor $R_p(n)$ (equation (2.5)). It can be shown that

$$\hat{Q}_p(n) \begin{bmatrix} R_p^{-T}(n-1) \\ \underline{0}^T \end{bmatrix} = \begin{bmatrix} R_p^{-T}(n) \\ \underline{\kappa}_p^T(n) \end{bmatrix} \tag{5.1}$$

where the matrix $\hat{Q}_p(n)$ is that in equation (2.9) and $\underline{\kappa}_p(n)$ is related to the Kalman gain vector[1]:

$$\underline{k}_p(n) = X_p^T(n)X_p(n)\underline{x}_p(n) = \gamma_p^{-1}(n)\underline{\kappa}_p(n). \tag{5.2}$$

Further more the matrix $\hat{Q}_p(n)$ can be reconstructed entirely from knowledge of the vector

$$\underline{\tilde{e}}_p(n) = R_p^{-T}(n-1)\underline{x}_p(n) \tag{5.3}$$

since

$$\hat{Q}_p(n)\begin{bmatrix} -\underline{\tilde{e}}_p(n) \\ 1 \end{bmatrix} = \begin{bmatrix} \underline{0} \\ \gamma_p^{-1}(n) \end{bmatrix} \tag{5.4}$$

Thus given $R_p^{-T}(n-1)$ and the new data at time n ($\underline{x}_p(n)$), we may use equation (5.3) to calculate the vector $\underline{\tilde{e}}_p(n)$ and hence infer the value of $\hat{Q}_p(n)$ and so update the matrix $R_p^{-T}(n-1)$.

It is possible to use the relationship of the vector $\underline{\kappa}_p(n)$ with the Kalman gain vector, in the conventional way, to construct a time update equation for the optimum coefficients [**A**]:

$$\underline{\omega}_p(n) = \underline{\omega}_p(n-1) - \gamma_p^{-1}(n)\underline{\kappa}_p(n)e_y(n, n-1) \tag{5.5}$$

where $e_y(n, n-1)$ is the 'apriori estimation error':

$$e_y(n, n-1) = y(n) + \underline{x}_p^T(n)\underline{\omega}_p(n-1) \tag{5.6}$$

However, an alternative method of calculating the optimum coefficients exists[**SMH**]; consider the following augmented data matrix:

$$X_y(n) = \begin{bmatrix} X_p(n) & \underline{y}(n) \end{bmatrix} \tag{5.7}$$

The upper-triangular matrix resulting from its QR decomposition will then have the form:

$$R_y(n) = \begin{bmatrix} R_p(n) & \underline{u}_p(n) \\ \underline{0}^T & \epsilon_p(n) \end{bmatrix} \tag{5.8}$$

where $\epsilon_p(n)$ is the square-root of the filtering error power for $y(n)$ (see [**SMH**]). Furthermore, it is easy to show that the inverse of this upper triangular matrix is another upper triangular matrix:

$$R_y^{-1}(n) = \begin{bmatrix} R_p^{-1}(n) & \epsilon_p^{-1}(n)\underline{\omega}_y(n) \\ \underline{0}^T & \epsilon_p^{-1}(n) \end{bmatrix} \tag{5.9}$$

where $\underline{\omega}_y(n) = R_p^{-1}(n)\underline{u}_p(n)$ is the least squares weight vector for the estimation of $y(n)$ by $\underline{x}_p(n)$. Thus we can obtain the weight vector $\underline{\omega}_y(n)$ directly from the right-hand column of the inverse augmented triangular matrix $R_y^{-1}(n)$ (or equivalently, the bottom row of $R_y^{-T}(n)$).

A SFG for the inverse QRD-RLS algorithm is shown in figure 9, for the particular case p=3, along with the tapped delay-line required for the time series adaptive filtering problem. The basic cells for the SFG are shown in figure 10. As can be seen they are essentially based on the standard QRD-RLS cells of figure 2. In figure 9 the elements of the matrix $R_y^{-T}(n)$ are explicitly shown but for notational

---

[1] See Haykin [**H**] but note that the quantity denoted there by $\gamma(n)$ is the square of that here.

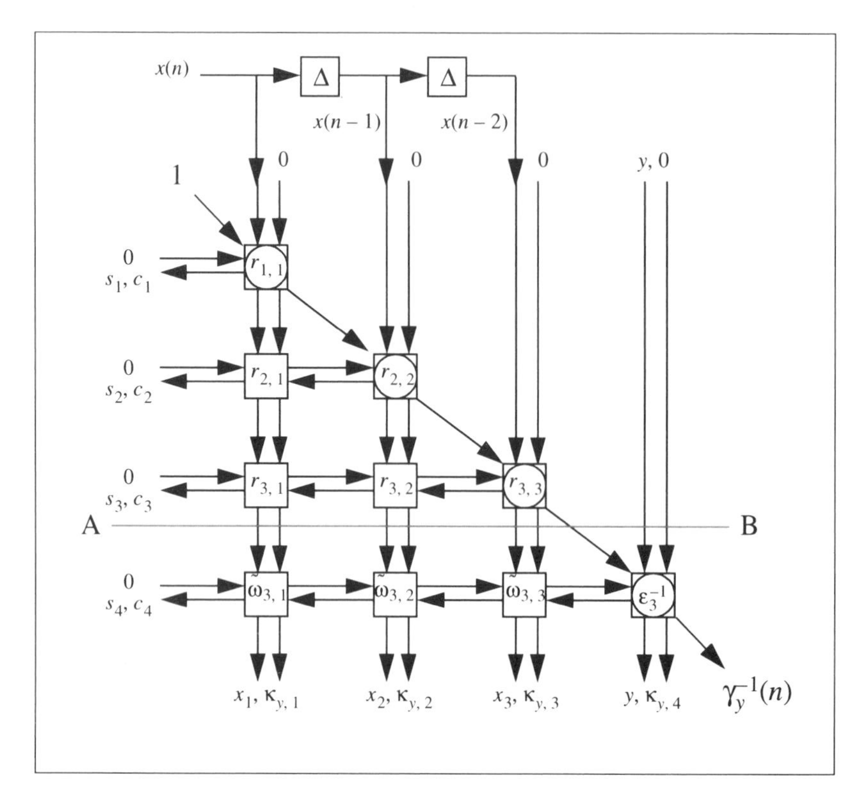

FIGURE 9. SFG for the inverse QRD-RLS Operator as Applied to Time-Series Adaptive Filtering

convenience its elements are denoted by $r_{i,j}$. Furthermore, we define the energy normalised weight vector $\tilde{\omega}_y$ by

$$\tilde{\omega}_y^T(n) = \epsilon_p^{-1}(n)\omega_y^T(n) \tag{5.10}$$

The algorithm operates as follows: the input data vector $\underline{x}_y(n)$ (cf. equation 5.7) flows from the top of the SFG to the bottom allowing the vector $R_y^{-T}(n-1)\underline{x}_y(n)$ to be calculated as the initially zero inputs on the left hand side move to the right (cf. equation 5.3). Once this data reaches the cells on the diagonal, the sines and cosines of the elementary 2×2 rotations that make up $\hat{Q}_y(n)$ can be calculated (cf. equation 5.4). These are then sent to the left of the SFG where they are used to update the stored matrix $R_y^{-T}(n-1)$ in time.

The SFG in figure 9 can be simplified by defining some new, more complex operators: see figure 11. Clearly the algorithm shown in figure 11 requires $O(p^2)$ operations per time sample. This can be seen from the fact that the triangular operator, labelled R_p^{-T} constitutes a p-th order inverse QRD-RLS operator and hence from figure 9 will require $O(p^2)$ operations. In order to develop a fast algorithm it is again necessary to focus on the tapped delay line which will in turn lead us to the two linear prediction problems. Our goal will be to replace this triangular operator by another operator that only requires $O(p)$ operations.

Just as in the previous section, we consider what happens when the inputs to the p-th order inverse QRD-RLS operator are permuted. First however, figure 12

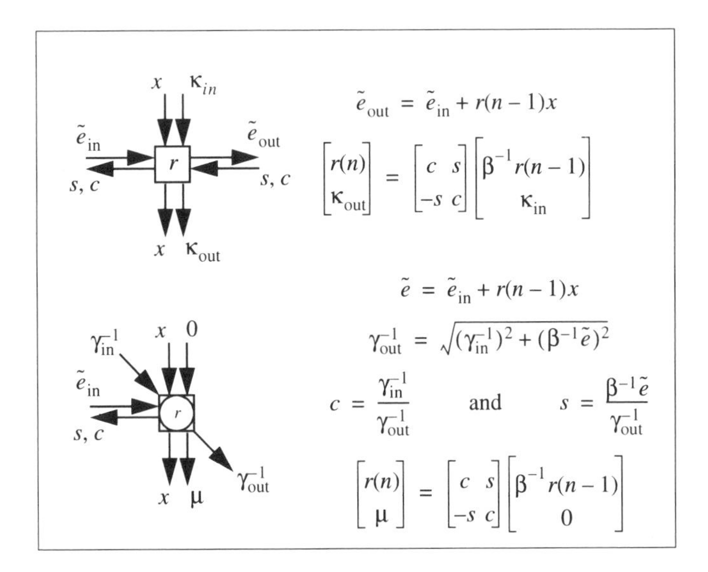

FIGURE 10. Basic cells for inverse QRD-RLS Operator

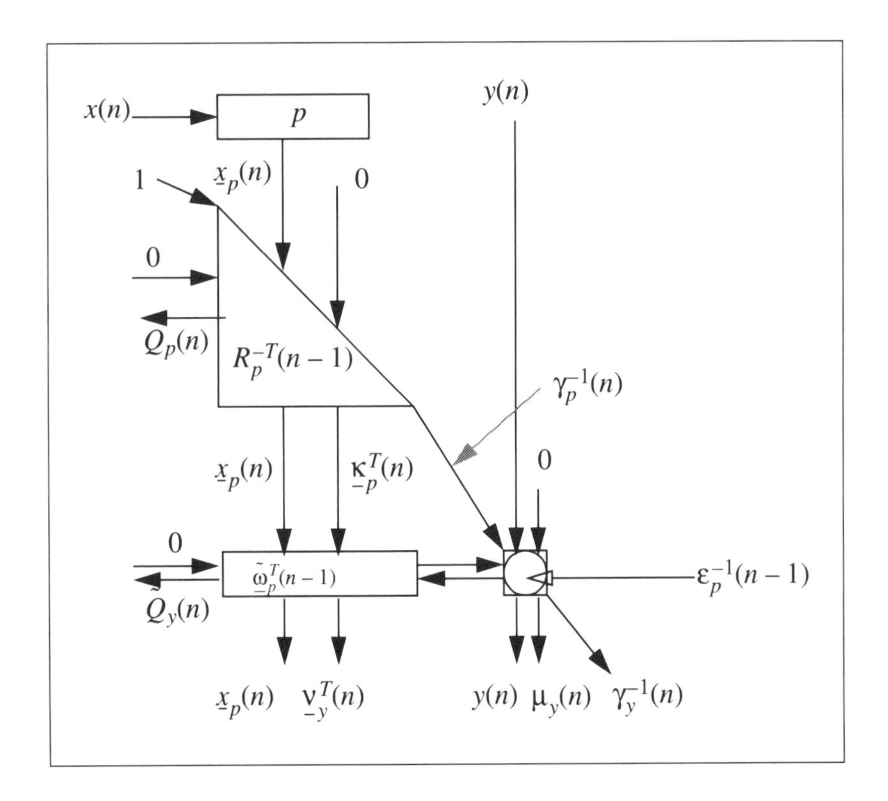

FIGURE 11. Simplified SFG for the $O(p2)$ Adaptive Filtering Algorithm.

shows the basic p-th order inverse QRD-RLS operator with the right most input shown separate from the others. Given the regularity of the inverse QRD-RLS

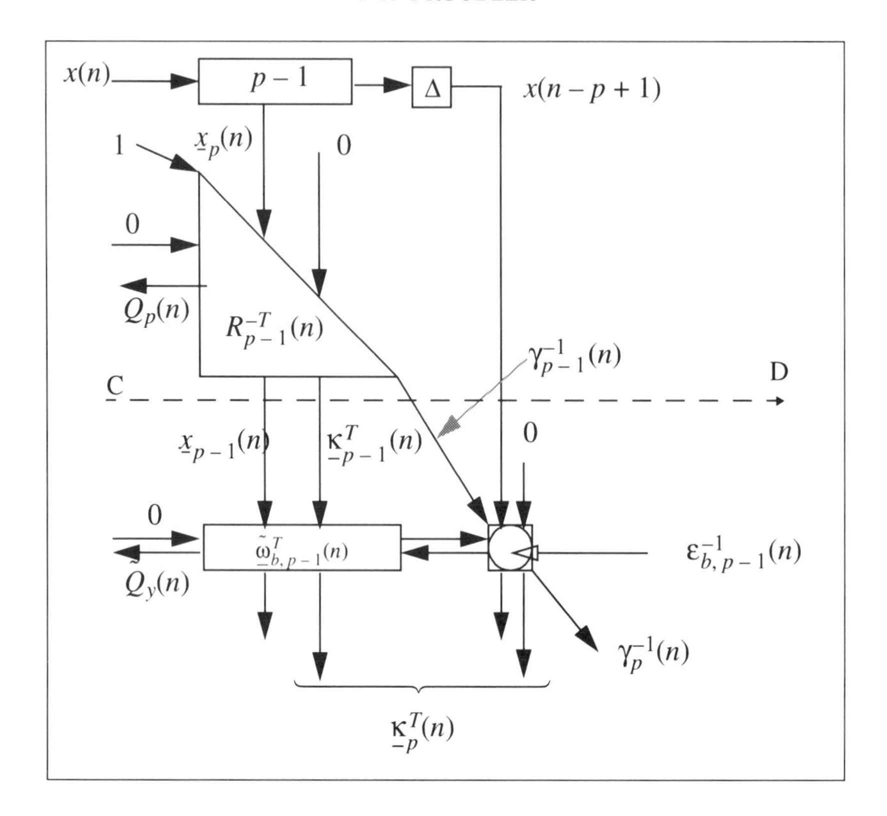

FIGURE 12. SFG for the p-th order Linear Prediction Operator

operator (figure 10) it is easy to see that the effect of this is to create a $(p-1)$-th order inverse QRD-RLS operator (above the line CD) connected to the bottom row of the original p-th order inverse QRD-RLS operator. By comparison with figure 9 with $y(n) = x(n-p+1)$, it should be clear that this last row contains the normalised (p-1)th order backward prediction weights and that the data flowing between these two operator is the vector $\underline{\kappa}^T_{p-1}(n)$ (see equation (5.1)). In figure 13 the inputs have now been permuted so that the undelayed input now appears to the right of the others. The tapped delay line has been split in two parts (one of length 1 and the other $(p-1)$) to reflect this. We now have a lower order inverse QRD-RLS operator (above the line EF) and a new bottom row. From figure 9 with $y(n) = x(n+1)$, this last row contains the normalised (p-1)th order forward prediction weights.

Just as in the previous (lattice) example, given that the RLS problem should essentially be invariant to a permutation of inputs, it is relevant to ask what relation does the new gain vector output (the composite vector $\left[\underline{\nu}^T_{fp}(n) \quad \mu_{fp}(n)\right]^T$) have to the original one ($\underline{\kappa}_p(n)$ in figure 12). Although, perhaps, not as obvious as in the lattice example, it can be shown that the new gain vector is just a permuted version of the original. This can be seen as follows: since the least squares residual $y(n) + \underline{x}^T(n)\underline{\omega}(n)$ is invariant to a permutation of the input data, the weight vector $\underline{\omega}$ will itself only be permuted. Then given equation (5.5) it is clear that the Kalman gain vector will also be permuted. Hence if

$$\Pi^T \underline{x}_p(n) = \left[\underline{x}_{p-1}(n-1) \quad x(n)\right]^T \tag{5.11}$$

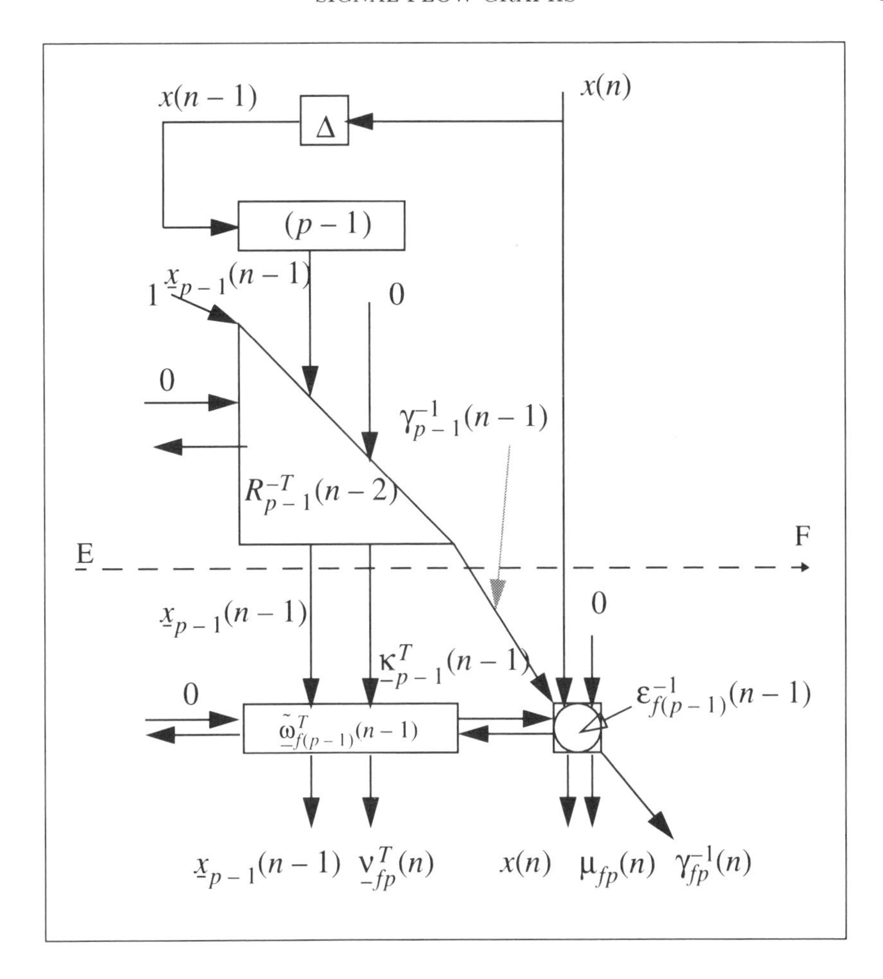

FIGURE 13. Transformed SFG for the p-th order Linear Prediction Operator

where Π^T is a cyclic permutation matrix that moves the first element of $\underline{x}_p(n)$ to the p-th position and moves all the others up one place then

$$\underline{\kappa}_p(n) = \Pi \left[\underline{\nu}^T_{fp}(n) \quad \mu_{fp}(n)\right]^T = \left[\mu_{fp}(n) \quad \underline{\nu}^T_{fp}(n)\right]^T \tag{5.12}$$

In addition, it is easy to show [**Pr**] that the likelihood variable $\gamma_{fp}(n)$ is invariant to a permutation of the input vector hence

$$\gamma_{fp}(n) = \gamma_p(n). \tag{5.13}$$

Given these observations, it is possible to show that the two linear prediction problems and the filtering problem share a common intermediate variable (in fact this is exactly the Kalman gain vector) - see figure 14.

The basis of the fast algorithm for inverse QRD-RLS time series filtering can now be seen from figure 14: the forward linear prediction operator (labelled FLP) and the backward linear prediction (BLP) operator produce the same outputs but whereas the backward prediction operator has as inputs quantities at time n (say) the forward prediction operator has as inputs quantities at time $n-1$. Thus if the backward prediction operator (BLP) could be inverted (i.e. given its outputs infer

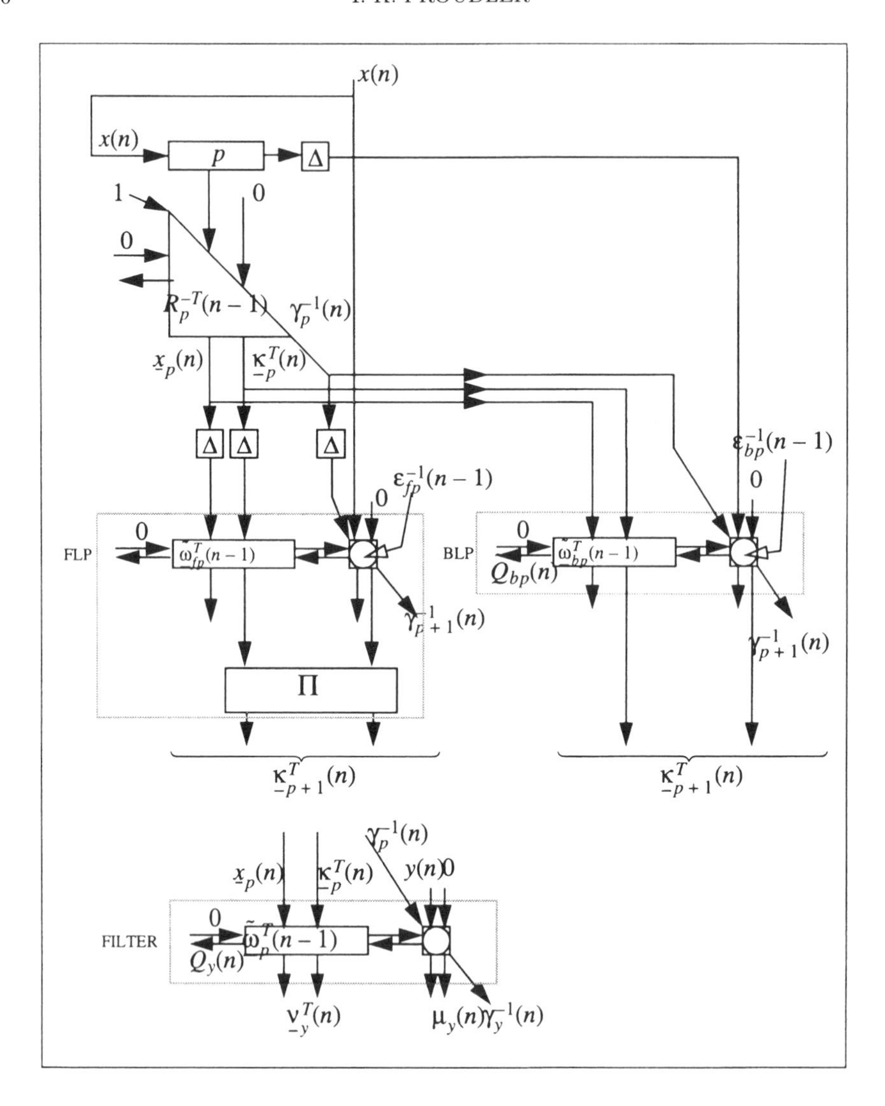

FIGURE 14. SFG for the Combined Linear Prediction Problems.

its inputs) it can use the outputs from the forward linear prediction operator to generate the next set of inputs for the forward prediction operator. The common $p \times p$ inverse QRD-RLS operator would then be superfluous and could be removed to leave a SFG of complexity $O(p)$ - as shown in figure 15. The operators for the new backward linear prediction section are the inverse of those in figure 14 in the sense that the latter perform an update operation using Givens rotations whereas the former are downdate operators which involve hyperbolic rotations.The resulting downdate cells are shown in figure 16. For completeness, figure 15 also shows the operator that solves the adaptive filtering problem.

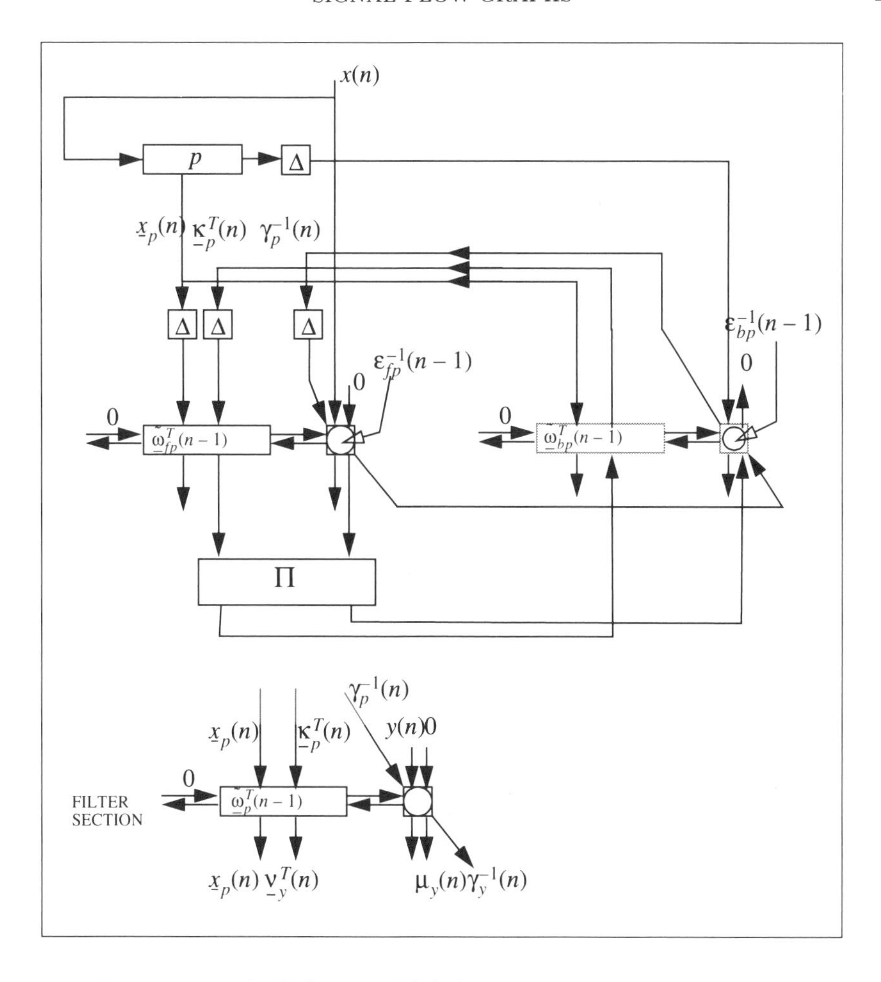

FIGURE 15. SFG for the O(p) Adaptive Filtering Algorithm.

6. Toeplitz-Block-Toeplitz

In section 2 it was mentioned that in image processing one can have a linear prediction problem where the data or covariance matrix has a 'Toeplitz-block-Toeplitz' structure described by equations (2.15) and (2.16) respectively. Such a situation can also arise in radar signal processing in the so-called wideband space-time processing problem. It is easy to see that such a problem can be solved in $O(t^2q^2)$ operations without taking advantage of the structure since there are tq weights to be found. It can also be considered to be a block-Toeplitz problem by ignoring one of the two Toeplitz structures and hence solved in $O(tq^2)$ or $O(t^2q)$ operations. It has been speculated that there may exist an algorithm that exploits both Toeplitz structures and thus have an operation count of $O(tq)$ or at least not quadratic in either t or q. Thus far no such algorithm has been found. Further more a proof that an efficient algorithm is not possible has not been forthcoming leading to continued interest in attempting the solve the problem. In the following, an attempt to generate such an algorithm using SFG manipulation will be described. Although

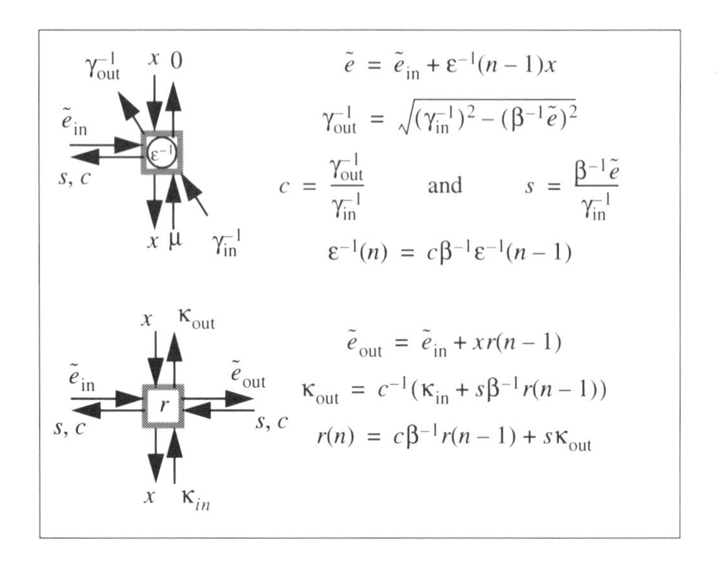

FIGURE 16. Downdate Cells

it does not produce an $O(tq)$ algorithm it constitutes another interesting example of the use of SFG manipulations.

In section 3 we saw that the combination of one or more tapped delay line feeding a QRD-RLS operator (figure 1) is equivalent to the lattice operation (figure 8). Now it is easy to see that, for the Toeplitz-block-Toeplitz problem, the $O(t^2q^2)$ algorithm can be described in SFG form by having two concatenated tapped delay lines feeding a QRD-RLS operator (figure 17a). One of the tapped delay lines produces copies of the input signal delayed by several sample instants (e.g. the image width d - see section 2) whilst the other produces the required number of consecutive samples (t in section 2). In figure 17 the former type of tapped delay line is distinguished from those seen so far by the legend '$n \times d$' where n is the number of taps and d is the size of each delay (cf. figure 3 with each delay stage Δ seconds replaced by one of $d\Delta$ seconds). It is therefore immediately clear that we can transform this SFG to that in figure 17b by replacing the QRD-RLS operator and the second set of tapped delay lines by a multichannel lattice.

Now given the internal structure of the lattice SFG (figure 18), it is possible to commute the remaining tapped delay line operator with the delay operator in the lattice (see figure 8) and, duplicating it where necessary, move the tapped delay line to the input to the two small QRD-RLS operators and the operators represented by the square blocks in figure 18. Once again the combination of the tapped delay lines and the (small) QRD-RLS operators can be replaced by a (small) lattice (figure 19). The critical question relates to the square blocks: is it possible to commute these with the tapped delay line? If these two operators do commute, then the above process can continue until all of the small QRD-RLS operators in the SFG can be replaced by (small) lattices and a reduction in computational load would result.

The square blocks represent the update of a matrix U (say) via the application of the $\hat{Q}$ matrix from the adjacent QRD-RLS operator and its input (the output of

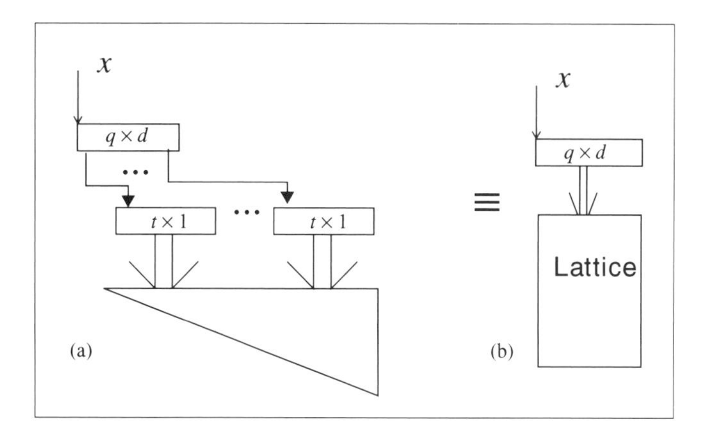

FIGURE 17. Toeplitz-block-Toeplitz SFG Transformation

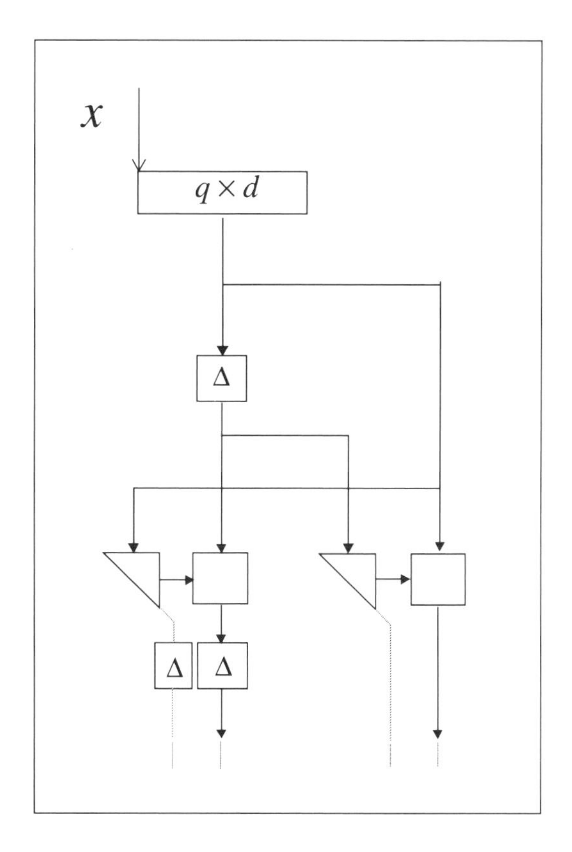

FIGURE 18. Detail of First Section of the Lattice SFG

the tapped delay line) (cf. equation (4.1)) i.e.

$$\hat{Q}^T \begin{bmatrix} U(n-1) \\ \underline{v}^T(n) \end{bmatrix} = \begin{bmatrix} U(n) \\ \underline{\alpha}^T(n) \end{bmatrix} \tag{6.1}$$

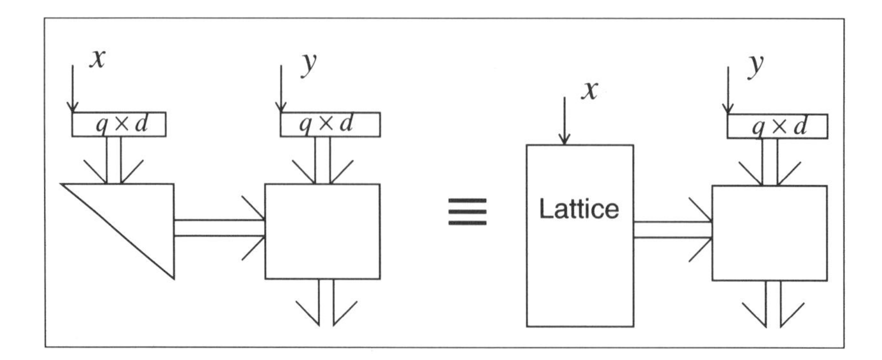

FIGURE 19. Transformation of Small QRD-RLS Operator

Unfortunately even if $\underline{v}^T(n)$ was produced by a tapped delay line, the transformed vector $\underline{\alpha}^T(n)$ does not in general have a time shift structure. Hence it appears that the tapped delay line and the square operator in figure 19 do not commute. This means that this approach to generating a reduced computation algorithm stops at this point and therefore, unfortunately, fails to achieve its objective.

7. Summary

The utility of representing an algorithm as a signal flow graph has been presented by discussing the development of two fast RLS adaptive filtering algorithms. The SFG representation allows the algorithm to be transformed in a rigorous manner into new variants that can have more desirable properties. In particular two fast RLS algorithms have been presented: the RLS lattice and the FTF algorithms.

This SFG manipulation techniques has also been used to transform the basic inverse QRD-RLS algorithm into two other functionally equivalent forms that have the property that they can be turned in to pipelined circuits. It can be seen from figure 9 that the inverse QRD-RLS algorithm has data flowing from top to bottom and both from left to right and right to left. This is in contrast to the QRD-RLS algorithm in figure 1 which only has data flowing one way in the horizontal direction (left to right). This contraflow of data makes it impossible to build an efficiently pipelined circuit since it is necessary to wait for the one calculation to finish before the next can start. However by transforming the algorithm's SFG[**MP94, PMMH**] it proves possible to generate two further algorithms in which the contraflow is broken and the new circuit can be pipelined. These techniques have also been used to good effect in the teaching of adaptive filtering to undergraduates[**M**].

It was shown that some adaptive filtering problems involve a Toeplitz-block-Toeplitz matrix. Despite much effort, as yet no way of exploiting this structure to reduce the amount of computation required to solve the problem has been found.

8. Acknowledgements

The author would like to thank Prof. John McWhirter and Dr. Marc Moonen for their contribution to the work discussed here.

References

[A] S T Alexander and A L Ghirnikar, *A Method for Recursive Least Squares Filtering based upon an Inverse QR Decomposition*, IEEE Trans. SP **41**, (1993), 20–30.

[CMK] G. Carayannis, D. G. Manolakis, and N. Kalouptsidis, *A Fast Sequential Algorithm for Least Squares Filtering and Prediction*, IEEE Trans. ASSP **31**, (1983), 1394–1402.

[ChK] J. Chun and T. Kailath, *Generalised displacement structure for Block-Toeplitz, Toeplitz-block and Toeplitz-derived matrices*, in *Numerical Linear Algebra, Digital signal Processing and Parallel Algorithms*, Eds. G.H. Golub and P. Van Dooren, Springer-Verlag, NATO ASI series F **70**, 1989.

[CiK] J.M. Cioffi and T.Kailath, *Fast, Recursive Least Squares Transversal Filters for Adaptive Filtering*, IEEE trans. ASSP **32**, (1984), 304–37

[F] D D Falconer and L Ljung, *Application of Fast Kalman Estimation to Adaptive Equalization*, IEEE Trans. Communications, **26**, (1987), 1439–1446.

[GeK] W.M.Gentleman and H.T.Kung, *Matrix Triangularisation by Systolic Arrays*, Proc. SPIE, **298**, *Real Time Signal Processing IV*, (1981), 19–26.

[GlK1] G-O. A. Glentis and N. Kalouptsidis, *Efficient Order Recursive Algorithms for Multichannel Least Squares Filtering*, IEEE Trans. SP, **40**, (1992), 1354–1374.

[GlK2] G-O. A. Glentis and N. Kalouptsidis, *Fast Adaptive Algorithms for Multichannel Filtering and System Identification*, IEEE Trans. SP, **40**, (1992), 2433–2458.

[HSMP] M. Harteneck, R.W. Stewart, J.G. McWhirter and I.K. Proudler, *An Algorithmically Engineered Fast Multichannel Adaptive Filter based on the QR-RLS*, IEE Proc. VIS **146**(1999), 7–14.

[H] S Haykin, *Adaptive Filter Theory*, 2nd Edition, Prentice-Hall, Englewood Cliffs, NJ, USA, 1991.

[Ka] T. Kailath, *A theorem of I. Schur and its impact on modern signal processing*, Operator Theory: Advances and Applications, **18**, 1986 Birkhauser Verlag, Basel, 9–30.

[Ku] S.Y.Kung, *VLSI Array Processors*, Prentice Hall (Information and Systems Science Series), 1988

[LD] The Levinson-Durbin Algorithm, S. Haykin, *Adaptive Filter Theory*, 2nd Edition, Prentice-Hall, Englewood Cliffs, New Jersey, USA, 1991, Section 6.3.

[Le] P.S. Lewis, *QR-based Algorithms for Multichannel Adaptive Least Squares Lattice Filters*, IEEE Trans. ASSP, **38**, (1990), 421–32.

[Li] F. Ling, *Givens Rotation Based Least Squares Lattice and Related Algorithms*, IEEE Trans. SP, **39**, (1991), 1541– 1552.

[McW] J.G.McWhirter, *Recursive Least Squares Minimisation using a Systolic Array*, Proc. SPIE, **431**, *Real Time Signal Processing VI*, (1983), 105–112.

[MP93] JG McWhirter and IK Proudler, *Orthogonal Lattice Algorithms for Adaptive Filtering and Beamforming*, Integration: the VLSI journal, **14**, (1993), 231–247.

[MP94] J.G. McWhirter and I. K. Proudler, A Systolic Array for Recursive Least Squares Estimation By Inverse Updates, Proc. CONTROL'94, Warwick, UK, March 1994.

[M] M. Moonen, *Adaptive filtering Course Notes*, KU Leuven, http://www.esat.kuleuven.ac.be/~moonen/asp_course.html

[N] J Nagy and R J Plemmons, *An Inverse Factorization Algorithm for Linear Prediction*, Linear Algebra and Its Applications, **172**, (1992), 169–195.

[PP] C T Pan and R J Plemmons, *Least Squares Modifications with Inverse Factorization: Parallel Implementations*, J. Comput. and Applied Maths., **27**, (1989), 109–127.

[PMS] I K Proudler, J G McWhirter and T J Shepherd, *Computationally Efficient, QR Decomposition Approach to Least Squares Adaptive Filtering*, IEE Proceedings Pt. F, **138**, (1991), 341–353.

[Pr] I. K. Proudler, *A Fast Time-Series Adaptive Filtering Algorithm based on the QRD Inverse-Updates Method*, IEE Proceedings VIS **141** (1994), 325–333.

[PMMH] I. K. Proudler, J.G. McWhirter, M. Moonen and G. Hekstra, *The Formal Derivation of a Systolic Array for Recursive Least Squares Estimation*, IEEE trans. CAS II, **43**, (1996), 247–254.

[SMH] T J Shepherd, J G McWhirter and J Hudson, *Parallel Weight Extraction from a Systolic Adaptive Beamformer*, Mathematics in Signal Processing II, J G McWhirter (Ed), Clarendon Press, Oxford, 1990, 775–790.

[Y] B. Yang and J.F. Bohme, *On a Parallel Implementation of the Adaptive Multichannel Least Squares Lattice Filter*, Proc. Int. Symp. on Signals, Systems and Electronics, Erlangen, Sept 1989.

DERA, Room E506, St Andrews Road, Malvern, Worcestershire WR14 3PS, UK.
E-mail address: proudler@signal.dera.gov.uk
URL: http://spg.dera.gov.uk/

Contemporary Mathematics
Volume **280**, 2001

The Structured Total Least Squares problem

Nicola Mastronardi, Philippe Lemmerling, and Sabine Van Huffel

ABSTRACT. In this paper we consider the Structured Total Least Squares (STLS) problem. The latter is a natural extension of the ordinary Total Least Squares (TLS) problem and is also used to determine the parameter vector of a linear model given some noisy measurements.
We consider structure on two different levels. First of all, the STLS problem is a structured problem, meaning that it extends the TLS formulation by including matrix structure constraints. The structure at this first level is often introduced to obtain statistically more accurate solutions of the parameter vector. Secondly, the matrices involved in algorithms for solving the STLS problem partly inherit the structure imposed by the previously mentioned constraints. Therefore, the structure at this level can be exploited to develop computationally more efficient algorithms.
We focus on two particular types of STLS problems using different matrix structure constraints. For both of them fast algorithms are developed by exploiting the low displacement rank of the involved matrices and the sparsity of the corresponding generators. The higher statistical accuracy and the higher computational efficiency are illustrated on two different examples: a system identification example and a novel speech compression scheme.

1. Introduction

Linear models play an important role in a broad class of disciplines and related applications such as signal processing, control theory, system theory, statistics, economics, biomedical applications etc. These models can be described in the following way:

$$\alpha(1)x(1) + \alpha(2)x(2) + \ldots + \alpha(n)x(n) = \beta \tag{1.1}$$

1991 *Mathematics Subject Classification.* 65F99.

Philippe Lemmerling is funded by the KULeuven.

Sabine Van Huffel is a senior Research Associate with the F.W.O. (Fund for Scientific Research-Flanders). This paper presents research results of the Belgian Programme on Interuniversity Poles of Attraction (IUAP P4-02 and P4-24), initiated by the Belgian State, Prime Minister's Office - Federal Office for Scientific, Technical and Cultural Affairs, of the European Community TMR Programme, Networks, project CHRX-CT97-0160, of the Brite Euram Programme, Thematic Network BRRT-CT97 -5040 'Niconet', of the Concerted Research Action (GOA) projects of the Flemish Government, entit led MIPS (Model-based Information Processing Systems) and MEFISTO-666 (Mathematical Engineering for In formation and Communication Systems Technology) and of the FWO "Krediet aan navorsers" G.0326.98. The scientific responsibility is assumed by its authors.

with $x \in \mathbb{R}^{n\times 1}$ the parameter vector and $\alpha \in \mathbb{R}^{n\times 1}$, $\beta \in \mathbb{R}$ the observed variables. In an experiment, typically many measurements are performed, say m, all of which are supposed to satisfy relation (1.1). This leads to an overdetermined set of equations

$$\sum_{j=1}^{n} A(i,j)x(j) \approx b(i), i = 1,\dots,m \tag{1.2}$$

with $A \in \mathbb{R}^{m\times n}$, $b \in \mathbb{R}^{m\times 1}$ and $m > n$ or in matrix form:

$$Ax \approx b. \tag{1.3}$$

There exist many approaches for solving (1.3). One of them is the well-known Least Squares (LS) method. It determines the parameter vector x by solving the following problem:

$$\min_x \|Ax - b\|_2. \tag{1.4}$$

The latter can be seen as applying a correction to the observations in b in order to make the system of equations (1.3) consistent. However, in most real-life applications, both the observations in A and b are contaminated with noise and thus it makes more sense to use the Total Least Squares (TLS) approach [**GV2, V**]:

$$\min_{\Delta A, \Delta b, x} \|[\Delta A \ \Delta b]\|_F^2 \tag{1.5}$$
$$\text{such that } (A + \Delta A)x = b + \Delta b.$$

Properties and algorithms for the TLS approach are described in [**V**]. One of those properties is the fact that the TLS approach yields a ML estimate for the parameter vector under certain assumptions. In order to specify these assumptions, a statistical framework is now presented. Let A and b contain measurements of the true unobservable values in $A_0 \in \mathbb{R}^{m\times n}$ and $b_0 \in \mathbb{R}^{n\times 1}$. Furthermore let

$$A_0 x_0 = b_0$$
$$A = A_0 + \Delta A_n$$
$$b = b_0 + \Delta b_n,$$

in which $\Delta A_n \in \mathbb{R}^{m\times n}$ and $\Delta b_n \in \mathbb{R}^{n\times 1}$ contain the "measurement" errors on the elements of A_0 and b_0. In [**V**] it is then shown that x at the solution of (1.5) is a ML estimator for x_0 when the rows of $[\Delta A_n \ \Delta b_n]$ are independently identically distributed (i.i.d.) with common zero mean vector and common covariance matrix that is (an unknown) multiple of the identity matrix $I_{(n+1)\times(n+1)}$. In reality, these noise conditions are often violated when there exist linear relations between the true unobserved variables represented by the entries $[A_0 \ b_0]$ as well as between the observed variables represented by the entries of $[A \ b]$. Mostly the linear relations between the entries of $[A_0 \ b_0]$ are the same as those between the entries of $[A \ b]$. Thus, at least intuitively, it is clear that –in case the *different* entries of $[\Delta A_n \ \Delta b_n]$ are i.i.d. zero mean noise of equal (unknown) variance– a ML estimate of x can only be obtained when the entries of $[\Delta A \ \Delta b]$ in (1.5) obey the same linear relations as the entries of $[A \ b]$ and the entries of $[A + \Delta A \ b + \Delta b]$. An example of such linear relations occurs when elements of a signal vector s are stored in a Toeplitz matrix (such that the signal can be read from the first row and the first column): the linear relations are a mathematical way of expressing that all the elements on a diagonal are equal, since they represent one and the same element of the signal vector s. In matrix terms these linear relations can be represented by a matrix structure constraint. Adding the latter matrix structure constraint yields the Structured Total Least Squares (STLS) approach:

$$
\begin{aligned}
(1.6) \qquad & \min_{\Delta A, \Delta b, x} \|[\Delta A \; \Delta b]\|_F^2 \\
& \text{such that } (A + \Delta A)x = b + \Delta b, \\
& \text{and } [A + \Delta A \; b + \Delta b] \text{ has the same linear structure as } [A \; b].
\end{aligned}
$$

Summarizing we can say that in the STLS formulation matrix structure constraints are added to the TLS problem formulation in order to increase the statistical accuracy of the determined parameter vector. Notice that this is a first level at which we exploit structure in order to obtain statistically more accurate results. Further on we will take advantage of the structure present on a second level, namely the level of algorithms for solving STLS problems: the involved matrices inherit partly the structure that appears in the previously mentioned matrix structure constraint. Furthermore, it should be noted that the STLS problem is related to operator theory [**AD, Dy**]. However, in this paper we focus on specific numerical algorithms for solving the STLS problem.

The paper is structured as follows. In the next section we describe the mathematical formulation of the STLS problem. We present the two matrix structure constraints that we will consider in this paper and the outline of two algorithms for solving the STLS problems involving the presented matrix structure constraints. Many improvements for solving structured sets of equations using different approaches [**GO, O, OP, OS**] exist. However, we are not aware of superfast algorithms for solving our STLS problems. Sections 3 to 5 describe fast implementations of the two algorithms described in section 2 for the two types of matrix structure constraints. The increased computational efficiency is obtained by exploiting the low displacement rank of the involved matrices and the sparsity of the corresponding generators. In section 6 we consider two applications. A system identification example is used to illustrate the increased statistical accuracy obtained by exploiting structure at the first level, i.e. by imposing matrix structure constraints. The increased computational performance obtained by taking advantage of the structure at the second level, i.e. the structure of the matrices that appear in the proposed algorithm, is demonstrated on a novel speech compression scheme.

2. Problem formulation

Before giving the mathematical formulation of the STLS problem, we introduce some notations that will be used throughout the paper. To refer to specific entries or parts of a matrix or a vector, we adopt a Matlab-like [**M**] notation:

- A(i,j): the entry in the jth column of the ith row of A.
- A(i,:): the ith row of A.
- A(:,j): the jth column of A.
- A(p:q,r:s): the $(q-p+1) \times (s-r+1)$ submatrix of A containing the entries that belong to rows p till q **and** to columns r till s.
- b(i): the entry on the ith row of column vector b.
- b(p:q): the $(q-p+1) \times 1$ subvector of b containing the entries of row p till row q.

To indicate that a vector or a matrix is structured, we use a matrix function notation. E.g. when the matrix $A \in \mathbb{R}^{m \times n}$ is linearly structured and the vector

$s \in \mathbb{R}^{k\times 1}$ contains the different entries of A, the structure can be denoted by specifying a linear matrix function $\mathbf{M}(s) : \mathbb{R}^{k\times 1} \to \mathbb{R}^{m\times n} : s \to \mathbf{M}(s)$ and stating that $A = \mathbf{M}(s)$.
A more general formulation of the STLS problem than the one described in (1.6) is the following:

$$\begin{aligned} &\min_{\Delta s, x} 0.5\Delta s^T W \Delta s \\ &\text{such that } (A + \mathbf{M}(\Delta s))x = b + \mathbf{m}(\Delta s), \end{aligned} \tag{2.1}$$

where $\Delta s \in \mathbb{R}^{k\times 1}$ contains the k *different* elements of $[\Delta A\ \Delta b]$ and $W \in \mathbb{R}^{k\times k}$ is a weighting matrix. When comparing (1.6) to (2.1) we see first of all that in the latter case the matrix structure constraint of (1.6) becomes obsolete through the use of matrix functions $\mathbf{M}(\Delta s) : \mathbb{R}^{k\times 1} \to \mathbb{R}^{m\times n}$ and $\mathbf{m}(\Delta s) : \mathbb{R}^{k\times 1} \to \mathbb{R}^{m\times 1}$. Secondly, formulation (1.6) is a special case of formulation (2.1). To see this, construct a diagonal matrix $D \in \mathbb{R}^{k\times k}$ such that $D(i,i)$ represents the number of times that $\Delta s(i)$ occurs in $[\Delta A\ \Delta b]$ and set $W = 2D$.
Since we only consider linearly structured matrices it is obvious to see that we can find a matrix function $\mathbf{X}(x) : \mathbb{R}^{n\times 1} \to \mathbb{R}^{m\times k}$ such that

$$\mathbf{X}(x)\Delta s = [\mathbf{M}(\Delta s)\ \mathbf{m}(\Delta s)] \begin{bmatrix} x \\ -1 \end{bmatrix}. \tag{2.2}$$

Furthermore define the matrix function $\mathbf{r}(\Delta s, x)$ as follows

$$\mathbf{r}(\Delta s, x) = (A + \mathbf{M}(\Delta s))x - (b + \mathbf{m}(\Delta s)) = [A\ b] \begin{bmatrix} x \\ -1 \end{bmatrix} + \mathbf{X}(x)\Delta s.$$

Using the latter formalism, we can rewrite problem (2.1) as follows

$$\begin{aligned} &\min_{\Delta s, x} 0.5\Delta s^T W \Delta s \\ &\text{such that } [A\ b] \begin{bmatrix} x \\ -1 \end{bmatrix} + \mathbf{X}(x)\Delta s = 0. \end{aligned} \tag{2.3}$$

As can be seen from (2.3), the STLS problem is a constrained optimization problem with a quadratic objective function and nonlinear constraints. This problem can be solved in many ways, using standard constrained optimization software. However, our goal is to develop fast algorithms for the STLS problem. In order to obtain algorithms that exploit the structure of the STLS problem as much as possible, we avoid techniques that try to eliminate the constraints, since often this elimination means a loss of the original structure present in the matrices. To avoid this, we apply the Newton method for unconstrained optimization to the Lagrangian L of problem (2.1):

$$L(\Delta s, x, \gamma) = 0.5\Delta s^T W \Delta s - \gamma^T (b - Ax - \mathbf{X}(x)\Delta s),$$

where $\gamma \in \mathbb{R}^{m\times 1}$ is a vector of Lagrange multipliers. Straightforward application of the Newton method (see e.g. [**F, GMW**]) on the Lagrangian L yields the following algorithm:

Algorithm STLS1

Input: $[A\ b] \in \mathbb{R}^{m\times(n+1)}$
Output: the parameter vector $x \in \mathbb{R}^{n\times 1}$ and $\Delta s \in \mathbb{R}^{k\times 1}$ (i.e. the minimal representation of the structured matrix $[\Delta A\ \Delta b]$) that solve (2.3)

Step 1: Initialize Δs,x and γ
Step 2: while *stopcriterion* not satisfied
Step 2.1: Solve the following system of equations:

$$\begin{bmatrix} H & J^T \\ J & 0 \end{bmatrix} \begin{bmatrix} \Delta\tilde{s} \\ \Delta\tilde{x} \\ \Delta\tilde{\gamma} \end{bmatrix} = - \begin{bmatrix} g + J^T\gamma \\ \mathbf{r}(\Delta s, x) \end{bmatrix}$$

Step 2.2: $\Delta s \leftarrow \Delta s + \Delta\tilde{s}$
$x \leftarrow x + \Delta\tilde{x}$
$\gamma \leftarrow \gamma + \Delta\tilde{\gamma}$

end

where $g = \begin{bmatrix} W & 0 \\ 0 & 0 \end{bmatrix} \begin{bmatrix} \Delta s \\ x \end{bmatrix} \in \mathbb{R}^{(k+n)\times 1}$ is the gradient of the objective function in (2.3) w.r.t. $v = [\Delta s^T \; x^T]^T$ and $J = [\mathbf{X}(x) \; (A + \mathbf{M}(\Delta s))]$ is the Jacobian of the constraints $\mathbf{r}(\Delta s, x) = 0$ w.r.t. v. Furthermore, for optimal convergence rate (superlinear) the matrix H should be set to $\nabla^2_{vv} L(\Delta s, x, \gamma)$. It is straightforward to show that

$$\nabla^2_{vv} L = \begin{bmatrix} W & 0 \\ 0 & 0 \end{bmatrix} - \sum_{i=1}^{m} \gamma(i) \nabla^2_{vv} \mathbf{r}(\Delta s, x). \tag{2.4}$$

As shown in e.g. [**F**], H can also be chosen to be a positive definite approximation of $\nabla^2_{vv} L$, without changing the final solution of problem (2.3). Inclusion of the second term in (2.4) often renders the structure of H rather complicated. We therefore only retain the first term in (2.4):

$$H = \begin{bmatrix} W & 0 \\ 0 & 0 \end{bmatrix}.$$

By using this positive definite approximation of $\nabla^2_{vv} L$, we loose some of the convergence speed (linear instead of superlinear), but this is largely compensated by the fact that one iteration can be implemented in a very fast way, exploiting the low displacement rank structure of the matrices involved in Step 2.1 of Algorithm STLS1.

The first type of structure we consider in this paper is the case in which $[A \; b]$ and $[\Delta A \; \Delta b]$ are Toeplitz matrices. This can be expressed by the following matrix functions:

$$\begin{aligned} \mathbf{M}(s) &: \mathbb{R}^{(m+n)\times 1} \rightarrow \mathbb{R}^{m\times n} : s \rightarrow \mathit{toeplitz}(s(n : m + n - 1), s(n : -1 : 1)) \\ \mathbf{m}(s) &: \mathbb{R}^{(m+n)\times 1} \rightarrow \mathbb{R}^{m\times 1} : s \rightarrow [s(m+n) \; s(1 : m - 1)^T]^T, \end{aligned} \tag{2.5}$$

where $\mathit{toeplitz}(c, r)$ is a toeplitz matrix specified by its first column c and its first row r. For this particular STLS problem we have that

$$\begin{aligned} &\mathbf{X}(x) : \mathbb{R}^{n\times 1} \rightarrow \mathbb{R}^{m\times(m+n)} : x \rightarrow \\ &\qquad \mathit{toeplitz}([x(n) \;\; -1 \; 0 \dots 0]^T, [x(n : -1 : 1)^T \; 0 \dots 0 \;\; -1]^T) \end{aligned} \tag{2.6}$$

Before we introduce the second type of structure that we will consider in this paper, note that (2.2) can be rewritten as follows

$$\mathbf{X}(x)\Delta s = [\mathbf{X_a}(x) \; \mathbf{X_b}(x)] \begin{bmatrix} \Delta s_a \\ \Delta s_b \end{bmatrix} = [\Delta A \; \Delta b] \begin{bmatrix} x \\ -1 \end{bmatrix}, \tag{2.7}$$

with $\mathbf{X_a}(x) : \mathbb{R}^{n\times 1} \rightarrow \mathbb{R}^{m\times(k-m)}$, $\mathbf{X_b}(\mathbf{x}) : \mathbb{R}^{n\times 1} \rightarrow \mathbb{R}^{m\times m}$, $\Delta s_a \in \mathbb{R}^{(k-m)\times 1}$, $\Delta s_b \in \mathbb{R}^{m\times 1}$ and $\Delta s^T = [\Delta s_a^T \; \Delta s_b^T]$. If $\mathbf{X_b}(x)$ is not rank-deficient, it is clear that we can write the constraints of problem (2.3) as follows:

$$\Delta s_b = -\mathbf{X_b}^{-1}(x)([A\; b]\begin{bmatrix} x \\ -1 \end{bmatrix} + \mathbf{X}_a(x)\Delta s_a).$$

Noticing that the objective function in (2.3) equals

$$0.5\|W^{1/2}[\Delta s_b^T \; \Delta s_a^T]^T\|_2^2,$$

it is obvious that the STLS problem (2.3) can be stated as an *unconstrained* optimization problem:

$$\min_{\Delta s_a, x} 0.5\|\mathbf{f_2}(\Delta s_a, x)\|_2^2, \tag{2.8}$$

where

$$\mathbf{f_2}(\Delta s_a, x) = W^{1/2}[-(\mathbf{X_b}^{-1}(x)([A\; b]\begin{bmatrix} x \\ -1 \end{bmatrix} + \mathbf{X_a}(x)\Delta s_a))^T \;\; \Delta s_a^T]^T.$$

Problem (2.8) is a nonlinear LS (NLLS) problem and can easily be solved by e.g. the Gauss-Newton algorithm [**F, GMW**]. We did not choose this solution strategy for the general STLS problem, since as mentioned before this elimination of the constraints leads to a complicated and unstructured Jacobian of $\mathbf{f_2}$. The latter means that in general there is no structure to exploit at the algorithmic level if this strategy is followed. For some structures however, formulation (2.8) can be greatly simplified. This will be the case for the second structure we consider. The matrix A is Toeplitz again but b is unstructured. The matrix functions used in formulation (2.1) thus become:

$$\begin{aligned} \mathbf{M}(s) &: \mathbb{R}^{(2m+n-1)\times 1} \rightarrow \mathbb{R}^{m\times n} : s \rightarrow \mathit{toeplitz}(s(n : m+n-1), s(n : -1 : 1)) \\ \mathbf{m}(s) &: \mathbb{R}^{(2m+n-1)\times 1} \rightarrow \mathbb{R}^{m\times 1} : s \rightarrow s(m+n : 2m+n-1). \end{aligned} \tag{2.9}$$

Furthermore the matrix functions defined in (2.2) and (2.7) are as follows:

$$\begin{aligned} \mathbf{X}(x) &: \mathbb{R}^{n\times 1} \rightarrow \mathbb{R}^{m\times(2m+n-1)} : x \rightarrow \\ &\quad \mathit{toeplitz}([x(n)\; 0\ldots 0]^T, [x(n : -1 : 1)^T\; 0\ldots 0\; -1\; \underbrace{0\ldots 0}_{m-1}]^T) \\ \mathbf{X_a} &: \mathbb{R}^{n\times 1} \rightarrow \mathbb{R}^{m\times(m+n-1)} : x \rightarrow \\ &\quad \mathit{toeplitz}([x(n)\; 0\ldots 0]^T, [x(n : -1 : 1)^T\; 0\ldots 0]^T) \\ \mathbf{X_b} &: \mathbb{R}^{n\times 1} \rightarrow \mathbb{R}^{m\times m} : x \rightarrow -I_{m\times m}. \end{aligned} \tag{2.10}$$

As can be seen, problem formulation (2.8) is drastically simplified for the second structure we consider and it is possible to apply a simple Gauss-Newton algorithm without destroying the structure present at the algorithmic level:

Algorithm STLS2

Input: $[A\; b] \in \mathbb{R}^{m\times(n+1)}$

Output: the parameter vector $x \in \mathbb{R}^{n\times 1}$ and $\Delta s \in \mathbb{R}^{k\times 1}$ (i.e. the minimal representation of the structured matrix $[\Delta A\; \Delta b]$) that solve (2.3)

Step 1: Initialize Δs and x

Step 2: while *stopcriterion* not satisfied

Step 2.1: Solve the following LS problem

$$\min_{\Delta\tilde{s}_a,\Delta\tilde{x}} \|J_2 \begin{bmatrix} \Delta\tilde{s}_a \\ \Delta\tilde{x} \end{bmatrix} - f_2(\Delta s_a, x)\|_2$$

Step 2.2: $\Delta s_a \leftarrow \Delta s_a + \Delta\tilde{s}_a$
$x \leftarrow x + \Delta\tilde{x}$

end

where $\mathbf{f_2}$ is defined in (2.8) and J_2 is the Jacobian of $\mathbf{f_2}$ w.r.t. $[\Delta s_a^T\ x^T]^T$. Note that for the second type of structure we consider

$$J_2 = \begin{bmatrix} \mathbf{X_a}(x) & A + \mathbf{M}(\Delta s) \\ I_{(m+n-1)\times(m+n-1)} & 0 \end{bmatrix}.$$

Up til now, we left W in (2.3) unspecified. In many applications (see e.g. section 6) W is chosen to be the identity matrix. Therefore, the algorithms that will be developed in the next sections use $W = I$. Concerning the initialization of the algorithms we refer to [**L**], where several options are described. In the examples we describe in section 6, the initialization is simply based on LS estimates.

3. Fast LDL^T decomposition of the considered matrices

In this section we consider the structure at the second level i.e. the algorithmic level. Fast implementations will be developed for the kernel problems (i.e. Step 2.1) of algorithm STLS1 and STLS2. In the case of STLS1 and $[A\ b]$ Toeplitz (i.e. the first type of matrix structure constraint we consider), the kernel problem is a system of linear equations:

$$M_1 z = b_1, \tag{3.1}$$

with

$$M_1 = \begin{bmatrix} I_{(m+n)\times(m+n)} & 0_{(m+n)\times n} & X_1^T \\ 0_{n\times(m+n)} & 0_{n\times n} & \Lambda_1^T \\ X_1 & \Lambda_1 & 0_{m\times m} \end{bmatrix}, \tag{3.2}$$

where $X_1 = \mathbf{X}(x)$, $\Lambda_1 = A + \mathbf{M}(\Delta s) = toeplitz(\lambda(n : m+n-1), \lambda(n : -1 : 1))$ with $\lambda \in \mathbb{R}^{(m+n-1)\times 1}$ and the matrix functions as defined in (2.5)-(2.6).
The kernel problem of algorithm STLS2 for A Toeplitz and b unstructured (i.e. the second type of matrix structure constraint we consider) is the following LS problem

$$\min_y \|M_2 y + b_2\|_2, \tag{3.3}$$

with

$$M_2 = \begin{bmatrix} X_2 & \Lambda_2 \\ I_{(m+n-1)\times(m+n-1)} & 0_{(m+n-1)\times n} \end{bmatrix}, \tag{3.4}$$

where $X_2 = \mathbf{X_a}(x)$ and $\Lambda_2 = A + \mathbf{M}(\Delta s)$ with the matrix functions as defined in (2.9)-(2.10).

The solution of (3.1) can be gained computing the LDL^T factorization of M_1, where L is lower triangular and D is a signature matrix and solving the linear systems

$$\begin{array}{l} Lz_2 = b_1 \\ Dz_1 = z_2 \\ L^T z = z_1 \end{array}. \tag{3.5}$$

The solution of (3.3) can be gained computing first the QR decomposition of $M_2 = Q_2R_2$, then the solution of the seminormal equation

$$M_2^T(M_2y) = -M_2^Tb_2 \equiv R_2^T(R_2y) = -M_2^Tb_2. \tag{3.6}$$

Eventually, one step of iterative refinement can be considered. Despite the diversity of the problems, (3.1) and (3.3) can be solved in a similar way exploiting the displacement structure of the involved matrices $M_2^TM_2$ and M_1.

In fact, both matrices are Toeplitz–like–block matrices. Then an appropriate implementation of the generalized Schur algorithm can be considered to solve both problems in a fast way.

In the next section we will describe the generalized Schur algorithm.

4. The generalized Schur algorithm

In this section we introduce the generalized Schur algorithm to compute the LDL^T factorization of a symmetric matrix A, where L is an upper triangular matrix and D is a signature matrix. A more extensive description of the algorithm can be found in [**D, Mo, K**].

Given a strongly regular[1] $n \times n$ matrix A, and define

$$D_A = A - ZAZ^T,$$

we say that the displacement rank of A is α if $\operatorname{rank}(D_A) = \alpha$, where Z is a lower triangular matrix of order n. The choice of Z depends on the matrix A, e.g., if A is a Toeplitz matrix, Z is chosen equal to the shift matrix. If A is a block–Toeplitz matrix, Z is chosen equal to the block–shift matrix (for a more general choice of the matrix Z, see [**K**]). Clearly D_A will have a decomposition of the form

$$D_A = G^TJ_AG,$$

where

$$G = \begin{bmatrix} g_1^T \\ \vdots \\ g_p^T \\ g_{p+1}^T \\ \vdots \\ g_\alpha^T \end{bmatrix}, \quad J_A = I_p \oplus -I_q, \; q = \alpha - p.$$

The matrix $G \in \mathbb{R}^{\alpha \times n}$ and the vectors g_i, $i = 1, \ldots, \alpha$, are called the generator matrix and the generators of A, respectively. The generators $g_1, \ldots, g_p$ are said to be *positive*, the generators $g_{p+1}, \ldots, g_\alpha$ are said to be *negative*. The pair (p, q) is called the *displacement inertia* of D_A. A matrix Θ is said to be J_A–orthogonal if $\Theta^TJ_A\Theta = J_A$.

A generator matrix is not unique. In fact, if G is a generator matrix of A and Θ is a J_A–orthogonal matrix, then ΘG is a generator matrix of A, too. A generator matrix is said to be in *proper* form if its first nonzero colum has a single nonzero

[1] A square matrix A is said to be *strongly regular* if all its principal minors are different from zero.

entry, i.e.,

$$G = \begin{bmatrix} 0 & * & * & \cdots & * \\ \vdots & \vdots & \vdots & \cdots & \vdots \\ 0 & * & * & \cdots & * \\ * & * & * & \cdots & * \\ 0 & * & * & \cdots & * \\ \vdots & \vdots & \vdots & \cdots & \vdots \\ 0 & * & * & \cdots & * \end{bmatrix},$$

where the elements denoted by "$*$" are generally different from zero.

The number of iterations of the generalized Schur algorithm is equal to the order of the matrix A. Let $G_0 = G$. Denoted by G_{i-1} the generator matrix at the beginning of the i–th iteration, a J_A–orthogonal matrix Θ_i is chosen such that $H_{i-1} = \Theta_i G_{i-1}$ is in proper form. More precisely, denote by f_i the ith column of G_{i-1}, the index of the pivot has to be within $\{1, \ldots, p\}$ if $f_i^T J f_i > 0$ (positive step), within $\{p+1, \ldots, \alpha\}$ if $f_i^T J f_i < 0$ (negative step).

Denote this index by k. Then, the generator matrix G_i is updated in the following way:

$$\begin{aligned} &G_i(k,:) = H_{i-1}(1,:)Z^T \\ &G_i([1:k-1, k+1:\alpha],:) = H_{i-1}([1:k-1, k+1:\alpha],:). \end{aligned}$$

Furthermore, $H(k,:)^T$ becomes the ith column of L. If $f_i^T J f_i > 0$, we set $D(i,i) = 1$. If $f_i^T J f_i < 0$, we set $D(i,i) = -1$. Observe that the case $f_i^T J f_i = 0$ does not occur due to the strong regularity of A [**K**]. Since in general the matrix Θ_i is given by the product of a number of Givens and hyperbolic rotations proportional to α, the computational cost at the i–th iteration is $O(\alpha(n-i+1))$. Hence the computational cost of the generalized Schur algorithm is $O(\alpha n^2)$. In case A is symmetric positive definite, like the matrix $M_2^T M_2$, the signature matrix D is equal to the identity matrix, and L is the Cholesky factor of A, i.e., L^T is the R factor of the QR factorization of A, with all positive entries on the main diagonal.

4.1. The generalized Schur algorithm applied to M_1 and $M_2^T M_2$. Before applying the generalized Schur algorithm to M_1 and $M_2^T M_2$, we consider a permutation matrix P_1 in order to transform M_1 into the Toeplitz–block matrix

$$K = P_1 M_1 P_1^T = \begin{bmatrix} I_{(m+n)\times(m+n)} & 0_{(m+n)\times n} & X^T \\ 0_{n\times(m+n)} & 0_{n\times n} & \Lambda_1^T \\ X & \Lambda_1 & 0_{m\times m} \end{bmatrix}.$$

where

$$X = \begin{bmatrix} -1 & x_n & \cdots & x_1 & 0 & \cdots & \cdots & 0 \\ 0 & -1 & x_n & \cdots & x_1 & 0 & & \vdots \\ \vdots & & \ddots & \ddots & & \ddots & & \vdots \\ \vdots & & & \ddots & \ddots & & \ddots & 0 \\ 0 & \cdots & \cdots & 0 & -1 & x_n & \cdots & x_1 \end{bmatrix}.$$

Furthermore we observe that the matrix K is not strongly regular. In fact $\det(K(1:i, 1:i) = 0, i = m+n+1, \ldots, m+2n$. Hence a second permutation matrix P_2 is considered in order to transform K into the Toeplitz–block matrix $\tilde{K}$, i.e.,

$$\tilde{K} = P_2 K P_2^T = \begin{bmatrix} I_{(m+n)\times(m+n)} & X^T & 0_{(m+n)\times n} \\ X & 0_{m\times m} & \Lambda_1 \\ 0_{n\times(m+n)} & \Lambda_1^T & 0_{n\times n} \end{bmatrix}.$$

It is easy to prove that $\tilde{K}$ is strongly regular. Considering the Schur complement of $I_{(m+n)\times(m+n)}$ in $\tilde{K}$ we can obtain the following partial LDL^T decomposition of $\tilde{K}$ without any additional cost (of course the product XX^T is not explicitly computed),

$$\tilde{K} = \begin{bmatrix} I_{(m+n)\times(m+n)} & & \\ X & I & \\ 0_{n\times(m+n)} & & I \end{bmatrix} \begin{bmatrix} I_{(m+n)\times(m+n)} & \\ & \hat{K} \end{bmatrix} \begin{bmatrix} I_{(m+n)\times(m+n)} & X^T & 0_{(m+n)\times n} \\ & I & \\ & & I \end{bmatrix},$$

where the matrix

$$\hat{K} = \begin{bmatrix} -XX^T & \Lambda_1 \\ \Lambda_1^T & 0_{n\times n} \end{bmatrix}$$

of order $m+n$ is the Schur complement of $I_{(m+n)\times(m+n)}$ in the matrix $\tilde{K}$. Then the problem is reduced to compute the LDL^T decomposition of $\hat{K}$.

Let $Z^{(1)} = Z_m \oplus Z_n$ and $Z^{(2)} = Z_{m+n-1} \oplus Z_n$ be two shift-block matrices, where

$$Z_k = \begin{bmatrix} 0 & 0 & \cdots & 0 \\ 1 & 0 & \cdots & 0 \\ & \ddots & \ddots & \\ & & 1 & 0 \end{bmatrix} \in \mathbb{R}^{k\times k}.$$

Then the displacement rank of $\hat{K}$ with respect to $Z^{(1)}$ is 4, the displacement rank of $M_2^T M_2$ with respect to $Z^{(2)}$ is 5.

Denote by $v_1 = X(1,:)^T$, and $v = v_1/\|v_1\|_2$. Let $y = -Xv$ and $w = [\lambda(n), \lambda(n-1), \ldots, \lambda(1)]^T/\|v_1\|_2$. Then, the generators of $\hat{K}$ are defined in the following way.

$$\begin{aligned} g_1 &= [y^T, w^T]^T \\ g_2 &= [0, y(2:m)^T, w^T]^T \\ g_3 &= [0, \lambda(m+n-1), \lambda(m+n-2), \ldots, \lambda(n+1), .5, 0, \ldots, 0]^T \\ g_4 &= [0, \lambda(m+n-1), \lambda(m+n-2), \ldots, \lambda(n+1), -.5, 0, \ldots, 0]^T, \end{aligned}$$

where g_2 and g_3 are positive, g_1 and g_4 are negative.

Let $\tilde{w} = M_2(2:2m+n-1, m+n)/\|M_2(2:2m+n-1, m+n)\|_2$ and $t = M_2(2:2m+n-1, 2:m+2n-1)^T\tilde{w}$. Then the following vectors are generators of $M_2^T M_2$.

$$\begin{aligned} h_1 &= M_2(1,:)^T \\ h_2 &= e_1 \\ h_3 &= [0, t^T]^T \\ h_4 &= [0, t(1:m+n-2)^T, 0, t(m+n:m+2n-2)^T]^T \\ h_5 &= [0, M_2(m, 1:m+n-2)^T, 0, M_2(m, m+n:m+2n-2)^T]^T, \end{aligned}$$

where $e_1 = [1, \underbrace{0, \ldots, 0}_{m+2n-2}]^T$.

Since the orders of the matrices $\hat{K}$ and $M_2^T M_2$ are $m+n$ and $m+2n-1$, respectively, the computational cost of the generalized Schur algorithm should be proportional to $(m+n)^2$. In the next section we will show that, exploiting the particular structure of the generators of $\hat{K}$ and $M_2^T M_2$, the computational cost of the generalized Schur algorithm can be reduced to $O(mn + n^2)$.

As the technique to reduce the number of computations is similar for both the problems, for the sake of brevity we describe only the fast algorithm to compute the LDL^T factorization of $\hat{K}$. The implementation details of the algorithm for the problem (3.3) can be found in [**MLV**].

We observe that the matrix $\hat{K}$ is indefinite. However, analyzing the generators and the Schur complement of $-XX^T$ in $\hat{K}$ we are able to say a priori that the steps of the algorithm for $i = 1, \ldots, m$ are negative, the steps for $i = m+1, \ldots, m+n$, are positive.

REMARK 4.1. We observe that $g_1(1:m), g_2(1:m)$ are the generators for the symmetric negative definite Toeplitz matrix $-XX^T$. If we denote by $\hat{G}_1 = \begin{bmatrix} g_1^T(1:m) \\ g_2^T(1:m) \end{bmatrix}$, the generator matrix for $-XX^T$, by $\hat{G}_i$ and $\hat{f}_i$ the updated generator matrix and the ith column of $\hat{G}_i$, respectively, obtained at the i–th iteration for the computation of the LDL^T factorization of $-XX^T$, we have that $\hat{f}_i^T \begin{bmatrix} -1 & 0 \\ 0 & 1 \end{bmatrix} \hat{f}_i < 0$.

4.2. Description of the algorithm. As introduced in section 4, at each iteration i, we look for a J–orthogonal matrix Θ_i in order to eliminate all elements of f_i,the i–th column of G_i with exception of one element. This can be done by choosing J–orthogonal matrices Φ such that $\Phi \begin{bmatrix} f_i(j) \\ f_i(k) \end{bmatrix} = \begin{bmatrix} * \\ 0 \end{bmatrix}$. Φ can be either a Givens rotation (updating) if $\{j,k\} \in \{1,4\}$ or $\{j,k\} \in \{2,3\}$, or a hyperbolic rotation (downdating) elsewhere. Proceeding in this way we can eliminate all the entries of f_i with exception of a single pivot element. We choose the index of the pivot element equal to 1 if $f_i^T J f_i < 0$ and we set $D(i,i) = -1$ (negative step), equal to 2 in the other case and we set $D(i,i) = 1$ (positive step). We perform the downdating step by means of a *mixed* hyperbolic rotation [**BBVD, SV**].

We divide the algorithm in 4 phases:

- 1st phase: iteration `for` $i = 1$,
- 2nd phase: iteration `for` $i = 2, \ldots, m-n$,
- 3rd phase: iteration `for` $i = m-n+1, \ldots, m$,
- 4th phase: iteration `for` $i = m+1, \ldots, m+n$.

4.2.1. *1st phase: iteration for* $i = 1$. $g_1^{(0)}$ is the only vector with the first entry different from zero. Then we set $L(:,1) = g_1^{(0)}, D(1,1) = -1$, $g_1^{(1)}(2:m+n) = g_1^{(0)}(1:m+n-1), g_1^{(1)}(1) = 0, g_1^{(1)}(m+1) = 0$

4.2.2. *2nd phase: iterations for* $i = 2:m-n$. Before describing this phase we observe that the vectors $g_3^{(i-1)}$ and $g_4^{(i-1)}$ differ only for the $(m+1)$th entry. Then we will see that the updating of $g_1^{(i-1)}$ with $g_4^{(i-1)}$ and the downdating with $g_3^{(i-1)}$ modifies only the $(m+1)$th entry of $g_1^{(i-1)}$. Hence $g_1^{(i-1)}(1:m)$ and $g_2^{(i-1)}(1:m)$ continue to be the generator vectors at the beginning of the ith step for the LDL^T factorization of $-XX^T$. Thus each iteration of this phase is a negative step since $f_i^T J f_i < 0$. Now we describe how the generators are modified at each iteration of this phase. We have to update $g_1^{(i-1)}$ with $g_4^{(i-1)}$ and downdate with $g_3^{(i-1)}$. These vectors are

$$g_1^{(i-1)} = [\underbrace{0,\ldots,0}_{i-1}, \xi_i, \ldots, \xi_{n+i}, \underbrace{0,\ldots,0}_{m-n-i}, \xi_{m+1}, \ldots, \xi_{m+n}]^T \tag{4.1}$$

$$g_4^{(i-1)} = [\underbrace{0,\ldots,0}_{i-1}, \zeta_i, \ldots, \zeta_m, \zeta_{m+1}, \ldots, \zeta_{m+n}]^T$$

$$g_3^{(i-1)} = [\underbrace{0,\ldots,0}_{i-1}, \zeta_i, \ldots, \zeta_m, \mu_{m+1}, \ldots, \zeta_{m+n}]^T.$$

The Givens rotation used to update $g_1^{(i-1)}$ with $g_4^{(i-1)}$ is

$$G = \begin{bmatrix} c_G^{(i-1)} & s_G^{(i-1)} \\ -s_G^{(i-1)} & c_G^{(i-1)} \end{bmatrix}, \text{ with } c_G^{(i-1)} = \frac{\xi_i}{\sqrt{\xi_i^2 + \zeta_i^2}} \text{ and } s_G^{(i-1)} = \frac{\zeta_i}{\sqrt{\xi_i^2 + \zeta_i^2}}.$$

The updated vectors $\tilde{g}_1^{(i-1)}$ and $\tilde{g}_4^{(i-1)}$ are

$$\tilde{g}_1^{(i-1)} = c_G^{(i-1)} g_1^{(i-1)} + s_G^{(i-1)} g_4^{(i-1)} \tag{4.2}$$

$$\tilde{g}_4^{(i-1)} = -s_G^{(i-1)} g_1^{(i-1)} + c_G^{(i-1)} g_4^{(i-1)}, \tag{4.3}$$

with $\tilde{g}_1^{(i-1)} = [\underbrace{0,\ldots,0}_{i-1}, \tilde{\xi}_i, \ldots, \tilde{\xi}_{m+n}]^T \quad \left(\tilde{\xi}_i = \sqrt{\xi_i^2 + \zeta_i^2}\right)$. Moreover,

$$\tilde{g}_4^{(i-1)}(n+i+1:m) = c_G^{(i-1)} g_4^{(i-1)}(n+i+1:m), \tag{4.4}$$

since $g_1^{(i-1)}(j) = 0, j = n+i+1, \ldots, m$. The next step in the ith iteration is downdating $\tilde{g}_1^{(i-1)}$ with $g_3^{(i-1)}$ by means of a mixed hyperbolic rotation

$$H = \begin{bmatrix} 1 & 0 \\ \rho & \sqrt{1-\rho^2} \end{bmatrix} \begin{bmatrix} \frac{1}{\sqrt{1-\rho^2}} & 0 \\ 0 & 1 \end{bmatrix} \begin{bmatrix} 1 & \rho \\ 0 & 1 \end{bmatrix},$$

where ρ is such that $H[\tilde{\xi}_i, \zeta_i]^T = [\hat{\xi}_i, 0]^T$. Taking (4.2) into account, it is straightforward to see that

$$\rho = -s_G^{(i-1)} \text{ and } \sqrt{1-\rho^2} = c_G^{(i-1)}.$$

The downdated vectors $\hat{g}_1^{(i-1)}$ and $\tilde{g}_3^{(i-1)}$ are

$$\hat{g}_1^{(i-1)} = \frac{\tilde{g}_1^{(i-1)} - s_G^{(i-1)} g_3^{(i-1)}}{c_G^{(i-1)}} = g_1^{(i-1)} - s_G^{(i-1)} \frac{g_3^{(i-1)} - g_4^{(i-1)}}{c_G^{(i-1)}} \tag{4.5}$$

and

$$\tilde{g}_3^{(i-1)} = -s_G^{(i-1)} \hat{g}_1^{(i-1)} + c_G^{(i-1)} g_3^{(i-1)} \tag{4.6}$$

Hence,

$$\begin{aligned} \tilde{g}_3^{(i-1)} &= -s_G^{(i-1)} \hat{g}_1^{(i-1)} + c_G^{(i-1)} g_3^{(i-1)} \\ &= -s_G^{(i-1)} g_1^{(i-1)} + \frac{c_G^{(i-1)^2} g_3^{(i-1)} - s_G^{(i-1)^2} (g_4^{(i-1)} - g_3^{(i-1)})}{c_G^{(i-1)}} \\ &= -s_G^{(i-1)} g_1^{(i-1)} + \frac{g_3^{(i-1)} - (1 - c_G^{(i-1)^2}) g_4^{(i-1)}}{c_G^{(i-1)}} \end{aligned} \tag{4.7}$$

Hence, from (4.6) and (4.3), $\tilde{g}_3^{(1)}$ and $\tilde{g}_4^{(1)}$ continue to be equal, except for the $(m+1)$th entry. Furthermore, from (4.5), we observe that $g_1^{(i-1)}$ and $\hat{g}_1^{(i-1)}$ differ in their $(m+1)$th entry. Since $g_1^{(i-1)}(m+1) = 0$, then $\hat{g}_1^{(i-1)}(m+1) = -s_G^{(i-1)}(g_3^{(i-1)}(m+1) - g_4^{(i-1)}(m+1))/c_G^{(i-1)}$. We observe that is not necessary to compute the whole vector in (4.4) since, at the next iteration, the corresponding entries of $g_1^{(i)}(n+i+2 : m+1)$ are equal to 0. We need only to store the partial product

$$c_G^{(i-2)} \cdots c_G^{(2)} c_G^{(1)} \tag{4.8}$$

into a temporary variable, and multiply $g_4^{(i)}(n+i+1)$ with this variable at the beginning of the ith iteration.

To finish the iteration $\hat{g}_1^{(i-1)}$ has to be downdated with $g_2^{(i-1)}$. This computation does not distroy the structure of the vectors since

$$\begin{aligned} \hat{g}_1^{(i-1)} &= [\underbrace{0,\ldots,0}_{i-1}, \underbrace{*,\cdots,*,*}_{n+1}, \underbrace{0,\ldots,0}_{m-n-i}, \underbrace{*,\cdots,*}_{n}]^T \\ g_2^{(i-1)} &= [\underbrace{0,\ldots,0}_{i-1}, \underbrace{*,\cdots,*}_{n}, \underbrace{0,\ldots,0}_{m-n-i+1}, \underbrace{*,\cdots,*}_{n}]^T. \end{aligned}$$

Let $\tilde{H}$ be the stabilized hyperbolic rotation such that

$$\tilde{H} \begin{bmatrix} \hat{g}_1^{(i-1)}(i : m+n) \\ g_2^{(i-1)}(i : m+n) \end{bmatrix}^T = \begin{bmatrix} \breve{g}_1^{(i-1)}(i : m+n) \\ \breve{g}_2^{(i-1)}(i : m+n) \end{bmatrix}^T$$

with $\breve{g}_2^{(i-1)}(i) = 0$. Then $\breve{g}_1^{(i-1)}$ becomes the ith column of L, $D(i,i) = -1$, and, for the next iteration, the updated vectors are

$$\begin{aligned} g_1^{(i)} &= [0, \hat{g}_1^{(i-1)}(i+1 : m+n)],\ g_1^{(i)}(m+1) = 0. \\ g_2^{(i)} &= \breve{g}_2^{(i-1)} \\ g_4^{(i)} &= \tilde{g}_4^{(i-1)} \\ g_3^{(i)} &= [\tilde{g}_4^{(i-1)}(1:m); \gamma; \tilde{g}_4^{(i-1)}(m+2 : m+n)] \end{aligned},$$

where $\gamma = c_G^{(i-1)} g_3^{(i-1)}(m+1) - s_G^{(i-1)} \hat{g}_3^{(i-1)}(m+1)$. The number of flops of this phase is $18mn - 18n^2$.

4.2.3. *3rd phase: iterations for* $i = m-n+1 : m$. The iteration of this phase are very similar to those of the previous one. However we do not need to store the product of the Givens coefficients $c_G^{(i-1)}$ into a temporary variable, since $g_1^{(i-1)}(k) \neq 0, k = i, \ldots, m+n$. We recall that $g_3^{(i-1)}$ and $g_4^{(i-1)}$ continue to be equal with exception of the $(m+1)$th entry. Thus $g_1^{(i-1)}(i:m)$ and $g_2^{(i-1)}(i:m)$ are the generator vectors at the ith iteration of the LDL^T factorization of $-XX^T$. Hence each iteration of this phase continues to be a negative step ($D(i,i) = -1$). The number of flops of this phase is $13.5n^2$.

4.2.4. *4th phase: iterations for* $i = m+1 : m+n$. In this phase the vectors $g_3^{(i-1)}(m+1 : m+n)$ and $g_4^{(i-1)}(m+1 : m+n)$ are different. Now we observe that the vectors $g_i^{(m)}, i = 1, \ldots, 4$ are the generators for the Schur complement of $-XX^T$ in the matrix $\hat{K}$ [**KS**], that is, the generators for the $\Lambda^T(XX^T)^{-1}\Lambda$, a symmetric and positive definite matrix. Then each iteration of this phase is a positive step, that is $D(i,i) = 1$ and the ith column of L is $\breve{g}_2^{(i-1)}$. The number of flops of this phase is $9n^2$.

4.3. Stability of LDL^T factorization. The stability of the proposed generalized Schur algorithm is studied in [**MVV**]. The stability properties of the algorithms for the considered problems depend on the implementation of the hyperbolic rotations.

In [**MVV**] it is proved that the following results holds for the LDL^T factorization of $\hat{K}$, provided the hyperbolic rotations are implemented in a stable way [**BBVD, CS**].

THEOREM 4.2. *Let G be the generator matrix of $\hat{K}$. Let L and D be the matrices of the LDL^T factorization of $\hat{K}$ computed by means of the generalized Schur algorithm applying a sequence of Givens rotations and two mixed hyperbolic rotations per iteration. Then*

$$\|\hat{K} - LDL^T\|_F \leq 62(m+n-1)(m+n)\varepsilon\left(2\sqrt{m+n}\|\hat{K}\|_F + \|G_1\|_F^2\right).$$

A similar results holds for the $R^T R$ factorization of $M_2^T M_2$. Hence the proposed algorithms are weakly stable.

5. Solution of the linear systems

In this section we evaluate the computational cost of the solution of the linear systems (3.5). The solution of (3.6) has the same order of complexity. The details for this case can be found in [**MLV**].

Having computed the following factorization of $\tilde{K}$ in $O(mn+n^2)$ flops,

$$\begin{aligned}\tilde{K} &= \begin{bmatrix} I & & \\ X & I & \\ 0_{n\times(m+n)} & & I \end{bmatrix} \begin{bmatrix} I & \\ & R^T \end{bmatrix} \begin{bmatrix} I & \\ & D \end{bmatrix} \begin{bmatrix} I & \\ & R \end{bmatrix} \begin{bmatrix} I & X^T & 0_{(m+n)\times n} \\ & I & \\ & & I \end{bmatrix} \\ &= L_1 L_2 D_1 L_2^T L_1^T, \end{aligned}$$

we need now to solve 5 linear systems, with coefficient matrices $L_1, L_2, D_1, L_2^T, L_1^T$, respectively. The solution of the systems with coefficient matrix L_1 and L_1^T can be computed in $O(mn)$ flops. The solution of the linear system with coefficient matrix D_1 is obtained changing the sign of the entries $m+n+1, \ldots, 2m+n$, of b_1. Furthermore, the solution of the linear systems with coefficient matrix L_2 and L_2^T can be computed in $0(mn+n^2)$ flops since

$$R = \begin{bmatrix} * & * & * & * & & & * & * & * \\ & * & * & * & * & & * & * & * \\ & & * & * & * & * & * & * & * \\ & & & * & * & * & * & * & * \\ & & & & * & * & * & * & * \\ & & & & & * & * & * & * \\ & & & & & & * & * & * \\ & & & & & & & * & * \\ & & & & & & & & * \end{bmatrix},$$

where the first row is $R(1,:) = [\underbrace{*,\ldots,*}_{n+1},\underbrace{0,\ldots,0}_{m-n-1},\underbrace{*,\ldots,*}_{n},]^T$.

Hence the solution of the linear system (3.5) has the same computational complexity of the LDL^T factorization of $\hat{K}$.

6. Applications

6.1. System identification. In this section we present a time-domain system identification example. The starting point consists of some noisy input measurements $u(t)$ and some noisy output measurements $y(t)$ of a linear system. The goal is to determine the impulse response function that characterizes the linear system. If no noise were present and the system is of order n the following relation holds between the noiseless input u_0 and the noiseless output y_0:

$$y_0(t) = \sum_{i=0}^{n} x_0(i)u_0(t-i), \tag{6.1}$$

where x_0 is the true impulse response. In reality we only have noisy measurements u and y such that (6.1) is only satisfied approximately. By writing down (6.1) for several noisy measurements, we obtain a structured overdetermined system of equations that corresponds to the matrix structure constraint of the second type we considered. If we assume that the input and output measurements are contaminated by i.i.d. Gaussian noise, a ML estimate of x_0 can be obtained by solving the STLS problem (2.1) involving the second matrix structure constraint (A being Toeplitz and b unstructured).
To illustrate the improved statistical accuracy of STLS (obtained by exploiting the structure at the first level) compared to other methods such as TLS (that do not exploit the structure at this first level), we perform a Monte-Carlo study consisting of 100 runs of the following simulation experiment. We take the following true impulse response x_0:

$$[1.9\ 3.3\ 4.4\ 5.4\ 5.9\ 6.2\ \ 6.3\ 6.1\ 5.8\ 5.6\ 5.3\ 5.0\ 4.85\ 4.6\ 4.0\ 3.4\ 1.8\ 1.0\ 0.2\ 0.0]^T.$$

For u_0 we take i.i.d. Gaussian noise of standard deviation 1. Using relation (6.1) we calculate m values of y_0. Finally both u_0 and y_0 are contaminated with i.i.d. Gaussian noise of standard deviation σ_n. Using the noisy measurements y and u we use both the TLS and STLS method to obtain an estimate of the parameter vector, namely x_{TLS} and x_{STLS}. Finally we calculate for each run the following relative errors:

$$err_{TLS} = \frac{\|x_{TLS} - x_0\|_2}{\|x_0\|_2} \text{ and } err_{STLS} = \frac{\|x_{STLS} - x_0\|_2}{\|x_0\|_2}.$$

The results of this Monte-Carlo study are summarized in table 1. The statistical accuracy improves for this particular example with a factor 1.5.

6.2. Speech compression scheme.

6.2.1. *Description.* In this subsection we describe a subband based speech compression scheme. The incoming speech signal is processed on a frame by frame basis. In a first step a frame of the incoming signal is fed to a lowpass filter. This yields the low-frequency (LF) part. By subtracting the latter from the incoming signal, we obtain the high-frequency (HF) , non-smoothed part. The motivation behind splitting the speech signal into two bands is that the LF part of a speech signal can

TABLE 1. This table shows the increased statistical performance when structure is exploited at the first level: we compare the TLS and STLS estimates for the system identification example.

σ_n	err_{TLS}	err_{STLS}
$1e-8$	$1.22e-8$	$7.84e-9$
$1e-6$	$1.20e-6$	$7.94e-7$
$1e-4$	$1.19e-4$	$7.46e-5$

efficiently be modeled by means of a linear model (Auto Regressive (AR) model). Instead of determining a classical AR model, leading to a vector of model parameters and a residual vector [**LJU**], we apply a minimal correction to the measured data such that the corrected measurements exactly satisfy an AR model. As shown in [**LDV, L**], the latter procedure corresponds to a Toeplitz STLS problem (2.1) (with $W = I_{m+n}$) in which the observed data vector s corresponds to the LF part of the incoming frame (i.e. $A = \mathbf{M}(s)$, $b = \mathbf{m}(s)$ with $\mathbf{M}(s)$, $\mathbf{m}(s)$ as defined in (2.5), the corrections we apply correspond to the vector Δs and the model for that particular frame is represented by the vector x. The compression comes about as follows. Due to the Toeplitz structure of $[A + \Delta A \; b + \Delta b]$ it should be clear that we only need $[s(1) + \Delta s(1) \; s(2) + \Delta s(2) \dots s(n) + \Delta s(n)]^T \in \mathbb{R}^{n\times 1}$ and $x \in \mathbb{R}^{n\times 1}$ in order to reconstruct the entire frame $s + \Delta s \in \mathbb{R}^{(m+n)\times 1}$. Thus, loosely speaking, we obtain a compression ratio of $(m+n)/(2n)$. For further implementation details we refer to [**LDV, L**].

6.2.2. *Numerical tests.* In this subsection we compare the efficiency of two implementations of algorithm STLS1 for the first type of matrix structure constraints. The first implementation is the fast implementation of the STLS1 algorithm as described in previous sections. It will be referred to as STLS1f. We also consider a straightforward implementation of algorithm STLS1, referred to as STLS1s, in which Step 2.1 of algorithm STLS1 is solved by Gaussian elimination with partial pivoting. No use is made of the particular structure of the matrix involved in this system of equations. As shown in the previous section, the computational complexity of STLS1f is $O(mn + n^2)$. The other implementation, STLS1s, clearly has a complexity of $O((m+n)^3)$.

The speech signal (see left hand side of figure 1) we will compress is 15000 samples long. It is processed on a frame by frame basis, each frame being 500 samples long. As mentioned before each frame is split in a LF and a HF part. It is the LF part that is compressed using the STLS problem formulation. The LF part of the frame can be modelled sufficiently well using a 4th order model, thus yielding the following dimensions of the STLS problem (2.1): $m = 496$, $n = 5$. The right hand side of figure (1) shows the original 6th frame (i.e. the signal s, full line) and the reconstructed frame obtained after compression (i.e. the signal $s + \Delta s$, dashed line). The latter frame can be reconstructed using only the first 4 samples of $s + \Delta s$ and the parameter vector $x \in \mathbb{R}^{4\times 1}$.

Per frame, the two implementations STLS1f and STLS1s require the same (but varying) number of iterations. The important figure is thus the number of flops per iteration. For this particular problem these figures are displayed in table 2 (see first line of the table). We clearly see the drastically increased computational performance obtained with STLS1f. To investigate the dependence of the computational cost of the different implementations on the size of the problem, we change the parameters of the compression scheme. The results are shown on the second line of table 2. In this case frames of length 252 are used and the estimated order is still 4. This yields $m \times n = 248 \times 4$. Going from the second to the first line of the table, m is doubled. As proven theoretically, the number of flops for STLS1s is multiplied by 8 whereas the number of flops for STLS1f only doubles.

TABLE 2. This table shows the increased performance of the implementation STLS1f compared to the straightforward implementation STLS1s, for different problem sizes.

frame length	estimated order	$m \times n$	STLS1s	STLS1f
500	4	496×4	$6.744e8$	129756
252	4	248×4	$8.734e7$	65028

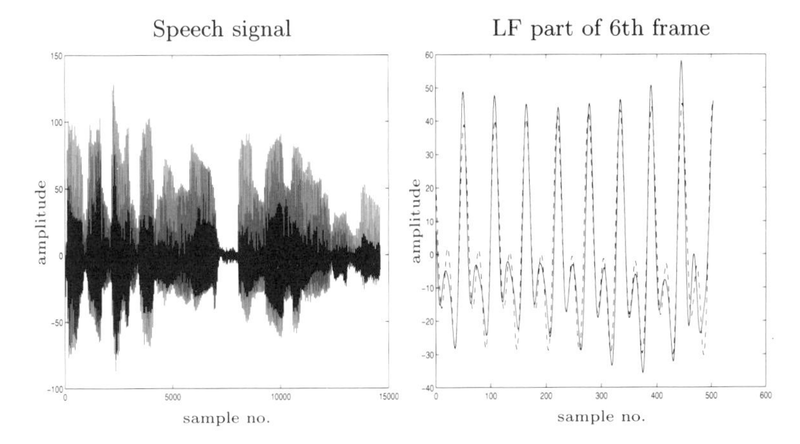

FIGURE 1. *The left hand side of this figure shows the original speech signal. The right hand side shows the* 6*th frame of the original signal (i.e.* s*, the full line) toghether with the reconstructed (after compression)* 6*th frame (dashed line).*

7. Conclusions

In this paper we described STLS problems involving two different types of matrix structures that often occur in applications. For both types of matrix structure constraints, we developed fast algorithms based on the low displacement rank of the involved matrices. An extra increase in computational efficiency is obtained by exploiting the sparsity of the corresponding generators. By means of a system identification example we have shown that the exploitation of structure at the first level (i.e. the level of the problem formulation) yields statistically more accurate results. The increased computational performance (obtained by taking advantage of the structure at the second level, i.e. the algorithmic level) is illustrated by means of a novel speech compression scheme.

References

[AD] D. Alpay and H. Dym, *On application of reproducing kernel spaces to the Schur algorithm and rational J unitary interpolation,* in: I.Schur Methods in Operator Theory and Signal Processing, (I.Gohberg, ed.), Oper. Theory: Adv. Appl., OT18, Birkhäuser-Verlag, Basel, 1986, pp. 89–159.

[BBVD] A.W. Bojanczyk, R.P. Brent, P. Van Dooren and F.R. De Hoog, *A note on downdating the Cholesky factorization,* SIAM J. Sci. Stat. Comput., **1** (1980), pp. 210–220.

[CS] S. Chandrasekaran and Ali H. Sayed, *Stabilizing the generalized Schur algorithm*, SIAM J. Matrix Anal. Appl., **17** (1996), pp. 950–983.

[CKL] J. Chun, T. Kailath and H. Lev–ari, *Fast parallel algorithms for QR and triangular factorization*, SIAM J. Sci. and Stat. Comp., **8** No. 6 (1987), pp. 899–913.

[D] J.M.Delosme, *Fast algorithms for finite shift-rank processes*, Ph.D.Thesis, Stanford University, Stanford, CA, 1982.

[Dy] H.Dym, *Structured Matrices, Reproducing Kernels and Interpolation,* This volume.

[F] R. Fletcher. *P*ractical Methods of Optimization. John Wiley & Sons, New York, 1987.

[GMW] P. E. Gill, W. Murray, and M. H. Wright. *P*ractical Optimization. Academic Press, London, 1981.

[GO] I.Gohberg and V.Olshevsky, *Fast state space algorithms for matrix Nehari and Nehari-Takagi interpolation problems*, Integral Equations and Operator Theory, **20, No. 1** (1994), 44-83.

[GV] G. H. Golub and C. F. Van Loan, *Matrix Computations,* Third ed., The John Hopkins University Press, Baltimore, MD, 1996.

[GV2] G. H. Golub and C.F. Van Loan. An analysis of the total least squares problem. *S*IAM J. Numer. Anal., 1980.

[K] T. Kailath, *Displacement structure and array algorithms,* in Fast Reliable Algorithms for Matrices with Structure, T. Kailath and A. H. Sayed, Ed., SIAM, Philadelpia, 1999.

[KC] T. Kailath, and J. Chun, *Generalized displacement structure for block–Toeplitz, Toeplitz–block and Toeplitz–derived matrices,* SIAM J. Matrix Anal. Appl., **15** (1994), pp. 114–128.

[KKM] T. Kailath, S. Kung and M. Morf, *Displacement ranks of matrices and linear equations*, J. Math. Anal. Appl., **68** (1979), pp. 395–407.

[KS] T. Kailath and A.H. Sayed, Displacement structure:theory and applications, SIAM Review, **37** (1995), pp. 297–386.

[LMV] P. Lemmerling, N. Mastronardi and S. Van Huffel, *Fast algorithm for solving the Hankel/Toeplitz structured total least squares problem,* Numerical Algorithms, to appear.

[L] P. Lemmerling. *S*tructured total least squares: analysis,algorithms and applications. PhD thesis, Dept. of Elect. Eng., ESAT-SISTA, Katholieke Universiteit Leuven, May 1999.

[LDV] P. Lemmerling, I. Dologlou, and S. Van Huffel. Variable rate speech compression based on exact modeling and waveform vector quantization. In *S*ignal Processing Symposium (SPS 98), pages 127–130, Katholieke Universiteit Leuven, Leuven, March 1998. IEEE Benelux Signal Processing Chapter.

[LJU] L. Ljung. *S*ystem identification: theory for the user. Prentice-Hall, Inc., 1987.

[MLV] N. Mastronardi, P. Lemmerling and S. Van Huffel, *Fast structured total least squares algorithm for solving the basic deconvolution problem,* ESAT-SISTA Report TR 98-89, ESAT Laboratory, K.U.Leuven, Belgium, September 1998, SIAM J. Matrix Anal., to appear.

[MVV] N. Mastronardi, P. Van Dooren and S. Van Huffel, *Stability of the generalized Shur algorithm*, ESAT-SISTA Report TR 99-82, ESAT Laboratory, K.U.Leuven, Belgium, October 1999.

[M] The MathWorks Inc. *M*ATLAB User's Guide. High-Performance Numeric Computation and Visualization Software, August 1992.

[Mo] M.Morf, *Fast algorithms for Multivariable Systems*, Ph.D.Thesis, Stanford University, Stanford, CA, 1974.

[O] V. Olshevsky, *Pivoting for structured matrices with Applications,* 1997, to appear in LAA, www.cs.gsu.edu/ matvro/papers.html

[OP] V.Olshevsky and V.Pan, *A unified superfast algorithm for boundary rational tangential interpolation problems and for inversion and factorization of dense structured matrices*, *Proc. of 39th Annual Symposium on Foundations of Computer Science (FOCS'98)*, IEEE Computer Society, Los Alamitos, CA, 1998, 192-201.

[OS] V.Olshevsky and A.Shokrollahi, *Fast matrix-vector multiplication algorithms for confluent Cauchy-like matrices with applications*, *Proceedings of the 31st Annual ACM Symposium on Theory of Computing (STOC'00)*, 2000.

[SV] M. Stewart and P. Van Dooren, *Stability issues in the factorization of structured matrices*, SIAM J. Matrix Anal. Appl., **18** (1997), pp. 104–118.

[V] S. Van Huffel and J. Vandewalle. *The* Total Least Squares Problem: computational aspects and analysis, volume 9. SIAM, Philadelphia, 1991.

DIPARTIMENTO DI MATEMATICA, UNIVERSITÀ DELLA BASILICATA, VIA N. SAURO, 85, 85100 POTENZA, ITALY

Current address: Department of Electrical Engineering, ESAT-SISTA/COSIC, Katholieke Universiteit Leuven, Kardinaal Mercierlaan 94, 3001 Heverlee, Belgium

E-mail address: nicola@esat.kuleuven.ac.be

DEPARTMENT OF ELECTRICAL ENGINEERING, ESAT-SISTA/COSIC, KATHOLIEKE UNIVERSITEIT LEUVEN, KARDINAAL MERCIERLAAN 94, 3001 HEVERLEE, BELGIUM

E-mail address: philippe.lemmerling@esat.kuleuven.ac.be

DEPARTMENT OF ELECTRICAL ENGINEERING, ESAT-SISTA/COSIC, KATHOLIEKE UNIVERSITEIT LEUVEN, KARDINAAL MERCIERLAAN 94, 3001 HEVERLEE, BELGIUM

E-mail address: sabine.vanhuffel@esat.kuleuven.ac.be

Contemporary Mathematics
Volume **280**, 2001

Exploiting Toeplitz structure in atmospheric image Restoration

W. K. Cochran, R. J. Plemmons, and T. C. Torgersen

Abstract. A phase-diversity-based approach is taken in this paper for developing numerical techniques and a software package for post-processing images taken through the atmosphere, where the blurring process is spatially invariant. This leads to a block Toeplitz with Toeplitz blocks (BTTB) blurring matrix, and enables Fast Fourier Transforms (FFTs) to be used in the reconstruction computations. The approach is based on a multiframe/multichannel formulation, where noisy, differently blurred short exposure images of the same object are available. Such situations occur when the same object is observed at successive time instants through the turbulent atmosphere, with a different transfer function at each instant. In general, the BTTB transfer functions differ primarily by changes in their phase. Phase-diversity image data is used to simultaneously estimate the object (i.e., the true image) and the phase, or wavefront profile. In our approach, regularization is applied and a resulting large-scale nonlinear optimization problem is solved efficiently using a limited memory quasi-Newton method with bound constraints and efficient application of FFTs to evaluate the objective function and its gradient. Test results in applying our software package to atmospheric image data are presented.

1. Introduction

Phase-diversity-based speckle (PDS) phase recovery and image reconstruction is an advanced imaging technique for restoring fine-resolution detail when imaging in the presence of phase aberrations such as atmospheric turbulence. The method, which in its basic form was developed by Gonsalves [**3**], has been applied using various numerical optimization techniques (see, e.g., [**7, 12, 15, 17**]), and its overall effectiveness for space-object identification has been well-accepted [**12, 15**]. The primary drawback to PDS in comparison the other image reconstruction schemes such as direct hardware implemented adaptive optics methods and deconvolution by wavefront sensing, has been the need for high performance optimization algorithm design and software implementation. Our purpose in this paper is take advantage of the spatial invariance and resulting Toeplitz structure of the atmospheric blurring process to address these two important issues.

1991 *Mathematics Subject Classification.* Primary: 94A08, Secondary: 15A57 .
Key words and phrases. Toeplitz structure, space object imaging, image reconstruction, phase-diversity, regularization, optimization.

The PDS method involves the simultaneous collection of several short-exposure images, some of which are conventional images that have been blurred by unknown aberrations. One or more additional images are collected in *separate channels*, by blurring the first image by a known amount, e.g., using a beam splitter and an out-of-focus lens, which generates a quadratic blur. Using the images collected by the phase-diversity method as data, one can set up a mathematical optimization problem (typically using a maximum likelihood formulation) for recovering the original image as well as the phase aberrations.

Light rays propagating through the earth's atmosphere are distorted because of variations in the index of refraction due to differences in air temperature, humidity, and other factors. This causes distortion, or blurring. A consequence of this blurring is the limited spatial resolution of space objects viewed through ground-based telescopes. Two approaches are often taken to remove these degradations. With adaptive optics [**9, 16**], a deformable mirror is used to restore, or phase conjugate, the distorted light rays to planarity in real time, prior to the formation of the image of the object. A second approach is to solve the inverse problem of estimating the object and certain features of the atmosphere, given observed image data and the spatially invariant image formation model.

Modeling the process of image formation is central to image reconstruction. In optical imaging it is generally assumed that quasi-monochromatic incoherent light energy propagates from object plane sources through an intervening medium and the telescope optical system, to an image plane. Let the object and image plane irradiance distributions be denoted by $f(\nu, \mu)$ and $d(x, y)$, respectively. A simple model of the spatially-invariant image formation process (see, e.g., [**6, 10**]) is the convolution integral equation

$$(1) \quad d(x,y) = \int\int s(x-\nu, y-\mu) f(\nu,\mu) d\nu d\mu \odot \eta(x,y) = (s \star f)(x,y) \odot \eta(x,y),$$

where $\odot$ indicates a point-by-point operation (e.g., addition or multiplication) with the noise process $\eta(x, y)$, and $\star$ represents convolution. Here d thus represents *image data*, f represents the *object*, or true image, and s is known as the *point spread function*, or PSF. The PSF is the image that would result from an idealized point object, and it characterizes atmospheric blurring effects which are spatially invariant in the immediate field of view. In this context, the inverse problem is to determine both the PSF s, whose matrix form is block Toeplitz with Toeplitz blocks (BTTB), see [**6**], as well as the object f, given the image data d.

This inverse problem, which is referred to as *blind deconvolution*, is ill-posed in several respects. First, since convolution is symmetric (i.e., $s \star f = f \star s$), one cannot uniquely determine both s and f from a "single" observed image d. Additional information is needed. Even if the non-uniqueness difficulty in the model (1) can be overcome, the additional difficulty of *instability* may arise, since we are solving an ill-posed inverse problem [**2**]. The term *regularization* is used for schemes which restore stability in a manner which provides a good approximation to the true solution.

The atmospheric imaging problem we consider here has the form (1) with important additional BTTB special structure. In the spatially invariant case generally assumed for atmospherically blurred images (as well as other applications), the blurring operates uniformly across the image of the object in question, i.e.,

$$s(i,j;k,l) = s(i-k, j-l)\,.$$

Let C denote the discretization of s written in matrix form. Then C is a banded block column circulant matrix with column circulant blocks. Moreover, the blocks themselves have the banded form. To form the PSF s, the rows of the blurred pixel image are stacked into a column vector, see e.g. [**6**]. The PSF matrix C can then be given in block form

$$C = \begin{bmatrix} C_0 & & & \\ C_1 & C_0 & & \\ \vdots & & \ddots & \\ C_\gamma & & & \ddots \\ & \ddots & & C_0 \\ & & \ddots & C_1 \\ & & \ddots & \vdots \\ & & & C_\gamma \end{bmatrix}, \tag{2}$$

with the discretized vector s as the first column. Since C is block circulant with circulant blocks, $C^T C$ is a BTTB square matrix [**6**]. This enables the use of 2-D FFTs in the blind deconvolution process.

Another important consideration in our work is that the PSF can be described in terms of a function known as the *phase*, or wavefront profile. A physical process known as *phase-diversity* [**3, 7, 12, 15, 17**] can then be applied to generate additional data to partially overcome the non-uniqueness difficulties in blind deconvolution. From several phase-diversity images, one then seeks to estimate both the object f *and* the phase.

The paper is outlined as follows. Our approach to PDS computations is described, regularization is discussed, and a derivation of the resulting cost functional is sketched in Section 2. A formula for the gradient of the cost functional is also given in this section, following the analysis in [**17**]. A fast limited memory quasi-Newton optimization method allowing bound constraints [**1**] for the minimization of our particular cost functional is the topic of Section 3. The method is a nonlinear minimization technique which combines low cost with rapid convergence. Only the cost functional and its gradient are computed at each iteration, allowing an efficient approximation to the Hessian based on updating. Our Fortran software package for multiframe PDS computations is discussed in Section 4. In Section 5, we present some preliminary computational results obtained in applying the package to simulated atmospheric space object image data based on representative phase screens for distributed atmospheres. Finally, some future directions of this work are listed in Section 6.

2. Multiframe Phase-Diversity

For simplicity we first describe the single frame phase-diversity formulation, based on a "pair" of images. We will assume that recorded image data can be accurately represented as an $n_x \times n_y$ arrays of pixel intensities with components

$$[d]_{ij} = (s \star f)(x_i, y_j) \odot \eta_{ij}, \quad 1 \le i \le n_x, \quad 1 \le j \le n_y, \tag{3}$$

where where $\odot$ indicates a point-by-point operation with the noise process $\eta(x, y)$, $\star$ represents convolution, f denotes the true image, or object, and s denotes the point spread function, or PSF. The PSF s quantifies the blurring effects of the atmosphere. In block matrix form the PSF can expressed as the BTTB matrix $T = C^T C$, where C is given by (2).

We will assume PSF dependence on the phase, or wavefront profile, $\phi(x, y)$ as given by

$$s[\phi] = |\mathcal{F}^{-1}\{pe^{\imath\phi}\}|^2, \tag{4}$$

where $\mathcal{F}$ denotes the 2-D Fourier transform, $\imath = \sqrt{-1}$, and $p = p(x, y)$ denotes the *pupil*, or aperture, function. For ground-based telescopes, p is an indicator function whose support is determined by the extent of the telescope mirror (see [**10**] for a detailed discussion of atmospheric imaging models).

The non-uniqueness and other difficulties for the model (3)-(4) can be at least partially resolved with phase-diversity data. In its simplest form, one collects a pair of images

$$\begin{aligned} d_1 &= s[\phi] \star f \odot \eta_1, \\ d_2 &= s[\phi + \theta] \star f \odot \eta_2, \end{aligned} \tag{5}$$

where θ represents a known phase perturbation (see [**7**]). In practice, this data is generated in hardware by splitting the beam of light collected from the telescope's primary mirror. From one beam, the conventional image d_1 is formed. An out-of-focus image d_2 is formed from the second beam. This corresponds to a quadratic phase perturbation, $\theta(x, y) = c(x^2 + y^2)$, where the constant c depends on the defocus length, generally chosen to have magnitude between π and 2π for space object imaging (see, e.g., [**12**]).

However, the single frame pair phase-diversity-based image recovery formulation described above will generally not lead to acceptable phase recoveries and image reconstructions [**7, 10, 12**]. The blind deconvolution problem (1) is still not adequately constrained. Fortunately, using multiple differently blurred frames of data is in itself a powerful constraint on the restored object.

PDS involves the collection of multiple short-exposure images of data, giving rise to a time series of recorded images. During short-exposure image acquisition, the turbulence structure of the atmosphere is effectively frozen, and the image is distorted by this instantaneous turbulence structure. These images are referred to as speckle because of their modulated appearance. Here, in spite of the severity of the image degradation, there is spatial-frequency information about the object up to the diffraction limit of the telescope system [**10**]. The high quality of PDS image reconstructions is illustrated in our tests reported in Section 5.

The multiframe/multichannel phase-diversity formulation can be described as follows. Let T denote the *number of time frames* of multiple short-exposure sets of data and K denote the *number of diversity channels for each time frame*. In applications relating to space object image recovery, T is generally a multiple of 16, since each Gemini record of data contains 16 frames (see [**11**]). Also, the number of diversity channels K for each time frame is usually taken to be 2 or 3, e.g., [**12**]. The phase-diversity data is then described by the equation

$$d_{tk} = s[\phi_t + \theta_k] \star f \odot \eta_{tk}, \quad t = 1, \ldots, T, \quad k = 1, \ldots, K, \tag{6}$$

where $k = 1$ involves no diversity, i.e., θ_1 is a zero array for each ϕ_t.

Assuming a Gaussian fit-to-data criterion is applied, then the PDS blind deconvolution problem is to compute the phase screens $\phi_1, \ldots, \phi_T$ and the image f to minimize

$$J[\phi, f] = \frac{1}{2} \sum_{t=1}^{T} \sum_{k=1}^{K} \|s_{tk} \star f - d_{tk}\|^2 + \frac{\gamma}{2} J_{reg}[f] + \frac{\alpha}{2} J_{reg}[\phi_1, \ldots, \phi_T], \tag{7}$$

where $s_{tk} = s[\phi_t + \theta_k]$.

In (7), $\alpha J_{reg}[\phi_1, \ldots, \phi_T]$ is a regularization functional, whose purpose is to establish stability with respect to perturbations in the phase screens ϕ_t. Similarly, the term $\gamma J_{reg}[f]$ establishes stability with respect to perturbations in f. Here, γ and α are positive regularization parameters. Regularization is discussed in more detail in the next section.

2.1. Regularization. The PDS approach itself can greatly reduce the non-uniqueness difficulties. But it should be noted that constant offsets in ϕ still cannot be resolved, since $s[\phi + c] = s[\phi]$ for any fixed c. Such ambiguities include phase wrapping. This is a consequence of the fact that $e^{\imath\phi} = e^{\imath(\phi+2\pi)}$. Planar offsets in ϕ (which imply a shift of the object) can be avoided by a pre-processing step which aligns the centroid of the each data image with the center pixel.

One must also deal with instability with respect to perturbations in the data. This can be overcome by Tikhonov regularization [**2**], or penalized least squares. Regularization methods yield solutions to ill-posed inverse problems which depend on regularization parameters, which quantify the tradeoff between error amplification due to instability and truncation due to regularization. Regularization functionals can be used to enforce smoothness constraints or prior information about the unknowns [**2, 4**].

Given the multiframe/multichannel data (6), one might minimize the *joint* cost functional (7), where γ and α are nonnegative scalar regularization parameters, to be multiplied times regularization, or penalty, functionals J_{reg}. These penalty functionals restore stability in a manner which incorporates a priori information about the object and phase. For the purpose of simplifying the PDS computations, the object regularization functional in PDS methods is generally taken to be

$$J_{reg}[f] = ||f||^2. \tag{8}$$

This corresponds to the minimal assumption that the object has finite intensity. It incorporates no prior smoothness assumptions.

There are various choices for the phase regularization functional. A study of three such choices is given in [**4**]. Define

$$J_{reg}[\phi_1, \ldots, \phi_T] = \sum_{t=1}^{T} ||L\phi_t||^2. \tag{9}$$

Then the choice of the operator L determines regularization functional for the phase. The choice $L = I$ corresponds to the minimal assumption that the phase has finite intensity. A smoothness constraint on the phase can be imposed by choosing L to be a differential operator, typically a Laplacian. However, if prior second order statistical information about the atmospheric turbulence is available, say from a wavefront sensor, then a sometimes more effective choice of L can be made [**4**]. If the phase ϕ is assumed to be a realization from a wide-sense stationary stochastic process whose covariance operator has a Von Karman spectrum [**10**] with covariance matrix A, then one might set

$$L = A^{-\frac{1}{2}}. \tag{10}$$

Choosing L as in (10) is called *MAP regularization*. Efficient numerical methods for computing A and the resulting L for this MAP regularization are given in [**8, 9**], where fast Hankel transforms are used. In fact, the covariance matrix A can be approximated as a sum of products of highly structured matrices, including discrete cosine, Toeplitz and Vandermonde matrices [**8**]. In ideal situations, the use of (10) for phase retrieval is recommended in the study by Irwan and Lane [**4**]. However, in working with real space object data, second order statistics may not be available, in which case the choice of L as the Laplacian may be most appropriate.

2.2. The Reduced Cost Functional and Its Gradient. Some of the notation and technical details in this section are taken from the paper by Vogel, Chan and Plemmons[**17**], which considered only the single frame phase-diversity case. Let upper case letters denote Fourier transformed variables. Then due to the BTTB matrix structure the discretized multiframe/multichannel cost functional (7) has a Fourier domain representation

$$J_{\gamma,\alpha}[\phi, F] = \frac{1}{2}\left(\sum_{t=1}^{T}\sum_{k=1}^{K}||S[\phi_t + \theta_k]F - D_{tk}||^2\right) + \frac{\gamma}{2}||F||^2 + \frac{\alpha}{2}J_{phase}[\phi_1, \ldots, \phi_T], \tag{11}$$

where $J_{phase}[\phi_1, \ldots, \phi_T]$ denotes the Fourier transform of the right-hand-side of (9). In order to simplify the notation, we set $S_{tk} = S[\phi_t + \theta_k]$ in the equations to follow, and in general, the subscript notation tk corresponds to a function evaluated at $\phi_t + \theta_k$. The exception is d_{tk} defined in (6), and its 2-D Fourier transform $D_{tk} = \mathcal{F}(d_{tk})$.

From (11) we thus obtain an unconstrained minimization problem with a very large number of unknowns. With $n_x \times n_y$ pixel image arrays, we have $(T+1)\, n_x\, n_y$ unknowns. As in [**3, 7, 17**], we eliminate the object and cut the number of unknowns by $n_x\, n_y$. Setting $\frac{\partial J}{\partial F} = 0$ yields

$$F[\phi_1, \ldots, \phi_T] = \frac{\sum_{t=1}^{T}\sum_{k=1}^{K} S_{tk}^* D_{tk}}{\gamma + \sum_{t=1}^{T}\sum_{k=1}^{K}|S_{tk}^*|^2}. \tag{12}$$

Here the superscript $*$ denotes complex conjugate, and $|\cdot|$ denotes component-wise magnitude of a complex quantity array. Given estimates for the T phase arrays $\phi_1, \ldots, \phi_T$, one can take the inverse Fourier transform in (12) to obtain an estimate for the object f. Note that the positive object regularization parameter γ in the denominator of (12) induces stability by preventing division by very small quantities or zero.

By substituting $F = F[\phi_1, \ldots, \phi_T]$ from (12) back into (11), one obtains the *reduced cost functional*

$$J[\phi_1, \ldots, \phi_T] = \tag{13}$$

$$\frac{1}{2}\left(\sum_{t=1}^{T}\sum_{k=1}^{K}||D_{tk}||^2 - \left|\left|\frac{\sum_{t=1}^{T}\sum_{k=1}^{K} S_{tk}^* D_{tk}}{\gamma + \sum_{t=1}^{T}\sum_{k=1}^{K}|S_{tk}|^2}\right|\right|^2\right) + \alpha J_{phase}[\phi_1, \ldots, \phi_T].$$

The gradient of the reduced cost functional then has a representation

(14)

$$(g[\phi_1,\ldots,\phi_T])_t = -2\sum_{k=1}^{K} \mathrm{Imag}(H_{tk}^* \mathcal{F}(\mathrm{Real}(h_{tk}\mathcal{F}^{-1}(V_{tk})))) + \alpha\ (g_{reg}[\phi_1,\ldots,\phi_T])_t\,,$$

for $t = 1,\ldots,T$, where $\theta_1 = 0$,

$$H_{tk} = pe^{\imath(\phi_t+\theta_k)}, \qquad h_{tk} = \mathcal{F}^{-1}(H_{tk}), \tag{15}$$

$$s_{tk} = |h_{tk}|^2, \qquad S_{tk} = \mathcal{F}(s_{tk}), \tag{16}$$

$$V_{tk} = F^* D_{tk} - |F|^2 S_{tk}, \tag{17}$$

and $g_{reg}[\phi_1,\ldots,\phi_T]$ is the gradient of the phase regularization functional $J_{phase}[\phi]$. The latter quantity simplifies according to the choice of L in (9).

3. Limited Memory Optimization

To minimize the reduced cost functional $J[\phi_1,\ldots,\phi_T]$ given in (13), we apply a limited memory quasi-Newton optimization method allowing bound constraints on the phase. The basic algorithm is given in [**1**] and has been incorporated into an optimization package at the Argonne National Laboratory in a highly efficient code.

Quasi-Newton methods yield approximations to a (local) minimizer u_* to the phase screens ϕ_t in (13) of the form

$$u_{i+1} = u_i + s_i, \quad i = 0, 1, \ldots,$$

where $s_i = \mu d_i$ updates the current vector iterate, μ is a positive step length parameter, and the quasi-Newton direction vector d_i solves

$$H_i d = -g[u_i], \tag{18}$$

with H_i a symmetric positive definite approximation to the true Hessian $H[u_i]$, and $g[u_i]$ is the gradient vector given in (14), evaluated at the current phase approximation. Positive definiteness guarantees that d_i is a descent direction for J for some $\mu > 0$, provided the gradient is nonzero (see, e.g., [**5**]).

If H_i in (18) is taken to be the true Hessian H, then one obtains Newton's method. This method has the advantage of quadratic convergence near a local minimizer. Unfortunately, the computation, storage, and inversion of the Hessian may be prohibitively expensive. Moreover, far from a local minimizer the Hessian need not be positive definite, and hence the Newton step need not be a descent direction.

An alternative to Newton's method is the BFGS method (see, e.g., [**1**]). Given an initial Hessian approximation H_0, it generates a sequence of Hessian approximates via the rank-two update

$$H_{i+1} = H_i + \frac{1}{y^T s} y y^T - \frac{1}{s^T H_i s}(H_i s)(H_i s)^T, \tag{19}$$

where $s = s_i = u_{i+1} - u_i$ is the current scaled step vector and $y = g[u_{i+1}] - g[u_i]$ is the difference between gradients. If H_i is positive definite and $y^T s$ is positive, then H_{i+1} is guaranteed to be positive definite. Under standard assumptions, i.e., J is smooth, u_0 is sufficiently close to a local minimizer u_*, H_0 is sufficiently close to $H[u_*]$, and $H[u_*]$ is positive definite, the BFGS method is rapidly convergent (see [**1**] for details).

A recursive formula for the inverses of the matrices in (19) is

$$H_{i+1}^{-1} = (I - \frac{1}{y^T s} s y^T) H_i^{-1} (I - \frac{1}{y^T s} y s^T) + \frac{1}{y^T s} s s^T. \tag{20}$$

This recursion can be used to easily solve equation (18) with a BFGS Hessian approximation H_i. To do so requires storage of i vectors s and y (i.e., all the previous steps and gradient differences), inner product computations involving the gradient $g[u_i]$ and the s and y vectors, and the computation of $H_0^{-1}v$ for some vector v. With the limited memory BFGS method, only a fixed number of the s and y vectors, say 5 pairs, are retained. As new vectors are added to storage, the oldest vectors are discarded. This can substantially reduce the storage requirements of the method and the cost of computing the quasi-Newton steps d.

The numerical optimization routine incorporated into our PDS software package is called L-BFGS-B, for limited memory BFGS with bound constraints. The basic code, in FORTRAN, is available, as is the paper [**1**], from the Argonne National Laboratory Optimization Technology Center on their web page at www-unix.mcs.anl.gov/neos/Server/solvers/BCO:L-BFGS-B/.

4. The Software Package

We have developed and tested a FORTRAN software package called **Multiframe Phase-Diversity Reconstruction** (MPDR), for the purpose of atmospheric image reconstruction using the techniques described in this paper. In the current version (v0.2), the user interface and the atmospheric simulation is based on Matlab 5.3. For efficiency, the restoration is performed by (compiled) FORTRAN code using Matlab's Mex interface to interconnect the restoration code and the user interface.

The data dependencies inherent in equations (13) through (17) imply that two passes over the phase data are required. A time/space trade-off exists between storing several fairly large intermediate values for use in the second pass versus re-computing those intermediate values during the second pass. Two versions of the restoration code have been developed. One optimizes for time, but assumes sufficient memory is available. Another optimizes for lower memory use but incurs an increased time cost of six FFTs per function evaluation instead of four. We call it the *small memory* version of MPDR. In both versions, we applied a compiler optimization technique known as *lifetime analysis* to optimize the re-use of temporary arrays.

MPDR uses the highly efficient FFTW (2.1.2) library (see www.fftw.org) for evaluation of the reduced cost functional (13), the gradient (14) and for other computations. This exploits the BTTB matrix structure the discretized multiframe/multichannel cost functional (7).

Wherever possible, MPDR also exploits the well-known Hermitian symmetry property for Fourier transforms of real data. Given $A \in \mathcal{R}^{m\times n}$, let $B = \mathcal{F}(A)$, where $\mathcal{F}$ denotes the 2-D discrete Fourier transform. Then, for $0 \le i < m$ and $0 \le j < n$

$$B_{i,j} = \overline{B}_{m-i \text{ modulo } m, n-j \text{ modulo } n} \,. \tag{21}$$

Note that this symmetry property described in (21) is preserved under addition, conjugation, and component-wise multiplication. MPDR also uses a modified version of the numerical optimization Fortran routine L-BFGS-B from the Argonne National Laboratory Optimization Technology Center to minimize the reduced phase cost functional given in (13).

Original Matlab m-files in the directory Simulate were primarily written by Brent Ellerbroek (AFRL and Gemini) and Curt Vogel (Montana State University), for the purpose of generating test phase screen data allowing for various atmospheric turbulence conditions. Some of the original Matlab m-files in the directory Utilities for manipulating the data, etc., were also written by Curt Vogel. In addition, Dave Tyler (University of New Mexico) provided phase screen data generated using FORTRAN code. He and Brent Ellerbroek also gave helpful advice on choosing certain atmospheric turbulence parameters. Some of the FORTRAN routines for cost functional and gradient evaluations were translated from Matlab code written by Curt Vogel (MSU). In computing restorations from real telescope data [**14**], all the production code in MPDR can be used independent of Matlab.

5. Simulation Tests

The numerical experiments we discuss in this section were performed on data produced by simulated multiple time frame, multiple channel phase-diversity data. The phase screens are generated according to a Von Karman turbulence model [**10**] using a phase screen generator primarily written by Brent Ellerbroek to simulate "seeing conditions" at the U.S. Air Force Starfire Optical Range in New Mexico. It can be used to generate phase screens for distributed atmospheres, including some or all of the following effects:

- Finite outer scale.
- Time series of phase screens with distinct wind velocities for each layer.
- Anisoplanatism (i.e., phase screens for point sources in different directions).
- Scintillation and diffraction effects.

According to Brent Ellerbroek, "the principal limitation of the generator is that the phase screens are periodic, so that (a) one cannot expect really long sequences to have the correct temporal statistics, (b) one must keep the aperture diameter less than about 0.5 times the width of the screen for reasonably correct higher-order turbulence statistics, and (c) one should keep the diameter greater than about 0.1 times the width of the screen for reasonable tilt-included statistics".

Simulated atmospheric phase was generated according to the multiframe/multichannel phase-diverse speckle model in equation (1). Simulated phase-diversity image data was then generated according to a discretization of the model. Although the pupil may lie within an $n \times n$ grid, a computational grid of size $2n \times 2n$ was used in the Fourier domain to avoid wrap and reduce other edge effects. Discrete Fourier transforms were computed using the 2-D fast Fourier transform package FFTW. To simulate instrument noise, Poisson and Gaussian errors were added to the generated phase data.

To solve the phase recovery and object reconstruction inverse problems in a stable manner, Tikhonov regularization with the regularization functionals described in Section 2.1 were applied. The reduction scheme of Section 2.2 was employed to eliminate the unknown object from the cost functional, thereby reducing the number of unknowns for the optimization routine to Tn^2 point values of the estimated phases. Each $n \times n$ estimated phase array corresponds to a moment in time t (with $1 \leq t \leq T$), at which one frame of K diversity images, each of size $n \times n$, is collected. The reduced cost functional was minimized using the implementation of the limited memory BFGS method described in Section 3. Sample true, blurred and noisy, and reconstructed objects are shown in Figure 1.

Notice in Figure 1 that restoring the object by using 32 frames of data leads to a considerable visual improvement over the restoration using only 1 frame. Since the tests given here are simulations and the true object is known, a more formal comparison can be made. In particular, we compare the relative error

$$\frac{\|f - \mathcal{F}^{-1}(F[\phi_1, \phi_2, ..., \phi_T])\|_2}{\|f\|_2}$$

where f is the true object, $F[\phi_1, \phi_2, ..., \phi_T]$ is defined in equation (12), and $\mathcal{F}^{-1}(F[\phi_1, \phi_2, ..., \phi_T])$ is the restored object. Using a single frame yields a relative error of 0.4343, but the relative error is only 0.1438 using 32 frames.

A table of representative execution times on a 250MHz Sun Ultra is given in Figure 2. The regularization and stopping parameters were chosen to illustrate typical values and the times shown are typical of those needed to achieve an acceptable restoration. A performance profile of the code shows that about 80% of the execution time is spent computing 2-D FFTs. Thus the block Toeplitz with Toeplitz blocks (BTTB) structure of the PSF matrices is critical.

The performance of the limited memory algorithm depends on the number of vectors s and y saved, as discussed in Section 3. If no vectors are saved, the algorithm reduces to a scaled version of the steepest descent method. On the other hand, if all the vectors are saved, the usual BFGS scheme results. As should be expected, performance improves as more vectors are saved. Significant improvement results in saving 5 vectors as opposed to saving none at all. However, relatively little is gained in going from 5 to more saved vectors, so the algorithm is quite memory and cost efficient.

6. Future Research Directions

Thus far, we have addressed the development of high performance optimization algorithms, and software implementations that are fast and yet storage-efficient on serial computers. Tests on simulated multiple time frame, multiple channel phase-diversity data confirm the effectiveness of our MPDR software package on realistic phase recovery and object reconstruction. However, several tasks remain to be undertaken before this project is fully complete. We hope to consider the following additional work:

- Further investigation of matrix structured regularization techniques (such as MAP) for the phase, and development of an effective phase unwrapping scheme in the context of wavefront phase recovery.
- Further tests of the algorithms and code using data from the Maui Space Surveillance Site in Hawaii to confirm the effectiveness of our software package on real atmospheric imaging problems.
- Investigation of alternate reduced formulations of the cost functional that avoid the standard Wiener filter restoration step for the image common to PDS methods.

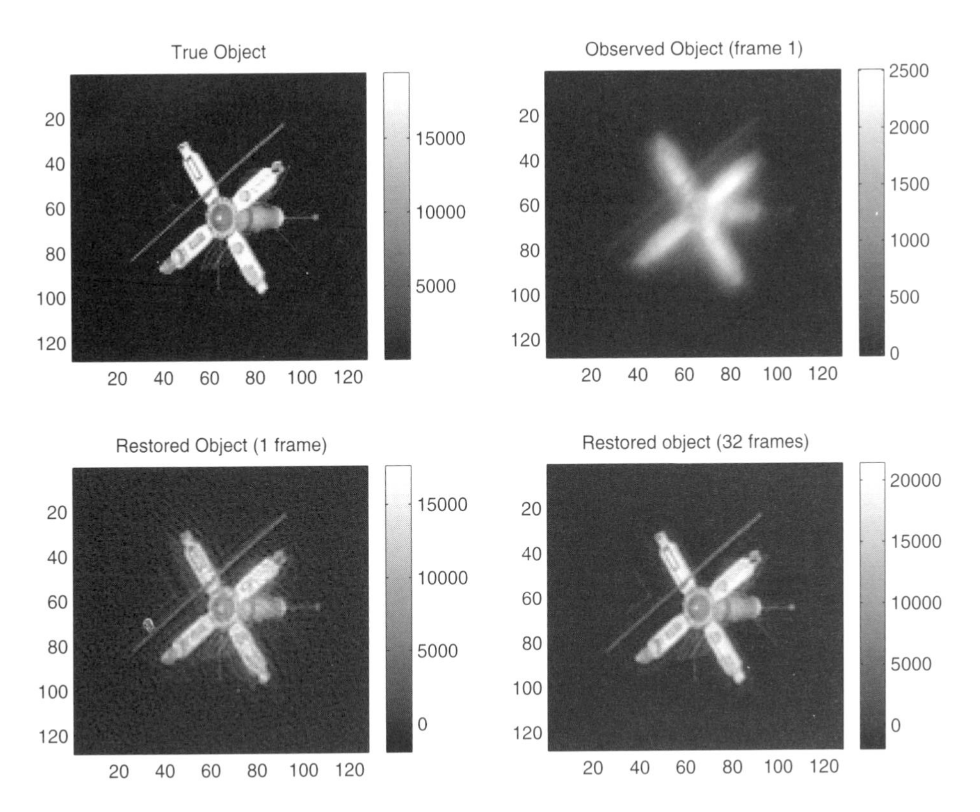

FIGURE 1. *Sample object recovery using* 128 × 128 *resolution. The upper left subplot shows the true object, the upper right subplot shows the blurred and noisy observed object, the lower left subplot shows a recovered image using a single frame of data, while the lower right subplot shows the recovered image using 32 frames.*

64 × 64 Problems			128 × 128 Problems		
Frames	Iterations	Elapsed Time	Frames	Iterations	Elapsed Time
1	22	2.9 secs.	1	24	20 secs.
16	36	1.15 min.	16	35	5.5 min.
32	35	2.2 min.	32	36	10.95 min.
64	35	4.4 min.	64	40	30.7 min.

FIGURE 2. Sample Execution Times with Two Diversity Channels. The small-memory version of the code, used for the 128×128, 64 frame problem, solves for more than 10^6 phase values at each nonlinear iteration.

- Parallelization of the codes on the Maui High Performance Computing Center IBM parallel system with 603 IBM SP nodes with: 256 gigaflops of processing power, 167 gigabytes of total memory, 2.1 terabytes of internal disk space, as well as incorporation of the MPDR software package into the Maui image manager and on-line systems archive (MIMOSA) at the DOD space surveillance facility (see [**11**]).

We envision that the project described in this paper will result in new technologies for atmospheric image reconstruction, in the form of robust and efficient algorithms as well as their implementation in our MPDR software package. Our techniques will be tested on applications from space object imaging. Packaging the results of our research into reliable software will hopefully facilitate the effective and timely transfer of new knowledge research laboratories and industry.

Acknowledgments. The authors wish to thank Brent Ellerbroek, Kathy Schulze, Dave Tyler, and Curt Vogel for providing very helpful advice during this study. Research by Robert J. Plemmons and Todd C. Torgersen on this project was supported in part by the AFOSR under grant F49620-00-1-0155, and by the NSF under grant CCR-9732070. Research by William K. Cochran was supported by Research Experiences for Undergraduates funds in the NSF grant. This project also received support from the HPCERC/Maui Project under AFRL Contract F29601-96-D-0128, Task Order 05, including accounts on the MHPCC IBM SP2 computing system.

References

[1] R.H. Byrd, P. Lu, J. Nocedal and C. Zhu, "A limited memory algorithm for bound constrained optimization," *SIAM J. Scientific Computing*, Vol. 21, pp. 1190-1208, 1996.

[2] H. Engl, M. Hanke, and A. Neubauer, *Regularization of Inverse Problems*, Kluwer Academic Publishers, Dordrecht, 1996.

[3] R.A. Gonsalves, "Phase retrieval and diversity in adaptive optics," *Optical Eng.* **21**, pp. 829-832, 1982.

[4] I. Irwan and R. Lane, "Phase retrieval with prior information," *J. Optical. Soc. Amer., A*, Vol. 15, pp. 2302-2311, 1998.

[5] C.T. Kelley, *Iterative Methods for Optimization*, SIAM Press, 1999.

[6] J. Nagy, P. Pauca R. Plemmons, and T. Torgersen "Degradation reduction in optics imagery using Toeplitz structure," *CALCOLO J. on Numerical Analysis and Theory of Computation*, Vol. 33, pp. 269-288, 1997.

[7] R.G. Paxman, T.J. Schulz, and J.R. Fineup, "Joint estimation of object and aberrations by using phase-diversity," *J. Optical Soc. Am., A*, Vol. 9, pp. 1072-1085, 1992.

[8] P. Pauca, B. Ellerbroek, N. Pitsianis, R. Plemmons, and X. Sun. "Performance modeling of adaptive-optics imaging systems using fast Hankel transforms," *Advanced Signal Processing Algorithms, Architectures, and Implementations VIII*, 3461:339–347, 1998.

[9] P. Pauca and R. Plemmons, "Some computational problems arising in adaptive optics imaging systems," *Computational and Applied Mathematics*, Vol. 123, pp. 467-487, 2000.

[10] M.C. Roggemann and B. Welsh, *Imaging Through Turbulence*, CRC Press, 1996.

[11] K.J. Schulze, "Maui image manager and on-line systems archive (MIMOSA) – Intelligent observatory data management and processing," preprint, 1999.

[12] J.H. Seldin, M.F. Reiley, R.G. Paxman, B.E. Stribling, B.L. Ellerbroek, and D.C. Johnston, "Space-object identification using phase-diverse speckle," in *Digital Image Recovery and Synthesis III, SPIE Proceedings Vol. 3170*, pp. 2-15, 1997.

[13] D.G. Sheppard, B.R. Hunt, and M.W. Marcellin, "Iterative multiframe superresolution algorithms for atmospheric-turbulence-degraded imagery," *J. Optical. Soc. Amer., A*, Vol. 15, pp. 978-992, 1998.

[14] T.C. Torgersen and D.W. Tyler, "Practical problems in restoring images from real phase diverse speckle data," preprint, November, 2000.

[15] D.W. Tyler, S.D. Ford, B.R. Hunt, R.G. Paxman, M.C. Roggemann, J.C. Roundtree, T.J. Schulz, K.J. Schulze, D.G. Sheppard, J.H. Seldin, B.E. Stribling, W.C. Van Kampen and B.M. Welsh, "Comparison of image reconstruction algorithms using adaptive optics instrumentation," in *Adaptive Optical System Technologies, SPIE Kona Proc. Vol. 3353* , pp. 160-171, 1998.

[16] R.K. Tyson, *Principles of Adaptive Optics, 2nd Edition*, Academic Press, 1998.

[17] C.R. Vogel, T. Chan, and R.J. Plemmons, "Fast algorithms for phase-diversity-based blind deconvolution," in *Adaptive Optical System Technologies*, *SPIE Kona Proc. Vol. 3353*, pp. 994-1005, 1998.

Wake Forest University, Box 7388, Winston-Salem, NC 27109

E-mail address: plemmons@mthcsc.wfu.edu

E-mail address: torgerse@mthcsc.wfu.edu

PART III. Control Theory

Contemporary Mathematics
Volume **280**, 2001

A survey of model reduction methods for large-scale systems

A.C. Antoulas, D.C. Sorensen, and S. Gugercin

ABSTRACT. An overview of model reduction methods and a comparison of the resulting algorithms are presented. These approaches are divided into two broad categories, namely SVD based and moment matching based methods. It turns out that the approximation error in the former case behaves better globally in frequency while in the latter case the local behavior is better.

1. Introduction and problem statement

Direct numerical simulation of dynamical systems has been an extremely successful means for studying complex physical phenomena. However, as more detail is included, the dimensionality of such simulations may increase to unmanageable levels of storage and computational requirements. One approach to overcoming this is through model reduction. The goal is to produce a low dimensional system that has the same response characteristics as the original system with far less storage requirements and much lower evaluation time. The resulting reduced model might be used to replace the original system as a component in a larger simulation or it might be used to develop a low dimensional controller suitable for real time applications.

The *model reduction problem* we are interested in can be stated as follows. Given is a linear dynamical system in state space form:

$$\Sigma : \left\{ \begin{array}{l} \sigma x(t) = Ax(t) + Bu(t) \\ y(t) = Cx(t) + Du(t) \end{array} \right. \tag{1.1}$$

where σ is either the derivative operator $\sigma f(t) = \frac{d}{dt} f(t)$, $t \in \mathbb{R}$, or the shift $\sigma f(t) = f(t+1)$, $t \in \mathbb{Z}$, depending on whether the system is continuous- or discrete-time. For simplicity we will use the notation:

$$\Sigma = \left[\begin{array}{c|c} A & B \\ \hline C & D \end{array} \right] \in \mathbb{R}^{(n+p)\times(n+m)} \tag{1.2}$$

1991 *Mathematics Subject Classification.* Primary 93B11; Secondary 93B15,93B40.

This work was supported in part by the NSF Grant DMS-9972591 and by the NSF Grant CCR 9988393.

The *problem* consists in approximating $\mathbf{\Sigma}$ with:

$$\hat{\mathbf{\Sigma}} = \left[\begin{array}{c|c} \hat{A} & \hat{B} \\ \hline \hat{C} & \hat{D} \end{array}\right] \in \mathbb{R}^{(k+p)\times(k+m)} \tag{1.3}$$

where $k \ll n$ such that the following properties are satisfied:

1. The approximation error is *small*, and there exists a *global* error bound.
2. System properties, like *stability*, *passivity*, are preserved.
3. The procedure is *computationally stable* and *efficient*.

There are two sets of methods which are currently in use, namely

(a) **SVD based methods** and

(b) **moment matching based methods**.

One commonly used approach is the so-called Balanced Model Reduction first introduced by Moore [**23**], which belongs to the former category. In this method, the system is transformed to a basis where the states which are difficult to reach are simultaneously difficult to observe. Then, the reduced model is obtained simply by truncating the states which have this property. Two other closely related model reduction techniques are Hankel Norm Approximation [**24**] and the Singular Perturbation Approximation [**20**], [**22**]. When applied to stable systems, all of these approaches are guaranteed to preserve stability and provide bounds on the approximation error. Recently much research has been done to establish connections between Krylov subspace projection methods used in numerical linear algebra and model reduction [**9**], [**13**], [**14**], [**16**], [**17**], [**19**], [**29**], [**15**]; consequently implicit restarting [**27**] has been applied to obtain stable reduced models [**18**]. For recent surveys on this topic see [**10, 28**].

Issues arising in the approximation of large systems are: **storage**, **computational speed**, and **accuracy**. In general storage and computational speed are finite and problems are *ill-conditioned. In addition*: we need global error bounds and preservation of stability/passivity. SVD based methods provide error bounds and preserve stability, but are computationally not efficient. On the other hand, moment matching based methods can be implemented *iteratively*, which leads to numerically efficient algorithms, but these do not automatically preserve stability and have no global error bounds. To remedy this situation, the *approximate balancing* method was introduced in [**9**]. It attempts to combine the best properties of the above methods, by iteratively computing a reduced order approximately balanced system.

The paper is organized as follows. After the problem definition, the first part is devoted to approximation methods which are related to the SVD. Subsequently, moment matching methods are reviewed. The third part of the paper is devoted to a comparison of the resulting seven algorithms applied on six dynamical systems of low to moderate complexity. We conclude with some remarks on unifying features of these reduction methods, and complexity considerations.

2. Approximation in the 2-norm: SVD-based methods

2.1. The singular value decomposition: static systems. Given a matrix $A \in \mathbb{R}^{n\times m}$, its *Singular Value Decomposition* (SVD) is defined as follows:

$$A = U\Sigma V^*, \ \Sigma = \text{diag}(\sigma_1, \cdots, \sigma_n) \in \mathbb{R}^{n\times m},$$

where $\sigma_1(A) \geq \cdots \geq \sigma_n(A) \geq 0$, are the *singular values*, and $\sigma_1(A)$ is the 2-*induced norm* of A. The columns of the orthogonal matrices $U = (u_1\ u_2\ \cdots\ u_n)$, $UU^* = I_n$, $V = (v_1\ v_2\ \cdots\ v_m)$, $VV^* = I_m$, are the *left, right singular vectors* of A, respectively. Assuming that $\sigma_r > 0$, $\sigma_{r+1} = 0$, implies that the rank of A is r. Finally, the SVD induces a *dyadic decomposition* of A:

$$A = \sigma_1 u_1 v_1^* + \sigma_2 u_2 v_2^* + \cdots \sigma_r u_r v_r^*$$

A basic matrix approximation problem is as follows: given $A \in \mathbb{R}^{n\times m}$ with rank $A = r \leq n \leq m$, we seek to find $X \in \mathbb{R}^{n\times m}$ with rank $X = k < r$, such that the 2-norm of the *error* $E := A - X$ is minimized.

THEOREM 2.1. **Schmidt-Mirsky, Eckart-Young: Optimal approximation in the 2 norm**. *Provided that $\sigma_k > \sigma_{k+1}$, there holds:* $\min_{\text{rank} X\leq k} \| A - X \|_2 = \sigma_{k+1}(A)$. *A (non-unique) minimizer X_* is obtained by truncating the dyadic decomposition:* $X_* = \sigma_1 u_1 v_1^* + \sigma_2 u_2 v_2^* + \cdots \sigma_k u_k v_k^*$.

Example: Clown approximation. The above approximation method is applied to the *clown* image shown in figure 1, which can be downloaded from Matlab. In the black and white version, this is a 320×200 pixel image, each pixel having 64 levels of gray. First, the 200 singular values of this 2-dimensional array are computed (see upper right-hand side subplot of the figure); the singular values drop-off rapidly making a low-order approximation with small error, possible. The optimal approximants for rank $k = 1, 3, 10, 30$ are shown. Notice that the storage reduction of a rank k approximant is $(n + m + 1) * k$ compared to $n * m$ for the original image.

2.2. SVD methods applied to dynamical systems. There are different ways of applying the SVD to the approximation of dynamical systems. The table below summarizes the different approaches.

Nonlinear systems	Linear systems
POD methods	Hankel-norm approximation Balanced truncation Singular perturbation New method (Cross grammian)

TABLE 1. SVD based approximation methods

Its application to non-linear systems is known under the name *POD*, that is: Proper Orthogonal Decomposition. In the linear case we can make use of additional structure. The result corresponding to a generalization of the Schmidt-Mirsky theorem is known under the name of Hankel norm approximation. Closely related methods are approximation by balanced truncation and approximation by singular perturbation. Finally, the new method proposed in section 3.6 is based on the SVD in a different way.

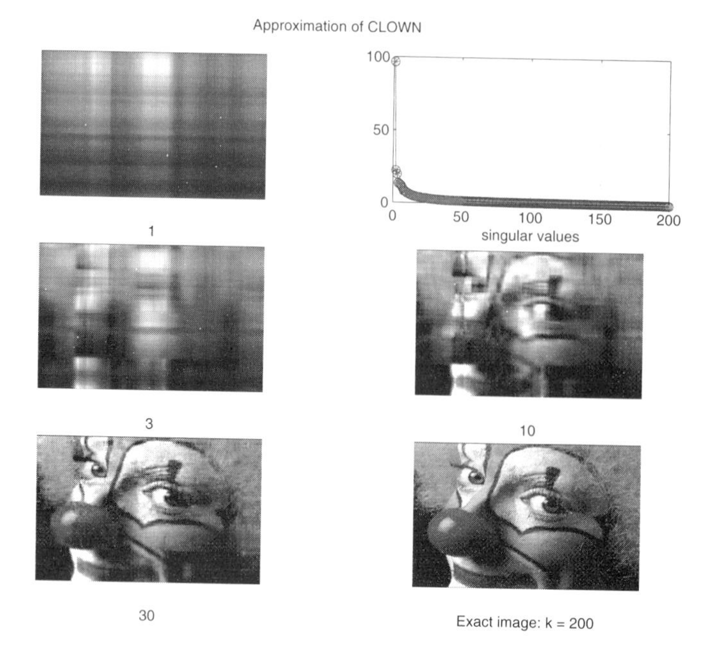

FIGURE 1. Approximation of the clown image

2.2.1. *Proper Orthogonal Decomposition (POD) methods.* Consider the nonlinear system described by $\dot{x}(t) = f(x(t), u(t))$; let

$$\mathcal{X} = [x(t_1)\ x(t_2)\ \cdots\ x(t_N)] \in \mathbb{R}^{n \times N}$$

be a collection of snapshots of the solution of this system. We compute the singular value decomposition and truncate depending on how fast the singular values decay:

$$\mathcal{X} = U\Sigma V^* \approx U_k \Sigma_k V_k^*,\ k \ll n$$

The state is now approximated by its projection on the space spanned by the dominant left singular vectors of the collection of snapshots $\mathcal{X}$: $x(t) \approx U_k \xi(t)$, $\xi(t) \in \mathbb{R}^k$. Thus the approximation $\xi(t)$ of the state $x(t)$ evolves in a *low-dimensional* space. This implies the reduced order state equation:

$$\dot{\xi}(t) = U_k^* f(U_k \xi(t), u(t))$$

Thus under certain assumptions, the original dynamical system can be replaced by one evolving in a low dimensional space. POD is a popular method in all applications involving the solution of PDEs (see e.g. [**11**]).

2.2.2. *Optimal approximation of linear systems.* Consider the following *Hankel operator*:

$$\mathcal{H} = \begin{pmatrix} \alpha_1 & \alpha_2 & \alpha_3 & \cdots \\ \alpha_2 & \alpha_3 & \alpha_4 & \cdots \\ \alpha_3 & \alpha_4 & \alpha_5 & \cdots \\ \vdots & \vdots & \vdots & \ddots \end{pmatrix} : \ \ell(\mathbb{Z}_+) \longrightarrow \ell(\mathbb{Z}_+)$$

It is assumed that rank $\mathcal{H} = n < \infty$, which is equivalent with the rationality of the (formal) power series:

$$\sum_{t>0} \alpha_t z^{-t} = \frac{\pi(z)}{\chi(z)} =: G_{\mathcal{H}}(z), \ \deg \chi = n > \deg \pi$$

It is well known that in this case $G_{\mathcal{H}}$ possesses a state-space realization denoted by (1.2):

$$G_{\mathcal{H}}(z) = \frac{\pi(z)}{\chi(z)} = C(zI - A)^{-1}B, \ A \in \mathbb{R}^{n \times n}, \ B, C^* \in \mathbb{R}^n$$

This is a discrete-time system; thus the eigenvalues of A (roots of χ) lie inside the unit disc if and only if $\sum_{t>0} \mid \alpha_t \mid^2 < \infty$.

The problem which arises now is to approximate $\mathcal{H}$ by a Hankel operator $\hat{\mathcal{H}}$ of lower rank, *optimally* in the 2-induced norm. The system-theoretic interpretation of this problem is to optimally approximate the linear system described by $G = \frac{\pi}{\chi}$, by a system of lower complexity, $\hat{G} = \frac{\hat{\pi}}{\hat{\chi}}$, $\deg \chi > \deg \hat{\chi}$. This is the problem of *approximation in the Hankel norm.*

First we note that $\mathcal{H}$ is bounded and compact and hence possesses a discrete set of non-zero singular values with an accumulation point at zero:

$$\sigma_1(\mathcal{H}) \geq \sigma_2(\mathcal{H}) \geq \cdots \geq \sigma_n(\mathcal{H}) > 0$$

These are called the *Hankel singular values* of the system $\mathbf{\Sigma} = \left[\begin{array}{c|c} A & B \\ \hline C & \end{array}\right]$, and σ_1 is its *Hankel norm*, denoted by $\sigma_1 = \|\mathbf{\Sigma}\|_H$. By the Schmidt-Mirsky, Eckart-Young result, any approximant $\mathcal{K}$, not necessarily structured, of rank $k < n$ satisfies:

$$\| \mathcal{H} - \mathcal{K} \|_2 \geq \sigma_{k+1}(\mathcal{H})$$

The question which arises is whether *there exist an approximant of rank k which has Hankel structure and achieves the lower bound.* In system-theoretic terms we seek a low order approximant $\hat{\mathbf{\Sigma}}$ to $\mathbf{\Sigma}$. The question has an affirmative answer.

THEOREM 2.2. **Adamjan, Arov, Krein (AAK).** *There exists a unique approximant $\hat{\mathcal{H}}$ of rank k, which has Hankel structure and attains the lower bound:* $\sigma_1(\mathcal{H} - \hat{\mathcal{H}}) = \sigma_{k+1}(\mathcal{H})$.

The above result holds for continuous-time systems as well. In this case the discrete-time Hankel operator introduced above is replaced by the continuous-time Hankel operator defined as follows: $y(t) = \mathcal{H}(u)(t) = \int_{-\infty}^{0} h(t-\tau)u(\tau)d\tau$, $t > 0$, where $h(t) = Ce^{At}B$ is the impulse response of the system; for details see e.g. [**7, 8**].

2.2.3. *Optimal and suboptimal approximation in the 2-norm.* It turns out that both *sub-optimal* and *optimal* approximants of MIMO (multi-input, multi-output) linear continuous- and discrete-time systems can be treated within the same framework. The problem is thus: Given a stable system $\mathbf{\Sigma}$, we seek approximants $\mathbf{\Sigma}_*$ satisfying

$$\sigma_{k+1}(\mathbf{\Sigma}) \leq \|\mathbf{\Sigma} - \mathbf{\Sigma}_*\|_H \leq \epsilon < \sigma_k(\mathbf{\Sigma}) \tag{2.1}$$

This is accomplished by the following construction.

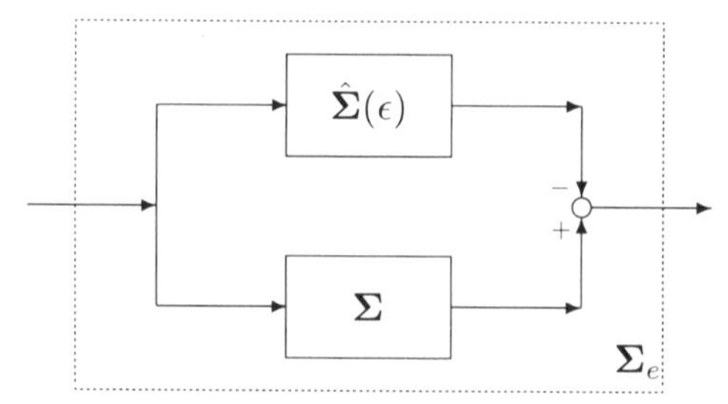

Given $\mathbf{\Sigma}$, construct $\hat{\mathbf{\Sigma}}$ such that $\mathbf{\Sigma}_e := \mathbf{\Sigma} - \hat{\mathbf{\Sigma}}$ has norm ϵ, and is all-pass; in this case $\hat{\mathbf{\Sigma}}$ is called an *ϵ-all-pass dilation* of $\mathbf{\Sigma}$. The following result holds.

THEOREM 2.3. *Let $\hat{\mathbf{\Sigma}}$ be an ϵ-all-pass dilation of the linear, stable, discrete- or continuous-time system $\mathbf{\Sigma}$, satisying (2.1). The stable part $\hat{\mathbf{\Sigma}}_+$ of $\mathbf{\Sigma}$ has exactly k stable poles and the inequalities (2.1) hold. Furthermore, if $\sigma_{k+1}(\mathbf{\Sigma}) = \epsilon$, $\sigma_{k+1}(\mathbf{\Sigma}) = \|\mathbf{\Sigma} - \hat{\mathbf{\Sigma}}\|_H$.*

Computation. The Hankel singular values of $\mathbf{\Sigma}$ given by (1.2), can be computed by solving two Lyapunov equations in finite dimensions. For continuous-time systems, let $\mathcal{P}$, $\mathcal{Q}$ be the system grammians:

$$AP + \mathcal{P}A^* + BB^* = 0, \;\; A^*\mathcal{Q} + \mathcal{Q}A + C^*C = 0 \tag{2.2}$$

It can be shown that [**8**]:

$$\sigma_i(\mathbf{\Sigma}) = \sqrt{\lambda_i(\mathcal{P}\mathcal{Q})} \tag{2.3}$$

The *error bound* for optimal approximants, in the 2-norm of the *convolution operator* is:

$$\sigma_{k+1} \leq \| \mathbf{\Sigma} - \hat{\mathbf{\Sigma}} \|_\infty \leq 2(\sigma_{k+1} + \cdots + \sigma_n) \tag{2.4}$$

Recall that the $\mathcal{H}_\infty$ norm of $\mathbf{\Sigma}$ is maximum of the largest singular value of the frequency response, or alternatively, the 2-induced norm of the *convolution operator*, namely: $y(t) = \mathcal{S}(u)(t) = \int_{-\infty}^{\infty} h(t-\tau)u(\tau)d\tau$, $t \in \mathbb{R}$, where $h(t) = Ce^{At}B$. For details on these issues we refer e.g. to [**6, 7, 8**].

2.2.4. *Approximation by balanced truncation.* A linear system $\mathbf{\Sigma}$ in state space form is called *balanced* if the solutions of the two grammians (2.2) are equal and diagonal:

$$\mathcal{P} = \mathcal{Q} = \Sigma = \text{diag}\,(\sigma_1, \cdots, \sigma_n) \tag{2.5}$$

It turns our that every reachable and observable system can be transformed to balanced form by means of a basis change $\hat{x} = Tx$. Let $\mathcal{P} = UU^*$ and $\mathcal{Q} = LL^*$ where U and L are upper and lower triangular matrices respectively. Let also $U^*L = Z\Sigma Y^*$ be the singular value decomposition (SVD) of U^*L. A the balancing transformation is $T = \Sigma^{\frac{1}{2}}Z^*U^{-1} = \Sigma^{-\frac{1}{2}}Y^*L^*$. Let $\mathbf{\Sigma}$ be balanced with grammians equal to $\Sigma = \text{diag}\,(\Sigma_1, \Sigma_2)$, where $\Sigma_1 \in \mathbb{R}^{k\times k}$, and Σ_2 contains all the small Hankel singular values. Partition conformally the system matrices:

$$\mathbf{\Sigma} = \left[\begin{array}{cc|c} A_{11} & A_{12} & B_1 \\ A_{21} & A_{22} & B_2 \\ \hline C_1 & C_2 & \end{array}\right] \quad \text{where } A_{11} \in \mathbb{R}^{k\times k},\; B_1 \in \mathbb{R}^{k\times m},\; C_1 \in \mathbb{R}^{p\times k} \tag{2.6}$$

The system $\hat{\mathbf{\Sigma}} := \left[\begin{array}{c|c} A_{11} & B_1 \\ \hline C_1 & \end{array}\right]$, is a reduced order system obtained by *balanced truncation*. This system has the following guaranteed properties: (a) stability is preserved, and (b) the same error bound (2.4) holds as in Hankel-norm approximation.

2.2.5. *Singular Perturbation Approximation.* A closely related approximation method, is the so-called *singular perturbation* approximation. It is based on the balanced form presented above. Thus, let (2.6) hold; the reduced order model is given by

$$\hat{\mathbf{\Sigma}} = \left[\begin{array}{c|c} \hat{A} & \hat{B} \\ \hline \hat{C} & \hat{D} \end{array}\right] = \left[\begin{array}{c|c} A_{11} - A_{12}A_{22}^{-1}A_{21} & B_1 - A_{12}A_{22}^{-1}B_2 \\ \hline C_1 - C_2A_{22}^{-1}A_{21} & D - C_2A_{22}^{-1}B_2 \end{array}\right]. \tag{2.7}$$

Again, the same guaranteed properties as for the approximation by balanced truncation, are satisfied.

3. Approximation by moment matching

Given a linear system $\mathbf{\Sigma}$ in state space form (1.2), its transfer function $G(s) = C(sI - A)^{-1}B + D$, is expanded in a Laurent series around a given point $s_0 \in \mathbb{C}$ in the complex plane:

$$G(s_0 + \sigma) = \eta_0 + \eta_1\sigma + \eta_2\sigma^2 + \eta_3\sigma^3 + \cdots$$

The η_t are called the *moments* of $\mathbf{\Sigma}$ at s_0. We seek a reduced order system $\hat{\mathbf{\Sigma}}$ as in (1.3), such that the Laurent expansion of the corresponding transfer function at s_0 has the form

$$\hat{G}(s_0 + \sigma) = \hat{\eta}_0 + \hat{\eta}_1\sigma + \hat{\eta}_2\sigma^2 + \hat{\eta}_3\sigma^3 + \cdots$$

where a certain number of moments ℓ is matched:

$$\eta_j = \hat{\eta}_j, \; j = 1, 2, \cdots, \ell$$

for $\ell \ll n$. If s_0 is infinity, the moments are called *Markov parameters*; the corresponding problem is known as *partial realization*, or *Padé approximation*; the solution of this problem can be found in [**1**], [**4**]. Importantly, the solution of these problems can be implemented in a numerically stable and efficient way, by means of the *Lanczos* and *Arnoldi* procedures. For arbitrary $s_0 \in \mathbb{C}$, the problem is known as *rational interpolation*, see e.g. [**2**], [**3**]. A numerically efficient solution is given by means of the *rational Lanczos/Arnoldi* procedures. For connections between moment matching and Lanczos see [**5**].

Recently, there has been renewed interest in moment matching and projection methods for model reduction in LTI systems. Three leading efforts in this area are Padé via Lanczos (PVL) [**13**], multi-point rational interpolation [**16**], and implicitly restarted dual Arnoldi [**19**].

The PVL approach exploits the deep connection between the (nonsymmetric) Lanczos process and classic moment matching techniques. The multi-point rational interpolation approach utilizes the rational Krylov method of Ruhe [**26**] to provide moment matching of the transfer function at selected frequencies and hence to obtain enhanced approximation of the transfer function over a broad frequency range. These techniques have proven to be very effective. PVL has enjoyed considerable success in circuit simulation applications. Rational interpolation achieves remarkable approximation of the transfer function with very low order models. Nevertheless, there are shortcomings to both approaches. In particular, since the methods are local in nature, it is difficult to establish rigorous error bounds. Heuristics have been developed that appear to work, but no global results exist. Secondly, the rational interpolation method requires selection of interpolation points. At present, this is not an automated process and relies on ad-hoc specification by the user.

In [**19**] an implicitly restarted dual Arnoldi approach is described. The dual Arnoldi method runs two separate Arnoldi processes, one for the reachability subspace, and the other for the observability subspace and then constructs an oblique projection from the two orthogonal Arnoldi basis sets. The basis sets and the reduced model are updated using a generalized notion of implicit restarting. The updating process is designed to iteratively improve the approximation properties of the model. Essentially, the reduced model is reduced further, keeping the best features, and then expanded via the dual Arnoldi processes to include new information. The goal is to achieve approximation properties similar to those of balanced truncation. Other related approaches [**12, 21, 25**] work directly with projected forms of the two Lyapunov equations (2.2) to obtain low rank approximations to the system Grammians.

In the sequel we will review the Lanczos and Arnoldi procedures. We will also review the concept of implicit restarting. For simplicity only the scalar (SISO) versions will be discussed.

3.1. The Lanczos procedure. Given is the scalar ($m = p = 1$) system $\mathbf{\Sigma}$ as in (1.2). We seek to find $\hat{\mathbf{\Sigma}}$ as in (1.3), $k < n$, such that the first $2k$ *Markov parameters* $\eta_i = CA^{i-1}B$, of $\mathbf{\Sigma}$, and $\hat{\eta}_i := \hat{C}\hat{A}^{i-1}\hat{B}$, of $\hat{\mathbf{\Sigma}}$, are matched: $\eta_i = \hat{\eta}_i$, $i = 1, \cdots, 2k$. We will solve this problem following a non-conventional path with system-theoretic flavor; the Lanczos factorization in numerical analysis is introduced using a different set of arguments. First, the *observability* matrix $\mathcal{O}_t$, and the *reachability* matrix $\mathcal{R}_t$ are defined:

$$\mathcal{O}_t = \begin{bmatrix} C \\ CA \\ \vdots \\ CA^{t-1} \end{bmatrix} \in \mathbb{R}^{t\times n}, \ \mathcal{R}_t = \begin{bmatrix} B & AB & \cdots & A^{t-1}B \end{bmatrix} \in \mathbb{R}^{n\times t}$$

Secondly, we define the $t \times t$ Hankel matrix, and its shift:

$$\mathcal{H}_t := \begin{bmatrix} \eta_1 & \eta_2 & \cdots & \eta_t \\ \eta_2 & \eta_3 & \cdots & \eta_{t+1} \\ \vdots & & \ddots & \\ \eta_t & \eta_{t+1} & \cdots & \eta_{2t-1} \end{bmatrix}, \ \sigma\mathcal{H}_t := \begin{bmatrix} \eta_2 & \eta_3 & \cdots & \eta_{t+1} \\ \eta_3 & \eta_4 & \cdots & \eta_{t+2} \\ \vdots & & \ddots & \\ \eta_{t+1} & \eta_{t+2} & \cdots & \eta_{2t} \end{bmatrix}$$

It follows that $\mathcal{H}_t = \mathcal{O}_t\mathcal{R}_t$ and $\sigma\mathcal{H}_t = \mathcal{O}_tA\mathcal{R}_t$. The *key step* is as follows: under the *assumption* that $\det\mathcal{H}_i \neq 0$, $i = 1, \cdots, k$, we compute the LU factorization of $\mathcal{H}_k$:

$$\mathcal{H}_k = LU$$

with $L(i,j) = 0$, $i < j$, $U(i,j) = 0$, $i > j$, and $L(i,i) = \pm U(i,i)$. Define the maps:

$$\pi_L := L^{-1}\mathcal{O}_k \text{ and } \pi_U := \mathcal{R}_kU^{-1} \tag{3.1}$$

Clearly, the following properties hold: (a) $\pi_L\pi_U = 1$, and (b) $\pi_U\pi_L$: (oblique) projection. The **reduced order system $\hat{\mathbf{\Sigma}}$**, is now defined as follows:

$$\hat{A} := \pi_LA\pi_U, \ \hat{B} = \pi_LB, \ \hat{C} = C\pi_U \tag{3.2}$$

THEOREM 3.1. *$\hat{\Sigma}$ as defined above matches $2k$ Markov parameters. Furthermore, $\hat{A}$ is tridiagonal, and $\hat{B}, \hat{C}^*$ are multiples of the unit vector e_1.*

3.2. The Arnoldi procedure. As in the Lanczos case, we will derive the Arnoldi factorization following a non-conventional path, which is different from the path usually adopted in numerical analysis in this case. Let $\mathcal{R}_n = [B\ AB\ \cdots\ A^{n-1}B]$ with $A \in \mathbb{R}^{n\times n}$, $B \in \mathbb{R}^n$. Then:

$$A\mathcal{R}_n = \mathcal{R}_n F \text{ where } F = \begin{pmatrix} 0 & 0 & \cdots & 0 & -\alpha_0 \\ 1 & 0 & \cdots & 0 & -\alpha_1 \\ 0 & 1 & \cdots & 0 & -\alpha_2 \\ & & \ddots & & \\ 0 & 0 & \cdots & 1 & -\alpha_{n-1} \end{pmatrix}$$

and $\chi_A(s) = \det(sI - A) = s^n + \alpha_{n-1}s^{n-1} + \cdots + \alpha_1 s + \alpha_0$. The *key step* in this case consists in computing the QR factorization of $\mathcal{R}_n$:

$$\mathcal{R}_n = VU$$

where V is orthogonal and U is upper triangular. It follows that $AVU = VUF$, $AV = VUFU^{-1}$, which in turn with $\bar{F} = UFU^{-1}$, implies that $AV = V\bar{F}$; since U, U^{-1} are upper triangular, and F is upper Hessenberg, it follows that $\bar{F}$ is *upper Hessenberg*. The first k columns of this relationship yield the k-step Arnoldi factorization:

$$[AV]_k = \left[V\bar{F}\right]_k \Rightarrow A[V]_k = [V]_k\bar{F}_{kk} + fe_k^*$$

where f is a multiple of the $(k+1)$-st column of V; $\bar{F}_{kk}$, is still upper Hessenberg, and the columns of $[V]_k$ provide an orthonormal basis for the space spanned by the first k columns of $\mathcal{R}_n$.

Recall that $\mathcal{H}_k = \mathcal{O}_k\mathcal{R}_k$; a projection π can be attached to the QR factorization of $\mathcal{R}_k$:

$$\pi := \mathcal{R}_k U^{-1} = V,\ V \in \mathbb{R}^{n\times k},\ V^*V = I_k,\ U: \text{ upper triangular} \tag{3.3}$$

The *reduced order system* $\hat{\Sigma}$ is now defined as follows:

$$\hat{A} := \pi^* A\pi,\ \hat{B} = \pi^* B,\ \hat{C} = C\pi \tag{3.4}$$

THEOREM 3.2. *$\hat{\Sigma}$ matches k Markov parameters: $\hat{\eta}_i = \eta_i$, $i = 1, \cdots, k$. Furthermore, $\hat{A}$ is in Hessenberg form, and $\hat{B}$ is a multiple of e_1.*

Remarks. **(a)** Number of operations needed to compute $\hat{\Sigma}$ using Lanczos or Arnoldi is $O(k^2n)$, vs. $O(n^3)$ operations needed for the other methods. The procedure is *iterative* and only *matrix-vector* multiplications are required as opposed to matrix factorizations and/or inversions.

(b) Drawback of Lanczos: it breaks down if $\det \mathcal{H}_i = 0$, for some $1 \le i \le n$. The remedy in this case are *look-ahead* methods. Arnoldi breaks down if $\mathcal{R}_i$ does not have full rank; this happens less frequently.

(c) $\hat{\Sigma}$ tends to approximate the high frequency poles of Σ. Hence the steady-state error may be significant. Remedy: match expansions around other frequencies. This leads to *rational Lanczos*.

(d) $\hat{\Sigma}$ may not be stable, even if Σ is stable. Remedy: *implicit restart* of Lanczos and Arnoldi. ■

3.3. The algorithms.

3.3.1. *The Lanczos algorithm: recursive implementation.* **Given**: the triple $A \in \mathbb{R}^{n\times n}$, $B, C^* \in \mathbb{R}^n$, **find**: $V, W \in \mathbb{R}^{n\times k}$, $f, g \in \mathbb{R}^n$, and $K \in \mathbb{R}^{k\times k}$, such that

$$AV = VK + fe_k^*, \;\; A^*W = WK^* + ge_k^*, \text{ where}$$
$$K = W^*AV, \;\; V^*W = I_k, \;\; W^*f = 0, \;\; V^*g = 0$$

where e_k denotes the kth unit vector in $\mathbb{R}^n$. The projections π_L, π_U defined above are given by W^*, V.

Two-sided Lanczos algorithm

(1) $\beta_1 := \sqrt{|CB|}$, $\gamma_1 := \operatorname{sgn}(CB)\beta_1$
$v_1 := B/\beta_1$, $w_1 := C^*/\gamma_1$
(2) For $j = 1, \cdots, k$, set
  (a) $\alpha_j := w_j^* A v_j$
  (b) $r_j := Av_j - \alpha_j v_j - \gamma_j v_{j-1}$, $q_j := A^* w_j - \alpha_j w_j - \beta_j w_{j-1}$
  (c) $\beta_{j+1} = \sqrt{|r_j^* q_j|}$, $\gamma_{j+1} = \operatorname{sgn}(r_j^* q_j)\beta_{j+1}$
  (d) $v_{j+1} = r_j/\beta_{j+1}$, $w_{j+1} = q_j/\gamma_{j+1}$

The following relationships hold: $V_k = (v_1 \; v_2 \; \cdots \; v_k)$, $W_k = (w_1 \; w_2 \; \cdots \; w_k)$, where $AV_k = V_k K_k + \beta_{k+1} v_{k+1} e_k^*$, $A^*W_k = W_k K_k^* + \gamma_{k+1} w_{k+1} e_k^*$, and $K_k = \begin{pmatrix} \alpha_1 & \gamma_2 & & \\ \beta_2 & \alpha_2 & \ddots & \\ & \ddots & \ddots & \gamma_k \\ & & \beta_k & \alpha_k \end{pmatrix}$, $r_k \in \mathcal{R}_{k+1}(A, B)$, $q_k^* \in \mathcal{O}_{k+1}(C, A)$.

3.3.2. *The Arnoldi algorithm: recursive implementation.* **Given**: the triple $A \in \mathbb{R}^{n\times n}$, $B, C^* \in \mathbb{R}^n$, **find**: $V \in \mathbb{R}^{n\times k}$, $f \in \mathbb{R}^n$, and $K \in \mathbb{R}^{k\times k}$, such that

$$AV = VK + fe_k^*, \text{ where}$$
$$K = V^*AV, \;\; V^*V = I_k, \;\; V^*f = 0$$

where K is in *upper Hessenberg* form. The projection π defined above is given by V.

The Arnoldi algorithm

(1) $v_1 := \frac{v}{\|v\|}$, $w := Av_1$; $\alpha_1 := v_1^* w$
$f_1 := w - v_1\alpha_1$; $V_1 := (v_1)$; $K_1 := (\alpha_1)$
(2) For $j = 1, 2, \cdots, k-1$
  $\beta_j := \| f_j \|$, $v_{j+1} := \frac{f_j}{\beta_j}$
  $V_{j+1} := (V_j \; v_{j+1})$, $\hat{K}_j = \begin{pmatrix} K_j \\ \beta_j e_j^* \end{pmatrix}$
  $w := Av_{j+1}$, $h := V_{j+1}^* w$, $f_{j+1} = w - V_{j+1} h$
  $K_{j+1} := \left(\hat{K}_j \; h \right)$

Remarks. **(a)** Let $f_j := Av_j - V_j h_j$ be the residual of the Arnoldi process. It can be shown that at each step h_j is chosen so that the norm $\| f_j \|$ of this residual is minimized. It turns out that $V_j^* h_j = 0$, and $h_j = V_j^* A v_j$, where $w = Av_j$, that is $f_j = (I - V_j V_j^*) A v_j$.

(b) If A is symmetric, then H_j is tridiagonal, and the Arnoldi algorithm coincides with the Lanczos algorithm. ■

3.4. Implicitly restarted Arnoldi and Lanczos methods. The concept of *implicit restarting* (IR) of the Lanczos and Arnoldi factorizations was introduced by Sorensen [**27**]. Its goal is to get a better approximation of some desired set of preferred eigenvalues, for example, those eigenvalues that have

- Largest modulus
- Largest real part
- Positive or negative real part

Let A have eigenvalues in the left half-plane, and let the approximant K_m obtained through Lanczos or Arnoldi, have an eigenvalue μ in the right half-plane. To eliminate this unwanted eigenvalue the reduced order system obtained at the m-th step $AV_m = V_m K_m + f_m e_m^*$, is projected onto an $(m-1)$-st order system as follows. First, compute the QR-factorization of $K_m - \mu I_m = Q_m R_m$:

$$A\bar{V}_m = \bar{V}_m \bar{K}_m + f_m e_m^* Q_m \text{ where } \bar{V}_m = V_m Q_m,\ \bar{K}_m = Q_m^* K_m Q_m$$

We now truncate the above relationship to contain $m-1$ columns; let $\bar{K}_{m-1}$ denote the principal submatrix of $\bar{K}_m$, containing the leading $m-1$ rows and columns.

THEOREM 3.3. *Given the above set-up, $\bar{K}_{m-1}$ can be obtained through an $m-1$ step Arnoldi process with A unchanged, and the new starting vector $\bar{B} := (\mu I_n - A)B$: $A\bar{V}_{m-1} = \bar{V}_{m-1}\bar{K}_{m-1} + \bar{f}e_{m-1}^*$.*

This process can be repeated to eliminate other unwanted eigenvalues (poles) from the reduced order system. The price we have to pay for this is that the original moments are no longer exactly macthed.

3.5. The Rational Krylov Method. The rational Krylov Method is a generalized version of the standard Arnoldi and Lanczos methods. Given a dynamical system Σ, a set of interpolation points $w_1, \cdots, w_l$, and an integer N, the Rational Krylov Algorithm produces a reduced order system Σ_k that matches N moments of Σ at $w_1, \cdots, w_l$. The reduced system is not guaranteed to be stable and no global error bounds exist. Moreover the selection of interpolation points which determines the reduced model is not an automated process and has to be figured out by the user using trial and error.

3.6. A new approach: the cross grammian. The approach to model reduction proposed below is related to the implicitly restarted dual Arnoldi approach developed in [**19**]; although it is not a moment matching method it belongs to the general set of Krylov projection methods. Its main feature is that it is based on *one Sylvester equation* instead of *two Lyapunov equations*. One problem with prior attempts at working with the two Lyapunov equations separately and then applying dense methods to the reduced equations, is consistency. One cannot be certain that the two separate basis sets are the ones that would have been selected if the full system Grammians had been available. Since our method actually provides the best rank k approximations to the system Grammians with a computable error bound, we are assured to obtain a valid approximation to the balanced reduction.

Given $\mathbf{\Sigma}$ as in (1.2) with $m = p = 1$, the *cross grammian* $X \in \mathbb{R}^{n \times n}$ is the solution to the following Sylvester equation:

$$AX + XA + BC = 0 \tag{3.5}$$

Recall the definition of the reachability and observability grammians (2.2). The relationship between these three grammians is:

$$X^2 = \mathcal{P}\mathcal{Q} \tag{3.6}$$

Moreover, the eigenvalues of X are equal to the non-zero Hankel singular values of $\mathbf{\Sigma}$. If A is stable

$$X = \int_0^\infty e^{At}BCe^{At}dt = \frac{1}{2\pi}\int_{-\infty}^{\infty}(j\omega - A)^{-1}BC(-j\omega - A)^{-1}\,d\omega$$

Furthermore, in this case the $\mathcal{H}_2$ norm of the system is given by $\| \mathbf{\Sigma} \|_{\mathcal{H}_2}^2 = CXB$. Often, the *singular values* of X drop off very rapidly and X can be well approximated by a low rank matrix. Therefore the idea is to capture most of the energy with X_k, the best rank k approximation to X:

$$\| \mathbf{\Sigma} \|_{\mathcal{H}_2}^2 = CX_kB + \mathcal{O}(\sigma_{k+1}(X))$$

The Approximate Balanced Method [**9**] solves a Sylvester Equation to obtain a reduced order almost balanced system iteratively without computing the full order balanced realization $\mathbf{\Sigma}$ in (2.6).

3.6.1. *Description of the solution.* We now return to the study of the *Sylvester equation* (3.5). It is well known that X is a solution iff

$$\begin{pmatrix} A & BC \\ 0 & -A \end{pmatrix}\begin{pmatrix} I & X \\ 0 & I \end{pmatrix} = \begin{pmatrix} I & X \\ 0 & I \end{pmatrix}\begin{pmatrix} A & 0 \\ 0 & -A \end{pmatrix}$$

This suggests that X can be computed using a projection method:

$$\begin{pmatrix} A & BC \\ 0 & -A \end{pmatrix}\begin{pmatrix} V_1 \\ V_2 \end{pmatrix} = \begin{pmatrix} V_1 \\ V_2 \end{pmatrix}H,\ V = \begin{pmatrix} V_1 \\ V_2 \end{pmatrix}$$

where V is orthogonal: $V_1^*V_1 + V_2^*V_2 = I$. If V_2 is non-singular, then $AV_1 + BCV_2 = V_1H$, $-AV_2 = V_2H$, which implies $A(V_1V_2^{-1}) + BC = (V_1V_2^{-1})\hat{H}$, $H = -A$, where $\hat{H} = V_2HV_2^{-1}$. Therefore the solution is:

$$X = V_1V_2^{-1}$$

The best rank k approximation to X is related to the *C-S decomposition.* Let $V = [V_1^*\ V_2^*]^* \in \mathbb{R}^{2n\times k}$, with $V^*V = I_k$; then we have $V_1 = U_1\Gamma W^*$, $V_2 = U_2\Delta W^*$, where U_1, U_2 are orthogonal, W nonsingular and $\Gamma^2 + \Delta^2 = I_k$. Assuming that A is stable, it readily follows that V_2 has full rank iff the eigenvalues of A include those of H. It follows that the SVD of X can be expressed as $X = U_1(\Gamma/\Delta)U_2^*$. To compute the best rank k approximation to X, we begin with the full (n-step) decomposition: $V_1, V_2 \in \mathbb{R}^{n\times n}$, V_2 full rank:

$$\begin{aligned} AV_1 + BCV_2 &= V_1H \\ -AV_2 &= V_2H \end{aligned}$$

Let $W_k := W(:, 1:k)$, $\Gamma_k := \Gamma(1:k, 1:k)$, $\Delta_k := \Delta(1:k, 1:k)$. Then

$$\begin{aligned} A(V_1W_k) + BC(V_2W_k) &= (V_1W)(W^*HW_k) \\ -A(V_2W_k) &= (V_2W)(W^*HW_k) \end{aligned}$$

Therefore

$$\begin{aligned} A(U_{1k}\Gamma_k) + BC(U_{2k}\Delta_k) &= (U_{1k}\Gamma_k)H_k + E_k \\ -U_{2k}^*AU_{2k}\Delta_k &= \Delta_kH_k \end{aligned}$$

where $U_{1k}^* E_k = 0$ and $H_k = W_k^* H W_k$. We thus obtain the *projected* Sylvester equation

$$U_{1k}^*(AX_k + X_k A + BC)U_{2k} = 0$$

This implies the error equation

$$AX_k + X_k A + BC = -A(X - X_k) - (X - X_k)A = \mathcal{O}(\gamma_{k+1}/\delta_{k+1})$$

where

$$X_k = U_{1k}(\Gamma_k/\Delta_k)U_{2k}^*$$

is the best rank k approximation of the cross grammian X. The *Reduced order system* is now defined as follows: let the SVD of the cross grammian be $X = U\Sigma V^*$, and the best rank k approximant be $X_k = U_k \Sigma_k V_k^*$. Then $\hat{\boldsymbol{\Sigma}}$ is given by

$$\hat{\boldsymbol{\Sigma}} = \left[\begin{array}{c|c} \hat{A} & \hat{B} \\ \hline \hat{C} & \end{array}\right] = \left[\begin{array}{c|c} V_k^* A V_k & V_k^* B \\ \hline C V_k & \end{array}\right] \tag{3.7}$$

A closely related alternative method of defining a reduced order model is the following. Let $\hat{X}$ be the best rank k approximation to X. Compute a partial eigenvalue decomposition of $\hat{X} = Z_k D_k W_k^*$ where $W_k^* Z_k = I_k$. Then an approximately balanced system is obtained as

$$\hat{\boldsymbol{\Sigma}} = \left[\begin{array}{c|c} \hat{A} & \hat{B} \\ \hline \hat{C} & \end{array}\right] = \left[\begin{array}{c|c} W_k^* A Z_k & W_k^* B_k \\ \hline C Z_k & \end{array}\right] \tag{3.8}$$

The advantage of this method which will be referred to as *Approximate Balancing* in the sequel, is that it computes an *almost* balanced reduced system *iteratively* without computing a balanced realization of the full order system first, and then truncating. Some details on the implementation of this algorithm are provided in section 3.6.3; more details can be found in [**9**].

Finally, a word about the MIMO case. Recall that (3.5) is not defined unless $m = p$. Hence, we to apply this method to MIMO systems, we proceed by embedding the system $\boldsymbol{\Sigma}$ in a system $\tilde{\boldsymbol{\Sigma}}$ which has the same order, is square, and symmetric:

$$\tilde{\boldsymbol{\Sigma}} = \left[\begin{array}{c|c} A & \tilde{B} \\ \hline \tilde{C} & \end{array}\right] \left[\begin{array}{c|c} A & JC^* \quad B \\ \hline C & \\ B^* J^{-1} & \end{array}\right], \quad J = J^* \tag{3.9}$$

The matrix J is symmetric and is chosen so that $AJ = JA^*$; consequently it is called *symmetrizer*. The remaining degrees of freedom of J are chosen so that $\lambda_i(\mathcal{PQ}) \approx \lambda_i(\tilde{\mathcal{P}}\tilde{\mathcal{Q}}) = \lambda(\tilde{X})^2$ where $\tilde{X}$ is the cross grammian of $\tilde{\boldsymbol{\Sigma}}$; for details see [**9**].

3.6.2. *A stopping criterion.* As *stopping criterion*, we look at the $\mathcal{L}_\infty$-norm of the following *residual*:

$$R(s) = BC - (sI - A)V_k(sI - \hat{A})^{-1}\hat{B}\hat{C}(-sI - \hat{A})^{-1}V_k^*(-sI - A)$$

Notice that *projected residual* is zero: $V_k^* R(s) V_k = 0$. Consider:

$$(sI - A)V(sI - V^*AV)^{-1}V^*B = (sI - A)VV^*(sI - AVV^*)^{-1}B$$

Assume that we change basis so that $V^* = [I_k \ 0]$; let in this basis

$$A = \begin{pmatrix} A_{11} & A_{12} \\ A_{21} & A_{22} \end{pmatrix}, \quad B = \begin{pmatrix} B_1 \\ B_2 \end{pmatrix}, \quad C = \begin{pmatrix} C_1 & C_2 \end{pmatrix}$$

Then this last expression becomes:

$$\begin{pmatrix} I \\ -A_{21}(sI_k - A_{11})^{-1} \end{pmatrix} B_1$$

Similarly, $CV(-sI - V^*AV)^{-1}V^*(-sI - A) = C_1[I \ (-sI - A_{11})^{-1}A_{12}]$. Hence

$$R(s) = \begin{pmatrix} B_1 \\ B_2 \end{pmatrix} \begin{pmatrix} C_1 & C_2 \end{pmatrix} - \begin{pmatrix} I \\ -A_{21}(sI_k - A_{11})^{-1} \end{pmatrix} B_1 C_1 \begin{pmatrix} I & (-sI - A_{11})^{-1}A_{12} \end{pmatrix}$$

In state space form we have

$$R = \left[\begin{array}{cc|cc} A_{11} & B_1C_1 & B_1C_1 & 0 \\ 0 & -A_{11} & 0 & A_{12} \\ \hline 0 & -B_1C_1 & 0 & B_1C_2 \\ A_{21} & 0 & B_2C_1 & B_2C_2 \end{array}\right] \in \mathbb{R}^{(2k+n)\times(2k+n)}$$

The implication is that the residual system is described by an $n \times n$ proper rational matrix $R(s)$, whose McMillan degree is $2k$; that is the residual system although it has many inputs and outputs, it has low complexity and therefore its $\mathcal{L}_\infty$-norm $\| R(s) \|_\infty$ can be readily computed.

3.6.3. *Computation of the solution.* The following points are important to keep in mind. First, we give up Krylov (Arnoldi requires special starting vector), and second, an iterative scheme related to the Jacobi-Davidson algorithm together with implicit restarting will be used. Given is

$$(-A)V = VH + F, \ V^*V = I, \ V^*F = 0$$

(1) Solve the projected Sylvester equation in the Controllability space and compute the SVD of the solution:

$$AY + BCV = YH \ \Rightarrow \ [Q, S, W] = \mathrm{svd}\,(Y)$$

(2) Project onto the space of the largest singular values

$$\left|\begin{array}{l} S_k = S(1:k, 1:k) \\ Q_k = Q(:, 1:k) \\ W_k = W(1:k, :) \end{array}\right. \qquad \left|\begin{array}{l} V \leftarrow VW_k \\ \bar{H} = Q_k^* A Q_k \\ H \leftarrow W_k^* H W_k \end{array}\right.$$

(3) Correct the projected Sylvester equation in the observability space:

$$E := A^*VW_kS_k + VW_kS_k\bar{H}^* + C^*B^*Q_k$$

Solve $A^*Z + Z\bar{H}^* = -E$.

(4) Adjoin Correction and project:

$$\left|\begin{array}{l} [V, R] = \mathrm{qr}\,([VW_k, \ Z]) \\ H \leftarrow -(V^*AV) \\ F \leftarrow (-I + VV^*)AV \end{array}\right.$$

Remark. It should be stressed that the equations in 1. and 3. above

$$\begin{pmatrix} A & BCV \\ 0 & -H \end{pmatrix}\begin{pmatrix} V_1 \\ V_2 \end{pmatrix} = \begin{pmatrix} V_1 \\ V_2 \end{pmatrix} R, \ \begin{pmatrix} A^* & E \\ 0 & -H^* \end{pmatrix}\begin{pmatrix} W_1 \\ W_2 \end{pmatrix} = \begin{pmatrix} W_1 \\ W_2 \end{pmatrix} Q$$

are solved by IRAM (Implicitly restrated Arnoldi method). No inversions are required. However convergence may be accelerated with a single sparse direct factorization. ∎

4. Application of the reduction algorithms

In this section we apply the algorithms mentioned above to six different dynamical systems: a structural model, a building model, a heat transfer model, a CD player model, a clamped beam model, and a low-pass butterworth filter. We reduce the order of models with a *tolerance*[1] value, ρ, of 1×10^{-3}. Table-2 shows the order of the systems, n; the number of inputs, m; and outputs, p; and the order of reduced system, k. Moreover, the normalized[2] Hankel singular values of each model are depicted in Figure 2-a and 2-b. To make a better comparison between the systems, in Figure 3 we also show relative degree reduction $\frac{k}{n}$ vs a given error tolerance $\frac{\sigma_k}{\sigma_1}$. This figure shows how much the order can be reduced for the given tolerance: the lower the curve, the easier the system to approximate. It can be seen from Figure 3 that among all models for a fixed tolerance value less than 1.0×10^{-1}, the building model is the hardest one to approximate. One should notice that specification of the tolerance value ρ determines the reduced model completely in all of the methods, except the rational Krylov method. The order of the reduced model and the eigenvalue placements are completely automatic. On the other hand, in the rational Krylov method, one has to choose the interpolation points and the integer N which determines the number of moments matched per point.

	n	m	p	k
Structural Model	270	3	3	37
Building Model	48	1	1	31
Heat Model	197	2	2	5
CD Player	120	1	1	12
Clamped Beam	348	1	1	13
Butterworth Filter	100	1	1	35

TABLE 2. The systems used for comparing the model reduction algorithms

In each subsection below, we briefly describe the system and then apply the algorithms. The following quantities are plotted in each case: (i) the largest singular value of the frequency response of the full and reduced order models, (ii) the same quantity for the corresponding error systems, (iii) the nyquist plots of the full and reduced order systems. Moreover, the relative $\mathcal{H}_\infty$ and $\mathcal{H}_2$ norms of the error systems are tabulated. Since balanced reduction and approximate balanced reduction approximants were *almost* the same for all the models except the heat model, we show and tabulate results just for the former for those cases.

4.1. Structural model. This is a model of component 1r (Russian service module) of the International Space Station. It has 270 states, 3 inputs and 3 outputs. The real part of the pole closest to imaginary axis is -3.11×10^{-3}. The normalized Hankel singular values of the system are shown in Figure 2-a. We approximate the system with reduced models of order 37. Since the system is MIMO, the Arnoldi and Lanczos algorithms are not applied. The resultant reduced order systems are shown in figure 4-a. As seen from this figure, all reduced models

[1]The *tolerance* corresponding to a kth order reduced system is given by the ratio $\frac{\sigma_k}{\sigma_1}$, where σ_1 and σ_k are the largest and k^{th} largest singular value of the system respectively.

[2]For comparison, we normalize the highest Hankel Singular Value of each system to 1.

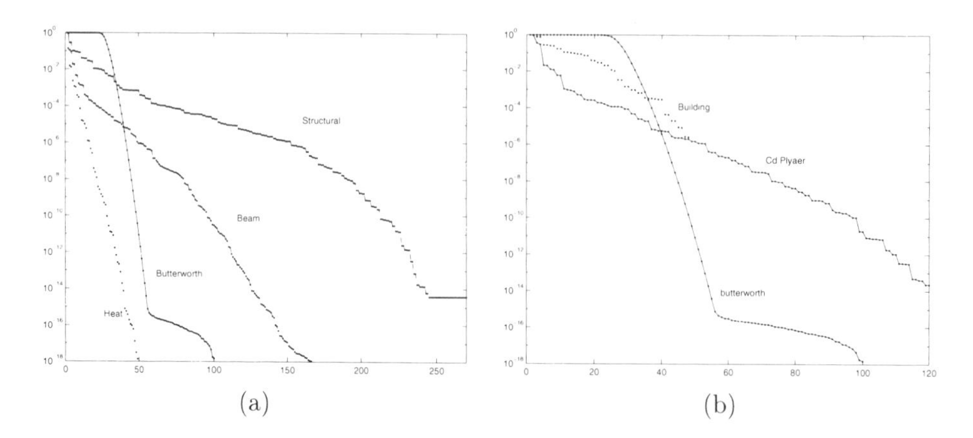

FIGURE 2. Normalized Hankel singular values of (a) heat model, Butterworth filter , clamped beam and structural model; (b) Butterworth filter, building, CD player.

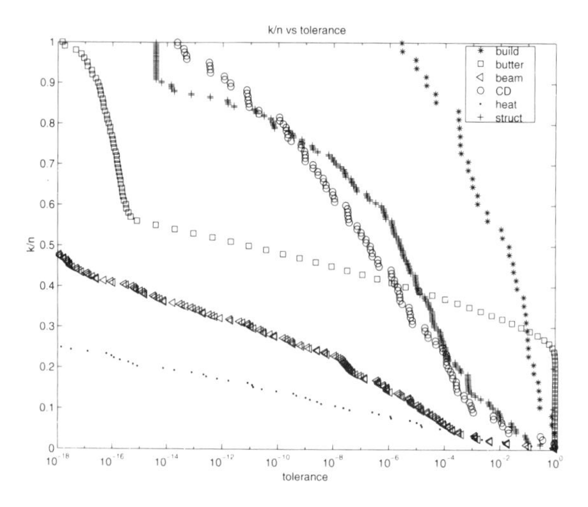

FIGURE 3. Relative degree reduction $\frac{k}{n}$ vs error tolerance $\frac{\sigma_k}{\sigma_1}$

work quite well. The peaks, especially the ones at the lower frequencies, are well approximated. Figure 4-b shows the largest singular value $\sigma_{\max}$ of the frequency response of the error systems. Rational Krylov does a perfect job at the lower and higher frequencies. But for the moderate peak frequency levels, it has the highest error amplitude. This is because the selection of interpolation points is not an automated process and relies on ad-hoc specification by the user. Singular perturbation approximation is the worst for low and higher frequencies. Table 3 lists the relative[3] $\mathcal{H}_\infty$ and $\mathcal{H}_2$ norms of the errors system. As seen from the figure, rational Krylov has the highest errors. Considering both the relative $\mathcal{H}_\infty$, $\mathcal{H}_2$ norms error norms and the whole frequency range, balanced reduction is the best. The

[3]To find the relative error, we divide the norm of the error system with the corresponding norm of the full order system

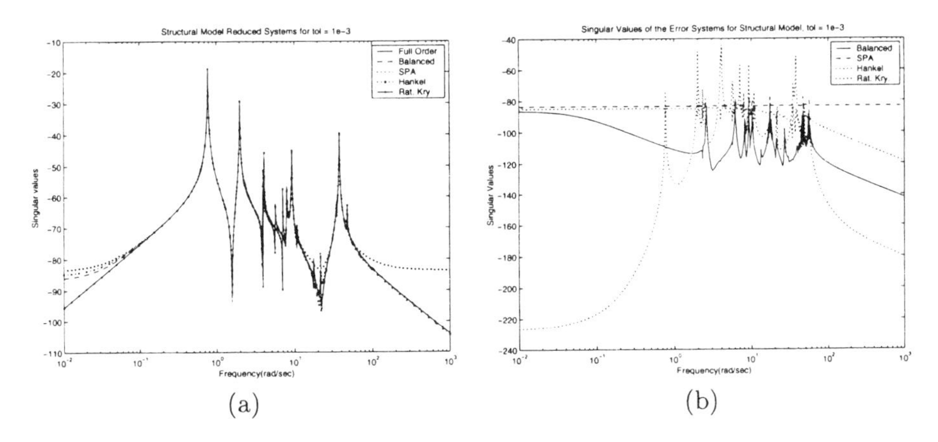

(a) (b)

FIGURE 4. $\sigma_{\max}$ of the frequency response of the (a) reduced and (b) error systems of the structural model

	$\mathcal{H}_\infty$ norm	$\mathcal{H}_2$ norm
Balanced	6.93×10^{-4}	5.70×10^{-3}
Hankel	8.84×10^{-4}	1.98×10^{-2}
Sing. Pert.	1.08×10^{-3}	3.66×10^{-2}
Rat. Krylov	4.46×10^{-2}	1.33×10^{-1}

TABLE 3. Relative Error Norms for Structural Model

nyquist plots of the full order and the reduced order systems are shown in figure 5-a. Notice that all the approximants match the full order model very well except for rational Krylov which deviates around the origin.

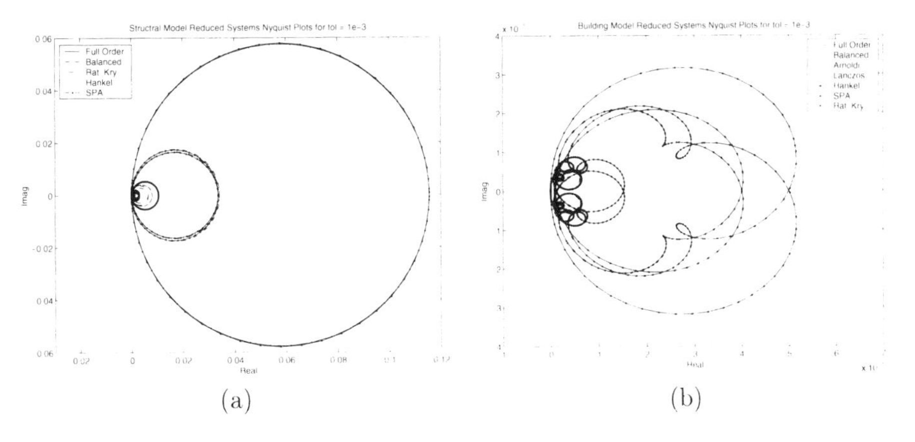

(a) (b)

FIGURE 5. Nyquist plots of the full and reduced order models for the (a) structral model, (b) building

4.2. Building model. This is the model of a building (the Los Angeles University Hospital) with 8 floors each having 3 degrees of freedom, namely displacements in x and y directions, and rotation. Hence we have 24 variables with the following type of second order differential equation describing the dynamics of the system:

$$M\ddot{q}(t) + D\dot{q}(t) + Kq(t) = v\,u(t) \tag{4.1}$$

where $u(t)$ is the input. (4.1) can be put into state-space form by defining $x^* = [\,q^*\ \dot{q}^*\,]^*$:

$$\dot{x}(t) = \begin{bmatrix} 0 & I \\ -M^{-1}K & -M^{-1}D \end{bmatrix} x(t) + \begin{bmatrix} 0 \\ M^{-1}v \end{bmatrix} u(t)$$

We are mostly interested in the motion in the first coordinate $q_1(t)$. Hence, we choose $v = [1\ 0\ \cdots\ 0]^*$ and the output $y(t) = \dot{q}_1(t) = x_{25}(t)$.

The state-space model has order 48, and is single input and single output. For this example, the pole closest to the imaginary axis has real part equal to -2.62×10^{-1}. The chosen tolerance yields a reduced order system of order 31. The largest singular value $\sigma_{\max}$ of the frequency response of the reduced order

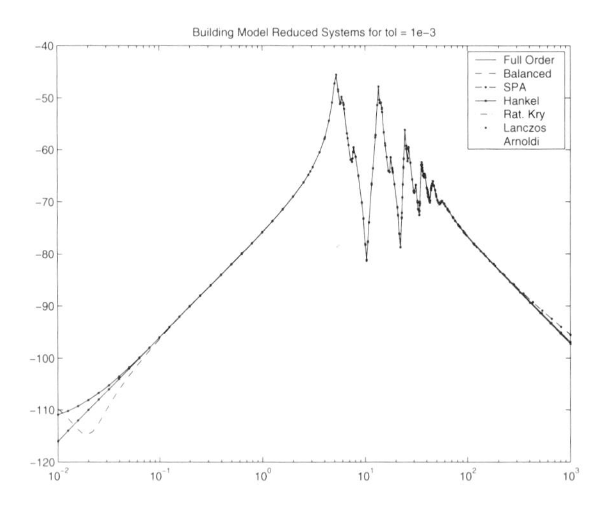

FIGURE 6. $\sigma_{\max}$ of the frequency response of the reduced systems of the building model

system and of the error systems are shown in figures 6 and 7 respectively. Since the expansion of the transfer function $G(s)$ around $s_0 = \infty$ results in unstable reduced systems for Arnoldi and Lanczos, we made use of the shifted version of these two methods with $s_0 = 1$. The effect of choosing s_0 as a low frequency point can be observed in figure 7-b (Arnoldi and Lanczos result in very good approximants for the low frequency range). The same is valid for rational Krylov methods as well, with $s_0 = 1$ chosen as one of the interpolation points. When compared to SVD based methods, the moments matching methods are much better in the low frequency range. Among the SVD based methods, singular perturbation and balanced reduction methods are the best for the low frequency and high frequency range respectively. When we consider the whole frequency range, balancing and singular perturbation are closer to the original model. But in terms of relative $\mathcal{H}_\infty$ error norm, Hankel norm approximation is the best. As expected rational Krylov, Arnoldi and Lanczos result in high relative errors due to their local nature. Among them, rational Krylov is the best. Figure 5-b illustrates the nyquist plots of the

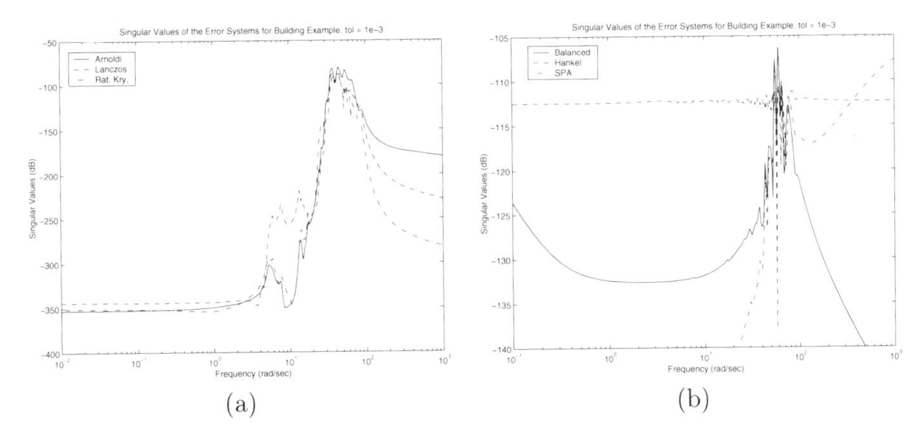

FIGURE 7. (a)-(b) $\sigma_{\max}$ of the frequency response of the error systems of building model

full order and the reduced order systems. The figure shows that all approximants match the nyquist plots of the full order model quite well.

	$\mathcal{H}_\infty$ norm of error	$\mathcal{H}_2$ norm of error
Balanced	$9,64 \times 10^{-4}$	2.04×10^{-3}
Hankel	5.50×10^{-4}	6.25×10^{-3}
Sing. Pert.	9.65×10^{-4}	$2,42 \times 10^{-2}$
Rat. Krylov	7.51×10^{-3}	1.11×10^{-2}
Lanczos	7.86×10^{-3}	1.26×10^{-2}
Arnoldi	1.93×10^{-2}	3.33×10^{-2}

TABLE 4. Relative error norms building model

4.3. Heat diffusion model. The original system is a plate with two heat sources and two points of measurements, described by the heat equation. A model of order 197 is obtained by spatial discretization. The real part of the pole closest to the imaginary axis is -1.52×10^{-2}. It is observed from figure 2-a that this system is rather easy to approximate since the Hankel singular values decay very rapidly. The chosen tolerance of 10^{-3} leads to reduced models of order 5. Since this is a MIMO system, Lanczos and Arnoldi were not applied. As expected, all methods generate satisfactory approximants matching the full order model through the whole frequency range (see figure 8). Only the rational Krylov method has some problems for moderate frequencies due to the unautomated choice of interpolation points. The nyquist plots of the full order and the reduced order systems are shown in figure 9-a. The figure reveals that as in the structural model, rational Krylov has a problem matching the full order system around the origin. With the exception of the rational Krylov approximant, the nyquist plots of the approximants are close to that of the full order model.

4.4. CD Player. This system describes the dynamics between the lens actuator and the radial arm position of a portable CD player. The model has 120 states with a single input and a single output. The pole closest to the imaginary axis has

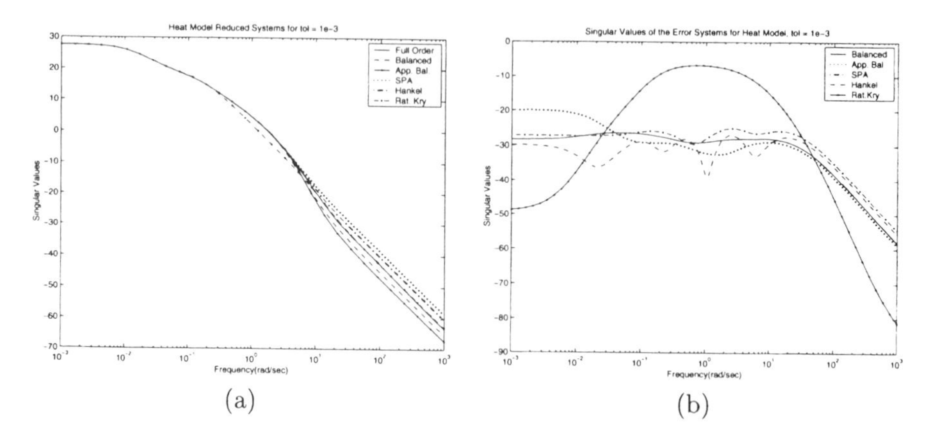

(a) (b)

FIGURE 8. $\sigma_{\max}$ of the frequency response of the (a) reduced and (b) error systems of heat diffusion model

	$\mathcal{H}_\infty$ norm of error	$\mathcal{H}_2$ norm of error
Balanced	2.03×10^{-3}	5.26×10^{-2}
Approx. Bal.	4.25×10^{-3}	4.68×10^{-2}
Hankel	1.93×10^{-3}	6.16×10^{-2}
Sing. Pert.	2.39×10^{-3}	7.39×10^{-2}
Rat. Krylov	1.92×10^{-2}	2.01×10^{-1}

TABLE 5. Relative error norms of the heat model approximants

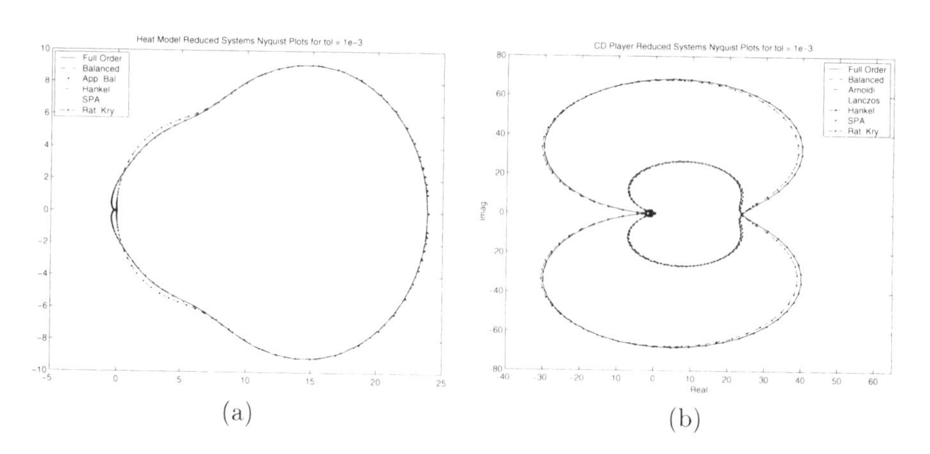

(a) (b)

FIGURE 9. Nyquist plots of the full and reduced order models for the (a) heat model, (b) CD player

real part equal to -2.43×10^{-2}. Approximants have order 12. The first moment of the system is zero. Hence, instead of expanding the transfer function around $s = \infty$, we expand it around $s_0 = 200$ rad/sec. This overcomes the breakdown of the Lanczos procedure. We also use the shifted version of the Arnoldi procedure with $s_0 = 200$ rad/sec. Figure 10-a illustrates the largest singular value of the frequency response of the reduced order models together with that of the full order model. One should notice that only rational Krylov catches the peaks around the frequency range $10^4 - 10^5$ rad/sec. No SVD based method matches those peaks. Among the SVD based ones, Hankel norm approximation is the worst around $s = 0$, and also around $s = \infty$. The largest singular values of the frequency response of the error systems in figure 10-b reveal that the SVD based methods are better when we consider the whole frequency range. Despite doing a perfect job at $s = 0$ and $s = \infty$, rational Krylov has the highest relative $\mathcal{H}_\infty$ and $\mathcal{H}_2$ error norms as listed in table 6. But one should notice that rational Krylov is superior to the Arnoldi and Lanczos procedures except the frequency range $10^2 - 10^3$ rad/sec. When we consider the whole frequency range, balanced reduction is again the best. Figure 9-b illustrates the nyquist plots of the full order and the reduced order systems. Only the rational Krylov approximant has some (small) deviation from the full order model.

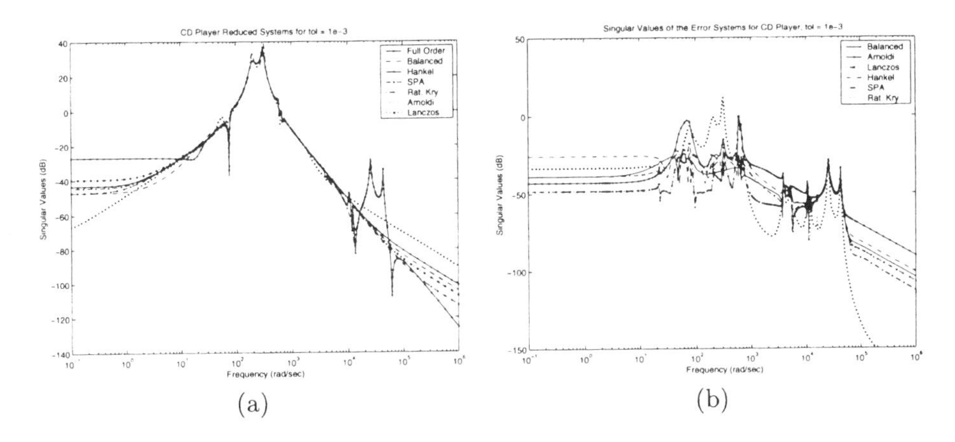

FIGURE 10. $\sigma_{\max}$ of the frequency response of the (a) reduced and (b) error systems of the CD player

	$\mathcal{H}_\infty$ norm of error	$\mathcal{H}_2$ norm of error
Balanced	9.74×10^{-4}	3.92×10^{-3}
Approx. Bal.	9.74×10^{-4}	3.92×10^{-3}
Hankel	9.01×10^{-4}	4.55×10^{-3}
Sing. Pert.	1.22×10^{-3}	4.16×10^{-3}
Rat. Krylov	5.60×10^{-2}	4.06×10^{-2}
Arnoldi	1.81×10^{-2}	1.84×10^{-2}
Lanczos	1.28×10^{-2}	1.28×10^{-2}

TABLE 6. Relative error norms for the CD player

4.5. Clamped Beam Model. The clamped beam model has 348 states and is SISO. It is again obtained by spatial discretization of an appropriate partial differential equation. The input represents the force applied to the structure at the free end, and the output is the resulting displacement. For this example, the real part of the pole closest to imaginary axis is -5.05×10^{-3}. We approximate the system with a model of order 13 (tolerance 10^{-3}). The plots of the largest singular value of the frequency response of the approximants and error systems are shown in figure 11-a and 11-b respectively. Since $CB = 0$, in order to prevent the breakdown of Lanczos, we expand the transfer function $G(s)$ of the original system around $s_0 = 0.1$ instead of $s = \infty$; we also use the shifted Arnoldi method with $s_0 = 0.1$ rad/sec. Rational Krylov is again the best for both $s = 0$ and $s = \infty$. Indeed except for the frequency range between 0.6 and 30 rad/sec, this method gives the best approximant among all the methods. Lanczos and Arnoldi procedures also lead to good approximants especially for the frequency range $0 - 1$ rad/sec. This is due to the choice of s_0 as a low frequency point. Balanced model reduction is the best one among the SVD methods after $s = 1$ rad/sec. In terms of error norms, SVD based methods are better than moment matching based methods, but the differences are not as high as for the previous examples. Again, rational Krylov is the best among moment matching methods. The nyquist plots of the full order and the reduced order systems are shown in figure 12-a. This figure shows that all the approximants match the the nyquist plots of the full order model well. Indeed, this is the best match of nyquist plots among all the six examples.

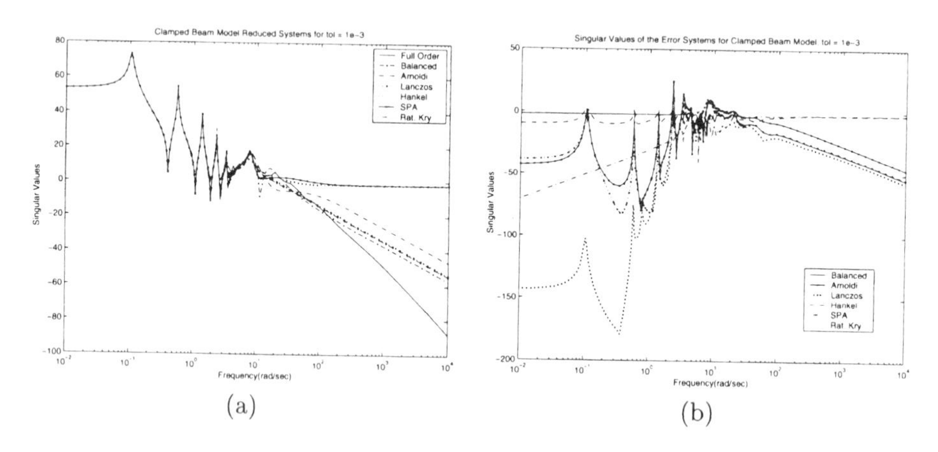

(a) (b)

FIGURE 11. $\sigma_{\max}$ of the frequency response of the (a) reduced and (b) error systems of clamped beam

	$\mathcal{H}_\infty$ norm of error	$\mathcal{H}_2$ norm of error
Balanced	2.14×10^{-4}	7.69×10^{-3}
Hankel	2.97×10^{-4}	8.10×10^{-3}
Sing. Pert.	3.28×10^{-4}	4.88×10^{-2}
Rat. Krylov	5.45×10^{-4}	8.88×10^{-3}
Arnoldi	3.72×10^{-3}	1.68×10^{-2}
Lanczos	9.43×10^{-4}	1.67×10^{-2}

TABLE 7. Relative error norms for the clamped beam

4.6. Low-pass Butterworth filter. The full order model is a low-pass Butterworth filter of order 100 with the cutoff frequency at 1 rad/sec. The normalized

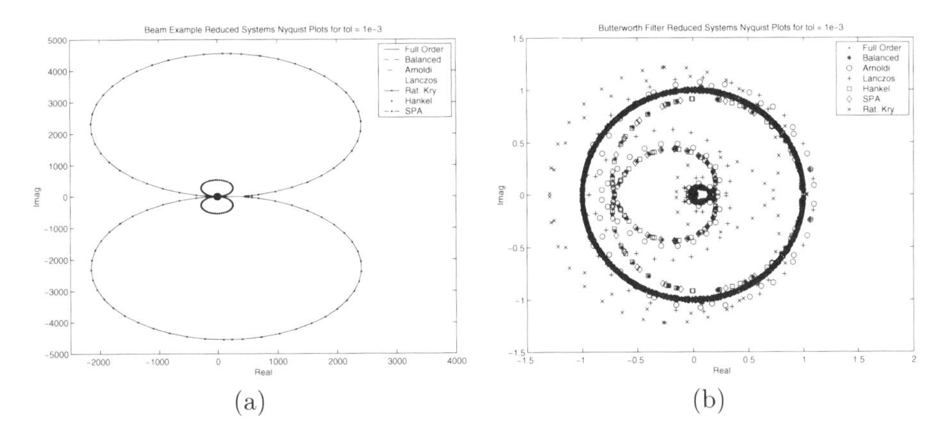

FIGURE 12. Nyquist plots of the full and reduced order models for the (a) clamped beam, (b) Butterworth filter

Hankel singular values are shown in figure 2-a and figure 2-b. It should be noticed that unlike the other systems, the initial 25 singular values are equal. Therefore, this system cannot be reduced below order 25. The approximants in this case have order 35. One should notice that the transfer function of this example has no zeros. Thus Arnoldi and Lanczos procedures do not work if we expand the transfer function $G(s)$ around $s = \infty$. Instead, we expand $G(s)$ around $s_0 = 0.1$. As figure 13-a illustrates, all moment matching based methods have difficulty especially around the cutoff frequency. Among them, Lanczos and Arnoldi show very similar results and are better than rational Krylov. On the other hand, SVD based methods work without problems producing good approximants for the whole frequency range. Among the SVD based methods, Hankel norm approximation is the best in terms of $\mathcal{H}_\infty$ norm, while it is the worst in terms of $\mathcal{H}_2$ norm. Singular perturbation methods and balanced reduction show very close behaviors for frequencies less than 1 rad/sec; subsequently, balanced reduction is better. Figure 12-b depicts the nyquist plots of the full order and the reduced order systems; notice that moment matching methods are far from matching the full order model. SVD based methods do not yield very good approximants, but compared to the former, they do a better job.

	$\mathcal{H}_\infty$ norm of error	$\mathcal{H}_2$ norm of error
Balanced	6.29×10^{-4}	5.19×10^{-4}
Approx. Bal.	6.29×10^{-4}	5.19×10^{-4}
Hankel	5.68×10^{-4}	1.65×10^{-3}
Sing. Pert.	6.33×10^{-4}	5.21×10^{-4}
Rat. Krylov	1.02×10^{0}	4.44×10^{-1}
Arnoldi	1.02×10^{0}	5.38×10^{-1}
Lanczos	1.04×10^{0}	3.68×10^{-1}

TABLE 8. Relative error norms for the Butterworth filter

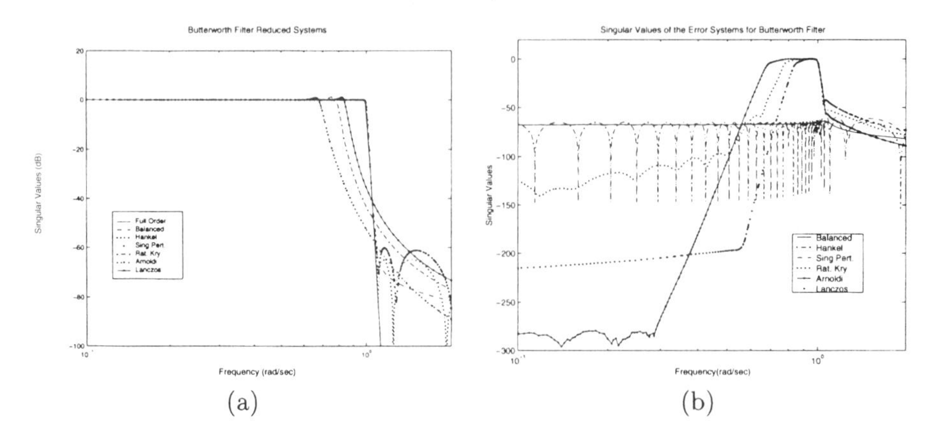

(a) (b)

FIGURE 13. $\sigma_{\max}$ of the frequency response of the (a) reduced and (b) error systems of Butterworth filter

5. Projectors and computational complexity

The unifying feature of all model reduction methods presented above is that they are obtained by means of *projections*. Let $\pi = VW^*$, $V, W \in \mathbb{R}^{n\times k}$, $k \leq n$, be a projection, that is $W^*V = I_k$. The corresponding reduced order model $\hat{\mathbf{\Sigma}}$ in (1.3) is obtained as follows:

$$\begin{array}{rcl} \sigma\hat{x} & = & (W^*AV)\hat{x} + (W^*B)u \\ \hat{y} & = & (CV)\hat{x} \end{array} \tag{5.1}$$

The quality of the approximant is measured in terms of the frequency response $G(j\omega) = C(j\omega I - A)^{-1}B$. Optimal Hankel norm and balancing emphasize energy of grammians

$$\mathcal{P} = \frac{1}{2\pi}\int_{-\infty}^{+\infty} (j\omega I - A)^{-1}BB^*(j\omega I - A^*)^{-1}d\omega,$$

$$\mathcal{Q} = \frac{1}{2\pi}\int_{-\infty}^{+\infty} (j\omega I - A^*)^{-1}C^*C(j\omega I - A)^{-1}d\omega$$

Krylov methods adapt to frequency response and emphasize relative contributions of $C(j\omega I - A)^{-1}B$. The new method emphasizes the energy of the cross grammian

$$X = \frac{1}{2\pi}\int_{-\infty}^{+\infty} (j\omega I - A)^{-1}BC(-j\omega I - A)^{-1}d\omega$$

The *choices of projectors* for the different methods are as follows.

(1) **Balanced truncation**. Solve: $A\mathcal{P}+\mathcal{P}A^*+BB^* = 0$, $A^*\mathcal{Q}+\mathcal{Q}A+C^*C = 0$, and *project* onto the dominant eigenspace of $\mathcal{P}\mathcal{Q}$.
(2) **Optimal Hankel norm approximation**. Solve for the grammians. Embed in a lossless transfer function and project onto its stable eigenspace.
(3) **Krylov-based approximation**. Project onto reachability and/or observability spaces.
(4) **New method**. Project onto the space spanned by the dominant right singular vectors or eigenvectors of the cross grammian.

The *complexity* of these methods using *dense* decompositions taking into account only dominant terms of the total cost, is as follows:

(1) **Balanced truncation**. Compute grammians $\approx 70n^3$ (QZ algorithm); perform balancing $\approx 30n^3$ (eigendecomposition).
(2) **Optimal Hankel norm approximation**. Compute grammians $\approx 70n^3$ (QZ algorithm); perform balancing and embedding $\approx 60n^3$.
(3) **Krylov approximation**. Approximately k^2n operations.

The *complexity* using *approximate* and/or *sparse* decompositions, is as follows. Let α be the average number of non-zero elements per row in A, and let k be the number of expansion points. Then:

(1) **Balanced truncation**. Grammians $\approx c_1\alpha kn$; balancing $\mathcal{O}(n^3)$.
(2) **Optimal Hankel norm approximation**. Grammians $\approx c_1\alpha kn$; embedding $\mathcal{O}(n^3)$.
(3) **Krylov approximation**. Approximately $c_2k\alpha n$ operations

6. Conclusions

In this note we presented a comparative study of seven algorithms for model reduction, namely: balanced model reduction, approximate balanced reduction, singular perturbation method, Hankel norm approximation, Arnoldi procedure, Lanczos procedure, and rational Krylov method. These algorithms have been applied to six different dynamical systems. The first four make use of the Hankel singular values and the latter three are based on matching of the moments; i.e. the coefficients of the Laurent expansion of the transfer function around some point of the complex plane. The results show that balanced reduction and approximate balanced reduction are the best over the whole frequency range. Between these two, approximate balancing has the advantage that it computes an *almost* balanced reduced system *iteratively* without obtaining a balanced realization of the full order system first, and subsequently truncating, thus reducing the computational cost and storage requirements. Although it has the lowest $\mathcal{H}_\infty$ error norm in most of the cases, it leads to the highest $\mathcal{H}_2$ error norm. Being local in nature moment matching methods always lead to higher error norms than SVD based methods; but they reduce the computational cost and storage requirements remarkably. Among them, rational Krylov gives better results due to the flexibility of the selection of interpolation points. However, the selection of these points which determines the reduced model is not an automated process and has to be specified by the user, with little guidance from the theory on how to choose these points. By contrast, in the other methods a given error tolerance value determines everything.

References

[1] A.C. Antoulas, *On recursiveness and related topics in linear systems*, IEEE Transactions on Automatic Control, **AC-31**: 1121-1135 (1986).
[2] A.C. Antoulas, J.A. Ball, J. Kang, and J.C. Willems, *On the solution of the minimal rational interpolation problem*, Linear Algebra and Its Applications, Special Issue on Matrix Problems, **137/138**: 511-573 (1990).
[3] A.C. Antoulas and J.C. Willems, *A behavioral approach to linear exact modeling*, IEEE Transactions on Automatic Control, **AC-38**: 1776-1802 (1993).
[4] A.C. Antoulas, *Recursive modeling of discrete-time time series*, IMA volume on Linear Algebra for Control, P. van Dooren and B.W. Wyman Editors, Springer Verlag, **62**: 1-20 (1993).

[5] A.C. Antoulas, E.J. Grimme and D.C. Sorensen, *On behaviors, rational interpolation, and the Lanczos algorithm*, Proc. 13th IFAC Triennial World Congress, San Francisco, Pergamon Press (1996).

[6] A.C. Antoulas, *Approximation of linear operators in the 2-norm*, Special Issue of Linear Algebra and Applications on *Challenges in Matrix Theory*, **278**: 309-316, (1998).

[7] A.C. Antoulas, *Approximation of linear dynamical systems*, in the Wiley Encyclopedia of Electrical and Electronics Engineering, edited by J.G. Webster, volume **11**: 403-422 (1999).

[8] A.C. Antoulas, *Lectures on the approximation of large-scale dynamical systems*, Draft, SIAM Press, to appear (2001).

[9] A.C. Antoulas and D.C. Sorensen, *Projection methods for balanced model reduction*, Technical Report ECE-CAAM Depts, Rice University, September 2000.

[10] A.C. Antoulas and P.M. van Dooren, *Short course on the approximation of large-scale dynamical systems*, SIAM Annual Meeting, San Juan, Puerto Rico, July (2000).

[11] P. Holmes, J.L. Lumley, and G. Berkooz, *Turbulence, coherent structures, dynamical systems and symmetry*, Cambridge Momographs on Mechanics, Cambridge University Press (1996).

[12] D.L. Boley, *Krylov space methods on state-space control models*, Circuits, Systems, and Signal Processing, **13**: 733-758 (1994).

[13] P. Feldman and R.W. Freund, *Efficient linear circuit analysis by Padé approximation via a Lanczos method*, IEEE Trans. Computer-Aided Design, **14**, 639-649, (1995).

[14] R.W. Freund, *Reduced-order modeling techniques based on Krylov subspaces and their use in circuit simulation*, Numerical Analysis Manuscript 98-3-02, Bell Laboratories, February (1988).

[15] W.B. Gragg and A. Lindquist, *On the partial realization problem*, Linear Algebra and Its Applications, Special Issue on Linear Systems and Control, **50**: 277-319 (1983).

[16] E.J. Grimme,*Krylov Projection Methods for Model Reduction*, Ph.D. Thesis, ECE Dept., U. of Illinois, Urbana-Champaign,(1997).

[17] K. Gallivan, E.J. Grimme, and P. Van Dooren, *Asymptotic waveform evaluation via a restarted Lanczos method*, Applied Math. Letters, **7**: 75-80 (1994).

[18] E.J. Grimme, D.C. Sorensen, and P. Van Dooren, *Model reduction of state space systems via an implicitly restarted Lanczos method*, Numerical Algorithms, **12**: 1-31 (1995).

[19] I.M. Jaimoukha, E.M. Kasenally, *Implicitly restarted Krylov subspace methods for stable partial realizations*, SIAM J. Matrix Anal. Appl., **18**: 633-652 (1997).

[20] P. V. Kokotovic, R. E. O'Malley, P. Sannuti, *Singular perturbations and order reduction in control theory - an overview*, *Automatica*, **12**: 123-132, 1976.

[21] J. Li, F. Wang, J. White, *An efficient Lyapunov equation-based approach for generating reduced-order models of interconnect*, Proc. 36th IEEE/ACM Design Automation Conference, New Orleans, LA, (1999).

[22] Y. Liu and B.D.O. Anderson, *Singular perturbation approximation of balanced systems*, Int. J. Control, **50**: 1379-1405 (1989).

[23] B. C. Moore, *Principal component analysis in linear system: controllability, observability and model reduction*, IEEE Transactions on Automatic Control, **AC-26**: 17-32 (1981).

[24] K. Glover, *All optimal Hankel-norm approximations of linear mutilvariable systems and their L^∞-error bounds*, Int. J. Control, **39**: 1115-1193 (1984).

[25] T. Penzl, *Eigenvalue decay bounds for solutions of Lyapunov equations: The symmetric case*, Systems and Control Letters, to appear (2000).

[26] A. Ruhe, *Rational Krylov algorithms for nonsymmetric eigenvalue problems II: matrix pairs*, Linear Alg. Appl., **197**: 283-295 (1984).

[27] D.C. Sorensen, *Implicit application of polynomial filters in a k-step Arnoldi method*, SIAM J. Matrix Anal. Applic., **13**: 357-385 (1992).

[28] P. Van Dooren, *Gramian based model reduction of large-scale dynamical systems*, SIAM Annual Meeting, San Juan, Puerto Rico, July (2000).

[29] P. Van Dooren, *The Lanczos algorithm and Padé approximations*, Short Course, Benelux Meeting on Systems and Control (1995).

DEPTARTMENT OF ELECTRICAL AND COMPUTER ENGINEERING, MS 380, RICE UNIVERSITY, HOUSTON, TEXAS 77251-1892.
E-mail address: aca@rice.edu

DEPARTMENT OF COMPUTATIONAL AND APPLIED MATHEMATICS, MS 134, RICE UNIVERSITY, HOUSTON, TEXAS 77251-1892.
E-mail address: sorensen@rice.edu

DEPTARTMENT OF ELECTRICAL AND COMPUTER ENGINEERING, MS 380, RICE UNIVERSITY, HOUSTON, TEXAS 77251-1892.
E-mail address: serkan@rice.edu

Contemporary Mathematics
Volume **280**, 2001

Theory and Computations of Some Inverse Eigenvalue Problems for the Quadratic Pencil

Biswa N. Datta and Daniil R. Sarkissian

Abstract. The paper presents a "direct and partial-modal" approach to solutions of two important inverse eigenvalue problems; namely, the partial eigenvalue assignment and the partial eigenstructure assignment problems, arising in control and feedback stabilization of vibrating systems modeled by systems of matrix second-order differential equations. The approach is direct, since it is obtained completely in second-order setting without any reformulation to a first-order (state-space) system and it is partial-modal, since it uses only a partial knowledge of eigenvalues and eigenvectors of the associated quadratic matrix pencil. An orthogonality relation between the eigenvectors of the quadratic pencil derived in this paper plays an important role in obtaining these solutions. The approach is practical for feedback stabilization and supression of unwanted vibrations in vibrating structures such as automobiles, high-rise buildings, aircrafts and large-space structures.

1. Introduction

1.1. Problem Statements and Motivation. An inverse eigenvalue problem for a matrix A is the problem of finding A given the complete or a part of the spectrum and/or eigenvectors. There are many different forms of inverse eigenvalue problems and they arise in various applications (see the recent review paper of [**MTC**]).

In this paper, we focus on certain types of inverse eigenvalue problems associated with a quadratic matrix pencil arising in feedback control of a matrix second-order system

$$M\ddot{x}(t) + D\dot{x}(t) + Kx(t) = f(t), \tag{1.1}$$

1991 *Mathematics Subject Classification.* Primary 34A55, 93B55; Secondary 93B52, 70Q05.

Based on an invited presentation at the AMS Research Conference on "*Structured Matrices in Operator Theory, Numerical Analysis, Control, Signal and Image Processing*", Boulder, Colorado, July 1999 and at the first workshop on "*Inverse Problems and Eigenvalue Localization*", Antofagasta, Chile, December 1999. The research was partially supported by National Science Foundation grant under contract no. ECS-0074411. The work of B.N. Datta was also partially supported by Proyecto de Cooperación Internacional, FONDECYT # 7990011/1999.

The paper is comprised of joint work of the authors with S. Elhay and Y. Ram.

where

$M = M^T$ = *mass* or *inertia* matrix
$K = K^T$ = *stiffness* matrix
$D = D^T$ = *viscous damping* matrix

and $x = x(t) \in \mathbb{R}^n$ is a vector representing the *displacement* of the masses in the lumped mass model. The vectors $\dot{x}(t)$ and $\ddot{x}(t)$ represent, respectively, the *velocities* and *accelerations*. The vector $f = f(t) \in \mathbb{R}^n$ represents the *external force* to the system. In many cases, the matrices M and K are positive definite and positive semidefinite, respectively.

Upon separation of variables, the system gives rise to the quadratic eigenvalue problem for the quadratic matrix pencil

$$P(\lambda) = \lambda^2 M + \lambda D + K. \tag{1.2}$$

The pencil (1.2) has $2n$ eigenvalues which are the roots of the equation

$$\det(P(\lambda)) = 0$$

and $2n$ corresponding eigenvectors. Each eigenvalue of $P(\lambda)$ is related to one of the *natural frequencies* of the homogeneous system:

$$M\ddot{x}(t) + D\dot{x}(t) + Kx(t) = 0, \tag{1.3}$$

and the eigenvectors are referred to as the *modes* of vibration of the system (see [**BNDa**], [**DJI**]).

Dangerous oscillations (called *resonance*) occur when one or more eigenvalues of the pencil (1.2) became equal or close to the frequency of the external force $f(t)$. A natural way to control such unwanted vibrations is to apply the techniques of feedback control in control theory. The examples of feedback control range from automobiles to high-rise buildings and large space structures (see [**DJI**]). Suppose that a control force of the form $Bu(t)$, where B is an $n \times m$ matrix, called the *control matrix*, is applied to the structure modeled by (1.1). We then have the control system:

$$M\ddot{x}(t) + D\dot{x}(t) + Kx(t) = Bu(t) + f(t), \tag{1.4}$$

Assume that the displacement vector $x(t)$ and the velocity vector $\dot{x}(t)$ can be measured and the control vector $u(t)$ is chosen in the form

$$u(t) = F^T\dot{x}(t) + G^T x(t), \tag{1.5}$$

where F and G are constant matrices. Then the system (1.4) becomes

$$M\ddot{x}(t) + (D - BF^T)\dot{x}(t) + (K - BG^T)x(t) = f(t). \tag{1.6}$$

In feedback control literature, the quadratic matrix pencil

$$P_c(\lambda) = \lambda^2 M + \lambda(D - BF^T) + K - BG^T \tag{1.7}$$

associated with the system (1.6) is called a *closed-loop pencil* and the matrices F and G are called *feedback matrices*. In analogy, the pencil $P(\lambda) = \lambda^2 M + \lambda D + K$ is called an *open-loop pencil*. The eigenvalues of the pencil $P(\lambda)$ and $P_c(\lambda)$ are called, *open-loop* and *closed-loop eigenvalues*, respectively.

To control the unwanted vibrations in a structure, the problem, therefore, will be to choose the feedback matrices F and G such that none of the natural frequencies is in the frequency range of the external force.

Another important use of feedback control technique is to stabilize the response of a system. The homogeneous system (1.3) is said to be *asymptotically stable* if

$$||x(t)|| \to 0 \text{ as } t \to \infty.$$

A well-known result in theory of stability is that the *system (1.3) is asymptotically stable if and only if all the eigenvalues of $P(\lambda)$ have negative real parts.*

Thus, to stabilize a system using feedback, the problem will be to choose the feedback matrices F and G such that the eigenvalues of the closed-loop pencil $P_c(\lambda)$ have negative real parts.

In a realistic situation, however, only a few eigenvalues are "troublesome"; so it makes more sense to alter only those "troublesome" eigenvalues, while keeping the rest of the spectrum invariant. This leads to the following inverse eigenvalue problem, known as the *partial eigenvalue assignment problem* for the pencil (1.2).

PROBLEM 1.1. Given

(1) Real $n \times n$ matrices $M = M^T > 0$, $D = D^T$, $K = K^T$.
(2) The $n \times m$ $(m \leq n)$ control matrix B.
(3) The self-conjugate subset $\{\lambda_1, \dots, \lambda_p\}$, $p < n$ of the open-loop spectrum $\{\lambda_1, \dots, \lambda_p; \lambda_{p+1}, \dots, \lambda_{2n}\}$ and the corresponding eigenvector set $\{x_1, \dots, x_p\}$.
(4) The self-conjugate set $\{\mu_1, \dots, \mu_p\}$ of numbers.

Find *real* feedback matrices F and G such that the spectrum of the closed-loop pencil (1.7) is $\{\mu_1, \dots, \mu_p; \lambda_{p+1}, \dots, \lambda_{2n}\}$.

While Problem 1.1 is important in its own right, it is to be noted that, if the system response needs to be altered by feedback, both eigenvalue assignment as well as eigenvector assignment should be considered. *This is because, the eigenvalues determine the rate at which system response decays or grows while the eigenvectors determine the shape of the response.* Such a problem is called the *eigenstructure assignment problem.* Unfortunately, the eigenstructure assignment problem, in general, is not solvable if the matrix B is given apriori (see [**IK**]). This consideration leads to the following more tractable (but practical) inverse eigenstructure assignment problem for the quadratic pencil (1.2), known as the *partial eigenstructure assignment problem* for the pencil (1.2).

PROBLEM 1.2. Given

(1) Real $n \times n$ matrices $M = M^T > 0$, $D = D^T$, $K = K^T$.
(2) The self-conjugate subset $\{\lambda_1, \dots, \lambda_p\}$, $p < n$ of the open-loop spectrum $\{\lambda_1, \dots, \lambda_p; \lambda_{p+1}, \dots, \lambda_{2n}\}$ and the corresponding eigenvector set $\{x_1, \dots, x_p\}$.
(3) The self-conjugate sets of numbers and vectors $\{\mu_1, \dots, \mu_p\}$ and $\{y_1, \dots, y_p\}$, such that $\mu_j = \overline{\mu_k}$ implies $y_j = \overline{y_k}$.

Find a *real* control matrix B of order $n \times m$ $(m < n)$, and *real* feedback matrices F and G of order $n \times m$ such that the spectrum of the closed-loop pencil (1.7) is $\{\mu_1, \dots, \mu_p; \lambda_{p+1}, \dots, \lambda_{2n}\}$ and the eigenvector set is $\{y_1, \dots, y_p; x_{p+1}, \dots, x_{2n}\}$, where $x_{p+1}, \dots, x_{2n}$ are the eigenvectors of (1.2) corresponding to $\lambda_{p+1}, \dots, \lambda_{2n}$.

1.2. Current Solution Techniques and their Computational and Engineering Difficulties. An obvious solution approach for any of the above problems is to recast the problem in terms of a first-order reformulation and then apply one of the many well-established techniques developed specifically for a first-order state-space system (see e.g. [**BNDb**]). There are some computational difficulties with this approach. Let's assume in this section for the sake of simplicity, that $f(t) = 0$. Then the standard first-order reformulation of the control system (1.4) is:

$$\dot{z}(t) = \begin{pmatrix} 0 & I \\ -M^{-1}K & -M^{-1}D \end{pmatrix} z(t) + \begin{pmatrix} 0 \\ M^{-1}B \end{pmatrix} u(t), \quad \text{where} \quad z(t) = \begin{pmatrix} x(t) \\ \dot{x}(t) \end{pmatrix}.$$

The obvious numerical difficulty with this approach is that the matrix M has to be inverted, and, if it is ill-conditioned, the state matrix will not be computed accurately. *Furthermore, all the exploitable properties such as definiteness, sparsity, bandness, etc. of the coefficient matrices M, D, and K, usually offered by a practical problem, will be completely destroyed.*

The use of a nonstandard first-order reformulation, such as

$$\begin{pmatrix} M & 0 \\ 0 & M \end{pmatrix} \dot{z}(t) = \begin{pmatrix} 0 & M \\ -K & -D \end{pmatrix} z(t) + \begin{pmatrix} 0 \\ B \end{pmatrix} u(t)$$

will give rise to a *descriptor system* of the form $E\dot{z}(t) = Az(t) + \hat{B}u(t)$, and *the eigenvalue and eigenstructure assignment methods for the descriptor systems, especially, when the matrix E is ill-conditioned, are not well developed.*

A second approach, popularly known in the engineering literature as the *independent modal space control* (IMSC) *approach* (see [**DJI**]), also suffers from some serious computational difficulties and it is almost impossible to implement this approach in practice. The basic idea here is to decouple the problem into a set of n independent problems, solve each of these independent problems separately, and then piece the individual solutions together to obtain a solution of the given problem. This is done using simultaneous diagonalization of the matrices M, D and K as follows.

Let S be the matrix of eigenvectors of the linear pencil $K - \lambda M$ and let Λ_K be a diagonal matrix containing its eigenvalues. Then, since $M = M^T > 0$ and $K = K^T$, we have

$$S^T MS = I\,, \; S^T KS = \Lambda_K\,.$$

The same transforming matrix S also diagonalizes the matrix D as well; that is

$$S^T DS = \Lambda_D$$

if and only if

$$KM^{-1}D = DM^{-1}K. \tag{1.8}$$

With the change in coordinates $x(t) = Sy(t)$, the equations

$$\begin{aligned} M\ddot{x}(t) + D\dot{x}(t) + Kx(t) &= f(t) + Bu(t)\,, \\ u(t) &= F^T\dot{x}(t) + G^T x(t) \end{aligned}$$

become

$$I\ddot{y}(t) + (\Lambda_D - S^T BF^T S)\dot{y}(t) + (\Lambda_K - S^T BG^T S)y(t) = S^T f(t)\,.$$

These differential equations will decouple into n independent equations if and only if

$$BF^T M^{-1}D = DM^{-1}BF^T \quad \text{and} \quad BG^T M^{-1}D = KM^{-1}BG^T. \tag{1.9}$$

The commutativity relations (1.8) and (1.9) are almost impossible to satisfy in practice. Indeed, it is remarked in [**DJI**]: "This puts very stringent requirements on the locations and number of sensors and actuators". Furthermore, computing the matrices Λ_K, Λ_D and S amounts to finding the complete spectrum and the associated eigenvectors of the pencil $P(\lambda) = \lambda^2 M + \lambda D + K$. Unfortunately, numerical methods for finding the complete spectrum of the quadratic pencil are not well developed, especially for large and sparse matrices. The state-of-the-art techniques are capable of computing only a few extremal eigenvalues and eigenvectors (see [**PC**] and [**SBFV**]).

The current engineering practice is to solve these problems using a small number of eigenvalues and the corresponding eigenvectors hoping that the remaining large number of eigenvalues that are not to be reassigned remain invariant; that is, *spill-over* does not occur. Spill-over, however, invariably occurs with such an *ad-hoc* practice.

1.3. Contributions of the paper. In view of the above considerations, a practical approach to solutions of Problem 1.1 and Problem 1.2 should be to solve these problems

(i) without reformulation to a first-order system
(ii) using only those few eigenvalues and eigenvectors that can either be computed or measured in a vibration laboratory
(iii) guaranteeing that no spill-over will occur.

A solution technique of this type will be called a *direct partial modal approach*. It is "direct", because the solution is obtained directly in the second-order setting without any types of reformulations. It is "partial modal", because only a part of the spectral data is needed for the solution.

In [**DERa**], a direct and partial modal approach to Problem 1.1 was proposed in the single-input case (that is when the control matrix B is a vector b), under a rather strict condition that all the $2n$ eigenvalues of the pencil $P(\lambda)$ are distinct. One of the three *orthogonality relations* (see Theorem 2.1 below) between the eigenvectors of the pencil $P(\lambda)$, which was also proved under the above condition, played a key role in deriving the solution.

In this paper, a slightly different orthogonality relation (relation (2.9)) is proved under a weaker condition; namely that the p eigenvalues $\lambda_1, \ldots, \lambda_p$ that need to be reassigned are different from the remaining $2n - p$ eigenvalues of $P(\lambda)$, and then a solution to the Problem 1.1 in the single-input case (Theorem 4.1) is derived using this new orthogonality relation. The proposed solution is constructive and leads to Algorithm 4.2 for the single-input partial eigenvalue assignment problem. A generalization of Algorithm 4.2 to the multi-input case is then stated (Algorithm 4.3). We remark that though there exist at least two other solutions to the multi-input problem [**DERa, DS**], *the solution presented here is new and more practical from computational point of view.* Indeed, the solutions to both the single-input and multi-input problems are obtained by solving a $p \times p$ small linear system. Furthermore, mathematical results are proved guaranteeing that there will be no spill-over (Theorem 4.1, part (i)).

Next, a solution to Problem 1.2 adapted from the recent paper of [**DERS**] is described here again under the weaker condition that the eigenvalues $\lambda_1, \ldots, \lambda_p$ are different from the remaining eigenvalues. The matrix Z_1 defined by 1.10) below plays an important role in the solutions of both Problem 1.1 and Problem 1.2.

Thus, a unified approach to the solutions of both the problems is presented in this paper via the above single matrix Z_1.

1.4. Organization and Notation. The organization of the paper is as follows:

In Section 2, we derive three orthogonality relations between the eigenvectors of $P(\lambda)$; one of them is used to obtain solutions to Problem 1.1. However, these relations are of independent interests and are important contributions in their own rights in the literature of linear algebra.

In Section 3, we state a modal criterion of controllability for a second-order system which will be used later.

In Section 4, we present our solutions to Problem 1.1 : Theorem 4.1 and Algorithm 4.2 for the single-input problem and Algorithm 4.3 for the multi-input problem.

Section 5 is on the solution of Problem 1.2. The solution is presented in Theorem 5.1 and its constructive proof leads to Algorithm 5.2.

Numerical results illustrating Algorithms 4.3 and 5.2 are presented in Section 6.

The following notations will be used in our presentation throughout the paper.

$\Lambda = \text{diag}(\lambda_1, \ldots, \lambda_{2n})$	–	matrix of the open-loop eigenvalues
$X = (x_1, \ldots, x_{2n})$	–	matrix of the open-loop eigenvectors
$\Lambda_1 = \text{diag}(\lambda_1, \ldots, \lambda_p)$	–	matrix of eigenvalues to be reassigned
$X_1 = (x_1, \ldots, x_p)$	–	matrix of eigenvectors to be reassigned
$\Lambda_2 = \text{diag}(\lambda_{p+1}, \ldots, \lambda_{2n})$	–	matrix of eigenvalues to remain unaltered
$X_2 = (x_{p+1}, \ldots, x_{2n})$	–	matrix of eigenvectors to remain unaltered
$\Lambda_1' = \text{diag}(\mu_1, \ldots, \mu_p)$	–	matrix of the new eigenvalues
$Y_1 = (y_1, \ldots, y_p)$	–	matrix of the new eigenvectors

$$Z_1 = \Lambda_1' Y_1^T M X_1 \Lambda_1 - Y_1^T K X_1. \tag{1.10}$$

2. Orthogonality Relations of the Eigenvectors of Quadratic Matrix Pencil

In this section, we derive three orthogonality relations (due to [**DERb**]) between the eigenvectors of a symmetric definite quadratic pencil. These results generalize the well-known results on orthogonality between the eigenvectors of a symmetric matrix and those of a symmetric definite linear pencil (see [**BNDa**]) of the form $K - \lambda M$.

THEOREM 2.1. (Orthogonality of the Eigenvectors of Quadratic Pencil). *Let* $P(\lambda) = \lambda^2 M + \lambda D + K$, *where* $M = M^T > 0$, $D = D^T$, *and* $K = K^T$. *Assume that the eigenvalues* $\lambda_1, \ldots, \lambda_n$ *are all distinct and different from zero. Then there exist diagonal matrices* D_1, D_2, *and* D_3 *such that*

$$\Lambda X^T M X \Lambda - X^T K X = D_1 \tag{2.1}$$

$$\Lambda X^T D X \Lambda + \Lambda X^T K X + X^T K X \Lambda = D_2 \tag{2.2}$$

$$\Lambda X^T M X + X^T M X \Lambda + X^T D X = D_3 \tag{2.3}$$

Furthermore

$$D_1 = D_3\Lambda \tag{2.4}$$
$$D_2 = -D_1\Lambda \tag{2.5}$$
$$D_2 = -D_3\Lambda^2. \tag{2.6}$$

PROOF. By definition, the pair (X, Λ) must satisfy the $n \times 2n$ system of equations (called the *eigendecomposition* of the pencil $P(\lambda) = \lambda^2 M + \lambda D + K$):

$$MX\Lambda^2 + DX\Lambda + KX = 0. \tag{2.7}$$

Isolating the term in D, we have from above

$$-DX\Lambda = MX\Lambda^2 + KX.$$

Multiplying this on the left by ΛX^T gives

$$-\Lambda X^T DX\Lambda = \Lambda X^T MX\Lambda^2 + \Lambda X^T KX$$

Taking the transpose gives

$$-\Lambda X^T DX\Lambda = \Lambda^2 X^T MX\Lambda + X^T KX\Lambda$$

Now, subtracting the latter from the former gives, on rearrangement,

$$\Lambda X^T MX\Lambda^2 - X^T KX\Lambda = \Lambda^2 X^T MX\Lambda - \Lambda X^T KX$$

or

$$(\Lambda X^T MX\Lambda - X^T KX)\Lambda = \Lambda(\Lambda X^T MX\Lambda - X^T KX). \tag{2.8}$$

Thus, the matrix $\Lambda X^T MX\Lambda - X^T KX$ which we denote by D_1, must be diagonal since it commutes with a diagonal matrix, the diagonal entries of which are distinct. We thus have the *first orthogonality relation (2.1):*

$$\Lambda X^T MX\Lambda - X^T KX = D_1$$

Similarly, isolating the term in M of the eigendecomposition equation, we get

$$-MX\Lambda^2 = DX\Lambda + KX,$$

and multiplying this on the left by $\Lambda^2 X^T$ gives

$$-\Lambda^2 X^T MX\Lambda^2 = \Lambda^2 X^T DX\Lambda + \Lambda^2 X^T KX.$$

Taking the transpose, we have

$$-\Lambda^2 X^T MX\Lambda^2 = \Lambda X^T DX\Lambda^2 + X^T KX\Lambda^2.$$

Subtracting the last equation from the previous one and adding $\Lambda X^T KX\Lambda$ to both sides gives, after some rearrangement,

$$\Lambda(\Lambda X^T DX\Lambda + \Lambda X^T KX + X^T KX\Lambda) = (\Lambda X^T DX\Lambda + \Lambda X^T KX + X^T KX\Lambda)\Lambda$$

Again, this commutativity property implies, since Λ has distinct diagonal entries, that

$$\Lambda X^T DX\Lambda + \Lambda X^T KX + X^T KX\Lambda = D_2$$

is a diagonal matrix. This is the *second orthogonality relation (2.2).*

The first and second orthogonality relations together easily imply the *third orthogonality relation (2.3):*

$$\Lambda X^T MX + X^T MX\Lambda + X^T DX = D_3$$

To prove (2.4) we multiply the last equation on the right by Λ giving

$$\Lambda X^T M X \Lambda + X^T M X \Lambda^2 + X^T D X \Lambda = D_3 \Lambda,$$

which, using the eigendecomposition equation, becomes

$$\Lambda X^T M X \Lambda + X^T(-KX) = D_3 \Lambda.$$

So, from the first orthogonality relation (2.1) we see that

$$D_1 = D_3 \Lambda$$

Next, using the eigendecomposition equation (2.7), we rewrite the second orthogonality relation (2.2) as

$$\begin{aligned} D_2 &= \Lambda X^T(DX\Lambda + KX) + X^T K X \Lambda \\ &= \Lambda X^T(-MX\Lambda^2) + X^T K X \Lambda = (-\Lambda X^T M X \Lambda^+ X^T K X)\Lambda \end{aligned}$$

By the first orthogonality relation we then have

$$D_2 = -D_1 \Lambda$$

Finally, from $D_1 = D_3\Lambda$ and $D_2 = -D_1\Lambda$ we have

$$D_2 = -D_3 \Lambda^2.$$

We remind the reader that matrix and vector transposition here does not mean conjugation for complex quantities. □

COROLLARY 2.2. *Let the sets* $\{\lambda_1, \ldots, \lambda_p\}$ *and* $\{\lambda_{p+1}, \ldots, \lambda_{2n}\}$ *be disjoint. Then*

$$\Lambda_1 X_1^T M X_2 \Lambda_2 - X_1^T K X_2 \quad = \quad 0, \tag{2.9}$$

PROOF. The equation (2.8) can be written as

$$N\Lambda = \Lambda N, \text{ where } N = (n_{ij}) = \Lambda X^T M X \Lambda - X^T K X,$$

then $n_{ij}\lambda_i = \lambda_j n_{ij}$ implies $n_{ij} = 0$ if $1 \le i \le p < j \le 2n$ or $1 \le j \le p < i \le 2n$.
In matrix notation, this implies relation (2.9). □

3. Modal Criterion of Controllability

In this section, we state a well-known criterion of controllability for the control system (1.4). The criterion is a generalization of the Hautus criterion of controllability [**MLJH**] to the second-order system.

DEFINITION 3.1. The system

$$M\ddot{x}(t) + D\dot{x}(t) + Kx(t) \quad = \quad Bu(t) \tag{3.1}$$

is *controllable* if for any initial $x(0)$ and $\dot{x}(0)$, there exists a finite time $0 \le t_f < \infty$ and a control $u_0(t)$ such that $x(t_f) = \dot{x}(t_f) = 0$.

THEOREM 3.2 (see [**LA**]). (Modal Criterion of Controllability).
The system (3.1) or, equivalently, the pair $(P(\lambda), B)$ *is* controllable *if and only if*

$$\operatorname{rank}(\lambda^2 M + \lambda D + K, B) = n \textit{ for all } \lambda \in \mathbb{C} \tag{3.2}$$

Since each mode of the system can be checked individually for controllability, it makes sense to talk about controllability of the pair $(P(\lambda), B)$ with respect to a given mode.

Then, we define partial controllability as:

DEFINITION 3.3. The pair $(P(\lambda), B)$ is *partially controllable* with respect to the set Ω, if it is controllable with respect to each member of the Ω.

We can then state the following criterion for partial controllability.

COROLLARY 3.4. (Modal Criterion of Partial Controllability).
The system (3.1) or the pair $(P(\lambda), B)$ *is* partially controllable *with respect to the set* $\Omega \subset \mathbb{C}$ *if and only if*

$$\operatorname{rank}(\lambda^2 M + \lambda D + K, B) = n \text{ for all } \lambda \in \Omega \tag{3.3}$$

4. Solution to Problem 1.1

In this section, we present a direct partial modal approach for the solution of Problem 1.1. We consider the single-input case first.

4.1. Case 1. Single-input Case. In the single-input case, Problem 1.1 reduces to the following problem:
Given $n \times n$ matrices $M = M^T > 0$, $D = D^T$, $K = K^T$, the $n \times 1$ $(m = 1)$ control vector b; a part of the spectrum $\{\lambda_1, \dots, \lambda_p\}$ of the open-loop pencil $P(\lambda)$, and the set $\{\mu_1, \dots, \mu_p\}$, both closed under complex conjugation, find real feedback vectors f and g such that the spectrum of $P_c(\lambda) = \lambda^2 M + \lambda(D - bf^T) + K - bg^T$ is precisely the set $\{\mu_1, \mu_2, \dots, \mu_p, \lambda_{p+1}, \dots, \lambda_{2n}\}$, where $\lambda_{p+1}, \dots, \lambda_{2n}$ are the remaining eigenvalues of the pencil $P(\lambda)$.
The direct and partial-modal approach for a solution of the single-input problem is contained in the following theorem.

THEOREM 4.1. (Solution to Single-input Partial Eigenvalue Assignment Problem for a Quadratic Pencil).
If $\{\lambda_1, \dots, \lambda_p\} \cap \{\lambda_{p+1}, \dots, \lambda_{2n}\} = \emptyset$ *then*

(i) *For any arbitrary vector* β, *the feedback vectors* f *and* g *defined by*

$$f = MX_1\Lambda_1\beta \tag{4.1}$$

$$\text{and } g = -KX_1\beta. \tag{4.2}$$

are such that $2n - p$ *eigenvalues* $\lambda_{p+1}, \dots, \lambda_{2n}$ *of the closed-loop pencil* $P_c(\lambda) = \lambda^2 M + \lambda(D - bf^T) + K - bg^T$ *are the same as those of the open-loop pencil* $P(\lambda) = \lambda^2 M + \lambda D + K$.

(ii) *If pair* $(P(\lambda), B)$ *is partially controllable with respect to* $\Omega_p = \{\lambda_1, \dots, \lambda_p\}$, $0 \notin \Omega_p$ *and* $\{\mu_1, \dots, \mu_p\} \cap \{\lambda_1, \dots, \lambda_{2n}\} = \emptyset$ *then Problem 1.1 has unique solution in the form (4.1)-(4.2) with* β *given by the solution of the algebraic linear system*

$$Z_1\beta = (1, 1, \dots, 1)^T, \tag{4.3}$$

where Z_1 *is defined by (1.10).*

PROOF.

(i). In terms of the eigenvalue and eigenvector matrices, proving Part (i) amounts to showing that

$$MX_2\Lambda_2^2 + (D - bf^T)X_2\Lambda_2 + (K - bg^T)X_2 = 0.$$

To do so, we consider the eigendecomposition equation again:

$$MX\Lambda^2 + DX\Lambda + KX = 0. \tag{4.4}$$

From this, we obtain

$$\begin{aligned}&MX_2\Lambda_2^2 + (D - bf^T)X_2\Lambda_2 + (K - bg^T)X_2\\&= MX_2\Lambda^2 + DX_2\Lambda_2 + KX_2 - b\beta^T(\Lambda_1 X_1^T MX_2\Lambda_2 - X_1^T KX_2)\\&= -b\beta^T(\Lambda_1 X_1^T MX_2\Lambda_2 - X_1^T KX_2).\end{aligned}$$

Indeed, (4.4) implies $MX_2\Lambda_2^2 + DX_2\Lambda_2 + KX_2 = 0$ and, furthermore, since $\{\lambda_1, \dots, \lambda_p\} \cap \{\lambda_{p+1}, \dots, \lambda_{2n}\} = \emptyset$, then by (2.9) in Corollary 2.2 we have

$$\Lambda_1 X_1^T MX_2\Lambda_2 - X_1^T KX_2 = 0.$$

Thus $MX_2\Lambda_2^2 + (D - bf^T)X_2\Lambda_2 + (K - bg^T)X_2 = 0$.

(ii). The proof of (ii) comes in two stages. In stage 1, we show that if the vector β is chosen satisfying (4.3), then the feedback vectors f and g defined by (4.1) and (4.2) are such that the eigenvalues $\lambda_1, \dots, \lambda_p$ will be assigned to $\mu_1, \dots, \mu_p$. In state 2, we show that the feedback vectors f and g determined this way are real.

To prove stage 2, we note that if there exists a vector β that moves the eigenvalues $\{\lambda_j\}_{j=1}^p$ to the eigenvalue $\{\mu_j\}_{j=1}^p$, then there exists an eigenvector matrix Y_1 of order $n \times p$ consisting of the eigenvectors $y_1, \dots, y_p$ corresponding to the eigenvalues $\mu_1, \dots, \mu_p$ such that

$$MY_1(\Lambda_1')^2 + (D - bf^T)Y_1\Lambda_1' + (K - bg^T)Y_1 = 0,$$

where

$$Y_1 = (y_1, y_2, \dots, y_p),\ y_j \neq 0,\ j = 1, 2, ..., m.$$

Substituting for f, g and rearranging, we have

$$\begin{aligned}MY_1(\Lambda_1')^2 + DY_1\Lambda_1' + KY_1 &= b\beta^T(\Lambda_1 X_1^T MY_1\Lambda_1' - X_1^T KY_1)\\&= b\beta^T Z_1^T = bc^T,\end{aligned}$$

where Z_1 is given by (1.10) and $c = Z_1\beta$ is a vector that will depend on the scaling chosen for the eigenvectors in Y_1.

Since the condition (ii) ensures the existence of the set of p vectors $\{y_1, \dots, y_p\}$ such that for each $k = 1, 2, \dots, p$

$$\begin{pmatrix} y_k \\ 1 \end{pmatrix} \in \text{null}(\mu_k^2 M + \mu_k D + K, -b).$$

we can solve for each of the eigenvectors y_i using the equations

$$(\mu_j^2 M + \mu_j D + K)y_j = b, \qquad j = 1, 2, ..., p \tag{4.5}$$

to obtain Y_1.

This corresponds to choosing the vector $c = (1, 1, ..., 1)^T$, so, having computed the eigenvectors we could solve the $p \times p$ linear system

$$Z_1\beta = (1, 1, ..., 1)^T$$

for β, and hence determine the vectors f and g. Indeed, to prove that Z_1 is nonsingular we subtract

$$\mu_k y_k^T(Mx_j\lambda_j^2 + Dx_j\lambda_j + \mu_k y_k^T Kx_j) = 0$$

from

$$(My_k\mu_k^2 + Dy_k\mu_k + Ky_k)^T x_j\lambda_j = b^T x_j\lambda_j$$

and obtain that

$$(\mu_k - \lambda_j)(\mu_k y_k^T M x_j \lambda_j - y_k K x_j) = b^T x_j \lambda_j.$$

Thus the k, j-th element of the matrix Z_1 is given by

$$z_{kj} = \frac{b^T x_j \lambda_j}{\mu_k - \lambda_j}$$

and

$$Z_1 = C \operatorname{diag}(b^T x_1 \lambda_1, \ldots, b^T x_p \lambda_p)$$

where the Cauchy matrix $C = \left(\frac{1}{\mu_k - \lambda_j}\right)_{k,j=1}^{p}$ is nonsingular when $\{\mu_1, \ldots, \mu_p\} \cap \{\lambda_1, \ldots, \lambda_p\} = \emptyset$ and $b^T x_j \neq 0$ for all $j = 1, \ldots, p$ since the pair $(P(\lambda), b)$ is controllable with respect to each λ_j.

We now prove stage 2. Since the set $\{\lambda_1, \ldots, \lambda_p\}$ is self-conjugate and the coefficient matrices M, D and K of the open-loop pencil $P(\lambda)$ are real, we know that $\lambda_j = \overline{\lambda_k}$ implies that $x_j = \overline{x_k}$ (conjugate eigenvectors correspond to conjugate eigenvalues). Therefore, there exists a nonsingular permutation matrix T such that

$$\overline{X_1} = X_1 T \text{ and } \overline{X_1 \Lambda_1} = X_1 \Lambda_1 T.$$

Similarly, there is a permutation matrix T' such that

$$\overline{Y_1} = Y_1 T' \text{ and } \overline{Y_1 \Lambda_1'} = Y_1 \Lambda_1' T'.$$

Thus, conjugating (1.10), we obtain

$$\overline{Z_1} = (T')^T \Lambda_1' Y_1^T M X_1 \Lambda_1 T - (T')^T Y_1^T K X_1 T \quad = \quad (T')^T Z_1 T,$$

and conjugation of (4.3) gives

$$\overline{Z_1 \beta} = ((T')^T Z_1 T)\overline{\beta} = (T')^T (1, 1, \ldots, 1)^T,$$

implying that $\overline{\beta} = T^T \beta$.

Therefore,

$$\overline{f} = M(X_1 \Lambda_1 T)(T^T \beta) = f \text{ and } \overline{g} = -K(X_1 T)(T^T \beta) = g$$

which shows that f and g are real vectors. □

Based on Theorem 4.1 we can state the following algorithm.

ALGORITHM 4.2. *(An Algorithm For the Single-input Partial Eigenvalue Assignment Problem for the Quadratic Pencil).*

Inputs:

1. The $n \times n$ matrices M, K, and D; $M = M^T > 0$, $D = D^T$ and $K = K^T$.
2. The $n \times 1$ control (input) vector b.
3. The set $\{\mu_1, \cdots, \mu_p\}$, closed under complex conjugation.
4. The self-conjugate subset $\{\lambda_1, \ldots, \lambda_p\}$ of the open-loop spectrum $\{\lambda_1, \ldots, \lambda_p; \lambda_{p+1}, \ldots, \lambda_{2n}\}$ and the associated eigenvector set $\{x_1, \ldots, x_p\}$.

Outputs: The feedback vectors f and g such that the spectrum of the closed-loop pencil $P_c(\lambda) = \lambda^2 M + \lambda(D - bf^T) + (K - bg^T)$ is $\{\mu_1, \ldots, \mu_p, \lambda_{p+1}, \ldots, \lambda_{2n}\}$.

Assumptions:

1. The quadratic pencil is partially controllable with respect to the eigenvalues $\Omega_p = \{\lambda_1, \ldots, \lambda_p\}$.
2. $\Omega_p \cap \{\lambda_{p+1}, \ldots, \lambda_{2n}\} = \emptyset$, $0 \notin \Omega_p$, $\{\mu_1, \ldots, \mu_p\} \cap \{\lambda_1, \ldots, \lambda_{2n}\} = \emptyset$.

Step 1. Form $\Lambda_1 = \text{diag}(\lambda_1, \ldots, \lambda_p)$ and $X_1 = (x_1, \ldots, x_p)$.

Step 2. Solve for $y_1, \ldots, y_p$:

$$(\mu_j^2 M + \mu_j D + K)y_j = b, \; j = 1, \ldots, p.$$

Step 3. Form

$$Z_1 = \Lambda_1' Y_1^T M X_1 \Lambda_1 - Y_1^T K X_1,$$

where $Y_1 = (y_1, \ldots, y_p)$ and $\Lambda_1' = \text{diag}(\mu_1, \ldots, \mu_p)$. If Z_1 is ill-conditioned, then warn the user that the problem is ill-posed.

Step 4. Solve for β:

$$Z_1 \beta = (1, 1, \cdots, 1)^T.$$

Step 5. Form

$$\begin{aligned} f &= M X_1 \Lambda_1 \beta \\ g &= -K X_1 \beta. \end{aligned}$$

4.2. Case 2. Multi-input case. In the multi-input case, we obtain the following generalization of Algorithm 4.2. Proof of the algorithm will appear elsewhere.

ALGORITHM 4.3. *(An Algorithm For the Multi-input Partial Eigenvalue Assignment Problem for the Quadratic Pencil).*

Inputs:

1. The $n \times n$ matrices M, K, and D; $M = M^T > 0$, $K = K^T$ and $D = D^T$.
2. The $n \times m$ control (input) matrix B.
3. The set $\{\mu_1, \cdots, \mu_p\}$, closed under complex conjugation.
4. The self-conjugate subset $\{\lambda_1, \ldots, \lambda_p\}$ of the open-loop spectrum $\{\lambda_1, \ldots, \lambda_p; \lambda_{p+1}, \ldots, \lambda_{2n}\}$ and the associated eigenvector set $\{x_1, \ldots, x_p\}$.

Outputs: The feedback matrices F and G such that the spectrum of the closed-loop pencil $P_c(\lambda) = \lambda^2 M + \lambda(D - BF^T) + (K - BG^T)$ is $\{\mu_1, \ldots, \mu_p, \lambda_{p+1}, \ldots, \lambda_{2n}\}$.

Assumptions:

1. The quadratic pencil is (partially) controllable with respect to the eigenvalues $\Omega_p = \{\lambda_1, \ldots, \lambda_p\}$.
2. $\Omega_p \cap \{\lambda_{p+1}, \ldots, \lambda_{2n}\} = \emptyset$, $0 \notin \Omega_p$, $\{\mu_1, \ldots, \mu_p\} \cap \{\lambda_1, \ldots, \lambda_{2n}\} = \emptyset$.

Step 1. Form $\Lambda_1 = \text{diag}(\lambda_1, \ldots, \lambda_p)$ and $X_1 = (x_1, \ldots, x_p)$.

Step 2. Choose arbitrary vectors $\gamma_1, \ldots, \gamma_p$ in such a way that $\mu_j = \overline{\mu_k}$ implies $\gamma_j = \overline{\gamma_k}$ and solve for $y_1, \ldots, y_p$:

$$(\mu_j^2 M + \mu_j D + K)y_j = B\gamma_j, \; j = 1, \ldots, p.$$

Step 3. Form

$$Z_1 = \Lambda_1' Y_1^T M X_1 \Lambda_1 - Y_1^T K X_1,$$

where $Y_1 = (y_1, \ldots, y_p)$ and $\Lambda_1' = \text{diag}(\mu_1, \ldots, \mu_p)$. If Z_1 is ill-conditioned, then return to Step 2 and select different vectors $\gamma_1, \ldots, \gamma_p$.

Step 4. Form $\Gamma = (\gamma_1, \gamma_2, \ldots, \gamma_p)$ and solve for Φ:

$$\Phi Z_1^T = \Gamma.$$

Step 5. Form

$$\begin{aligned} F &= MX_1\Lambda_1\Phi^T \\ G &= -KX_1\Phi^T. \end{aligned}$$

5. Solution to Problem 1.2

THEOREM 5.1. (Solution to the Partial Eigenstructure Assignment for a Quadratic Pencil).

(i) *The triplet $(\hat{B}, \hat{F}, \hat{G})$ defined by*

$$\begin{aligned} \hat{B} &= MY_1(\Lambda_1')^2 + DY_1\Lambda_1' + KY_1, \\ \hat{F} &= MX_1\Lambda_1 Z_1^{-1}, \text{ and} \\ \hat{G} &= -KX_1Z_1^{-1} \end{aligned}$$

constitutes a (possibly complex) solution to the Problem 1.2, provided that Z_1 is nonsingular.

(ii) *A solution with real B, F, and G is obtained from the triplet $(\hat{B}, \hat{F}, \hat{G})$ by taking the economy size QR factorization or the singular value decomposition (SVD) of the real matrix $H = \hat{B}\left(\hat{F}^T, \hat{G}^T\right)$.*

If QR factorization is used and if $LR = H$ is the economy QR factorization of H, then

$$\begin{aligned} B &= L, \\ F^T &= (r_1, r_2, \ldots, r_n) \text{ and} \\ G^T &= (r_{n+1}, r_{n+2}, \ldots, r_{2n}); \end{aligned}$$

where $R = (r_1, \ldots, r_{2n})$.

If SVD is used and if $H = U\Sigma V^T$ is used then the above formulae could be used either with

$$L = U,\ R = \Sigma V^T,$$

or with

$$L = U\Sigma,\ R = V^T.$$

PROOF.

(i). Suppose that the triplet $(\tilde{B}, \tilde{F}, \tilde{G})$ is a solution. Then

$$MY_1(\Lambda_1')^2 + DY_1\Lambda_1' + KY_1 = \tilde{B}\left(\tilde{F}^T Y_1\Lambda_1' + \tilde{G}^T Y_1\right). \tag{5.1}$$

Note that if $\tilde{B}$, $\tilde{F}$, and $\tilde{G}$ constitute a solution to Problem 1.2, then for any invertible W, $\hat{B} = \tilde{B}W$, $\hat{F} = \tilde{F}W^{-T}$, and $\hat{G} = \tilde{G}W^{-T}$ also constitute a solution; because $\tilde{B}\tilde{F}^T = \hat{B}\hat{F}^T$ and $\tilde{B}\tilde{G}^T = \hat{B}\hat{G}^T$. Denote

$$W = \tilde{F}^T Y_1\Lambda_1' + \tilde{G}^T Y_1. \tag{5.2}$$

Then, provided that W is invertible, $\hat{B} = \tilde{B}W$ is admissible for some $\hat{F}$ and $\hat{G}$. Thus we can take

$$\hat{B} = MY_1(\Lambda_1')^2 + DY_1\Lambda_1' + KY_1 \tag{5.3}$$

by virtue of (5.1) and (5.2). Relations (5.3) imply that

$$\hat{F}^T Y_1 \Lambda_1' + \hat{G}^T Y_1 = I. \tag{5.4}$$

Generalizing the result in Part (*i*) of Theorem 4.1 (the multi-input case) it can be shown that for any *Phi* , the matrices

$$\hat{F} = M X_1 \Lambda_1 \Phi^T \quad \text{and} \quad \hat{G} = -K X_1 \Phi^T \tag{5.5}$$

satisfy

$$M X_2 \Lambda_2^2 + (D - \hat{B}\hat{F}^T) X_2 \Lambda_2 + (K - \hat{B}\hat{G}^T) X_2 = 0.$$

Substituting (5.5) into (5.4), we obtain

$$\Phi = \left(\Lambda_1 X_1^T M Y_1 \Lambda_1' - X_1^T K Y_1\right)^{-1} = \left(Z_1^T\right)^{-1}$$

from which $\hat{F}$ and $\hat{G}$ can be determined.

(ii). Since $\mu_j = \overline{\mu_k}$ implies $y_j = \overline{y_k}$, it follows that there exist permutation matrices T and T' such that

$\overline{X_1} = X_1 T$, $\overline{X_1 \Lambda_1} = X_1 \Lambda_1 T$, $\overline{Y_1} = Y_1 T'$, $\overline{Y_1 \Lambda_1'} = Y_1 \Lambda_1' T'$ and $\overline{Y_1 (\Lambda_1')^2} = Y_1 (\Lambda_1')^2 T'$.

Thus, $\overline{Z_1} = (T')^T Z_1 T$ and using (5.3) and (5.5) we obtain

$$\begin{aligned} \overline{\hat{B}} = M Y_1 (\Lambda_1')^2 T' + D Y_1 \Lambda_1' T' + K Y_1 T' &= \hat{B} T', \\ \overline{\hat{B}\hat{F}^T} = \hat{B} T' (M X_1 \Lambda_1 T (T^T Z_1^{-1} T'))^T = \hat{B} M X_1 \Lambda_1 Z_1^{-1} &= \hat{B}\hat{F}^T \text{ and} \\ \overline{\hat{B}\hat{G}^T} = \hat{B} T' (-K X_1 T (T^T Z_1^{-1} T'))^T = -\hat{B} K X_1 Z_1^{-1} &= \hat{B}\hat{G}^T, \end{aligned}$$

which implies that both $\hat{B}\hat{F}^T$ and $\hat{G}\hat{G}^T$ are real matrices.

Define now the real $n \times 2n$ matrix

$$H = \hat{B}\left(\hat{F}^T, \hat{G}^T\right)$$

and let $LR = H$ be a factorization of H; where L and R are, respectively, of order $n \times m$ and $m \times 2n$. Then we can take B to be L, the first n columns of R to be F^T, and the last n columns of R to be G^T. Either the economy size QR factorization of H or the economy singular value decomposition of H can be used to compute B, F, and G (see [**GVL**] or [**BNDa**]). □

Based on the Theorem 5.1 we can state the following algorithm.

ALGORITHM 5.2. *(An Algorithm For the Partial Eigenstructure Assignment Problem for a Quadratic Matrix Pencil).*

Inputs:

1. The $n \times n$ matrices M, K, and D; $M = M^T > 0$, $K = K^T$, $D = D^T$.
2. The set of p numbers $\{\mu_1, \cdots, \mu_p\}$ and the set of p vectors $\{y_1, \ldots, y_p\}$, both closed under complex conjugation.
4. The self-conjugate subset $\{\lambda_1, \ldots, \lambda_p\}$ of the open-loop spectrum $\{\lambda_1, \ldots, \lambda_p; \lambda_{p+1}, \ldots, \lambda_{2n}\}$ and the associated eigenvector set $\{x_1, \ldots, x_p\}$.

Outputs: The feedback matrices F and G such that the spectrum of the closed-loop pencil $P_c(\lambda) = \lambda^2 M + \lambda(D - BF^T) + (K - BG^T)$ is $\{\mu_1, \ldots, \mu_p, \lambda_{p+1}, \ldots, \lambda_{2n}\}$ and the eigenvectors corresponding to $\mu_1, \ldots, \mu_p$ are $y_1, \ldots, y_p$, respectively.

Assumptions:

1. $\{\lambda_1, \ldots, \lambda_p\} \cap \{\lambda_{p+1}, \ldots, \lambda_{2n}\} = \emptyset$.

2. $\mu_j = \overline{\mu_k}$ implies $y_j = \overline{y_k}$ for all $1 \le j, k \le p$.

3. Matrix Z_1 obtained in Step 3 is nonsingular.

Step 1. Obtain the first p eigenvalues $\lambda_1, \ldots, \lambda_p$ that need to be reassigned and the corresponding eigenvectors $x_1, \ldots, x_p$.

Form $\Lambda_1 = \text{diag}(\lambda_1, \ldots, \lambda_p)$, $\Lambda_1' = \text{diag}(\mu_1, \ldots, \mu_p)$, $X_1 = (x_1, \ldots, x_p)$ and $Y_1 = (y_1, \ldots, y_p)$.

Step 2. Form the matrices $\hat{B}$ and Z_1

$$\begin{aligned} \hat{B} &= MY_1(\Lambda_1')^2 + DY_1\Lambda_1' + KY_1, \\ Z_1 &= \Lambda_1' Y_1^T M X_1 \Lambda_1 - Y_1^T K X_1. \end{aligned}$$

Step 3. Solve for $\hat{H} \in \mathbb{C}^{p\times 2n}$

$$Z_1^T \hat{H} = (\Lambda_1^T X_1^T M, -X_1^T K)$$

and form $H = \hat{B}\hat{H}$.

(If the system does not have a soultion, the given set of eigenvectors can not be assigned).

Step 4. Compute the economy size QR decomposition of $H = BR$.

Step 5. Partition $R \in \mathbb{R}^{m\times 2n}$ to get $F, G \in \mathbb{R}^{m\times n}$

$$R = (F^T, G^T).$$

(This step can also be implemented using the SVD of H, as shown in Theorem 5.1).

NOTE 5.3. MATLAB codes for Algorithm 4.3 and 5.2 are available from the authors upon request.

6. Illustrative Numerical Examples

EXAMPLE 6.1. We illustrate Algorithm 4.3 for the quadratic pencil $P(\lambda) = \lambda^2 M + \lambda D + K$ with random matrices M, D, K and B given by

$$M = \begin{pmatrix} 1.4685 & 0.7177 & 0.4757 & 0.4311 \\ 0.7177 & 2.6938 & 1.2660 & 0.9676 \\ 0.4757 & 1.2660 & 2.7061 & 1.3918 \\ 0.4311 & 0.9676 & 1.3918 & 2.1876 \end{pmatrix}, D = \begin{pmatrix} 1.3525 & 1.2695 & 0.7967 & 0.8160 \\ 1.2695 & 1.3274 & 0.9144 & 0.7325 \\ 0.7967 & 0.9144 & 0.9456 & 0.8310 \\ 0.8160 & 0.7325 & 0.8310 & 1.1536 \end{pmatrix}$$

$$K = \begin{pmatrix} 1.7824 & 0.0076 & -0.1359 & -0.7290 \\ 0.0076 & 1.0287 & -0.0101 & -0.0493 \\ -0.1359 & -0.0101 & 2.8360 & -0.2564 \\ -0.7290 & -0.0493 & -0.2564 & 1.9130 \end{pmatrix} \text{ and } B = \begin{pmatrix} 0.3450 & 0.4578 \\ 0.0579 & 0.7630 \\ 0.5967 & 0.9990 \\ 0.2853 & 0.3063 \end{pmatrix}$$

The open-loop eigenvalues of $P(\lambda)$, computed via MATLAB [**MRG**], are

$$-0.0861 \pm 1.6242i, -0.1022 \pm 0.8876i, -0.1748 \pm 1.1922i \text{ and } -0.4480 \pm 0.2465i.$$

We will solve Problem 1.1, reassigning only the most unstable pair of the open-loop eigenvalues; namely, $-0.0861 \pm 1.6242i$ to the locations $-0.1 \pm 1.6242i$. That is, we want the closed-loop pencil $P_c(\lambda) = \lambda^2 M + \lambda(D - BF^T) + K - BG^T$ to have the spectrum

$$\{-0.1 \pm 1.6242i, -0.1022 \pm 0.8876i, -0.1748 \pm 1.1922i, -0.448 \pm 0.2465i\}. \tag{6.1}$$

The *random* choices of γ_1 and γ_2 produce matrix Z_1 with the condition number $\text{Cond}_2(Z_1) = ||Z_1||_2||Z_1^{-1}||_2 = 1.64$ and the feedback matrices

$$F = \begin{pmatrix} 3.3599 & -2.4691 \\ -2.5437 & 1.2692 \\ 10.3080 & -6.8494 \\ -8.0702 & 5.6643 \end{pmatrix} \text{ and } G = \begin{pmatrix} -4.1868 & 1.4318 \\ -0.3352 & 0.4506 \\ -6.4921 & 1.2369 \\ 7.2495 & -2.2698 \end{pmatrix}$$

with the norms $||F||_2 = 16.6$ and $||G||_2 = 10.99$ such that the spectrum of the closed-loop pencil $P_c(\lambda)$ is precisely (6.1).

Next, we apply a method, essentially similar to the method 2/3 in [**KNVD**], that uses the freedom in choosing vectors γ_1 and γ_2 in order to improve the condition number of Z_1. The method converges after 3 steps, producing the matrix $Z_1^{(\text{robust})}$ with the condition number $\text{Cond}_2(Z_1^{(\text{robust})}) = 1.1$ and the feedback matrices

$$F^{(\text{robust})} = \begin{pmatrix} -0.6532 & 0.4781 \\ 0.7079 & -0.4183 \\ -2.2620 & 1.5349 \\ 1.6636 & -1.1733 \end{pmatrix} \text{ and } G^{(\text{robust})} = \begin{pmatrix} 1.3988 & -0.7501 \\ -0.0075 & -0.0285 \\ 2.5185 & -1.2554 \\ -2.4964 & 1.3185 \end{pmatrix}$$

with the $||F^{(\text{robust})}||_2 = 3.6$ and $||G^{(\text{robust})}||_2 = 4.3$ such that the spectrum of the "robust" closed-loop pencil

$$P_c^{(\text{robust})}(\lambda) = \lambda^2 M + \lambda\left(D - B\left(F^{(\text{robust})}\right)^T\right) + \left(K - B\left(G^{(\text{robust})}\right)^T\right)$$

is precisely the set (6.1).

We call the last closed-loop pencil "robust" because, aside from the mere reduction in the norm of the feedback matrices, our numerical experiments suggest that the eigenvalues of $P_c^{(\text{robust})}(\lambda)$ are less affected by the random perturbations of feedback matrices. This is illustrated in Figure 1 that plots the convex hulls of the closed-loop eigenvalues, when the feedback matrices F, G, $F^{(\text{robust})}$ and $G^{(\text{robust})}$ are perturbed, respectively, by ΔF, ΔG, $\Delta F^{(\text{robust})}$ and $\Delta G^{(\text{robust})}$, such that

$$||\Delta F||_2 < 0.01||F||_2, \; ||\Delta G||_2 < 0.01||G||_2$$

and

$$||\Delta F^{(\text{robust})}||_2 < 0.01||F^{(\text{robust})}||_2, \; ||\Delta G^{(\text{robust})}||_2 < 0.01||G^{(\text{robust})}||_2,$$

using 200 random perturbations.

EXAMPLE 6.2. Example 6.1 is now used to illustrate Algorithm 5.2. We will solve Problem 1.2, reassigning again the most unstable pair of the open-loop eigenvalues; namely, $-0.0861 \pm 1.6242i$ to the same locations $-0.1 \pm 1.6242i$. That is, we want the closed-loop pencil to have the spectrum (6.1). Let the matrix of vectors to be assigned be:

$$Y_1 = \begin{pmatrix} 1.0000 & 1.0000 \\ 0.0535 + 0.3834i & 0.0535 - 0.3834i \\ 0.5297 + 0.0668i & 0.5297 - 0.0668i \\ 0.6711 + 0.4175i & 0.6711 - 0.4175i \end{pmatrix}.$$

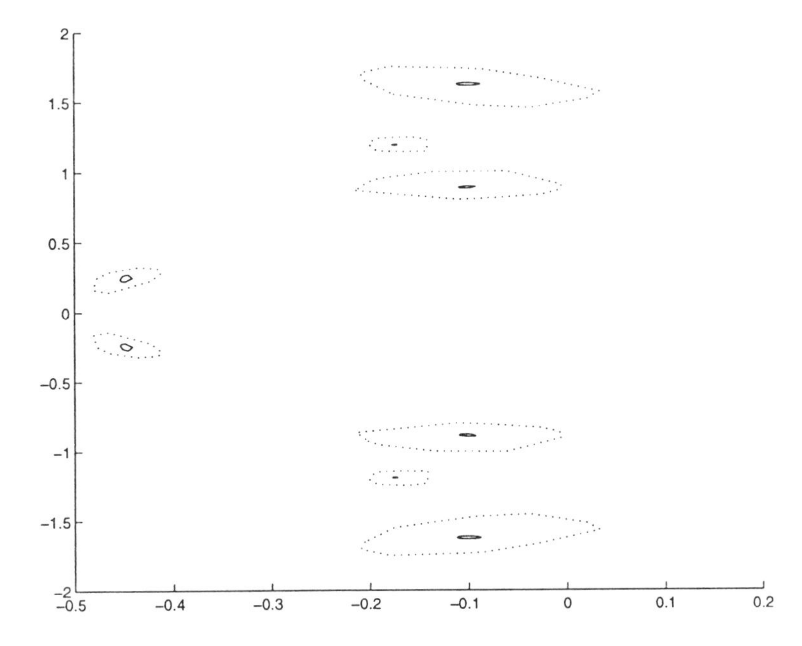

FIGURE 1. The convex hulls of closed-loop eigenvalues under 200 random 1% perturbation of the feedback matrices for quadratic pencils $P_c^{(\text{robust})}(\lambda)$ (solid lines) and $P_c(\lambda)$ (dashed lines).

Algorithm 5.2 produces the control matrix

$$B = \begin{pmatrix} -0.3814 & -0.5751 \\ -0.5555 & -0.4821 \\ -0.5191 & 0.4311 \\ -0.5258 & 0.5010 \end{pmatrix}$$

with $||B||_2 = 1$ and the feedback matrices

$$F = \begin{pmatrix} -8.1693 & -0.2320 \\ 2.5326 & -0.4130 \\ -20.6466 & 0 \\ 18.0019 & 0.2962 \end{pmatrix} \text{ and } G = \begin{pmatrix} 0.1688 & -1.3245 \\ 2.0584 & 0.2236 \\ -5.7223 & -3.0184 \\ 0.9821 & 2.4989 \end{pmatrix}$$

with the norms $||F||_2 = 28.6976$ and $||G||_2 = 7.07673$. The spectrum of the closed-loop pencil $P_c(\lambda)$ with above F and G is precisely (6.1) and the columns of Y_1 are the eigenvectors, corresponding to the eigenvalues $-0.1 \pm 1.6242i$.

REMARK 6.3. Examples 5.1 and 5.2 are purely illustrative ones. A *real-life example* involving an 211×211 quadratic pencil with sparse matrices has been solved in [**DS**], using Algorithm 4.3.

REMARK 6.4. In our numerical experiements on Algorithm 4.3, we have tried to achieve the robustness of the closed-loop eigenvalues by choosing the parametric vectors γ_1 and γ_2 appropriately.

Another idea to achieve robustness would be to select the poles appropriately in a certain region of the complex plane. In recent papers Calvetti et al. (CLRa,

CLRb), have shown that the idea works quite well in the single-input case for a first-order state-space system; we leave this to be done for future research.

7. Conclusions and Future Research

An uniform treatment of solutions, both theoretical and algorithmic, is presented for two important inverse eigenvalue problems for the quadratic pencil (1.2). The two problems are the problems of partial eigenvalue assignment and the partial eigenstructure assignment arising in feedback control of the matrix second-order control systems (1.1). The solutions have the following important practical features:

(1) They are "*direct*" in the sense that they are obtained directly in the matrix second-order settings without resorting to a first-order reformulation, so that the important structures, such as sparsity, definiteness, symmetry, etc. can be exploited.
(2) They are "*partial modal*", meaning that only a part of the spectrum (in fact, only that part that needs to be reassigned) and the corresponding eigenvectors are required.
(3) No *spill-over* occurs; that is, the eigenvalues and eigenvectors that are not required to be altered do not become affected by application of feedback.

In view of the above attractive practical features of our direct and partial-modal approach for the partial eigenvalue and eigenstructure assignment problems of the second-order model, we would like to extend the approach to the distributed parameter systems. In this context, we note that the second-order model (1.2) is just a discretized approximation (say, by the finite element method) of a distributed parameter system; thus, in spite of the fact that a second-order model is much used in practice for convenience, it has some severe limitations. For example, suppose that starting with a distributed parameter system, first a second-order model is obtained by discretization and then Problems 1.1 and 1.2 are solved using the direct partial-modal approach of this paper. Even though our results guarantee there will be no *spill-over* of the $2n-p$ eigenvalues that are not required to be reassigned, there still remains obvious uncertainty with the remaining infinite number of eigenvalues of the infinite-order system.

It is, therefore, desirable (though extremely hard) to obtain solutions directly from the distributed model without going through a discretization procedure. Some attempts, however, have been made already in this direction. Generalizing the results of [**DERb**], a solution to a single-input version of Problem 1.1 for a distributed gyroscopic system entirely in terms of the distributed parameters has been recently obtained in [**DRS**] (see also [**YMR**]). Specifically, the following problem has been solved:

Given the self-adjoint positive definite operators M and K, a gyroscopic operator G, and a self-conjugate set $\Omega = \{\mu_1, \ldots, \mu_p\}$, find feedback functions $f(x)$ and $g(x)$ such that each member of Ω is an eigenvalue of the closed-loop operator system

$$M\frac{\partial^2\nu(t,x)}{\partial t^2} + G\frac{\partial\nu(t,x)}{\partial t} + K\nu(t,x) \quad = \quad b(x)(f(x), \frac{\partial\nu(t,x)}{\partial t}) + b(x)(g(x), \nu(t,x)),$$

where $(\cdot,\cdot)$ is a scalar product, and the remaining infinite number of eigenvalues $\lambda_{p+1}, \lambda_{p+2}, \ldots$ remain the same as those of the open-loop operator system

$$M\frac{\partial^2 u(t,x)}{\partial t^2} + G\frac{\partial u(t,x)}{\partial t} + Ku(t,x) \quad = \quad 0.$$

The solution has been obtained in terms of the quantities given and entirely in the distributed parameter setting (that is, without any use of the discretization technique). The results obtained in this paper are the first and only results available for inverse eigenvalue problems for a quadratic operator pencil. Clearly, much remains to be done in this area.

Acknowledgement

The authors would like to thank the editor Vadim Olshevsky for his interest in this paper and for making some constructive suggestions that improved the readability of the paper.

References

[CLRa] D. Calvetti, B. Lewis and L. Reichel, *On the solution of the single input pole placement problem*, in Mathematical Theory of Networks and Systems, eds. A. Beghi, L. Finesso and G. Picci, Il Poliografo, Padova (1998), 585–588.

[CLRb] ———, *On the selection of the poles in the single input pole placement problem*, Linear Alg. Appl. (special issue dedicated to Hans Schneider), bf 302/**303** (1999), 331–345.

[CD] E. K. Chu and B. N. Datta, Numerically Robust Pole Assignment for the Second-Order Systems, Int. J. Control **4** (1996), 1113–1127.

[MTC] M. T. Chu, *Inverse Eigenvalue Problems*, SIAM Rev. **40** (1998), no. 1, 1–39.

[BNDa] B. N. Datta, *Numerical Linear Algebra and Applications*, Brook/Cole Publishing Co., Pacific Grove, California (1998).

[BNDb] B. N. Datta, *Numerical Methods for Linear Control Systems Design and Analysis*, Academic Press: New York, to appear.

[DERa] B. N. Datta, S. Elhay and Y. M. Ram, *An algorithm for the partial multi-input pole assignment problem of a second-order control system*, Proceedings of the IEEE Conference on Decision and Control (1996), 2025–2029.

[DERb] ———, *Orthogonality and Partial Pole Assignment for the Symmetric Definite Quadratic Pencil*, Linear Algebra and its Applications **257** (1997), 29–48.

[DERS] B. N. Datta, S. Elhay, Y. M. Ram and D. R. Sarkissian, *Partial Eigenstructure Assignment for the quadratic Pencil*, Journal of Sound and Vibration, **230** (2000), no. 1, 101–110.

[DRS] B. N. Datta, Y. M. Ram and D. R. Sarkissian, *Spectrum modification for gyroscopic systems*, to be submitted for publication (1999).

[DR] B. N. Datta and F. Rincón, *Feedback Stabilization of the Second-Order Model: A Nonmodal Approach*, Lin. Alg. Appl. **188** (1993), 138–161.

[DS] B. N. Datta and D. R. Sarkissian, *Multi-input Partial Eigenvalue Assignment for the Symmetric Quadratic Pencil*, Proceedings of the American Control Conference (1999), 2244–2247.

[GVL] G. Golub and C. Van Loan, *Matrix Computations*, The Johns Hopkins University Press, Baltimore (1984) & 3rd edition (1996).

[MLJH] M. L. J. Hautus, *Controllability and observability conditions of linear autonomous systems*, Proc. Kon. Ned. Akad. Wetensch., ser. A, **72** (1969), 443–448.

[DJI] D. J. Inman, *Vibrations: Control, Measurement and Stability*, Prentice Hall (1989).

[IK] D. J. Inman and A. Kress, *Eigenstructure Assignment via inverse eigenvalue methods*, AIAA J. Guidance, Control and Dynamics **18** (1995), 625–627.

[KNVD] J. Kautsky, N. K. Nichols and P. van Dooren, *Robust pole assignment in linear state feedback*, Int. J. Contr. **41** (1985), no. 5, 1129–1155.

[KFB] L. H. Keel, J. A. Fleming, S. P. Bhattacharyya, *Minimum norm pole assignment via Sylvester equation*, Contemporary Mathematics **47** (1985), 265–272.

[LA] A. J. Laub and W. F. Arnold, *Controllability and Observability Criteria for Multivariable Linear Second-Order Models*, IEEE Trans. Automatic Control **29** (1984) no. 2, 163–165.

[MRG] *MATLAB Reference Guide*, The Math Works, Inc. (1984-1996).

[PC] B. N. Parlett and H. C. Chen, *Use of indefinite pencils for computing damped natural modes*, Lin. Alg. Appl. **140** (1990), 53–88.

[YMR] Y. M. Ram, *Pole assignment for the vibrating rod*, Quarterly Journal of Mechanics and Applied Mathematics **51** (1998), no. 3, 461–476.

[YS] Y. Saad, *A projection method for partial pole assignment in linear state feedback*, IEEE Trans. Auto. Control **33** (1988), no. 3, 290–297.

[SBFV] G. L. G. Sleijpen, A. G. L. Booten, D.R. Fokkema and H. A. van der Vorst, *Jacobi-Davidson type methods for generalized eigenproblems and polynomial eigenproblems*, BIT **36** (1996), no. 3, 595–633.

Department of Mathematical Sciences, Northern Illinois University, DeKalb, IL, 60115

E-mail address: dattab@math.niu.edu

Department of Mathematical Sciences, Northern Illinois University, DeKalb, IL, 60115

E-mail address: sarkiss@math.niu.edu

Contemporary Mathematics
Volume **280**, 2001

Partial Eigenvalue Assignment for Large Linear Control Systems

D. Calvetti, B. Lewis, and L. Reichel

ABSTRACT. The eigenvalue assignment problem is a classical problem in Control Theory. This paper presents new algorithms for the stabilization of large single-input time-invariant control systems by partial eigenvalue assignment. Our algorithms are based on the implicitly restarted Arnoldi method, and are well suited for control systems that require the reassignment of a few eigenvalues.

1. Introduction

Consider the single-input time-invariant linear control system

$$\frac{d}{dt}x(t) = Ax(t) + bu(t), \qquad x(0) = x_0, \qquad t \geq 0, \tag{1.1}$$

where $A \in \mathbb{R}^{n\times n}$ is a large, nonsymmetric, possibly sparse, matrix, $b, x_0 \in \mathbb{R}^n$, $x(t)$ is a vector-valued function with values in $\mathbb{R}^n$ and $u(t)$ is a real-valued function. Let $\lambda(A) = \{\lambda_j\}_{j=1}^n$ denote the spectrum of A and introduce the set $\mathbb{P} = \{\psi_j\}_{j=1}^m$ of $m \leq n$ complex numbers. We refer to the ψ_j as poles. This paper is concerned with the problem of determining a vector $f \in \mathbb{R}^n$ with the property that

$$\lambda(A - bf^T) = \mathbb{P} \cup \{\lambda_j\}_{j=m+1}^n, \tag{1.2}$$

when such a vector exists. This problem is known as the *partial eigenvalue assignment problem.* The vector f in (1.2) is referred to as the feedback gain vector, because substituting $u(t) = -f^T x(t)$ into (1.1) yields a closed-loop system with solution

$$x(t) = \exp((A - bf^T)t)x_0, \qquad t \geq 0. \tag{1.3}$$

1991 *Mathematics Subject Classification.* Primary 93B55, 65F15.

Key words and phrases. Implicitly restarted Arnoldi process, eigenvalue assignment, control system.

Research supported in part by NSF grant DMS-9806702.

Research supported in part by NSF grant DMS-9806413.

Research supported in part by NSF grant DMS-9806413.

Assume that all but m eigenvalues of the matrix A have nonnegative real part. For definiteness, let

$$\mathrm{Re}(\lambda_j) \geq 0, \qquad 1 \leq j \leq m, \tag{1.4}$$

$$\mathrm{Re}(\lambda_j) < 0, \qquad m+1 \leq j \leq n. \tag{1.5}$$

If $f = 0$ in (1.3), then there are initial vectors x_0, such that the closed-loop solution $x(t)$ becomes unbounded as t increases. This is generally undesirable. Instead, we would like to choose f so that the closed-loop solution satisfies

$$\lim_{t \to \infty} x(t) = 0 \tag{1.6}$$

for any choice of initial vector x_0. In order to achieve this, let the poles ψ_j have negative real part and choose a feedback gain vector f, so that (1.2) holds. Then all eigenvalues of the matrix $A - bf^T$ have negative real part and (1.6) holds independently of the choice of x_0.

To simplify our exposition, we consider throughout this paper the problem of replacing a set of a few eigenvalues $\{\lambda_j\}_{j=1}^m$ of A with nonnegative real part by a set of poles $\mathbb{P}$ with negative real part. However, the methods presented can be used to replace other subsets of eigenvalues as well. The methods of this paper are designed for the solution of partial eigenvalue assignment problems (1.2) with m, the number of eigenvalues to be assigned, much smaller than n, the order of the matrix A.

The eigenvalue assignment problem (where $\mathbb{P}$ consists of n poles) has received considerable attention in the literature. Analyses of the sensitivity of this problem to perturbations can be found in [**7, 8**] and discussions on how to choose the poles ψ_j are presented in [**3, 9**]. Several numerical methods are available for small to medium-sized problems; see, e.g., [**1, 2, 5, 10, 11**] and references therein. Although these methods can be applied to the partial eigenvalue assignment problem for large systems, they are not computationally economical. These methods typically require $O(n^3)$ floating point operations and storage of up to n vectors in $\mathbb{R}^n$. Numerical methods for large-scale partial eigenvalue assignment problems have, so far, received less attention. Saad [**12**] described a projection method based on the Arnoldi process. We review this algorithm in Section 2 and present a modification thereof based on the implicitly restarted Arnoldi method. Both Saad's method and our modification require the evaluation of matrix-vector products with the matrix A^T, but not with A. Section 3 describes a variant of the partial eigenvalue assignment problem that allows both the vectors f and b to be chosen in (1.2). The possibility to choose both f and b reduces the sensitivity of the partial eigenvalue assignment problem to perturbations. Two algorithms are presented, one that requires the evaluation of matrix-vector products with the matrix A^T and one with the matrix A. Section 4 displays a few computed examples and Section 5 contains concluding remarks.

Finally, we remark that an eigenvalue assignment problem in which the vector f is fixed and b is to be determined so that (1.2) holds is discussed in [**4**]. The numerical method proposed is analogous to the method discussed in Section 2.

2. The partial eigenvalue assignment problem

Throughout this paper, we assume that the sets $\mathbb{P}$ and $\{\lambda_j\}_{j=m+1}^n$ in (1.2) are invariant under complex conjugation. It can be shown that the desired feedback gain vector f, if it exists, then has real-valued entries only.

Saad [**12**] solves the partial eigenvalue assignment problem (1.2) by computing the left invariant subspace of A associated with the set of eigenvalues $\{\lambda_j\}_{j=1}^m$. Let the columns of the matrix $V \in \mathbb{R}^{n\times m}$ form an orthonormal basis for this subspace. Then

$$A^T V = VH, \qquad V^T V = I_m, \tag{2.1}$$

for some matrix $H \in \mathbb{R}^{m\times m}$ with spectrum

$$\lambda(H) = \{\lambda_j\}_{j=1}^m. \tag{2.2}$$

Here I_m denotes the $m \times m$ identity matrix. We remark that the decomposition (2.1) is not unique; if the matrix pair $\{H, V\}$ satisfies (2.1) and (2.2), then so does the pair $\{U^T HU, VU\}$ for any orthogonal matrix $U \in \mathbb{R}^{m\times m}$.

Let the matrix pair $\{H, V\}$ satisfy (2.1) and (2.2). Introduce the vector $\tilde{b} = V^T b$ and consider the problem of determining a vector $\tilde{f} \in \mathbb{R}^m$, such that

$$\lambda(H - \tilde{f}\tilde{b}^T) = \mathbb{P}. \tag{2.3}$$

We refer to this eigenvalue assignment problem as the *projected eigenvalue assignment problem.* The matrix-vector pair $\{H^T, \tilde{b}\}$ is said to be controllable if

$$\operatorname{rank}(H^T - zI_m, \tilde{b}) = m, \qquad \forall z \in \mathbb{C}. \tag{2.4}$$

It is known that the projected eigenvalue assignment problem (2.3) has a solution $\tilde{f}$ for any set of poles $\mathbb{P}$, if and only if the matrix-vector pair $\{H^T, \tilde{b}\}$ is controllable; see, e.g., Wonham [**14**, Section 2.2].

The following result is a slight modification of Theorem 2.1 of Saad [**12**]. We present the proof because related arguments will be employed to show other results below.

THEOREM 2.1. *Let V and H satisfy equations (2.1) and (2.2), and let $\tilde{b} = V^T b$, where the vector b is given by (1.1). The partial eigenvalue assignment problem (1.2) has a solution $f \in range(V)$ if and only if the projected eigenvalue assignment problem (2.3) has a solution $\tilde{f} \in \mathbb{R}^m$.*

PROOF. Extend V to an orthogonal matrix $X = (V, W) \in \mathbb{R}^{n\times n}$. The decomposition (2.1) and

$$W^T V = 0 \tag{2.5}$$

yield $V^T A^T V = H$ and $W^T A^T V = W^T V H = 0$. These identities show that the matrix A^T is similar to the block-triangular matrix

$$X^T A^T X = \begin{pmatrix} H & V^T A^T W \\ 0 & W^T A^T W \end{pmatrix}.$$

It follows from (2.2) that

$$\lambda(W^T A^T W) = \{\lambda_j\}_{j=m+1}^n. \tag{2.6}$$

Assume that $f \in \text{range}(V)$ satisfies the partial eigenvalue assignment problem (1.2) and introduce $\tilde{f} = V^T f$. The equations (2.1), (2.5) and $W^T f = 0$ yield

$$\begin{aligned} V^T(A - bf^T)^T V &= H - \tilde{f}\tilde{b}^T, \\ W^T(A - bf^T)^T V &= W^T A^T V - W^T f b^T V^T = 0, \\ W^T(A - bf^T)^T W &= W^T A^T W - W^T f b^T W = W^T A^T W \end{aligned}$$

and therefore

$$X^T(A - bf^T)^T X = \begin{pmatrix} H - \tilde{f}\tilde{b}^T & V^T A^T W - \tilde{f} b^T W \\ 0 & W^T A^T W \end{pmatrix}. \tag{2.7}$$

It follows from $\lambda(X^T(A-bf^T)^T X) = \mathbb{P} \cup \{\lambda_j\}_{j=m+1}^n$ and (2.6) that $\lambda(H - \tilde{f}\tilde{b}^T) = \mathbb{P}$. Thus, $\tilde{f}$ satisfies the projected eigenvalue assignment problem (2.3).

Conversely, assume that $\tilde{f} \in \mathbb{R}^m$ satisfies the projected eigenvalue assignment problem (2.3), where the matrix-pair $\{H, V\}$ satisfies (2.1) and (2.2). Define $f = V\tilde{f}$. Substituting (2.1) and f into $X^T(A - bf^T)^T X$ yields the right-hand side of (2.7). Equations (2.3) and (2.6) yield (1.2). ∎

Saad [**12**] proposed to use the Arnoldi process to determine a decomposition of the form (2.1). This approach is attractive for large-scale problems, because the matrix A is only used to evaluate matrix-vector products with A^T. In particular, neither A nor A^T have to be factored or transformed. Application of m steps of the Arnoldi process to the matrix A^T with initial unit vector v_1 yields the Arnoldi decomposition

$$A^T V_m - V_m H_m = \eta_m v_{m+1} e_m^T, \tag{2.8}$$

where $V_m \in \mathbb{R}^{n \times m}$, $V_m^T V_m = I_m$, $V_m e_1 = v_1$ and $H_m \in \mathbb{R}^{m \times m}$ is an upper Hessenberg matrix with positive subdiagonal entries. The scalar η_m in (2.8) is nonnegative. When $\eta_m > 0$, we require the vector $v_{m+1} \in \mathbb{R}^n$ to satisfy $v_{m+1}^T v_{m+1} = 1$ and $V_m^T v_{m+1} = 0$. Moreover, $e_m = (0, \ldots, 0, 1)^T \in \mathbb{R}^m$. We assume for now that m is small enough so that the decomposition (2.8) with the stated properties exists.

Assume for the moment that the matrix H_m and scalar η_m in the Arnoldi decomposition (2.8) satisfy

$$\lambda(H_m) = \{\lambda_j\}_{j=1}^m, \qquad \eta_m = 0. \tag{2.9}$$

Then the matrix pair $\{H_m, V_m\}$ in (2.8) satisfies (2.1). Let $\tilde{b} = V_m^T b$, where b is given by (1.1), and assume that $\tilde{f}$ solves the projected eigenvalue assignment problem $\lambda(H_m - \tilde{f}\tilde{b}^T) = \mathbb{P}$. It then follows from Theorem 2.1 that $f = V_m \tilde{f}$ solves the partial eigenvalue assignment problem (1.2). Thus, given an Arnoldi decomposition (2.8) such that (2.9) holds, we can solve the partial eigenvalue assignment problem (1.2), if a solution exists, by computing a solution to the projected eigenvalue assignment problem (2.3).

However, given an arbitrary initial vector v_1, application of m steps of the Arnoldi process to the matrix A^T is unlikely to produce a decomposition (2.8) such that (2.9) is satisfied. Several restarting strategies for the Arnoldi process have been proposed in the literature with the aim of iteratively determining an initial vector v_1 that gives an Arnoldi decomposition (2.8) with the property (2.9). The Implicitly Restarted Arnoldi (IRA) method, proposed by Sorensen [**13**], is one of the most effective restarting strategies. This method combines the Arnoldi process with the implicitly shifted QR algorithm to determine an Arnoldi decomposition (2.8) that satisfies (2.9).

We outline the IRA method. Let the eigenvalues λ_j of A satisfy (1.4) and (1.5), and assume for the moment that m is known. We would like to determine the invariant subspace of A^T associated with the set of eigenvalues $\{\lambda_j\}_{j=1}^m$. Apply $2m$ steps of the Arnoldi process to the matrix A^T with initial unit vector v_1 with normally distributed randomly generated entries to determine the decomposition

$$A^T V_{2m} = V_{2m} H_{2m} + \eta_{2m} v_{2m+1} e_{2m}^T. \tag{2.10}$$

Compute the spectrum $\lambda(H_{2m}) = \{\mu_j\}_{j=1}^{2m}$ and assume that

$$\mathrm{Re}(\mu_1) \leq \mathrm{Re}(\mu_2) \leq \ldots \leq \mathrm{Re}(\mu_\ell) < 0 \leq \mathrm{Re}(\mu_{\ell+1}) \leq \ldots \leq \mathrm{Re}(\mu_{2m}).$$

Define $k = \min\{m, \ell\}$. Sorensen [**13**] describes how the decomposition (2.10) can be updated without evaluating any matrix-vector products with the matrix A^T to give the Arnoldi decomposition

$$A^T \hat{V}_{2m-k} = \hat{V}_{2m-k} \hat{H}_{2m-k} + \hat{\eta}_{2m-k} \hat{v}_{2m-k+1} e_{2m-k}^T, \tag{2.11}$$

with initial vector

$$\hat{v}_1 = \hat{V}_{2m-k} e_1 = \frac{\prod_{j=1}^k (A^T - \mu_j I) v_1}{\| \prod_{j=1}^k (A^T - \mu_j I) v_1 \|}, \tag{2.12}$$

where v_1 is the initial vector in the Arnoldi decomposition (2.10). Throughout this paper $\| \cdot \|$ denotes the Euclidean vector norm or the associated induced matrix norm.

Thus, the new initial vector (2.12) is obtained by multiplying the original initial vector v_1 by a polynomial in A^T. The zeros of this polynomial are Ritz values of A^T with negative real part. The purpose of multiplying v_1 by this polynomial is to remove, or at least reduce, eigenvector components of A^T in v_1 associated with eigenvalues of A^T with negative real part, and thereby force the new initial vector (2.12) into the invariant subspace of A^T associated with the set of eigenvalues $\{\lambda_j\}_{j=1}^m$ with nonnegative real part. If range$(\hat{V}_{2m-k})$ does not contain a sufficiently good approximation of this subspace, then we apply k steps of the Arnoldi process to obtain the Arnoldi decomposition

$$A^T \hat{V}_{2m} = \hat{V}_{2m} \hat{H}_{2m} + \hat{\eta}_{2m} \hat{v}_{2m+1} e_{2m}^T$$

from (2.11). This decomposition is of the same form as (2.10), and we update it in the same fashion as (2.10). We proceed in this manner until a sufficiently accurate approximation of the invariant subspace of A^T associated with the set of eigenvalues $\{\lambda_j\}_{j=1}^m$ has been determined. Since, in general, the number m of eigenvalues with positive real part is not known a priori, we also seek to determine m during the iterations with the IRA method.

A detailed discussion of the IRA method is given by Sorensen [**13**] and more recently by Lehoucq et al. [**6**]. In our numerical examples of Section 4, we used a simple implementation based on the description of the IRA method in [**13**]. The application of the IRA method for partial eigenvalue assignment is outlined in the following algorithm.

Algorithm 1 *Partial eigenvalue assignment using the IRA method.*

(1) Apply the IRA method to the matrix A^T to compute an Arnoldi decomposition of the form (2.8), such that $\eta_m \geq 0$ is tiny and $\lambda(H_m)$ is a good approximation of the set $\{\lambda_j\}_{j=1}^m$ of eigenvalues of A. The range of the matrix V_m in the Arnoldi decomposition provides an approximation of the invariant subspace of A^T associated with $\{\lambda_j\}_{j=1}^m$.
(2) Select a set of poles $\mathbb{P} = \{\psi_j\}_{j=1}^m$. It may be attractive to use the pole selection method presented in [**3**]. The selection of poles will be addressed further in Section 4.
(3) Let $\tilde{b} = V_m^T b$ and solve the projected eigenvalue assignment problem (2.3). Denote the solution by $\tilde{f}$.
(4) Let $f = V_m \tilde{f}$. The vector f is the computed solution of the partial eigenvalue assignment problem (1.2).

The computation of $\tilde{f}$ in Step 3 of Algorithm 1 can be accomplished by any numerical method for eigenvalue assignment for small matrices. In the numerical examples of Section 4 we used the method of Datta [**5**].

Algorithm 1 is analogous to a projection algorithm proposed by Saad [**12**]. It differs from the latter in its use of the IRA method. The IRA method is often very effective at determining the eigenvalues $\{\lambda_j\}_{j=1}^m$ and the associated invariant subspace of A^T.

However, it is possible that the IRA method converges too quickly, in the sense that it determines an invariant subspace associated with some, but not all, of the eigenvalues in the set $\{\lambda_j\}_{j=1}^m$. We address this situation by restarting the IRA method with a new random initial vector, that is orthogonalized against the invariant subspace of A^T already determined. In fact, in order to avoid convergence to the invariant subspace already determined, we orthogonalize all columns of the matrices V_m generated after a restart against the invariant subspace of A^T already determined. The IRA method is restarted repeatedly in this manner until all eigenvalues in the set $\{\lambda_j\}_{j=1}^m$, characterized by (1.4), and the associated invariant subspace of A^T have been computed to sufficient accuracy.

We remark that this approach does not require the number of eigenvalues of A with nonnegative real part be known a priori; we restart the IRA method repeatedly until no more eigenvalues with nonnegative real part are found. This approach has reliably determined all eigenvalues with nonnegative real part in a large number of numerical experiments.

3. The modified partial eigenvalue assignment problem

In our experience Algorithm 1 performs very well for many large partial eigenvalue assignment problems. Nevertheless, the approach described in Section 2 can fail because the partial eigenvalue assignment problem (1.2) is not solvable for certain vectors b. This section introduces a variant of the partial eigenvalue assignment problem that avoids this difficulty by allowing both vectors f and b to be chosen. Thus, we seek to determine vectors $f, b \in \mathbb{R}^n$, such that

$$\lambda(A - bf^T) = \mathbb{P} \cup \{\lambda_j\}_{j=m+1}^n. \tag{3.1}$$

We refer to this problem as the *modified partial eigenvalue assignment problem*. The following theorem establishes that this assignment problem has a solution.

THEOREM 3.1. *Let $V \in \mathbb{R}^{n\times m}$ and the upper Hessenberg matrix $H \in \mathbb{R}^{m\times m}$ with nonvanishing subdiagonal entries satisfy (2.1) and (2.2). Then there are vectors $f, b \in$ range(V) that solve the modified partial eigenvalue assignment problem (3.1).*

PROOF. Theorem 2.1 established that the existence of a solution $f \in \text{range}(V)$ is equivalent to the existence of a solution $\tilde{f}$ of the projected eigenvalue assignment problem (2.3) with $\tilde{b} = V^T b$. We will show that for a suitable choice of $\tilde{b}$, the projected eigenvalue assignment problem (2.3) has a solution $\tilde{f}$. The proof shows how the vectors $\tilde{f}$ and $\tilde{b}$ can be computed. The construction used in the proof to determine the vectors $\tilde{f}$ and $\tilde{b}$ has previously been employed by Datta [**5**].

Let $r_1 = e_1$ and define the vectors

$$r_{j+1} = (H - \psi_j I)r_j, \qquad 1 \le j \le m. \tag{3.2}$$

The matrix $R = (r_1, r_2, \ldots, r_m)$ is upper triangular with diagonal entries

$$e_k^T R e_k = \prod_{j=1}^{k-1} h_{j+1,j}, \qquad 1 \le k \le m,$$

where $e_1^T R e_1 = 1$. Since H has nonvanishing subdiagonal entries, the matrix R is nonsingular. Its inverse R^{-1} is upper triangular with diagonal entries

$$e_k^T R^{-1} e_k = \prod_{j=1}^{k-1} h_{j+1,j}^{-1}, \qquad 1 \le k \le m.$$

Introduce the lower bidiagonal matrix

$$B = \begin{pmatrix} \psi_1 & 0 & 0 & \cdots & 0 \\ 1 & \psi_2 & 0 & \cdots & 0 \\ 0 & 1 & \psi_3 & & 0 \\ \vdots & \ddots & \ddots & \ddots & \vdots \\ 0 & 0 & 0 & 1 & \psi_m \end{pmatrix}.$$

with spectrum $\lambda(B) = \mathbb{P}$. The recurrence formula (3.2) yields

$$HR - RB = r_{m+1} e_m^T$$

and, therefore,

$$H - r_{m+1} e_m^T R^{-1} = RBR^{-1}.$$

Thus,

$$\lambda(H - r_{m+1} e_m^T R^{-1}) = \mathbb{P}. \tag{3.3}$$

Note that $e_m^T R^{-1} = (e_m^T R e_m)^{-1} e_m^T$. Letting $\tilde{f} = (e_m^T R e_m)^{-1} r_{m+1}$ and $\tilde{b} = e_m$ transforms (3.3) into the form (2.3). Finally, it follows from Theorem 2.1 that $f = V\tilde{f}$ and $b = V\tilde{b}$ satisfy (1.2). ∎

We remark that since the subdiagonal entries of the upper Hessenberg matrix H in Theorem 3.1 are nonvanishing, it is easy to verify that the matrix-vector pair $\{H^T, e_m\}$ in the proof of the theorem is controllable; cf. (2.4). Also, note that the solution $\{f, b\}$ of the modified partial eigenvalue assignment problem is not unique.

THEOREM 3.2. *Let the matrices H_m and V_m, the scalar η_m and the vector v_{m+1} make up an Arnoldi decomposition (2.8). Determine the vector $\tilde{f} \in \mathbb{R}^m$, so that $\lambda(H_m - \tilde{f}e_m^T) = \mathbb{P}$. Let $b = V_m e_m$ and $f = V_m\tilde{f} + \eta_m v_{m+1}$. Then*

$$\mathbb{P} \subset \lambda(A - bf^T). \tag{3.4}$$

PROOF. Since H_m is an upper Hessenberg matrix with nonvanishing subdiagonal elements, a vector $\tilde{f}$ with the desired property can be computed as described in the proof of Theorem 3.1. Extend V_m to an orthogonal matrix $X = (V_m, W) \in \mathbb{R}^{n\times n}$. We obtain with the specified choices of f and b that

$$\begin{aligned} V_m^T(A - bf^T)^T V_m &= H_m - \tilde{f}e_m^T \\ W^T(A - bf^T)^T V_m &= W^T(A^T V_m - fb^T V_m) \\ &= W^T(V_m H_m + \eta_m v_{m+1} e_m^T - (V_m\tilde{f} + \eta_m v_{m+1})e_m^T) = 0, \end{aligned}$$

and therefore

$$X^T(A - bf^T)X = \begin{pmatrix} H_m - \tilde{f}e_m^T & V_m^T A^T W \\ 0 & W^T A^T W \end{pmatrix}.$$

This shows (3.4). ■

Note that the inclusion (3.4) holds regardless of the size of $\eta_m \geq 0$, although the remaining eigenvalues of $A - bf^T$ may vary as η_m increases. This depends on that spectrum $\lambda(W^T AW)$ may depend on η_m. We recall that $\lambda(W^T AW) = \{\lambda_j\}_{j=m+1}^n$ when $\eta_m = 0$.

Algorithm 2 *Modified partial eigenvalue assignment using the IRA method applied to A^T.*

(1) Apply the IRA method to the matrix A^T to compute an Arnoldi decomposition of the form (2.8), such that $\eta_m \geq 0$ is small and $\lambda(H_m)$ is a good approximation of the set $\{\lambda_j\}_{j=1}^m$ of eigenvalues of A. The range of the matrix V_m in the Arnoldi decomposition provides an approximation of the invariant subspace of A^T associated with $\{\lambda_j\}_{j=1}^m$.
(2) Select a set of poles $\mathbb{P} = \{\psi_j\}_{j=1}^m$, e.g., by the method described in [**3**].
(3) Let $b = V_m e_m$, and compute $\tilde{f} \in \mathbb{R}^m$ so that $\lambda(H_m - \tilde{f}e_m^T) = \mathbb{P}$.
(4) Let $f = V_m\tilde{f} + \eta_m v_{m+1}$. The vector pair $\{f, b\}$ is the computed solution of the modified partial eigenvalue assignment problem (3.1).

The vector $\tilde{f}$ in Step 3 of Algorithm 2 can be computed by any numerical method for eigenvalue assignment for small matrices. We use the method by Datta [**5**], i.e., the method in the proof of Theorem 3.1, in the numerical examples of Section 4.

We have found that Algorithm 2 often gives higher accuracy than Algorithm 1 in the sense that $\lambda(A - bf^T)$ is closer to the set $\mathbb{P} \cup \{\lambda_j\}_{j=m+1}^n$ when f and b are computed by Algorithm 2 than when f is computed by Algorithm 1. A reason for this may be that when $\eta_m > 0$ and the vector f is computed by Algorithm 1, one cannot expect $\mathbb{P}$ to be a subset of the spectrum of $A - bf^T$. On the other hand, when both vectors f and b are computed by Algorithm 2, Theorem 3.2 guarantees that (3.4) holds independently of the size of η_m.

Both Algorithms 1 and 2 require the evaluation of matrix-vector products with the matrix A^T. However, for some problems it may be possible to evaluate matrix-vector products with the matrix A much faster than with A^T. This situation arises, for instance, when A is large and sparse, and the sparse storage scheme used to represent A allows much faster evaluation of matrix-vector products with A than with A^T.

Since Algorithm 2 determines both the vectors f and b, the algorithm can be modified to use an Arnoldi decomposition of the matrix A instead of A^T. The matrix-vector product evaluations with the matrix A^T of Algorithm 2 can then be replaced by matrix-vector product evaluations with the matrix A. Before describing the algorithm, we present analogues of Theorems 3.1 and 3.2.

THEOREM 3.3. *Let $V \in \mathbb{R}^{n\times m}$ and the upper Hessenberg matrix $H \in \mathbb{R}^{m\times m}$ with nonvanishing subdiagonal entries satisfy*

$$AV = VH, \qquad V^TV = I_m,$$

and (2.2). Then there are vectors $f, b \in range(V)$ that solve the modified partial eigenvalue assignment problem (3.1).

PROOF. The proof differs from the proof of Theorem 3.1 only in that the vectors $\tilde{f}$ and $\tilde{b}$, as well as the vectors f and b, are interchanged. ■

THEOREM 3.4. *Consider the Arnoldi decomposition*

$$AV_m - V_mH_m = \eta_m v_{m+1}e_m^T, \tag{3.5}$$

where $V_m \in \mathbb{R}^{n\times m}$, $V_m^TV_m = I_m$, $H_m \in \mathbb{R}^{m\times m}$ is an upper Hessenberg matrix with positive subdiagonal entries, and η_m is a nonnegative scalar. Determine the vector $\tilde{b} \in \mathbb{R}^m$, such that $\lambda(H_m - \tilde{b}e_m^T) = \mathbb{P}$. Let $f = V_me_m$ and $b = V_m\tilde{b} + \eta_m v_{m+1}$. Then

$$\mathbb{P} \subset \lambda(A - bf^T). \tag{3.6}$$

PROOF. The proof differs from the proof of Theorem 3.2 only in that the vectors $\tilde{f}$ and $\tilde{b}$, as well as the vectors f and b, are interchanged. ■

Algorithm 3 *Modified partial eigenvalue assignment using the IRA method applied to A.*

(1) Apply the IRA method to the matrix A to compute an Arnoldi decomposition of the form (3.5), such that $\eta_m \geq 0$ is small and $\lambda(H_m)$ is a good approximation of the set $\{\lambda_j\}_{j=1}^m$ of eigenvalues of A. The range of the matrix V_m in the Arnoldi decomposition provides an approximation of the invariant subspace of A associated with $\{\lambda_j\}_{j=1}^m$.
(2) Select a set of poles $\mathbb{P} = \{\psi_j\}_{j=1}^m$, e.g., by the method described in [**3**].
(3) Let $f = V_me_m$, and compute $\tilde{b} \in \mathbb{R}^m$ so that $\lambda(H_m - \tilde{b}e_m^T) = \mathbb{P}$.
(4) Let $b = V_m\tilde{b} + \eta_m v_{m+1}$. The vector pair $\{f, b\}$ is the computed solution of the modified partial eigenvalue assignment problem (3.1).

4. Numerical experiments

The computations reported in this section were carried out on an Intel Pentium workstation using Matlab 5.3 and floating point arithmetic with 16 significant digits. Our computed examples display the performance of Algorithms 1 and 2 when applied to a matrix $A \in \mathbb{R}^{2000\times 2000}$, which we define by its spectral factorization,

$$A = S\Lambda S^{-1}, \qquad \Lambda = \mathrm{diag}[\lambda_1, \lambda_2, \ldots, \lambda_{2000}].$$

The eigenvalues $\lambda_1, \lambda_2, \ldots, \lambda_5$ are distributed in the disk in the complex plane with center $\frac{1}{2}$ and radius $\frac{1}{2}$. Thus, they have nonnegative real part. The remaining eigenvalues are allocated in the open disk with center 1 and radius 1, so that their distances to the center of the disk are uniformly distributed and their angles with the positive real axis are uniformly distributed in the interval $[-\pi, \pi]$. These eigenvalues have negative real part. We require all eigenvalues to be real or appear in complex conjugate pairs. Thus, $m = 5$ in (1.4). The eigenvalues of A are marked by the symbol $\times$ in the figures of this section. The eigenvector matrix S of A has complex conjugate columns, with entries with real and imaginary parts uniformly distributed in the interval $[0, 1]$.

In Step 1 of Algorithms 1 and 2, we determine several Arnoldi decompositions of the form (2.10) and (2.11). These decompositions enable us to compute estimates of $\max\{t : t \in \mathrm{Re}(\lambda(A)), t < 0\}$ and $\max\{t : t \in \mathrm{Im}(\lambda(A))\}$. Denote the computed estimates by α and β, respectively. In Step 2 of Algorithms 1 and 2, we choose a set of poles $\mathbb{P} = \{\psi_j\}_{j=1}^m$ on an interval $\mathbb{I}_{\hat{\alpha},\beta} = [\hat{\alpha} + i\beta, \hat{\alpha} - i\beta]$ in the complex plane, for some $\hat{\alpha} \leq \alpha$, where $i = \sqrt{-1}$. The poles are allocated by using the formulas [**3**, (3.2)-(3.3)]. This choice of poles approximately minimizes the condition number of the eigenvector matrix of $H - \tilde{f}\tilde{b}^T$ over all sets of m distinct poles in the interval $\mathbb{I}_{\alpha,\beta}$. Available bounds for the sensitivity of the projected eigenvalue assignment problem (2.3) to perturbations grow with the condition number of the eigenvector matrix of $H_m - \tilde{f}\tilde{b}^T$; see [**3, 7, 9**]. It is therefore advantageous to allocate the poles so that this condition number is not very large.

The vector b in (1.1) used in our experiments has normally distributed random entries with zero mean, normalized so that $\|b\| = 1$. This vector is only required by Algorithm 1. However, we also use it as starting vector for the initial Arnoldi decomposition in Step 1 of Algorithm 2. Then Step 1 of Algorithm 1 and Step 1 of Algorithm 2 give the same Arnoldi decomposition.

We wish to determine how close the eigenvalues of the matrix $A - bf^T$ are to the set $\mathbb{P} \cup \{\lambda_j\}_{j=m+1}^n$. The difference between the sets $\lambda(A - bf^T)$ and $\mathbb{P} \cup \{\lambda_j\}_{j=m+1}^n$ is measured by the following metric, defined for compact sets $\mathbb{F}$ and $\mathbb{G}$ in the complex plane,

$$d(\mathbb{F}, \mathbb{G}) = \max\left\{\max_{z\in\mathbb{F}} \mathrm{dist}(z, \mathbb{G}), \max_{\zeta\in\mathbb{G}} \mathrm{dist}(\zeta, \mathbb{F})\right\},$$

where $\mathrm{dist}(z, \mathbb{G}) = \min_{\zeta\in\mathbb{G}} |z - \zeta|$.

Figure 1 illustrates the performance of Algorithm 2, which computes both of the vectors f and b. After two restarts in the IRA method, the set of poles $\mathbb{P}$ is a subset of $\lambda(A - bf^T)$ to graphing accuracy. Figure 1 shows that the eigenvalues of $A - bf^T$ with negative real part agree well with the eigenvalues of A with negative real part.

Figure 2 shows the eigenvalues of the matrix $A - bf^T$ computed by Algorithm 1 when the IRA method is restarted twice in Step 1. Figure 2 displays that the

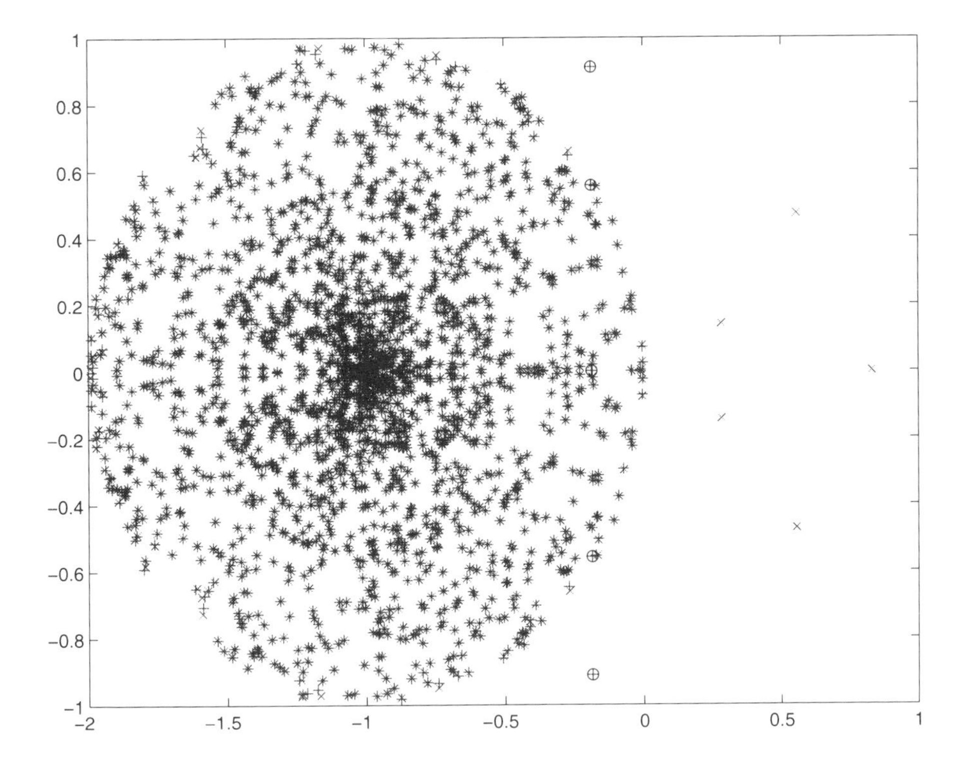

FIGURE 1. Solution determined by Algorithm 2 after two restarts in Step 1. The eigenvalues λ_j of A are marked by $\times$, the poles ψ_j by $\circ$, and the eigenvalues of $A - bf^T$ by $+$.

partial eigenvalue assignment problem is not solved successfully, in the sense that the computed vector f does not produce a matrix $A - bf^T$ with spectrum close to $\mathbb{P} \cup \{\lambda_j\}_{j=m+1}^n$. Moreover, the computed matrix $A - bf^T$ has eigenvalues with positive real part. Algorithm 1 required four restarts in the IRA method to determine a vector f, such that all eigenvalues of $A - bf^T$ have negative real part; see Figure 3. Comparing Figures 3 and 1 shows that four restarts in the IRA method in Step 1 of Algorithm 1 give a matrix $A - bf^T$ with a spectrum that is not as close to the set $\mathbb{P} \cup \{\lambda_j\}_{j=m+1}^n$ as the spectrum of the analogous matrix determined by Algorithm 2 with only two restarts in the IRA method in Step 1. Table 1 summarizes the numerical results.

For clarity of the presentation, we omitted a few details in the above description of the experiments, such as the determination of the constants α, $\hat{\alpha}$ and β. How these details are carried out may depend on the size of the matrix A. The matrix used for the computations reported in this section is small enough to allow storage of over 30 n-vectors in fast memory. We therefore in Step 1 of Algorithms 1 and 2 carried out 30 steps of the Arnoldi process to obtain an Arnoldi decomposition (2.8) with $m = 30$. The upper Hessenberg matrix H_{30} so determined had 25 eigenvalues $\mu_1, \mu_2, \ldots, \mu_{25}$ with negative real part and five eigenvalues $\mu_{26}, \mu_{27}, \ldots, \mu_{30}$ with

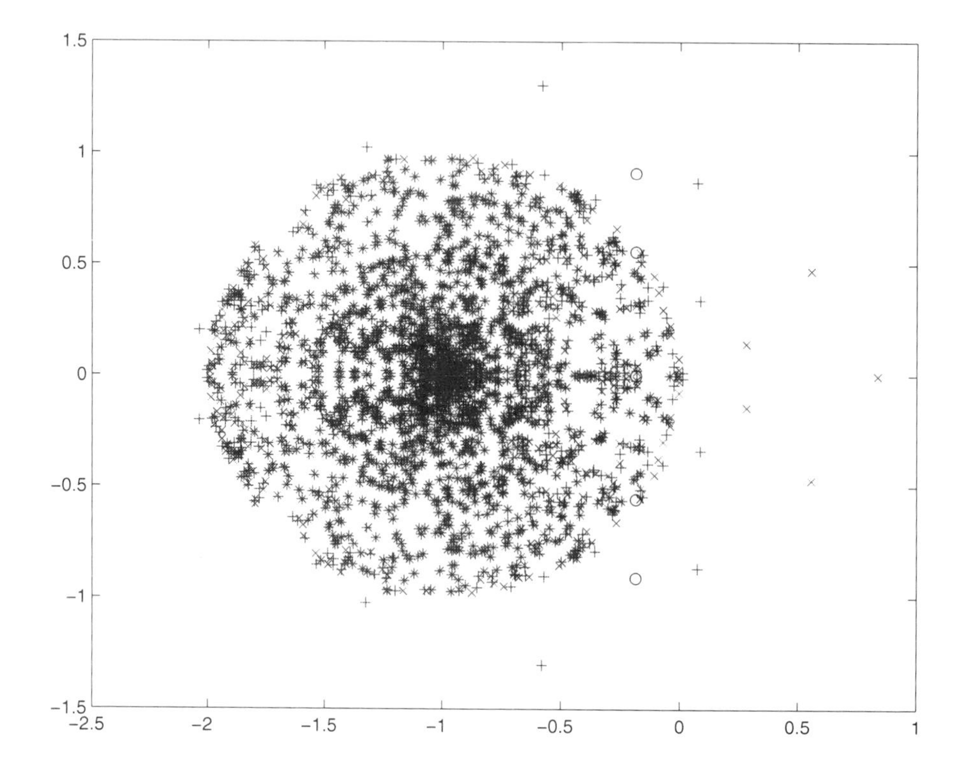

FIGURE 2. Solution determined by Algorithm 1 after two restarts in Step 1. The eigenvalues λ_j of A are marked by $\times$, the poles ψ_j by $\circ$, and the eigenvalues of $A - bf^T$ by $+$.

positive real part, and we let

$$\hat{\alpha} < \alpha = \max_{1 \leq j \leq 25} \mathrm{Re}(\mu_j), \qquad \beta = \max_{1 \leq j \leq 30} \mathrm{Im}(\mu_j).$$

The constant $\hat{\alpha}$, which together with β determines the interval $I_{\hat{\alpha},\beta}$ on which the poles were allocated, was chosen slightly smaller that α; see Figures 1-3. The value of $\hat{\alpha}$ was the same in all experiments.

We applied the eigenvalues $\mu_1, \mu_2, \ldots, \mu_{25}$ as shifts in the IRA method, as described in Section 2, to obtain an Arnoldi decomposition (2.8) with $m = 5$. The columns of the matrix V_5 in this Arnoldi decomposition did not span an invariant subspace of A associated with the eigenvalues $\lambda_1, \lambda_2, \ldots, \lambda_5$ with positive real part to desired accuracy. We therefore restarted the Arnoldi process as described in Section 2 several times until an invariant subspace associated with these eigenvalues had been determined sufficiently accurately. In each restart, we carried out five steps of the Arnoldi process to obtain an Arnoldi decomposition of the form (2.10) with $m = 5$ and then applied the eigenvalues with negative real part of the Hessenberg matrix in this decomposition as shifts.

The computations with Algorithms 1 and 2 reported in this section required a few minutes of execution time on our workstation. For comparison, we note that methods based on computing the Schur factorization of the matrix A used in our experiments require several hours of execution time on the same workstation.

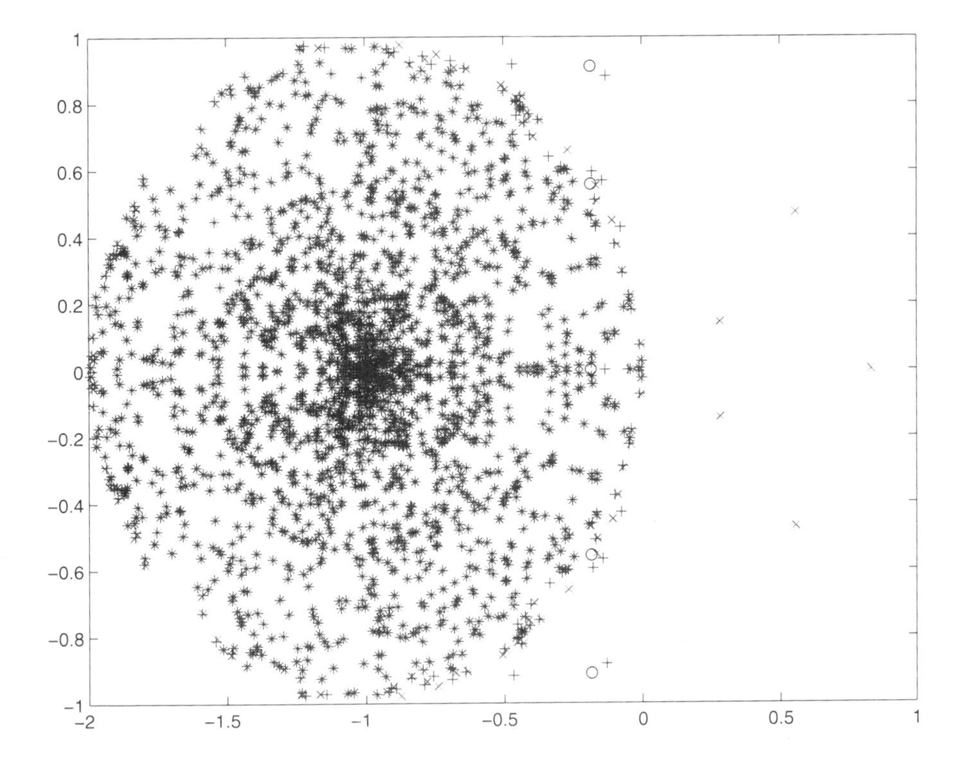

FIGURE 3. Solution determined by Algorithm 1 after four restarts in Step 1. The eigenvalues λ_j of A are marked by $\times$, the poles ψ_j by $\circ$, and the eigenvalues of $A - bf^T$ by $+$.

Quantity	Algorithm 2	Algorithm 1		
Restarts	2	2	3	4
$\|AV_m - V_mH_m\|$	$8.8 \cdot 10^{-2}$	$8.8 \cdot 10^{-2}$	$1.9 \cdot 10^{-2}$	$1.3 \cdot 10^{-4}$
$\mathrm{Re}(\lambda_{max})$	$-2.3 \cdot 10^{-3}$	$8.6 \cdot 10^{-2}$	$1.1 \cdot 10^{-2}$	$-1.7 \cdot 10^{-3}$
$d(\lambda(A - bf^T), \mathbb{P} \cup \{\lambda_j\}_{j=m+1}^n)$	$2.0 \cdot 10^{-2}$	$3.9 \cdot 10^{-1}$	$9.5 \cdot 10^{-2}$	$7.4 \cdot 10^{-2}$

TABLE 1. Comparison of Algorithms 1 and 2. The quantity λ_{max} denotes the eigenvalue of $A - bf^T$ with largest real part.

We remark that Algorithm 3 performs similarly as Algorithm 2, and we therefore do not show computed examples with the former algorithm.

5. Conclusion

This paper describes three algorithms for the partial eigenvalue assignment problem for a control system (1.1) with a large matrix A based on the implicitly restarted Arnoldi method. The algorithms do not require the matrix A to be stored or factored; it is only necessary to evaluate matrix-vector products with the matrices A or A^T.

The computed examples of Section 4 show the spectrum of the matrix $A - bf^T$ produced by Algorithm 2 to be closer to the set $\mathbb{P} \cup \{\lambda_j\}_{j=m+1}^n$ than the spectrum of the analogous matrix determined by Algorithm 1 with more restarts in the IRA method. Thus, Algorithm 2 computed a better matrix $A - bf^T$ and required fewer matrix-vector product evaluations with the matrix A^T than Algorithm 1. We found this to be typical for a large number of computed examples.

References

[1] M. Arnold and B. N. Datta, *Single-input eigenvalue assignment algorithms: A close look*, SIAM J. Matrix Anal., 19 (1998), pp. 444–467.

[2] D. Calvetti, B. Lewis and L. Reichel, *On the solution of the single input pole placement problem*, in Mathematical Theory of Netwcrks and Systems (MTNS 98), eds. A. Beghi, L. Finesso and G. Picci, Il Poliografo, Padova, 1998, pp. 585–588.

[3] ———, *On the selection of poles in the single input pole placement problem*, Linear Algebra Appl., 302-303 (1999), pp. 331–345.

[4] ———, *Partial eigenvalue assignment for large observer problems*, in Mathematical Theory of Networks and Systems (MTNS 2000), to appear.

[5] B. N. Datta, *An algorithm to assign eigenvalues in a Hessenberg matrix*, IEEE Trans. Autom. Control, AC-32 (1987), pp. 414–417.

[6] R. B. Lehoucq, D. C. Sorensen and C. Yang, *ARPACK Users' Guide: Solution of Large-Scale Eigenvalue Problems with Implicitly Restarted Arnoldi Methods*, SIAM, Philadelphia, 1998.

[7] V. Mehrmann and H. Xu, *An analysis of the pole placement problem* I. *The single-input case*, Elec. Trans. Numer. Anal., 4 (1996), pp. 89–105.

[8] ———, *An analysis of the pole placement problem* II. *The multi-input case*, Elec. Trans. Numer. Anal., 5 (1997), pp. 77–97.

[9] ———, *Choosing the poles so that the single-input pole placement is well-conditioned*, SIAM J. Matrix Anal., 19 (1998), pp. 664–681.

[10] G. Miminis and C. C. Paige, *An algorithm for pole assignment of time invariant linear systems*, Int. J. Contr., 35, 1981, pp. 341–354.

[11] G. Miminis and H. Roth, *Algorithm 747: A Fortran subroutine to solve the eigenvalue assignment problem for multiinput systems using state feedback*, ACM Trans. Math. Software, 21 (1995), pp. 299–326.

[12] Y. Saad, *Projection and deflation methods for partial pole assignment in linear state feedback*, IEEE Trans. Autom. Control, AC-33 (1988), pp. 290–297.

[13] D. C. Sorensen, *Implicit application of polynomial filters in a k-step Arnoldi method*, SIAM J. Matrix Anal. Appl., 13 (1992), pp. 357–385.

[14] W. M. Wonham, *Linear Multivariate Control: A Geometric Approach*, 3rd ed., Springer, New York, 1985.

Department of Mathematics, Case Western Reserve University, Cleveland, OH 44106

E-mail address: dxc57@po.cwru.edu

Department of Mathematics and Computer Science, Kent State University, Kent, OH 44242

E-mail address: blewis@mcs.kent.edu

Department of Mathematics and Computer Science, Kent State University, Kent, OH 44242

E-mail address: reichel@mcs.kent.edu

Contemporary Mathematics
Volume **280**, 2001

A Hybrid Method for the Numerical Solution of Discrete-Time Algebraic Riccati Equations

Heike Faßbender and Peter Benner

Abstract. A discrete-time algebraic Riccati equation (DARE) is a set of nonlinear equations. One of the oldest, best studied, numerical methods for solving it, is Newton's method. Finding a stabilizing starting guess which is already close to the desired solution is crucial. We propose to compute an approximate solution of the DARE by the (butterfly) SZ algorithm applied to the corresponding symplectic pencil where zero and infinity eigenvalues are removed using an iterative deflation strategy. This algorithm is a fast, reliable and structure-preserving algorithm for computing the stable deflating subspace of the symplectic matrix pencil associated with the DARE. From this, a stabilizing starting guess for Newton's method is easily obtained. The resulting method is very efficient and produces highly accurate results. Numerical examples demonstrate the behavior of the resulting hybrid method.

Keywords. discrete-time algebraic Riccati equation, Newton's method, SZ algorithm, symplectic matrix pencil.

1. Introduction

The standard (discrete-time) linear-quadratic optimization problem consists in finding a control trajectory $\{u(k), k = 0, 1, 2, \ldots\}$, minimizing the cost functional

$$\mathcal{J}(x_0, u) = \sum_{k=0}^{\infty} [x(k)^T Q x(k) + u(k)^T R u(k)]$$

in terms of u subject to the dynamical constraint

$$x(k+1) = Ax(k) + Bu(k), \qquad x(0) := x_0,$$

where $A \in \mathbb{R}^{n \times n}$, $B \in \mathbb{R}^{n \times m}$, $Q \in \mathbb{R}^{p \times p}$, and $R \in \mathbb{R}^{m \times m}$. Furthermore, we assume Q and R to be symmetric. Under certain conditions there is a unique control law,

$$u(k) = -\mathcal{K}(X_*)x(k), \qquad \mathcal{K}(X_*) := (R + B^T X_* B)^{-1} B^T X_* A,$$

minimizing $\mathcal{J}$ in terms of u subject to the dynamical constraint. The matrix X_* is the unique symmetric stabilizing solution of the algebraic matrix Riccati equation

$$0 = \mathcal{DR}(X) = Q - X + A^T X A - A^T X B (R + B^T X B)^{-1} B^T X A. \tag{1.1}$$

1991 *Mathematics Subject Classification.* Primary 65H10, 49N05; Secondary 15A24, 65F15.

That is, $X_* = X_*^T$ is the solution of (1.1) and all eigenvalues of $A - B\mathcal{K}(X_*)$ are inside the unit circle: $\sigma(A - B\mathcal{K}(X_*)) \subset \mathcal{D}_1(0)$, where $\mathcal{D}_1(0) = \{\zeta \in \mathbb{C}, |\zeta| < 1\}$. The equation (1.1) is usually referred to as discrete-time algebraic Riccati equation (DARE). It appears not only in the context presented, but also in numerous procedures for analysis, synthesis, and design of control and estimation systems with H_2 or H_∞ performance criteria, as well as in other branches of applied mathematics and engineering, see, e.g., [**1, 2, 3, 28, 22, 31**].

The DARE (1.1) can be considered as a set of nonlinear equations. Therefore, Newton's method has been one of the first methods proposed to solve DAREs [**19**]. Finding a stabilizing starting guess which is already close to the desired solution is crucial. It is well known that (under certain reasonable assumptions; see, e.g., Theorem 2.2) if X_0 is a stabilizing starting guess, then all iterates are stabilizing and converge globally quadratic to the desired solution X_* (see, e.g., [**19, 22, 25**]). Despite the ultimate rapid convergence, the iteration may initially converge slowly. This can be due to a large initial error $||X_* - X_0||$ or a disastrously large first Newton step resulting in a large error $||X_* - X_1||$. In both cases, it is possible that many iterations are required to find the region of rapid convergence.

Here we propose to compute an approximation $\widehat{X}$ to the stabilizing solution of $\mathcal{DR}(X)$ by the (butterfly) SZ algorithm. This solution is then used as a starting guess for Newton's method. In all our numerical experiments, roundoff errors never produced a computed $\widehat{X}$ that was not stabilizing. Hence, the computed $\widehat{X}$ always satisfied the assumptions for a starting guess needed in the convergence proof of Newton's method; see Theorem 2.2. The resulting hybrid method for solving (1.1) is a very efficient method and produces highly accurate results.

Assume R to be positive definite and define

$$L - \lambda M = \begin{bmatrix} A & 0 \\ Q & I \end{bmatrix} - \lambda \begin{bmatrix} I & -BR^{-1}B^T \\ 0 & A^T \end{bmatrix}. \tag{1.2}$$

Using furthermore the standard control-theoretic assumptions that

- (A, B) is stabilizable,
- (Q, A) is detectable,
- Q is positive semidefinite,

then $L - \lambda M$ has no eigenvalues on the unit circle and there exists a unique stabilizing solution X_* of the DARE (1.1); see, e.g., [**22**]. It is then easily seen that $L - \lambda M$ has precisely n eigenvalues in the open unit disk and n outside. Moreover, the Riccati solution X_* can be given in terms of the deflating subspace of $L - \lambda M$ corresponding to the n eigenvalues $\lambda_1, \ldots, \lambda_n$ inside the unit circle using the relation

$$\begin{bmatrix} A & 0 \\ Q & I \end{bmatrix} \begin{bmatrix} I \\ -X \end{bmatrix} = \begin{bmatrix} I & -BR^{-1}B^T \\ 0 & A^T \end{bmatrix} \begin{bmatrix} I \\ -X \end{bmatrix} \Lambda,$$

where $\Lambda \in \mathbb{R}^{n \times n}$, $\sigma(\Lambda) = \{\lambda_1, \ldots, \lambda_n\}$. Therefore, if we can compute $Y_1, Y_2 \in \mathbb{R}^{n \times n}$ such that the columns of $\left[\begin{smallmatrix} Y_1 \\ Y_2 \end{smallmatrix}\right]$ span the desired deflating subspace of $L - \lambda M$, then $X_* = -Y_2 Y_1^{-1}$ is the desired solution of the Riccati equation (1.1). See, e.g., [**22, 23, 25**], and the references therein.

It is worthwhile to note that $L - \lambda M$ of the form (1.2) is a symplectic matrix pencil. A symplectic matrix pencil $L - \lambda M, L, M \in \mathbb{R}^{2n \times 2n}$, is defined by the property

$$LJL^T = MJM^T,$$

where

$$J = \begin{bmatrix} 0 & I_n \\ -I_n & 0 \end{bmatrix},$$

and I_n is the $n \times n$ identity matrix. The nonzero eigenvalues of a symplectic matrix pencil occur in reciprocal pairs: If λ is an eigenvalue of $L - \lambda M$ with left eigenvector x, then $\overline{\lambda^{-1}}$ is an eigenvalue of $L - \lambda M$ with right eigenvector $(Jx)^H$. Hence, as we are dealing with real symplectic pencils, the finite generalized eigenvalues always occur in pairs if they are real or purely imaginary or in quadruples otherwise.

The numerical computation of a deflating subspace of a (symplectic) matrix pencil $L - \lambda M$ is usually carried out by an iterative procedure like the QZ algorithm. The QZ algorithm is numerically backward stable but it ignores the symplectic structure. Applying the QZ algorithm to a symplectic matrix pencil results in a general $2n \times 2n$ matrix pencil in generalized Schur form from which the eigenvalues and deflating subspaces can be read off. Sorting the eigenvalues in the generalized Schur form such that the eigenvalues inside the unit circle are contained in the upper left $n \times n$ block, this method results in the popular generalized Schur vector method for solving DAREs [**26**]. Due to roundoff errors unavoidable in finite-precision arithmetic, the computed eigenvalues will in general not come in pairs $\{\lambda, \lambda^{-1}\}$, although the exact eigenvalues have this property. Even worse, small perturbations may cause eigenvalues close to the unit circle to cross the unit circle such that the number of true and computed eigenvalues inside the open unit disk may differ. Moreover, the application of the QZ algorithm to $L - \lambda M$ is computationally quite expensive. The usual initial reduction to Hessenberg-triangular form requires about $70n^3$ flops plus $24n^3$ for accumulating the Z matrix; each iteration step requires about $88n^2$ flops for the transformations and $136n^2$ flops for accumulating Z; see, e.g., [**29**]. An estimated $40n^3$ flops are necessary for ordering the generalized Schur form. This results in a total cost of roughly $415n^3$ flops for computing a starting guess for Newton's method using the QZ algorithm, employing standard assumptions about convergence of the QZ iteration (see, e.g., [**17**]).

Here we propose to use the butterfly SZ algorithm for computing the deflating subspace of $L - \lambda M$. The butterfly SZ algorithm [**11, 16**] is a fast, reliable and efficient algorithm especially designed for solving the symplectic eigenproblem. It makes use of the fact that symplectic matrix pencils can be reduced to matrix pencils of the form

$$K - \lambda N = \begin{bmatrix} C & F \\ 0 & C^{-1} \end{bmatrix} - \lambda \begin{bmatrix} 0 & -I_n \\ I_n & T \end{bmatrix}, \tag{1.3}$$

where C, F are diagonal and T is symmetric tridiagonal. This form is determined by just $4n-1$ parameters. The symplectic matrix pencil $K - \lambda N$ is called a symplectic butterfly pencil. By exploiting this special reduced form and the symplecticity, the SZ algorithm is fast and efficient; in each iteration step only $O(n)$ arithmetic operations are required instead of $O(n^2)$ arithmetic operations for a QZ step. We thus save a significant amount of work. Of course, the accumulation of the Z matrix requires $O(n^2)$ arithmetic operations as in the QZ step. Moreover, by forcing the symplectic structure the above mentioned problems of the QZ algorithm are avoided. Using the so obtained solution as a starting guess for Newton's method, the resulting method for solving discrete-time algebraic Riccati equations is a very efficient method and produces highly accurate results. However, it should be noted that in the SZ algorithm non-orthogonal equivalence transformations have to be used. These are not as numerically stable as the orthogonal transformations used by the QZ algorithm. Therefore, the approximate DARE solution computed by the SZ algorithm is sometimes less accurate than the one obtained from using the QZ algorithm. Therefore one could expect that more Newton steps are necessary to refine the initial guess resulting from the SZ algorithm to maximum accuracy than needed when employing the QZ algorithm. But due to the quadratic convergence of Newton's method in a vicinity of the solution, this is rarely the case.

In Section 2 Newton's method is reviewed. Section 3 briefly describes the (butterfly) SZ algorithm for computing the deflating subspace of $L - \lambda M$. Combined with a strategy to deflate zero and infinity eigenvalues from the symplectic pencil in order to deal with discrete-time algebraic Riccati equations with singular A matrix, the hybrid method described in Section 4 consisting of the SZ algorithm followed by a few Newton iteration steps results in an efficient and accurate method for solving discrete-time algebraic Riccati equations. Numerical experiments are reported in Section 5.

2. Newton's method

The function $\mathcal{DR}(X)$ is a rational matrix function and $\mathcal{DR}(X) = 0$ defines a system of nonlinear equations. Hence it is straightforward to apply Newton's method to DAREs. Inspired by Kleinman's formulation of Newton's method for continuous-time algebraic Riccati equations [**20**], Hewer [**19**] proposed an analogous scheme for solving DAREs. A discussion of its convergence properties can be found in [**25, 22**].

Given a symmetric matrix X_0, the method can be given in algorithmic form as follows:

ALGORITHM 2.1.
FOR $k = 0, 1, 2, \ldots$
1. $K_k \leftarrow \mathcal{K}(X_k) = (R + B^T X_k B)^{-1} B^T X_k A$.
2. $A_k \leftarrow A - BK_k$.
3. $\mathcal{R}_k \leftarrow \mathcal{DR}(X_k)$.
4. Solve for N_k in the Stein equation

$$A_k^T N_k A_k - N_k = -\mathcal{R}_k. \tag{2.1}$$

5. $X_{k+1} \leftarrow X_k + N_k$.

END FOR

We have the following result for Algorithm 2.1 [**19, 25, 22**].

THEOREM 2.2. *If*

i) (A, B) *is stabilizable,*
ii) $R = R^T \geq 0$,
iii) *a unique stabilizing solution* X_* *of (1.1) exists such that* $R + B^T X_* B > 0$,
iv) X_0 *is stabilizing,*

then for the iterates produced by Algorithm 2.1 we have:

a) *All iterates* X_k *are stabilizing, i.e.,* $\sigma(A - B\mathcal{K}(X_k)) \subset \mathcal{D}_1(0)$ *for all* $k \in \mathbb{N}_0$.
b) $X_* \leq \ldots \leq X_{k+1} \leq X_k \leq \ldots \leq X_1$.

c) $\lim_{k\to\infty} X_k = X_*$.

d) *There exists a constant* $\gamma > 0$ *such that*

$$\|X_{k+1} - X_*\| \le \gamma \|X_k - X_*\|^2, \qquad k \ge 1,$$

i.e., the X_k *converge globally quadratic to* X_*.

The formulation of Algorithm 2.1 is analogous to the standard formulation of Newton's method as given, e.g., in [**15**, Algorithm 5.1.1] for the solution of nonlinear equations. Because of its robustness in the presence of rounding errors, we prefer to calculate the Newton step explicitly as in Algorithm 2.1 rather than to use the mathematically equivalent formulation of the Newton step [**19, 25, 22**],

$$A_k^T X_{k+1} A_k - X_{k+1} = -Q - K_k^T R K_k \tag{2.2}$$

which determines X_{k+1} directly. The coefficient matrices of the two Stein equations are the same, but the right-hand-sides are different. Define the Stein operator as the linear map

$$\Gamma_k : Z \longrightarrow (A - B\mathcal{K}(X_k))^T Z + Z(A - B\mathcal{K}(X_k)), \tag{2.3}$$

which is also the Fréchet derivative of $\mathcal{DR}(X)$ at X_k. Let us assume that the condition number of Γ_k permits us to solve the Stein equations (2.1) and (2.2) to ℓ correct significant digits. Loosely speaking this implies that when X_{k+1} is calculated directly as in (2.2), then its accuracy is limited to ℓ significant digits. On the other hand, in Algorithm 2.1, the accuracy of the computed Newton step N_k is limited to ℓ significant digits. Therefore, the sum $X_k + N_k$ has up to ℓ *more* correct digits than X_k. The accuracy of Algorithm 2.1 is ultimately limited only by the accuracy to which $\mathcal{K}(X_k)$, A_k, $\mathcal{DR}(X_k)$, and the sum $X_k + N_k$ are calculated.

The computational cost for Algorithm 2.1 mainly depends upon the cost for the numerical solution of the Stein equation (2.1). This can be done using the Bartels–Stewart algorithm [**7, 6**]. Then the cost for the solution of the Stein equation is about $32n^3$ flops. The computations for forming A_k and $\mathcal{R}_k$ can be arranged such that for $n \approx m$, they require $\approx 17n^3$ flops while for $m \ll n$, these matrices can be formed using only $\approx 3n^3$ flops. For an average value of $m = n/2$, the computational cost for one step of Algorithm 2.1 is about $42n^3$ flops.

One major difficulty is to find a stabilizing initial guess X_0. There exist stabilization procedures for discrete-time linear systems (see, e.g., [**4, 21, 29**]). But these may give large initial errors $\|X_* - X_0\|$ (see, e.g., [**8**]). The procedure suggested in [**21**] is even infeasible for numerical computations as it is based on explicitly summing up $A^k BB^T (A^T)^k$ for k up to n, thereby often causing overflow already for small values of n. This problem can be overcome in case A is stable. In that case, one can start from $X_0 = 0$.

Despite the ultimate rapid convergence indicated by Theorem 2.2 d), the iteration may initially converge slowly. This can be due to a large initial error $\|X_* - X_0\|$ or a disastrously large first Newton step resulting in a large error $\|X_* - X_1\|$. In both cases, it is possible that many iterations are required to find the region of rapid convergence. An ill-conditioned Stein equation makes it difficult to compute an accurate Newton step. An inaccurately computed Newton step can cause the usual convergence theory to break down in practice. Sometimes rounding errors or a poor choice of X_0 cause Newton's method to converge to a non-stabilizing solution.

For these reasons, Newton's method is usually not used by itself to solve DAREs. However, when it is used as a defect correction method or for iterative refinement of an approximate solution obtained by a more robust method, it is often able to squeeze out the maximum possible accuracy [**25**] after only one or two iterations. Therefore we propose here to find a stabilizing initial guess using the butterfly SZ algorithm.

An approach to overcome slow initial convergence is suggested in [**8, 9**]. There, a line search strategy is suggested that usually accelerates the convergence during the first iteration steps. The strategy is particularly successful if A is known to be stable and one can start from $X_0 = 0$. Still, for an unstable A matrix, this procedure relies on some initial stabilization and even for a stable A matrix, $X_0 = 0$ may result in a bad first step. Therefore this approach will also benefit from a good starting guess. Note, however, that most frequently a starting guess obtained from the SZ algorithm is so close to the desired solution that line search will usually not improve Newton's method significantly. We will therefore not discuss this topic here any further. Still, in case the problem of solving the DARE via the deflating subspace approach is ill-conditioned, line search may improve the convergence behavior.

3. The butterfly SZ algorithm

For simplicity let us assume at the moment that A in (1.2) is nonsingular. Premultiplying $L - \lambda M$ by $\left[\begin{smallmatrix} I & 0 \\ 0 & A^{-T} \end{smallmatrix}\right]$ results in a symplectic matrix pencil

$$L' - \lambda M' = \begin{bmatrix} A & 0 \\ A^{-T}Q & A^{-T} \end{bmatrix} - \lambda \begin{bmatrix} I & -BR^{-1}B^T \\ 0 & I \end{bmatrix}, \tag{3.1}$$

where L', M' are both symplectic, that is,

$$L'JL'^T = M'JM'^T = J,$$

as a matrix $X \in \mathbb{R}^{n \times n}$ is symplectic if $XJX^T = J$. In [**11, 16**] it is shown that for the symplectic matrix pencil $L' - \lambda M'$ there exist numerous symplectic matrices Z and nonsingular matrices S which reduce $L' - \lambda M'$ to a symplectic butterfly pencil $K - \lambda N$ (1.3)

$$S(L' - \lambda M')Z = K - \lambda N = \begin{bmatrix} C & F \\ 0 & C^{-1} \end{bmatrix} - \lambda \begin{bmatrix} 0 & -I \\ I & T \end{bmatrix},$$

where C and F are diagonal matrices, and T is a symmetric tridiagonal matrix. (More general, not only the symplectic matrix pencil in (3.1), but any symplectic matrix pencil $L' - \lambda M'$ with symplectic matrices L', M' can be reduced to a symplectic butterfly pencil). If T is an unreduced tridiagonal matrix, then the butterfly pencil is called unreduced. If any of the $n - 1$ subdiagonal elements of T are zero, the problem can be split into at least two problems of smaller dimension, but with the same symplectic butterfly structure.

Once the reduction to a symplectic butterfly pencil is achieved, the SZ algorithm is a suitable tool for computing the eigenvalues/deflating subspaces of the symplectic pencil $K - \lambda N$ [**11, 16**]. The SZ algorithm preserves the symplectic butterfly form in its iterations. It is the analogue of the SR algorithm (see [**10, 16**]) for the generalized eigenproblem, just as the QZ algorithm is the analogue of the QR algorithm for the generalized eigenproblem. Both are instances of the GZ algorithm [**30**].

Each iteration step begins with an unreduced butterfly pencil $K - \lambda N$. Choose a spectral transformation function q and compute a symplectic matrix Z_1 such that

$$Z_1^{-1} q(K^{-1}N)e_1 = \alpha e_1$$

for some scalar α. Then transform the pencil to

$$\widetilde{K} - \lambda \widetilde{N} = (K - \lambda N)Z_1.$$

This introduces a bulge into the matrices $\widetilde{K}$ and $\widetilde{N}$. Now transform the pencil to

$$\widehat{K} - \lambda \widehat{N} = S^{-1}(\widetilde{K} - \lambda \widetilde{N})\widetilde{Z},$$

where $\widehat{K} - \lambda \widehat{N}$ is of symplectic butterfly form. S and $\widetilde{Z}$ are symplectic, and $\widetilde{Z}e_1 = e_1$. This concludes the iteration. Under certain assumptions, it can be shown that the butterfly SZ algorithm converges cubically. The needed assumptions are technically involved and follow from the GZ convergence theory developed in [**30**]. The convergence theorem says roughly that if the eigenvalues are separated, and the shifts converge, and the condition numbers of the accumulated transformation matrices remain bounded, then the SZ algorithm converges. We refrain from listing the theorem here as it is not illuminating in the context of the proposed hybrid method. For a detailed discussion of the butterfly SZ algorithm see [**11, 16**].

Hence, in order to compute an approximate solution of the DARE (1.1) by the butterfly SZ algorithm, first the symplectic matrix pencil $L - \lambda M$ as in (1.2) has to be formed, then the symplectic matrix pencil $L' - \lambda M'$ as in (3.1) is computed. Next symplectic matrices Z_0 and S_0 are computed such that

$$\widehat{L} - \lambda \widehat{M} := S_0^{-1} L' Z_0 - \lambda S_0^{-1} M' Z_0$$

is a symplectic butterfly pencil. Using the butterfly SZ algorithm, symplectic matrices Z_1 and S_1 are computed such that

$$S_1^{-1} \widehat{L} Z_1 - \lambda S_1^{-1} \widehat{M} Z_1$$

is a symplectic butterfly pencil and the symmetric tridiagonal matrix $\widehat{T}$ in the lower right block of $S_1^{-1}\widehat{M}Z_1$ is reduced to quasi-diagonal form with 1×1 and 2×2 blocks on the diagonal. The eigenproblem decouples into a number of simple 2×2 or 4×4 generalized symplectic eigenproblems. Solving these subproblems, finally symplectic matrices Z_2, S_2 are computed such that

$$\check{L} = S_2^{-1} S_1^{-1} \widehat{L} Z_1 Z_2 = \begin{bmatrix} \phi_{11} & \phi_{12} \\ 0 & \phi_{22} \end{bmatrix},$$

$$\check{M} = S_2^{-1} S_1^{-1} \widehat{M} Z_1 Z_2 = \begin{bmatrix} \psi_{11} & \psi_{12} \\ 0 & \psi_{22} \end{bmatrix},$$

where the eigenvalues of the matrix pencil $\phi_{11} - \lambda \psi_{11}$ are precisely the n stable generalized eigenvalues. Let $Z = Z_0 Z_1 Z_2$. Partitioning Z conformably,

$$Z = \begin{bmatrix} Z_{11} & Z_{12} \\ Z_{21} & Z_{22} \end{bmatrix}, \tag{3.2}$$

the Riccati solution X_* is found by solving a system of linear equations:

$$X_* = -Z_{21} Z_{11}^{-1}. \tag{3.3}$$

This algorithm requires about $193n^3$ arithmetic operations in order to compute the desired deflating subspace of $L - \lambda M$ and is therefore cheaper than the QZ algorithm which requires about $415n^3$ arithmetic operations (neither flop count takes into account the cost for forming $L - \lambda M$ and the cost for solving the linear system (3.3) at the end, as these steps are the same for both algorithms). The cost of the different steps of the approach described above are given as follows. The computation of $A^{-T}Q$ and A^{-T} using an LR decomposition of A requires about $\frac{14}{3}n^3$ arithmetic operations. A careful flop count reveals that the initial reduction of $L' - \lambda M'$ to butterfly form $\widehat{L} - \lambda\widehat{M}$ requires about $75n^3$ arithmetic operations. For computing Z_0, an additional $28n^3$ arithmetic operations are needed. The butterfly SZ algorithm requires about $O(n^2)$ arithmetic operations for the computation of $\breve{L} - \lambda\breve{M}$ and additional $85n^3$ arithmetic operations for the computation of Z (this estimate is based on the assumption that $\frac{2}{3}$ iterations per eigenvalue are necessary as observed in [**11**]). Hence, the entire algorithm described, requires about $\frac{578}{3}n^3$ arithmetic operations.

Instead of generating the symplectic matrix Z as in (3.2), one can work with $n \times n$ matrices X, Y and T such that finally $X = -Z_{21}Z_{11}^{-1} = X_*$, $Y = Z_{11}^{-1}Z_{12}$, and $T = Z_{11}^{-1}$. Starting from $X = Y = 0, T = I$, this can be implemented without accumulating the intermediate symplectic transformations used in the butterfly SZ algorithm, just using the parameters that determine these transformations. As for every symplectic matrix Z written in the form (3.2), $Z_{21}Z_{11}^{-1}$ is symmetric, this approach guarantees that all intermediate (and the final) X are symmetric. Such an approach, called symmetric updating, was first proposed by Byers and Mehrmann [**14**] in the context of solving continuous-time algebraic Riccati equations via the Hamiltonian SR algorithm and has also been proposed for solving DAREs with an SR algorithm in [**5**]. A flop count shows that this approach is not feasible here. For the initial reduction of $L' - \lambda M'$ to butterfly form about $5n^4$ arithmetic operations are needed in order to compute X, Y, and Z. The butterfly SZ algorithm requires about $11n^4$ arithmetic operations for the symmetric updating. Hence, the symmetric updating requires $\mathcal{O}(n^4)$ arithmetic operations. Moreover, this increase in computational cost is not rewarded with a significantly more accurate computed solution X.

So far, we have assumed that A is nonsingular such that the symplectic pencil $L' - \lambda M'$ (3.1) can be formed. In case A is singular, the symplectic matrix pencil $L - \lambda M$ (1.2) has zero and infinity eigenvalues. Mehrmann proposes in [**25**, Algorithm 15.16] the following algorithm to deflate zero and infinite eigenvalues of $L - \lambda M$: Assume that $\operatorname{rank}(A) = k$. First, use the QR decomposition with column pivoting [**17**] to determine an orthogonal matrix $V \in \mathbb{R}^{n\times n}$, an upper triangular matrix $U \in \mathbb{R}^{n\times n}$ and a permutation matrix $P \in \mathbb{R}^{n\times n}$ such that

$$PA = UV = [0\ U_2]\begin{bmatrix} V_1 \\ V_2 \end{bmatrix},$$

where $U_2 \in \mathbb{R}^{n\times n-k}$ and $V_2 \in \mathbb{R}^{n-k\times n}$ have full rank. Then form

$$T = \begin{bmatrix} V^T & 0 \\ -QV^T & V^T \end{bmatrix}, \quad \text{and} \quad \widetilde{T} = \begin{bmatrix} V(I+GQ)^{-1} & 0 \\ U^TPQ(I+GQ)^{-1} & V \end{bmatrix},$$

where for ease of notation G denotes $BR^{-1}B^T$. Now as $AV^T = P^TU$, we obtain

$$\widetilde{T}(L - \lambda M)T =$$
$$= \begin{bmatrix} V(I+GQ)^{-1}P^TU & 0 \\ U^TPQ(I+GQ)^{-1}P^TU & I \end{bmatrix} - \lambda \begin{bmatrix} I & -V(I+GQ)^{-1}GV^T \\ 0 & U^TP(-Q(I+GQ)^{-1}G+I)V^T \end{bmatrix}.$$

Further, as $P^T U = P^T [0\ U_2]$,

$$\widetilde{T}(L - \lambda M)T = \left[\begin{array}{cc|cc} 0 & \widetilde{A}_1 & 0 & 0 \\ 0 & \widetilde{A} & 0 & 0 \\ \hline 0 & 0 & I & 0 \\ 0 & \widetilde{Q} & 0 & I \end{array}\right] - \lambda \left[\begin{array}{cc|cc} I & 0 & \widetilde{G}_{11} & \widetilde{G}_{12} \\ 0 & I & \widetilde{G}_{21} & \widetilde{G} \\ \hline 0 & 0 & 0 & 0 \\ 0 & 0 & \widetilde{A}_1^T & \widetilde{A}^T \end{array}\right]$$

where $\widetilde{A}_1, \widetilde{G}_{12} \in \mathbb{R}^{k \times n-k}$, $\widetilde{A}, \widetilde{Q}, \widetilde{G} \in \mathbb{R}^{n-k \times n-k}$. The first k columns of T span the right deflating subspace of $L - \lambda M$ corresponding to k zero eigenvalues and the rows $n+1, n+2, \ldots, n+k$ of $\widetilde{T}$ span the left deflating subspace corresponding to k infinity eigenvalues. We may therefore delete rows and columns $1, 2, \ldots, k$, $n+1, n+2, \ldots, n+k$ and proceed with the reduced pencil

$$\begin{bmatrix} \widetilde{A} & 0 \\ \widetilde{Q} & I \end{bmatrix} - \lambda \begin{bmatrix} I & \widetilde{G} \\ 0 & \widetilde{A} \end{bmatrix}.$$

There is no guarantee that $\widetilde{A}$ is nonsingular. Hence, the procedure described above has to be repeated until the resulting symplectic matrix pencil has no more zero and infinity eigenvalues and $\widetilde{A}$ is nonsingular. Note that neither the rank of A nor the number of zero eigenvalues of A determine the number of zero and infinity eigenvalues of the symplectic pencil. For instance, in Example 10 of [**12**], $n = 6$, $\operatorname{rank}(A) = 5$, A has three zero eigenvalues and $L - \lambda M$ as in (1.2) has two zero and infinite eigenvalues each.

All the computations in this algorithm can be carried out in a numerically reliable way. The solution of the linear systems with $I + GQ$ is well-conditioned, since Q and G are symmetric positive semidefinite.

If after the first iteration, $\widetilde{A}$ is nonsingular, then this process requires $(11n^2 + 6rn + 2r^2)n$ flops; the initial QR decomposition in order to check the rank of A costs $\frac{4}{3}n^3$ flops. Note that this initial decomposition is always computed. But in case A has full rank, then the symplectic pencil $L' - \lambda M'$ in (3.1) is formed using the original data in order not to introduce unnecessary rounding errors. In doing so, the QR decomposition of the A matrix should be used when computing $A^{-T}Q$ and A^{-T} instead of computing an LR decomposition of A.

Combined with a strategy to deflate zero and infinity eigenvalues from the symplectic pencil in order to deal with discrete-time algebraic Riccati equations with singular A matrix, the butterfly SZ algorithm is an efficient tool to compute a starting guess for Newton's method. Usually the same number of iterations is required as when refining an approximation computed by the generalized Schur vector method. Even if one or two iterations more are necessary due to the loss of accuracy caused by using non-orthogonal transformations, this is well compensated by the cheaper SZ iteration.

4. A hybrid method for DAREs

Combining the strategy to deflate zero and infinity eigenvalues from the symplectic pencil with the SZ algorithm followed by a few Newton iteration steps results in an efficient and accurate hybrid method.

Altogether, we propose the following algorithm to solve the DARE (1.1).

ALGORITHM 4.1.

Input: The coefficient matrices $A \in \mathbb{R}^{n \times n}$, $B \in \mathbb{R}^{n \times m}$, $Q = Q^T \in \mathbb{R}^{n \times n}$, and $R \in \mathbb{R}^{m \times m}$.

Output: An approximation $\tilde{X} = \tilde{X}^T \in \mathbb{R}^{n \times n}$ to the stabilizing solution of the DARE.

(1) Form the symplectic pencil $L - \lambda M$ as in (1.2).

(2) Use the procedure described in Section 3 to deflate all zero and infinite eigenvalues of $L - \lambda M$. That is, compute a nonsingular transformation matrix T_1 and a symplectic matrix S_1 such that

$$T_1(L - \lambda M)S_1 = \begin{bmatrix} 0 & \tilde{A}_1 & 0 & 0 \\ 0 & \tilde{A} & 0 & 0 \\ 0 & 0 & I_k & 0 \\ 0 & \tilde{Q} & 0 & I_{n-k} \end{bmatrix} - \lambda \begin{bmatrix} I_k & 0 & -\tilde{G}_{11} & -\tilde{G}_{12} \\ 0 & I_{n-k} & -\tilde{G}_{12}^T & -\tilde{G}_{22} \\ 0 & 0 & 0 & 0 \\ 0 & 0 & \tilde{A}_1^T & \tilde{A}^T \end{bmatrix}$$

with $\tilde{A}$ nonsingular and the first k columns of S_1 span the deflating subspace of $L - \lambda M$ corresponding to all zero eigenvalues.

(3) Apply the (quadruple shift) butterfly SZ algorithm described in Section 3 (and in detail in [**11**]) to the symplectic pencil

$$\tilde{L} - \lambda \tilde{M} := \begin{bmatrix} I_{n-k} & 0 \\ 0 & \tilde{A}^{-T} \end{bmatrix} \begin{bmatrix} \tilde{A} & 0 \\ \tilde{Q} & I_{n-k} \end{bmatrix} - \lambda \begin{bmatrix} I_{n-k} & -\tilde{G}_{22} \\ 0 & \tilde{A}^T \end{bmatrix}$$

such that

$$\tilde{T}_2(\tilde{L} - \lambda \tilde{M})\tilde{S}_2 = \begin{bmatrix} \phi_{11} & \phi_{12} \\ 0 & \phi_{22} \end{bmatrix} - \lambda \begin{bmatrix} \psi_{11} & \psi_{12} \\ 0 & \psi_{22} \end{bmatrix},$$

where the eigenvalues of $\phi_{11} - \lambda\psi_{11}$ are the stable nonzero eigenvalues of $L - \lambda M$.

(4) Partition $\tilde{S}_2 = \left[\begin{smallmatrix} S_{11} & S_{12} \\ S_{21} & S_{22} \end{smallmatrix}\right]$ where $S_{jj} \in \mathbb{R}^{n-k \times n-k}$, $j = 1, 2$. Set

$$Z := S_1 \begin{bmatrix} I_k & 0 & 0 & 0 \\ 0 & S_{11} & 0 & S_{12} \\ 0 & 0 & I_k & 0 \\ 0 & S_{21} & 0 & S_{22} \end{bmatrix}.$$

Then the first n columns of Z span the stable deflating subspace of $L - \lambda M$ and an approximate solution $\widehat{X}$ of the DARE can be computed as in (3.3).

(5) Use Newton's method (possibly endowed with a line search strategy as proposed in [**9**]) and starting guess $X_0 = \widehat{X}$ in order to iteratively refine the solution of the DARE to the highest achievable accuracy.

Note that all left transformation matrices need not be accumulated. The flop count for Algorithm 4.1 can be summarized assuming $k = 0$ and neglecting low-order terms: Steps 1. and 2. need $m\left(\frac{1}{2}n^2 + nm + \frac{1}{3}m^2\right)$ and $\frac{4}{3}n^3$ flops, respectively, Step 3. requires $\frac{576}{3}n^3$ flops (using the QR decomposition of A computed in Step 2.) while $\frac{8}{3}n^3$ flops are needed to solve the linear system in the 4. Step. One Newton iteration requires setting up the coefficient matrices of the Stein equation and then solving it. The coefficient matrices $A - BK_k$ and $\mathcal{DR}(X_k)$ are computed making use of common intermediate results. There are several options to implement these computations. Here we choose a version that amounts to $3n^3 + m\left(6n^2 + 5mn + \frac{1}{3}m^2\right)$ flops (see also [**8**]). When the Stein equation is solved using a discrete version of the Bartels-Stewart method [**7, 6**], the cost can be estimated as $32n^3$ flops based on flop counts for the QR algorithm given in [**17**]. In most applications, $m \ll n$ such that the cost of one Newton step is then about $35n^3$. If ℓ Newton steps are required for iterative refinement, the total cost of the hybrid method is given by

$$(196 + 3\ell)n^3 + m\left((6\ell + 0.5)n^2 + (5\ell + 1)mn + \frac{\ell + 1}{3}m^2\right) \text{ flops.}$$

Assuming $m \ll n$, this flop counts reveal that 5-6 Newton iterations are needed to reach the cost of the generalized Schur vector method. If compared to an iteratively refined generalized Schur vector method, this comparison becomes even more favorable. If the same number of Newton iterations is required, the hybrid method is almost twice as fast as the iteratively refined generalized Schur vector method.

Also note that as long as the SZ algorithm yields a stabilizing starting guess $X_0 = \widehat{X}$, the hybrid algorithm can be considered as numerically backward stable as it computes the solution to a nearby DARE; see [**25**, § 10]. This is due to the fact that the Newton iteration as formulated in Section 2 can be considered as a defect correction method for DAREs and the ultimate accuracy obtained by Newton's method is only limited by condition number of the DARE; see, e.g., [**27**].

5. Numerical Experiments

We have implemented Algorithm 4.1 as MATLAB[1] functions. The implementation of Newton's method for DAREs used here is described in [**8**]. As stopping criterion for Newton's method we used a criterion based on a normalized residual of the form

$$\|\mathcal{DR}(X_k)\|_F \leq n\varepsilon\|X_k\| \max\{\|A\|_F, \|B\|_F, \|R\|_F, \|Q\|_F\},$$

where ε is the machine precision.

In this section we compare the results obtained by Algorithm 4.1 with those obtained from the Schur vector method as proposed in [**26**], i.e., applying the QZ algorithm to the symplectic pencil $L - \lambda M$ from (1.2) and re-ordering the eigenvalues appropriately, followed by Newton's method in the same implementation as used in our hybrid method. In the tests, results of Newton's method without line search are reported. Only in Example 14, line search reduced the number of required Newton iterations by one in both methods.

The two algorithms are compared for the examples from the DARE benchmark collection [**12, 13**]. We do not give results for Examples 3 and 4 as there, R is singular and hence the approach via the symplectic pencil is not possible.

All computations were done using MATLAB Version 5.3 [**24**] under Solaris 7 on a Sun Ultra 10 Model 300 workstation with a 300 MHz Sun UltraSPARC-IIi CPU using IEEE double precision arithmetic. The machine precision was $\varepsilon \approx 2.2204 \times 10^{-16}$.

Table 1 reports the residual norms $\|\mathcal{DR}(X)\|_F$ for the solutions computed by the SZ and Schur vector algorithms before and after applying Newton's method and `NIT`, the number of Newton steps required to attain the final accuracy. We report absolute residual norms as any normalization applies in the same way to all results. Large absolute residuals (e.g., QZ solution in Example 12, SZ solution in Example 13) do not indicate failure of the method but can be explained by a large norm of the solution or any of the coefficients in the DARE. We therefore also

[1]MATLAB is a trademark of The MathWorks, Inc.

report the Frobenius norm of the exact solution, $\|X\|_F$. In case the exact solution is not known, the norm of the approximate solution computed by the hybrid method is reported. We do not report norms of the coefficient matrices as in all cases with large absolute residual norm, $\|X\|_F$ is large, too, which explains the large values. Moreover, the size n of the problem as well as the condition numbers of A and of the DARE are also provided. The condition K_{DARE} of the DARE is estimated using the method proposed in [**18**]. If cond $(A) = \|A\|_2 \|A^{-1}\|_2 = \infty$, then A is singular. In that case, the number in parentheses indicates the number of zero eigenvalues of the symplectic pencil $L - \lambda M$ in (1.2). Recall that these have to be deflated before applying the SZ algorithm. Table 2 gives information about relative errors for those examples where the exact solution is known. The information about the example and the number of iteration steps needed by Newton's method is not duplicated, though. In both tables X_* denotes the exact stabilizing solution, X_{SZ} and X_{QZ} are the initial guesses for Newton's method obtained from the butterfly SZ algorithm and the Schur vector method, respectively, and $X_{\texttt{NIT}}$ is the final approximation to X_* after `NIT` iteration steps of Newton's method.

From both tables it can be seen that for all examples, the final accuracy is similar for both methods. This is to be expected as the accuracy obtained by Newton's method is only limited by the conditioning of the DARE and not by the initial error $\|X_* - X_0\|_F$. The zero residuals in Examples 5, 12, 15 obtained by the hybrid method are due to the deflation of zero and infinite eigenvalues. In these cases no reduction to butterfly form and no SZ iteration are necessary to compute the solution. Moreover, comparing the residuals obtained by the SZ and the QZ algorithm, it can be seen that by using non-orthogonal transformations (in the SZ algorithm) the approximate solution may not be as accurate as when using only orthogonal transformations (in the QZ algorithm). Nevertheless, in most cases, the number of Newton steps needed to refine the computed approximate solutions of the SZ and QZ algorithms is equal or differs by one only. Only in Example 11, two Newton steps are needed by the hybrid method while none are required for the Schur vector solution. But even in this case, given the flop counts from the previous sections, Algorithm 4.1 still computes the solution faster than the Schur vector method. That is, the hybrid method is always faster than the Schur vector method with iterative refinement based on Newton's method. Note that for all examples with known exact solution, the attained accuracy is of order machine precision times condition of the DARE.

6. Concluding remarks

We have discussed the numerical solution of discrete-time Riccati equations. By initializing Newton's method with a starting guess computed by the butterfly SZ algorithm combined with a method for deflating zero and infinite eigenvalues to the corresponding symplectic matrix pencil, an efficient and accurate hybrid method is derived. This method is usually about twice as fast as the generalized Schur vector method with iterative refinement by Newton's method, while the accuracy obtained is in both cases only limited by the conditioning of the DARE and not by the SZ or QZ method.

Ex.	n	cond (A)	K_{DARE}	$\|X\|_F$	Algorithm 4.1			Schur vector method		
					SZ	Newton	NIT	QZ	Newton	NIT
					$\|\mathcal{DR}(X_{SZ})\|_F$	$\|\mathcal{DR}(X_{\mathtt{NIT}})\|_F$		$\|\mathcal{DR}(X_{QZ})\|_F$	$\|\mathcal{DR}(X_{\mathtt{NIT}})\|_F$	
1	2	1.2×10^{2}	18.9	21.0	1.9×10^{-13}	3.0×10^{-14}	1	4.8×10^{-14}	4.8×10^{-14}	0
2	2	1.1	4.7	5.2×10^{-2}	1.8×10^{-16}	1.8×10^{-16}	0	1.0×10^{-16}	1.0×10^{-16}	0
5	2	∞ (1)	1.9	5.2	0	0	0	1.8×10^{-15}	1.8×10^{-15}	0
6	4	1.0	30.6	43.1	6.9×10^{-12}	7.2×10^{-15}	1	1.7×10^{-13}	7.6×10^{-15}	1
7	4	19.9	7.9×10^{2}	2.8	2.4×10^{-12}	2.4×10^{-16}	1	1.7×10^{-15}	1.7×10^{-15}	0
8	4	3.8×10^{2}	5.1×10^{4}	65.8	1.2×10^{-12}	1.2×10^{-12}	0	1.7×10^{-13}	1.7×10^{-13}	0
9	5	23.5	1.0×10^{2}	75.4	2.5×10^{-9}	1.1×10^{-14}	1	2.1×10^{-13}	1.5×10^{-14}	1
10	6	∞ (2)	3.9	3.8	8.5×10^{-15}	8.5×10^{-15}	0	4.0×10^{-15}	4.0×10^{-15}	0
11	9	1.6×10^{6}	74.2	8.1×10^{2}	1.6×10^{-2}	1.2×10^{-13}	2	2.7×10^{-12}	2.7×10^{-12}	0
12	2	∞ (2)	2.7	1.0×10^{12}	0	0	0	7.8×10^{7}	0	1
13	3	∞ (1)	2.5	9.3×10^{6}	3.5	2.7×10^{-7}	1	5.2×10^{-8}	5.2×10^{-8}	0
14	4	∞ (3)	1.8×10^{8}	3.1×10^{7}	4.0×10^{-2}	2.2×10^{-8}	2	2.6×10^{-3}	0	2
15	100	∞ (100)	2.8×10^{2}	5.8×10^{2}	0	0	0	2.4×10^{-12}	2.4×10^{-12}	0

TABLE 1. Residual norms of SZ and Schur vector solutions before and after iterative refinement.

Ex.	Algorithm 4.1		Schur vector method	
	SZ	Newton	QZ	Newton
	$\frac{\|X_{SZ}-X_*\|_F}{\|X_*\|_F}$	$\frac{\|X_{\text{NIT}}-X_*\|_F}{\|X_*\|_F}$	$\frac{\|X_{QZ}-X_*\|_F}{\|X_*\|_F}$	$\frac{\|X_{\text{NIT}}-X_*\|_F}{\|X_*\|_F}$
1	9.4×10^{-15}	6.1×10^{-15}	4.5×10^{-16}	4.5×10^{-16}
5	0	0	3.4×10^{-16}	3.4×10^{-16}
12	0	0	7.8×10^{-5}	0
13	4.2×10^{-7}	3.2×10^{-14}	6.2×10^{-15}	6.2×10^{-15}
14	2.8×10^{-2}	1.7×10^{-8}	1.9×10^{-3}	1.7×10^{-9}
15	0	0	4.8×10^{-15}	4.8×10^{-15}

TABLE 2. Relative errors of SZ and Schur vector solution before and after iterative refinement.

References

[1] C.D. Ahlbrandt and A.C. Peterson. *Discrete Hamiltonian Systems: Difference Equations, Continued Fractions, and Riccati Equations.* Kluwer Academic Publishers, Dordrecht, NL, 1998.

[2] B.D.O. Anderson and J.B. Moore. *Optimal Filtering.* Prentice-Hall, Englewood Cliffs, NJ, 1979.

[3] B.D.O. Anderson and B. Vongpanitlerd. *Network Analysis and Synthesis. A Modern Systems Approach.* Prentice-Hall, Englewood Cliffs, NJ, 1972.

[4] E.S. Armstrong and G. T. Rublein. A stabilization algorithm for linear discrete constant systems. *IEEE Trans. Automat. Control*, AC–21:629–631, 1976.

[5] G. Banse. *Symplektische Eigenwertverfahren zur Lösung zeitdiskreter optimaler Steuerungsprobleme.* Dissertation, Fachbereich 3 – Mathematik und Informatik, Universität Bremen, Bremen, FRG, June 1995. In German.

[6] A. Y. Barraud. A numerical algorithm to solve $A^TXA - X = Q$. *IEEE Trans. Automat. Control*, AC-22:883–885, 1977.

[7] R.H. Bartels and G.W. Stewart. Solution of the matrix equation $AX + XB = C$: Algorithm 432. *Comm. ACM*, 15:820–826, 1972.

[8] P. Benner. *Contributions to the Numerical Solution of Algebraic Riccati Equations and Related Eigenvalue Problems.* Logos–Verlag, Berlin, Germany, 1997. *Also:* Dissertation, Fakultät für Mathematik, TU Chemnitz–Zwickau, 1997.

[9] P. Benner. Accelerating Newton's method for discrete-time algebraic Riccati equations. In A. Beghi, L. Finesso, and G. Picci, editors, *Mathematical Theory of Networks and Systems*, pages 569–572, Il Poligrafo, Padova, Italy, 1998.

[10] P. Benner and H. Faßbender. The symplectic eigenvalue problem, the butterfly form, the *SR* algorithm, and the Lanczos method. *Linear Algebra Appl.*, 275/276:19–47, 1998.

[11] P. Benner, H. Faßbender, and D.S. Watkins. *SR* and *SZ* algorithms for the symplectic (butterfly) eigenproblem. *Linear Algebra Appl.*, 287:41–76, 1999.

[12] P. Benner, A.J. Laub, and V. Mehrmann. A collection of benchmark examples for the numerical solution of algebraic Riccati equations II: Discrete-time case. Technical Report SPC 95_23, Fakultät für Mathematik, TU Chemnitz–Zwickau, 09107 Chemnitz, FRG, 1995. Available from `http://www.tu-chemnitz.de/sfb393/spc95pr.html`.

[13] P. Benner, A.J. Laub, and V. Mehrmann. Benchmarks for the numerical solution of algebraic Riccati equations. *IEEE Control Systems Magazine*, 7(5):18–28, 1997.

[14] R. Byers and V. Mehrmann. Symmetric updating of the solution of the algebraic Riccati equation. *Methods of Operations Research*, 54:117–125, 1985.

[15] J. Dennis and R. B. Schnabel. *Numerical Methods for Unconstrained Optimization and Nonlinear Equations.* Prentice Hall, Englewood Cliffs, New Jersey, 1983.

[16] H. Faßbender. *Symplectic Methods for Symplectic Eigenproblems.* Habilitationsschrift, Fachbereich 3 – Mathematik und Informatik, Universität Bremen, 28334 Bremen, (Germany), 1998.

[17] G.H. Golub and C.F. Van Loan. *Matrix Computations.* Johns Hopkins University Press, Baltimore, third edition, 1996.

[18] T. Gudmundsson, C. Kenney, and A.J. Laub. Scaling of the discrete-time algebraic Riccati equation to enhance stability of the Schur solution method. *IEEE Trans. Automat. Control*, 37:513–518, 1992.

[19] G.A. Hewer. An iterative technique for the computation of steady state gains for the discrete optimal regulator. *IEEE Trans. Automat. Control*, AC-16:382–384, 1971.

[20] D.L. Kleinman. On an iterative technique for Riccati equation computations. *IEEE Trans. Automat. Control*, AC-13:114–115, 1968.

[21] D.L. Kleinman. Stabilizing a discrete, constant, linear system with application to iterative methods for solving the Riccati equation. *IEEE Trans. Automat. Control*, AC-19:252–254, 1974.

[22] P. Lancaster and L. Rodman. *The Algebraic Riccati Equation.* Oxford University Press, Oxford, 1995.

[23] A.J. Laub. Algebraic aspects of generalized eigenvalue problems for solving Riccati equations. In C.I. Byrnes and A. Lindquist, editors, *Computational and Combinatorial Methods in Systems Theory*, pages 213–227. Elsevier (North-Holland), 1986.

[24] The Mathworks, Inc., 24 Prime Park Way, Natick, MA 01760-1500 (USA). *Using* MATLAB *Version 5*, June 1997.

[25] V. Mehrmann. *The Autonomous Linear Quadratic Control Problem, Theory and Numerical Solution.* Number 163 in Lecture Notes in Control and Information Sciences. Springer-Verlag, Heidelberg, July 1991.

[26] T. Pappas, A.J. Laub, and N.R. Sandell. On the numerical solution of the discrete-time algebraic Riccati equation. *IEEE Trans. Automat. Control*, AC-25:631–641, 1980.

[27] P.H. Petkov, N.D. Christov, and M.M. Konstantinov. *Computational Methods for Linear Control Systems.* Prentice-Hall, Hertfordshire, UK, 1991.

[28] A. Saberi, P. Sannuti, and B.M. Chen. *H_2 Optimal Control.* Prentice-Hall, Hertfordshire, UK, 1995.

[29] V. Sima. *Algorithms for Linear-Quadratic Optimization*, volume 200 of *Pure and Applied Mathematics.* Marcel Dekker, Inc., New York, NY, 1996.

[30] D.S. Watkins and L. Elsner. Theory of decomposition and bulge chasing algorithms for the generalized eigenvalue problem. *SIAM J. Matrix Anal. Appl.*, 15:943–967, 1994.

[31] K. Zhou, J.C. Doyle, and K. Glover. *Robust and Optimal Control.* Prentice-Hall, Upper Saddle River, NJ, 1995.

ZENTRUM MATHEMATIK, TECHNISCHE UNIVERSITÄT MÜNCHEN, D-80290 MÜNCHEN, GERMANY

E-mail address: fassbender@mathematik.tu-muenchen.de

ZENTRUM FÜR TECHNOMATHEMATIK, FACHBEREICH 3 - MATHEMATIK UND INFORMATIK, UNIVERSITÄT BREMEN, D-28334 BREMEN, GERMANY

E-mail address: benner@math.uni-bremen.de

PART IV. Spectral properties. Conditioning

Contemporary Mathematics
Volume **280**, 2001

Condition Numbers of Large Toeplitz-like Matrices

A. Böttcher and S. Grudsky

ABSTRACT. Let $T_n(a)$ be the $n \times n$ Toeplitz matrix generated by a rational function a without poles on the unit circle. We study the behavior of the spectral condition numbers

$$\kappa(T_n(a) + F_n) = \|T_n(a) + F_n\| \, \|(T_n(a) + F_n)^{-1}\|$$

for a natural class of perturbation matrices F_n. In particular, we admit the cases where F_n is the $n \times n$ truncation of a rationally generated Hankel matrix $H(b)$ or where F_n has constant blocks of finite dimensions in the upper left and lower right corners and zeros elsewhere. We show that generically the condition numbers $\kappa(T_n(a) + F_n)$ converge to a finite limit with a speed of at least $O(e^{-\gamma\sqrt{n}})$ with $\gamma > 0$, while in certain exceptional cases (including the case $F_n = 0$) the condition numbers $\kappa(T_n(a) + F_n)$ may converge to a finite limit more slowly, for example, as $O(n^{-2})$.

1. Introduction

Let $\mathcal{R}$ be the collection of all rational functions without poles on the complex unit circle $\mathbf{T}$. For $a \in \mathcal{R}$, we denote by $T(a)$ and $T_n(a)$ the Toeplitz matrices

$$T(a) = (a_{j-k})_{j,k=0}^{\infty}, \quad T_n(a) = (a_{j-k})_{j,k=0}^{n-1}$$

composed by the Fourier coefficients of a,

$$a_n = \frac{1}{2\pi} \int_0^{2\pi} a(e^{i\theta}) e^{-in\theta} = \frac{1}{2\pi i} \int_{\mathbf{T}} a(z) z^{-n-1} \, dz. \tag{1.1}$$

From (1.1) we infer that the Fourier coefficients a_n decay exponentially: if $a(z)$ has no poles in the annulus $\{z \in \mathbf{C} : r \leq |z| \leq R\}$ $(0 < r < 1 < R)$, then

$$|a_n| \leq \|a\|_\infty \, \varrho^{|n|}, \quad \varrho := \max(r, 1/R). \tag{1.2}$$

Given $\sigma \in (0, 1)$, we let $\mathcal{K}_\sigma$ stand for the set of all infinite matrices $K = (K_{jk})_{j,k=0}^{\infty}$ for which there exists a constant $C_K < \infty$ such that

$$|K_{jk}| \leq C_K \sigma^{j+k} \text{ for all } j, k \geq 0.$$

1991 *Mathematics Subject Classification.* Primary 47B35; Secondary 15A12, 65F35.

This work was supported by DFG-Kooperationsprojekt 436 RUS 113/426 for German and Russian scientists within the "Memorandum of Understanding" between DFG and RFFI. Grudsky also acknowledges support by RFFI Grants 98-01-01023 and 96-01-01195a.

We put $\mathcal{K} := \bigcup_{\sigma \in (0,1)} \mathcal{K}_\sigma$. The set $\mathcal{K}$ contains in particular all matrices with at most finitely many nonzero entries. Moreover, from (1.2) we see that if $b \in \mathcal{R}$, then the Hankel matrix

$$H(b) = (b_{j+k+1})_{j,k=0}^{\infty}$$

belongs to $\mathcal{K}$.

We define the operators P_n and W_n on $l^2 := l^2(\mathbf{Z}_+)$ by the formulas

$$P_n : (x_0, x_1, x_2, \dots) \mapsto (x_0, x_1, \dots, x_{n-1}, 0, 0, \dots),$$
$$W_n : (x_0, x_1, x_2, \dots) \mapsto (x_{n-1}, \dots, x_1, x_0, 0, 0, \dots).$$

A bounded linear operator A on l^2 is given by an infinite matrix $(A_{jk})_{j,k=0}^{\infty}$. In what follows we freely identify an operator A with its matrix (A_{jk}). Furthermore, we identify the operators $P_n A P_n$ and $W_n A W_n$ with the matrices

$$(A_{jk})_{j,k=0}^{n-1} \text{ and } (A_{n-1-k,n-1-j})_{j,k=0}^{n-1},$$

respectively. In dependence on the context, we think of $P_n A P_n$ and $W_n A W_n$ as matrices, as operators on l^2, or as operators on $l_n^2 := l^2(\{0, 1, \dots, n-1\})$.

The norm of an operator A on l^2 or l_n^2 will be denoted by $\|A\|$:

$$\|A\| = \sup_{x \neq 0} \|Ax\| / \|x\|.$$

The (spectral) condition number of A is defined by

$$\kappa(A) = \|A\| \, \|A^{-1}\|;$$

we make thorough use of the convention to put $\|A^{-1}\| = \infty$ in case A is not invertible.

In this paper we study the asymptotic behavior of the condition numbers

$$\kappa\big(T_n(a) + P_n K P_n + W_n L W_n\big) \text{ as } n \to \infty$$

under the assumption that $a \in \mathcal{R}$ and $K, L \in \mathcal{K}$. Thus, in the case $K = L = 0$ we are dealing with the condition numbers of pure Toeplitz matrices $T_n(a)$. If K and L have only a finite number of nonzero indices, then $T_n(a) + P_n K P_n + W_n L W_n$ may be viewed as a perturbation of $T_n(a)$ by the matrix K in the upper left corner and by (the "reverse" of) L in the lower right corner. In the case where $K = H(b)$ and $L = 0$, we encounter the $n \times n$ truncations of the Toeplitz plus Hankel matrix $T(a) + H(b)$.

Suppose $a \in \mathcal{R}$ and $K, L \in \mathcal{K}$. Put

$$A_n := T_n(a) + P_n K P_n + W_n L W_n.$$

Since K and L are compact and $W_n \to 0$ weakly, it follows that $W_n K W_n \to 0$ and $W_n L W_n \to 0$ strongly. This observation together with the identity

$$W_n T_n(a) W_n = T_n(\widetilde{a}), \quad \widetilde{a}(t) := a(1/t) \quad (t \in \mathbf{T})$$

implies that

$$A_n \to T(a) + K, \quad W_n A_n W_n \to T(\widetilde{a}) + L$$

strongly. Notice that $T_n(\widetilde{a})$ is just the transpose of $T_n(a)$. Let

$$M := \max\big(\|T(a) + K\|, \, \|T(\widetilde{a}) + L\|\big),$$
$$M^{(-1)} := \max\Big(\|\big(T(a) + K\big)^{-1}\|, \, \|\big(T(\widetilde{a}) + L\big)^{-1}\|\Big).$$

It turns out that the condition numbers $\kappa(A_n)$ approach a limit as $n \to \infty$ and that this limit is $MM^{(-1)}$:

$$\lim_{n\to\infty} \kappa(A_n) = MM^{(-1)}. \tag{1.3}$$

This was established in [**5**]. An alternative proof is in [**9**].

If one of the operators $T(a) + K$ and $T(\widetilde{a}) + L$ is not invertible, then (1.3) says that $\kappa(A_n) \to \infty$ as $n \to \infty$. In this paper we assume that $T(a) + K$ and $T(\widetilde{a}) + L$ are invertible and we consider the question about the speed with which $\kappa(A_n)$ converges to its limit $MM^{(-1)}$. We conjecture that the following is true.

CONJECTURE 1.1. *Let $M^{(-1)} < \infty$. We always have*

$$|\kappa(A_n) - MM^{(-1)}| = O\left(\frac{1}{n^2}\right), \tag{1.4}$$

and generically we even have the estimate

$$|\kappa(A_n) - MM^{(-1)}| = O(e^{-\gamma n}) \tag{1.5}$$

with some $\gamma > 0$ depending only on a, K, L.

The meaning of "generically" will be made precise later. Here is what we will be able to prove in this paper.

THEOREM 1.2. *Let $M^{(-1)} < \infty$. Then*

$$|\kappa(A_n) - MM^{(-1)}| = O\left(\frac{\log n}{n}\right), \tag{1.6}$$

and generically there is even a $\gamma > 0$ depending on a, K, L such that the stronger estimate

$$|\kappa(A_n) - MM^{(-1)}| = O(e^{-\gamma\sqrt{n}}) \tag{1.7}$$

holds.

2. Brief history

For pure Toeplitz matrices, $A_n = T_n(a)$, the story has its beginning in the sixties. Reich [**42**], Baxter [**3**], and Gohberg and Feldman [**19**] studied the following problem: if $T(a)$ is invertible, is there an n_0 such that the matrices $T_n(a)$ are invertible for all $n \geq n_0$ and $\sup_{n\geq n_0} \|T_n^{-1}(a)\| < \infty$? Since, obviously, $\|T_n(a)\| \to \|T(a)\|$, this is the question whether the condition numbers remain bounded as $n \to \infty$, that is, whether

$$\limsup_{n\to\infty} \kappa\big(T_n(a)\big) < \infty. \tag{2.1}$$

The mathematicians cited above showed that this is indeed true. A round theory of this and related problems was developed by Gohberg and Feldman in their book [**20**]. They even considered compactly perturbed Toeplitz matrices and proved that

$$\limsup_{n\to\infty} \kappa\big(T_n(a) + P_nKP_n\big) < \infty \tag{2.2}$$

if and only if $T(a) + K$ is invertible. Notice in this connection that the invertibility of $T(a) + K$ implies that $T(\widetilde{a})$, the transpose of $T(a)$, is also invertible.

In 1976, Widom [**63**] published the following beautiful formula for the product of two finite Toeplitz matrices:

$$T_n(a)T_n(b) = T_n(ab) - P_nH(a)H(\widetilde{b})P_n - W_nH(\widetilde{a})H(b)W_n. \tag{2.3}$$

Recall that $\widetilde{c}$ is always defined by $\widetilde{c}(t) := c(1/t)\,(t \in \mathbf{T})$ and that $H(c)$ and $H(\widetilde{c})$ stand for the infinite Hankel matrices

$$H(c) = (c_{j+k+1})_{j,k=0}^{\infty}, \quad H(\widetilde{c}) = (c_{-j-k-1})_{j,k=0}^{\infty}.$$

If a and b are continuous, which is the case for $a, b \in \mathcal{R}$, then the Hankel operators occurring in (2.3) are all compact, and hence

$$T_n(a)T_n(b) = T_n(ab) + P_nKP_n + W_nLW_n$$

with compact operators K and L. This formula with $b = a^{-1}$ was used by Widom [**62**] to find a nice asymptotic inverse for $T_n(a)$:

$$T_n^{-1}(a) = T_n(a^{-1}) + P_nXP_n + W_nYW_n + E_n \tag{2.4}$$

with the compact operators

$$X = T^{-1}(a) - T(a^{-1}), \quad Y = T^{-1}(\widetilde{a}) - T(\widetilde{a}^{-1})$$

and with a sequence E_n such that $\|E_n\| \to 0$.

At the turn to the eighties, Silbermann [**54**] realized the importance of formula (2.3) in connection with localization techniques for approximation methods for Toeplitz operators. As a by-product, he obtained that

$$\limsup_{n\to\infty} \kappa\big(T_n(a) + P_nKP_n + W_nLW_n\big) < \infty \tag{2.5}$$

if and only if $T(a) + K$ and $T(\widetilde{a}) + L$ are invertible.

In the nineties, the investigation of the asymptotic behavior of the ε-pseudospectra of Toeplitz matrices received some popularity from Reichel and Trefethen's paper [**43**]. We remark that the ε-pseudospectrum of an $n \times n$ matrix A_n is defined by

$$\operatorname{sp}_\varepsilon A_n = \big\{\lambda \in \mathbf{C} : \|(A_n - \lambda I)^{-1}\| \geq 1/\varepsilon\big\}.$$

Clearly, to understand the asymptotics of $\operatorname{sp}_\varepsilon A_n$ one needs rather precise information about the behavior of the norms $\|(A_n - \lambda I)^{-1}\|$. Motivated by this set of problems, one of the authors [**5**] was able to compute the upper limits in (2.1), (2.2), (2.5) and to show that actually

$$\lim_{n\to\infty} \kappa\big(T_n(a) + P_nKP_n + W_nLW_n\big) = MM^{(-1)}. \tag{2.6}$$

The proof of (2.6) given in [**5**] is based on C^*-algebra arguments and was essentially inspired by [**55**]. Another proof, which also works for operators on l^p spaces, is in [**9**].

The question about the convergence speed in (2.6) was first studied by Kozak and one of the authors [**25**]. In the latter paper, the operators are considered on the space l^1 and estimates are given for the condition numbers $\kappa(T_n(a))$ of pure Toeplitz matrices. These results were carried over to the spaces l^p, and, in particular, to l^2, in our papers [**10**] and [**11**] with Kozak and Silbermann. In [**10**] we gave estimates for the difference

$$\|T_n(a) + P_nKP_n + W_nLW_n\| - M, \tag{2.7}$$

and in [**11**] we combined our insights into the behavior of (2.7) with Widom's representation (2.4) in order to obtain estimates for

$$\|T_n^{-1}(a)\| - M^{(-1)}. \tag{2.8}$$

The purpose of this paper is to extend the results to

$$\|\left(T_n(a) + P_nKP_n + W_nLW_n\right)^{-1}\| - M^{(-1)}$$

in place of (2.8).

These are a few pieces of the story. In a sense, the story is fascinating: 30 years ago one was happy to know that the condition numbers $\kappa(A_n)$ remain bounded, 5 years ago one was able to compute the limit of $\kappa(A_n)$, now one is tackling the speed with which $\kappa(A_n)$ approaches its limit, and we are sure that some day people will deal with the constants hidden in the O's on the right of (1.4)–(1.7).

3. Some instructive examples

To get a feeling for the matter, we examine a few examples and cite some results from [**10**] and [**11**].

EXAMPLE 3.1. Let $a(t) = 3 - t - t^{-1}$ $(t \in \mathbf{T})$. Thus,

$$T(a) = \begin{pmatrix} 3 & -1 & 0 & \dots \\ -1 & 3 & -1 & \dots \\ 0 & -1 & 3 & \dots \\ \dots & \dots & \dots & \dots \end{pmatrix}$$

is a Hermitian tridiagonal Toeplitz matrix. The eigenvalues of $T_n(a)$ are

$$\lambda_j(T_n(a)) = 3 + 2\cos\frac{\pi j}{n+1} \qquad (j = 1, \dots, n).$$

Hence,

$$\|T_n(a)\| = \max\left(\lambda_j(T_n(a))\right) = 3 + 2\cos\frac{\pi}{n+1} = 5 - \frac{\pi^2}{(n+1)^2} + O\left(\frac{1}{n^4}\right),$$

$$\|T_n^{-1}(a)\| = \frac{1}{\min \lambda_j(T_n(a))} = \frac{1}{3 - 2\cos\frac{\pi}{n+1}} = 1 + \frac{\pi^2}{(n+1)^2} + O\left(\frac{1}{n^4}\right),$$

and as

$$\|T(a)\| = \max_{t\in\mathbf{T}} |a(t)| = 5, \quad \|T^{-1}(a)\| = 1/\min_{t\in\mathbf{T}} |a(t)| = 1,$$

we see that

$$\big|\,\|T_n(a)\| - \|T(a)\|\,\big| = \frac{\pi^2}{(n+1)^2} + O\left(\frac{1}{n^4}\right),$$

$$\big|\,\|T_n^{-1}(a)\| - \|T^{-1}(a)\|\,\big| = \frac{\pi^2}{(n+1)^2} + O\left(\frac{1}{n^4}\right),$$

$$\left|\kappa(T_n(a)) - MM^{(-1)}\right| = \left|\kappa(T_n(a)) - \kappa(T(a))\right| = \frac{4\pi^2}{(n+1)^2} + O\left(\frac{1}{n^4}\right).$$

The moral is as follows: although $T(a)$ is a very nice matrix, $\kappa(T_n(a))$ converges to its limit "rather slowly", namely, with the order $1/n^2$. □

It turns out that the slow convergence of the norms $\|T_n(a)\|$ to the limit $\|T(a)\|$ observed in Example 3.1 is typical for rationally generated Toeplitz matrices. Here we adopt the point of view that rational functions with constant modulus are no typical rational functions.

THEOREM 3.2. *Let $a \in \mathcal{R}$ and assume $|a|$ is not constant. Denote by $2\gamma \in \{2, 4, 6, \dots\}$ the maximal order of the zeros of $\|a\|_\infty - |a(t)|$ on $\mathbf{T}$. Then*

$$\frac{d_1}{n^{2\gamma}} \le \|T(a)\| - \|T_n(a)\| \le \frac{d_2}{n^{2\gamma}} \text{ for all } n \ge 1 \tag{3.1}$$

with certain constants $0 < d_1 < d_2 < \infty$ depending only on a. On the other hand, if $a \in \mathcal{R}$ and $|a|$ is constant, then

$$\|T(a)\|(1 - de^{-\delta n}) \le \|T_n(a)\| \le \|T(a)\| \text{ for all } n \ge 1, \tag{3.2}$$

where d and δ are positive constants that depend only on a.

PROOF. We begin with the proof of (3.2). The inequality $\|T_n(a)\| \le \|T(a)\|$ in (3.2) is obvious. We are left with the estimate from below in (3.2). Let $f := e_{[n/2]}$ where $[n/2]$ is the integral part of $n/2$ and $e_{[n/2]} \in l_n^2$ is the vector with a unit at the position $[n/2]$ and zeros elsewhere. The adjoint of $T(a)$ is $T^*(a) = T(\overline{a})$ with $\overline{a}(t) := \overline{a(t)}$ $(t \in \mathbf{T})$. From formula (2.3) we obtain

$$T_n(a)T_n(\overline{a}) = T_n(|a|^2) - P_nKP_n - W_nLW_n \tag{3.3}$$

where $K = H(a)H(\widetilde{\overline{a}})$ and $L = H(\widetilde{a})H(\overline{a})$ are positively semi-definite:

$$(Kx, x) \ge 0, \ (Lx, x) \ge 0 \text{ for all } x \in l^2. \tag{3.4}$$

From (3.3) we get

$$\begin{aligned} \|T_n(a)\|^2 &= \|T_n(a)T_n(\overline{a})\| = \|T_n(a)T_n(\overline{a})\| \, \|f\|^2 \\ &\ge |(T_n(a)T_n(\overline{a})f, f)| = |(\|a\|_\infty^2 f, f) - (P_nKP_nf, f) - (W_nLW_nf, f)| \\ &\ge \|a\|_\infty^2 - |(P_nKP_nf, f)| - |(W_nLW_nf, f)| \\ &= \|a\|_\infty^2 - K_{[n/2],[n/2]} - L_{n-1-[n/2],n-1-[n/2]}, \end{aligned}$$

and since $K, L \in \mathcal{K}_\sigma$ for some $\sigma \in (0, 1)$ (which easily follows from (1.2)), we obtain

$$K_{[n/2],[n/2]} = O(\sigma^n), \quad L_{n-1-[n/2],n-1-[n/2]} = O(\sigma^n).$$

Hence $\|a\|_\infty^2 - \|T_n(a)\|^2 = O(\sigma^n)$, which implies that

$$\|T(a)\| - \|T_n(a)\| = \|a\|_\infty - \|T_n(a)\| = O(\sigma^n) = O(e^{-\delta n}).$$

This completes the proof of (3.2).

For Toeplitz band matrices, estimate (3.1) was proved in [**10**]. The proof given there also works for rationally generated Toeplitz matrices. For the reader's convenience, we repeat pieces of it here.

From (3.3) and (3.4) we infer that $(T_n(a)T_n(\overline{a})x, x) \le (T_n(|a|^2)x, x)$ for all $x \in l^2$, whence

$$\|T_n(a)\|^2 = \|T_n(a)T_n(\overline{a})\| \le \|T_n(|a|^2)\|. \tag{3.5}$$

Estimate (3.1) is well known for Hermitian matrices (see Serra [**50**], [**51**]). Thus, there is a constant $d_1 > 0$ such that

$$\|T(|a|^2)\| - \|T_n(|a|^2)\| \ge \frac{2d_1}{n^{2\gamma}} \|a\|_\infty,$$

and since $\|T(|a|^2)\| = \| \, |a|^2 \|_\infty = \|a\|_\infty^2$, we obtain

$$\|T_n(|a|^2)\| \le \|a\|_\infty^2 \left(1 - \frac{2d_1}{n^{2\gamma}} \frac{1}{\|a\|_\infty}\right). \tag{3.6}$$

Combining (3.5) and (3.6) we get

$$\|T_n(a)\| \le \|a\|_\infty \left(1 - \frac{2d_1}{n^{2\gamma}} \frac{1}{\|a\|_\infty}\right)^{1/2} \le \|a\|_\infty \left(1 - \frac{d_1}{n^{2\gamma}} \frac{1}{\|a\|_\infty}\right) = \|a\|_\infty - \frac{d_1}{n^{2\gamma}}.$$

As $\|T_n(a)\| \le \|T(a)\| = \|a\|_\infty$, this implies that

$$\|T(a)\| - \|T_n(a)\| = \|a\|_\infty - \|T_n(a)\| \ge \frac{d_1}{n^{2\gamma}},$$

which is the lower estimate in (3.1).

We now proceed to the upper estimate in (3.1). Let $2\gamma \ge 2$ be the maximal integer for which there are $t_0 \in \mathbf{T}$ and $D \in (0, \infty)$ such that

$$\|a\|_\infty - |a(t)| \le D|t - t_0|^{2\gamma} \text{ for all } t \in \mathbf{T}. \tag{3.7}$$

Since

$$T_n(a(t/t_0)) = \operatorname{diag}(1, t_0^{-1}, \ldots, t_0^{-n+1})\, T_n(a) \operatorname{diag}(1, t_0, \ldots, t_0^{n-1}),$$

we can without loss of generality assume that $t_0 = 1$. From (3.7) we see that there is a constant $C \in (0, \infty)$ such that

$$\|a\|_\infty^2 - |a(e^{i\theta})|^2 \le C|\theta|^{2\gamma} \text{ for all } \theta \in (-\pi, \pi]. \tag{3.8}$$

If $g \in l_n^2$, then, by (3.3),

$$\begin{aligned} \|T_n(a)\|^2\|g\|^2 &= \|T_n(a)T_n(\bar{a})\|\|g\|^2 \ge |(T_n(a)T_n(\bar{a})g, g)| \\ &= |(T_n(|a|^2)g, g) - (P_nKP_ng, g) - (W_nLW_ng, g)| \\ &\ge \|a\|_\infty^2\|g\|^2 - |((\|a\|_\infty^2 - T_n(|a|^2))g, g)| - |(P_nKP_ng, g)| - |(W_nLW_ng, g)|. \end{aligned} \tag{3.9}$$

Let m be the integer satisfying $m(\gamma + 1) < n \le (m + 1)(\gamma + 1)$ and define $g \in l_n^2$ as the sequence of the Fourier coefficients of the function

$$G(e^{i\theta}) = (1 + e^{i\theta} + \ldots + e^{im\theta})^{\gamma+1}.$$

It is easily checked (see, e.g., [**7**]) that the l^2 norm of g admits the estimate

$$\|g\| \ge E_1 n^{\gamma+1/2}$$

with some $E_1 \in (0, \infty)$ independent of n, and in [**10**] we showed that if (3.8) holds, then

$$|((\|a\|_\infty^2 - T_n(|a|^2))g, g)| \le E_2 n$$

with some $E_2 \in (0, \infty)$ that does not depend on n. Hence, (3.9) gives

$$\|T_n(a)\|^2 \ge \|T(a)\|^2 - \frac{E_2 n}{E_1^2 n^{2\gamma+1}} - \frac{|(P_nKP_ng, g)|}{E_1^2 n^{2\gamma+1}} - \frac{|(W_nLW_ng, g)|}{E_1^2 n^{2\gamma+1}}. \tag{3.10}$$

As $K \in \mathcal{K}_\sigma$ with $\sigma \in (0, 1)$ (recall the proof of (3.2)), we get

$$\begin{aligned} |(P_nKP_ng, g)| &\le \sum_{i,k=0}^{n-1} |K_{ik}|\,|g_i|\,|g_k| \le \sum_{i,k=0}^{n-1} \sigma^{i+k}|g_i|\,|g_k| \\ &= \left(\sum_{i=0}^{n-1} \sigma^i|g_i|\right)\left(\sum_{k=0}^{n-1} \sigma^k|g_k|\right) = \left(\sum_{i=0}^{n-1} \sigma^i|g_i|\right)^2. \end{aligned}$$

Clearly, $\sum_{i=0}^{n-1} \sigma^i |g_i|$ is the sum of the absolute values of the Fourier coefficients of the function $t \mapsto G(\sigma t)$. Consequently,

$$\left(\sum_{i=0}^{n-1} \sigma^i |g_i|\right) \leq (1 + \sigma + \ldots + \sigma^m)^{\gamma+1} < \left(\frac{1}{1-\sigma}\right)^{\gamma+1},$$

whence

$$|(P_n K P_n g, g)| \leq \left(\frac{1}{1-\sigma}\right)^{2(\gamma+1)}.$$

Analogously,

$$|(W_n L W_n g, g)| \leq \left(\frac{1}{1-\sigma}\right)^{2(\gamma+1)}.$$

From (3.10) we therefore obtain that

$$\|T_n(a)\|^2 \geq \|T(a)\|^2 - \frac{E_3}{n^{2\gamma}}$$

with some $E_3 \in (0, \infty)$ independent of n. This implies the upper estimate in (3.1). □

EXAMPLE 3.3. The functions $a(t) = t^k$ ($k \in \mathbf{Z}$) are trivial examples of rational functions with constant modulus, and it is clear that $\|T_n(a)\| = \|T(a)\| = 1$ in this case. A nontrivial example of a unimodular function $a \in \mathcal{R}$ is

$$a(t) = \frac{t^{-1} - \alpha}{1 - \overline{\alpha} t^{-1}} \frac{t - \beta}{1 - \overline{\beta} t} \quad (t \in \mathbf{T}) \tag{3.11}$$

where $|\alpha| < 1$, $|\beta| < 1$. More generally, we can take $a = \overline{B}_1 B_2$ where B_1 and B_2 are finite Blaschke products.

Now let $a \in \mathcal{R}$, $|a(t)| = 1$ for all $t \in \mathbf{T}$, and suppose $T(a)$ is invertible. We have this case if, for instance, a is given by (3.11) with $|\alpha| < 1$ and $|\beta| < 1$ or if $a = \overline{B}_1 B_2$ where B_1 and B_2 are finite Blaschke products with the same numbers of zeros. Proposition 6.2 and Theorem 4.1 of [**11**] show that

$$\left| \, \|T_n^{-1}(a)\| - \|T^{-1}(a)\| \, \right| = O\left(e^{-\delta\sqrt{n}}\right)$$

with some $\delta > 0$. This in conjunction with the second part of Theorem 3.2 gives

$$|\kappa(T_n(a)) - \kappa(T(a))| = O\left(e^{-\gamma\sqrt{n}}\right)$$

with some $\gamma > 0$. Moral: Theorem 3.2 tells us that typically the condition numbers of pure rationally generated Toeplitz matrices do not converge faster to their limit than the inverse of some power of n; rational functions of constant modulus are the only exception from this rule. □

The following theorem describes another situation in which we have exponentially fast convergence.

THEOREM 3.4. *If $K, L \in \mathcal{K}_\sigma$ and if $I + K$ and $I + L$ are invertible, then*

$$\left|\kappa(I + P_n K P_n + W_n L W_n) - M M^{(-1)}\right| = O(\sigma^{n/2}),$$

where

$$M = \max\left(\|I + K\|, \|I + L\|\right), \quad M^{(-1)} = \max\left(\|(I + K)^{-1}\|, \|(I + L)^{-1}\|\right).$$

The proof is based on a simple lemma. We put $Q_n := I - P_n$, that is,

$$Q_n : (x_0, x_1, x_2, \ldots) \mapsto (0, \ldots, 0, x_n, x_{n+1}, \ldots).$$

LEMMA 3.5. *If $R \in \mathcal{K}_\sigma$ then $\|RQ_n\| = O(\sigma^n)$ and $\|Q_n R\| = O(\sigma^n)$.*

PROOF. Let $\|\cdot\|_F$ denote the Frobenius norm (= Hilbert-Schmidt norm). We have

$$\|RQ_n\|^2 \le \|RQ_n\|_F^2 = \sum_{j=0}^{\infty}\sum_{k=n}^{\infty} |R_{jk}|^2$$
$$= O\Big(\sum_{j=0}^{\infty}\sum_{k=n}^{\infty} \sigma^{2(j+k)}\Big) = O\Big(\sum_{j=0}^{\infty} \sigma^{2j} \sum_{k=n}^{\infty} \sigma^{2k}\Big) = O(\sigma^{2n}),$$

and passage to adjoints gives the result for $\|Q_n R\|$. □

PROOF OF THEOREM 3.4. Let $[x]$ stand for the integral part of a real number x. In what follows we denote by $\boldsymbol{O}(\sigma^n)$ a sequence $\{C_n\}$ of matrices such that $\|C_n\| = O(\sigma^n)$. Let $n \ge 2$.

From Lemma 3.5 we deduce that

$$\begin{aligned} &P_n K P_n \\ &= P_n Q_{[n/2]} K Q_{[n/2]} P_n + P_n Q_{[n/2]} K P_{[n/2]} + P_{[n/2]} K Q_{[n/2]} P_n + P_{[n/2]} K P_{[n/2]} \\ &= P_{[n/2]} K P_{[n/2]} + \boldsymbol{O}(\sigma^{n/2}), \end{aligned}$$

and analogously,

$$W_n L W_n = W_n P_{[n/2]} L P_{[n/2]} W_n + \boldsymbol{O}(\sigma^{n/2}).$$

Thus, with $K_n := P_{[n/2]} K P_{[n/2]}$ and $L_n := W_n P_{[n/2]} L P_{[n/2]} W_n$, we have

$$I + P_n K P_n + W_n L W_n = I + K_n + L_n + \boldsymbol{O}(\sigma^{n/2}).$$

The $n \times n$ matrix $I + K_n + L_n$ is block diagonal:

$$I + K_n + L_n = \begin{pmatrix} I + K_n & 0 & 0 \\ 0 & I & 0 \\ 0 & 0 & I + W_{[n/2]} L W_{[n/2]} \end{pmatrix}$$

with $[n/2] \times [n/2]$ blocks $I + K_n$ and $I + W_{[n/2]} L W_{[n/2]}$ (notice that the middle block I appears only if n is an odd number). This implies that

$$\begin{aligned} (3.12)\qquad &\|I + K_n + L_n + \boldsymbol{O}(\sigma^{n/2})\| \\ &= \max(\|I + K_n\|, \|I + W_{[n/2]} L W_{[n/2]}\|) + O(\sigma^{n/2}), \\ (3.13)\qquad &\|\big(I + K_n + L_n + \boldsymbol{O}(\sigma^{n/2})\big)^{-1}\| \\ &= \max\big(\|(I + K_n)^{-1}\|, \|(I + W_{[n/2]} L W_{[n/2]})^{-1}\|\big) + O(\sigma^{n/2}). \end{aligned}$$

Again by Lemma 3.5,

$$\|K_n - K\| = \|P_{[n/2]} K P_{[n/2]} - K\| = O(\sigma^{n/2}),$$
$$\|W_{[n/2]} L W_{[n/2]} - L\| = \|P_{[n/2]} L P_{[n/2]} - L\| = O(\sigma^{n/2}),$$

which, together with (3.12) and (3.13), gives

$$\|I + K_n + L_n + \boldsymbol{O}(\sigma^{n/2})\| = M + O(\sigma^{n/2}),$$
$$\|\big(I + K_n + L_n + \boldsymbol{O}(\sigma^{n/2})\big)^{-1}\| = M^{(-1)} + O(\sigma^{n/2}). \qquad \square$$

4. Generic and exceptional cases

Let $a \in \mathcal{R}$, $K \in K$, $L \in \mathcal{K}$, $A_n = T_n(a) + P_nKP_n + W_nLW_n$. We put

$$M := \max\Big(\|T(a) + K\|, \|T(\widetilde{a}) + L\|\Big), \quad M_0 := \|T(a)\|,$$
$$M^{(-1)} := \max\Big(\|\big(T(a) + K\big)^{-1}\|, \|\big(T(\widetilde{a}) + L\big)^{-1}\|\Big), \quad M_0^{(-1)} := \|T(a^{-1})\|.$$

The following is well known.

LEMMA 4.1. *We always have $M \geq M_0$ and $M^{(-1)} \geq M_0^{(-1)}$.*

PROOF. Define $V^{(-n)}$ and $V^{(n)}$ on l^2 by

$$V^{(-n)} : (x_0, x_1, x_2, \ldots) \mapsto (x_n, x_{n+1}, x_{n+2}, \ldots),$$
$$V^{(n)} : (x_0, x_1, x_2, \ldots) \mapsto (\underbrace{0, \ldots, 0}_{n}, x_0, x_1, x_2, \ldots).$$

It is clear that $V^{(-n)}T(a)V^n = T(a)$. Since $V^n \to 0$ weakly and K is compact, we see that $V^{(-n)}KV^n \to 0$ strongly. Hence,

$$\|T(a)\| \leq \liminf_{n\to\infty} \|V^{(-n)}\big(T(a) + K\big)V^n\| \leq \|T(a) + K\|.$$

Analogously, $\|T(a)\| = \|T(\widetilde{a})\| \leq \|T(\widetilde{a}) + L\|$. This proves that $M_0 \leq M$.

The analogue of (2.3) for infinite matrices is

$$T(a)T(a^{-1}) = I - H(a)H(\widetilde{a}^{-1}), \tag{4.1}$$

and the operator $U := H(a)H(\widetilde{a}^{-1})$ is compact. From (4.1) we get

$$\big(T(a) + K\big)T(a^{-1}) = I - U + KT(a^{-1}),$$

whence

$$\big(T(a) + K\big)^{-1} = T(a^{-1}) + \big(T(a) + K\big)^{-1}\big(U - KT(a^{-1})\big). \tag{4.2}$$

As $(T(a) + K)^{-1}(U - KT(a^{-1}))$ is compact together with U and K, we obtain as in the preceding paragraph that

$$\|T(a^{-1})\| \leq \|\big(T(a) + K\big)^{-1}\|.$$

In the same way we get $\|T(a^{-1})\| = \|T(\widetilde{a}^{-1})\| \leq \|(T(\widetilde{a}) + L)^{-1}\|$. This completes the proof of the inequality $M_0^{(-1)} \leq M^{(-1)}$. $\square$

We will see that $\|A_n\|$ converges very fast to M if $M_0 < M$ and that $\|A_n^{-1}\|$ conveges very fast to $M^{(-1)}$ provided $M_0^{(-1)} < M^{(-1)}$. Thus, we arrive at the question about when in Lemma 4.1 strict inequalities hold. We equip $\mathcal{R} \times \mathcal{K} \times \mathcal{K}$ with the norm

$$\|(a, K, L)\| := \max\big(\|a\|_\infty, \|K\|, \|L\|\big).$$

Let $G\mathcal{R}$ be the set of all functions in $\mathcal{R}$ which do not have zeros on $\mathbf{T}$. Note that if $a \in \mathcal{R} \setminus G\mathcal{R}$, then $M^{(-1)} = M_0^{-1} = \infty$.

THEOREM 4.2. *The sets* $\left\{(a, K, L) \in \mathcal{R} \times \mathcal{K} \times \mathcal{K} : M > M_0\right\}$ *and* $\left\{(a, K, L) \in G\mathcal{R} \times \mathcal{K} \times \mathcal{K} : M^{(-1)} > M_0^{(-1)}\right\}$ *are open and dense subsets of* $\mathcal{R} \times \mathcal{K} \times \mathcal{K}$.

A similar result was already established in [**10**]. There we proved the theorem with $\mathcal{K}$ replaced by the set $\mathcal{C}$ of all compact operators on l^2. To prove the theorem as it is stated, we need two auxiliary results.

LEMMA 4.3. *If* $a \in \mathcal{R}$ *and* $R, S \in \mathcal{K}$, *then*

$$RS \in \mathcal{K}, \quad RT(a) \in \mathcal{K}, \quad T(a)R \in \mathcal{K}.$$

PROOF. The first inclusion is obvious:

$$|(RS)_{jk}| = \left|\sum_{l=0}^{\infty} R_{jl} S_{lk}\right| = O\left(\sum_{l=0}^{\infty} \sigma^{j+l} \sigma^{l+k}\right) = O(\sigma^{j+k}).$$

To prove the other inclusions, let $\varrho \in (0,1)$ be as in (1.2) and suppose $R \in \mathcal{K}_\sigma$, $\sigma \in (0,1)$. Choose $\tau_0, \tau \in (0,1)$ so that $\max(\varrho, \sigma) < \tau_0 < \tau$. Then

$$\begin{aligned} \left|\left(RT(a)\right)_{jk}\right| &= \left|\sum_{l=0}^{\infty} R_{jl} a_{l-k}\right| = O\left(\sum_{l=0}^{\infty} \sigma^{j+l} \varrho^{|l-k|}\right) \\ &= O\left(\sum_{l=0}^{k-1} \sigma^{j+l} \varrho^{k-l} + \sum_{l=k}^{\infty} \sigma^{j+l} \varrho^{l-k}\right) \\ &= O\left(k \tau_0^j \tau_0^k + \tau_0^j \tau_0^{-k} \sum_{l=k}^{\infty} \tau_0^{2l}\right) \\ &= O(k\tau_0^{j+k} + \tau_0^{j+k}) = O(\tau^{j+k}), \end{aligned}$$

which shows that $RT(a) \in \mathcal{K}_\tau$. Passing to adjoints we get that $T(a)R \in \mathcal{K}_\tau$. □

For further referencing, we state the following classical result. We denote by $C = C(\mathbf{T})$ the set of all continuous functions on $\mathbf{T}$.

THEOREM 4.4. (GOHBERG [**18**]) *Let* $a \in C$. *The operator* $T(a)$ *is invertible modulo compact operators if and only if* a *has no zeros on* $\mathbf{T}$, *and* $T(a)$ *is invertible if and only if* a *has no zeros on* $\mathbf{T}$ *and the winding number of* $a(\mathbf{T})$ *about the origin is zero.*

A full proof is also in [**14**, Section 1.6]. □

LEMMA 4.5. *Let* $a \in \mathcal{R}$, $K \in \mathcal{K}$, *and suppose* $T(a) + K$ *is invertible. Then* $a \in G\mathcal{R}$ *and*

$$X := \left(T(a) + K\right)^{-1} - T(a^{-1}) \in \mathcal{K}.$$

PROOF. Theorem 4.4 implies that $a \in G\mathcal{R}$. From (4.2) we obtain that

$$X = \left(T(a) + K\right)^{-1}\left(H(a)H(\widetilde{a}^{-1}) - KT(a^{-1})\right), \tag{4.3}$$

and Lemma 4.3 tells us that

$$R := H(a)H(\widetilde{a}^{-1}) - KT(a^{-1}) \in \mathcal{K}_\tau$$

for some $\tau \in (0,1)$. Hence,

$$\begin{aligned} |X_{jk}| &= \left|\sum_{l=0}^{\infty} \left[\big(T(a)+K\big)^{-1}\right]_{jk} R_{lk}\right| = O\Big(\sum_{l=0}^{\infty} |R_{lk}|\Big) \\ &= O\Big(\sum_{l=0}^{\infty} \tau^{l+k}\Big) = O(\tau^k). \end{aligned}$$

In the same way we derived (4.3) we get

$$\begin{aligned} X &= \Big(I - T(a^{-1})\big(T(a)+K\big)\Big)\big(T(a)+K\big)^{-1} \\ &= \big(I - T(a^{-1})T(a) - T(a^{-1})K\big)\big(T(a)+K\big)^{-1} \\ &= \big(H(a^{-1})H(\widetilde{a}) - T(a^{-1})K\big)\big(T(a)+K\big)^{-1} := S\big(T(a)+K\big)^{-1} \end{aligned}$$

with $S \in \mathcal{K}_\tau$ for some $\tau \in (0,1)$. As above, this gives

$$|X_{jk}| = O\Big(\sum_{l=0}^{\infty} S_{jl}\Big) = O(\tau^j).$$

In summary, $|X_{jk}|^2 = O(\tau^{j+k})$, that is, $X \in \mathcal{K}_{\sqrt{\tau}}$. □

PROOF OF THEOREM 4.2. Put $\mathcal{G} := \{(a,K) \in \mathcal{R}\times\mathcal{K} : \|T(a)+K\| > \|T(a)\|\}$. It is clear that $\mathcal{G}$ is open. To show that $\mathcal{G}$ is dense in $\mathcal{R}\times\mathcal{K}$, suppose $\|T(a)+K\| = \|T(a)\|$ for some $a \in \mathcal{R}$ and $K \in \mathcal{K}$. Given any $\varepsilon > 0$, we can find a $\lambda \in \mathbf{C}$ such that $|\lambda| < \varepsilon$ and $\|a+\lambda\|_\infty > \|a\|_\infty$. Then, by Lemma 4.1,

$$\|T(a)+K+\lambda I\| = \|T(a+\lambda)+K\| \ge \|a+\lambda\|_\infty > \|a\|_\infty = \|T(a)\|,$$

and since $T(a)+K+\lambda P_n$ converges strongly to $T(a)+K+\lambda I$, we obtain

$$\liminf_{n\to\infty} \|T(a)+K+\lambda P_n\| \ge \|T(a)+K+\lambda I\| > \|T(a)\|.$$

Consequently, letting $K_0 := K+\lambda P_n$, we have $K_0 \in \mathcal{K}$, $\|T(a)+K_0\| > \|T(a)\|$ for all sufficiently large n, and

$$\|\big(T(a)+K_0\big) - \big(T(a)+K\big)\| = |\lambda|\,\|P_n\| = |\lambda| < \varepsilon.$$

This shows that $\mathcal{G}$ is dense in $\mathcal{R}\times\mathcal{K}$. Now it easily follows that the set $\{(a,K,L) : M > M_0\}$ is open and dense in $\mathcal{R}\times\mathcal{K}\times\mathcal{K}$.

Let $\mathcal{H} := \{(a,K) \in G\mathcal{R}\times\mathcal{K} : \|(T(a)+K)^{-1}\| > \|T(a^{-1})\|\}$. Again it is obvious that $\mathcal{H}$ is dense in $G\mathcal{R}\times\mathcal{K}$. Assume $\|(T(a)+K)^{-1}\| = \|T(a^{-1})\|$ for certain $a \in G\mathcal{R}$ and $K \in \mathcal{K}$. By Lemma 4.5,

$$\big(T(a)+K\big)^{-1} = T(a^{-1}) + X, \quad X \in \mathcal{K}.$$

The reasoning of the preceding paragraph yields a $\lambda \in \mathbf{C}$ such that $|\lambda| < \varepsilon$ and

$$\|T(a^{-1}) + X + \lambda P_n\| > \|T(a^{-1})\| \tag{4.4}$$

for all sufficiently large n. Put

$$K_n = \big(T(a^{-1}) + X + \lambda P_n\big)^{-1} - T(a).$$

From Lemma 4.5 we deduce that $K_n \in \mathcal{K}$, and (4.4) gives

$$\|\big(T(a)+K_n\big)^{-1}\| = \|T(a^{-1}) + X + \lambda P_n\| > \|T(a^{-1})\|.$$

Finally, we have

$$\|\big(T(a)+K_n\big)^{-1}-\big(T(a)+K\big)^{-1}\|=\|\lambda P_n\|=|\lambda|<\varepsilon.$$

This shows that $\mathcal{H}$ and thus also $\{(a,K,L): M^{(-1)}>M_0^{(-1)}\}$ are open and dense. □

By virtue of Theorem 4.2, we may say that the strict inequalities $M>M_0$ and $M^{(-1)}>M_0^{(-1)}$ are the generic case, while the equalities $M=M_0$ or $M^{(-1)}=M_0^{(-1)}$ represent exceptional cases.

Let us return to the examples cited in Section 3.

For pure Toeplitz matrices, $A_n=T_n(a)$, we have

$$M=M_0=\|T(a)\|,\quad M^{(-1)}=\|T^{-1}(a)\|,\quad M_0^{(-1)}=\|T(a^{-1})\|.$$

Thus, when considering the convergence of $\|T_n(a)\|$ to M, we are in the exceptional case. The situation is different for the convergence of $\|T_n^{-1}(a)\|$ to $M^{(-1)}$: the case $\|T^{-1}(a)\|>\|T(a^{-1})\|$ is generic (see [**11**, Theorem 1.2]). In [**11**] we also showed that $\|T^{-1}(a)\|=\|T(a^{-1})\|$ (exceptional case) if $T(a)$ is Hermitian, as in Example 3.1, or if $T(a)$ is triangular, while $\|T^{-1}(a)\|>\|T(a^{-1})\|$ (generic case) if a is a non-constant unimodular function, as in Example 3.3.

Finally, let $A_n=I+P_nKP_n+W_nLW_n$ be as in Theorem 3.4. In this case

$$M=\max(\|I+K\|,\|I+L\|),\quad M_0=1,$$
$$M^{(-1)}=\max\big(\|(I+K)^{-1}\|,\|(I+L)^{-1}\|\big),\quad M_0^{(-1)}=1.$$

Obviously, all the four equalities

$$\|I+K\|=1,\ \|I+L\|=1,\ \|(I+K)^{-1}\|=1,\ \|(I+L)^{-1}\|=1$$

are satisfied in very rare situations only. Thus, the very fast convergence of the condition numbers $\kappa(A_n)$ to $MM^{(-1)}$ observed in Theorem 3.4 is in accordance with what will be proved in Section 9: *generically* $\kappa(A_n)$ converges to $MM^{(-1)}$ with very high speed. That in the situation at hand $\kappa(A_n)$ goes to $MM^{(-1)}$ very fast (even exponentially) *in either case* is due to some special circumstances. Notice that we will prove that slow convergence *may* happen in the exceptional case only, but that we do not assert that it *must* necessarily occur in the exceptional case.

5. C^*-algebras and stable matrix sequences

The purpose of this section is to illustrate how a few simple C^*-algebra arguments yield pretty nice pieces of information about the limit of the condition numbers $\kappa(A_n)$. The main results, Corollaries 5.3 to 5.5, are well known. We remark that these results can also be proved without having any recourse to C^*-algebras (see [**9**]).

Let $\mathcal{B}:=\mathcal{B}(l^2)$ and $\mathcal{C}:=\mathcal{C}(l^2)$ stand for the sets of all bounded and compact linear operators on $l^2:=l^2(\mathbf{Z}_+)$, respectively. Let further $\mathcal{S}$ be the set of all sequences $\{B_n\}_{n=1}^\infty$ of $n\times n$ matrices B_n such that

$$\|\{B_n\}\|:=\sup_{n\geq 1}\|B_n\|<\infty, \tag{5.1}$$

where $\|B_n\|$ is the norm of B_n on $l_n^2:=l^2(\{0,1,\dots,n-1\})$. It is readily seen that $\mathcal{S}$ is a C^*-algebra under the natural algebraic operations and the norm (5.1). Denote by $\mathcal{A}$ the smallest closed subalgebra of $\mathcal{S}$ which contains all sequences $\{T_n(a)\}_{n=1}^{\infty}$ with $a \in \mathcal{R}$. Obviously, $\mathcal{A}$ is also a C^*-algebra. Finally, let $\mathcal{N}$ denote the sequences $\{C_n\} \in \mathcal{S}$ such that $\|C_n\| \to 0$ as $n \to \infty$.

THEOREM 5.1. *The C^*-algebra $\mathcal{A}$ is the set of all sequences $\{B_n\}_{n=1}^{\infty}$ such that*

$$B_n = T_n(b) + P_n U P_n + W_n V W_n + C_n \tag{5.2}$$

with $b \in C := C(\mathbf{T})$, $U \in \mathcal{C}$, $V \in \mathcal{C}$, $\{C_n\} \in \mathcal{N}$.

This theorem was established by Silbermann and one of the authors in [**12**] (proofs are also in [**13**, Proposition 7.27] and [**9**, Proposition 2.2]). Of course, a key ingredient to the understanding of this theorem is Widom's formula (2.3).

Let B_n be given by (5.2). Then

$$B_n \to B := T(b) + U, \quad W_n B_n W_n \to \widetilde{B} := T(\widetilde{b}) + V \tag{5.3}$$

strongly as $n \to \infty$. The set $\mathcal{N}$ is obviously a closed two-sided ideal of $\mathcal{A}$.

THEOREM 5.2. *The map* Sym *defined by*

$$\mathrm{Sym} : \mathcal{A}/\mathcal{N} \to \mathcal{B} \oplus \mathcal{B}, \ \{B_n\} + \mathcal{N} \mapsto (B, \widetilde{B})$$

is a C^*-algebra homomorphism that preserves spectra and norms.

It is easily verified that Sym is a C^*-algebra homomorphism. Since the only compact Toeplitz operator is the zero operator (see the proof of Lemma 4.1), it follows that Sym is injective. As injective homomorphisms of unital C^*-algebras preserve spectra and norms, we arrive at the conclusion of Theorem 5.2. We remark that this simple reasoning goes back to [**12**], [**13**, Theorem 7.11], and [**5**].

Here are three immediate consequences of Theorem 5.2.

A sequence $\{A_n\} \in \mathcal{A}$ is said to be stable if $\{A_n\} + \mathcal{N}$ is invertible in $\mathcal{A}/\mathcal{N}$. Equivalently, $\{A_n\} \in \mathcal{A}$ is stable if and only if

$$\limsup_{n\to\infty} \|A_n^{-1}\| < \infty.$$

COROLLARY 5.3. *Let $A_n = T_n(a) + P_n K P_n + W_n L W_n + C_n$ with $a \in C$, $K \in \mathcal{C}$, $L \in \mathcal{C}$, $\|C_n\| \to 0$ as $n \to \infty$. The sequence $\{A_n\}$ is stable if and only if*

$$A = T(a) + K \ \text{and} \ \widetilde{A} = T(\widetilde{a}) + L$$

are invertible. □

COROLLARY 5.4. *Let A_n be as in the previous corollary. If $T(a) + K$ and $T(\widetilde{a}) + L$ are invertible, then*

$$A_n^{-1} = T_n(a^{-1}) + P_n X P_n + W_n Y W_n + E_n \tag{5.4}$$

with

$$X := \big(T(a) + K\big)^{-1} - T(a^{-1}) \in \mathcal{C}, \quad Y := \big(T(\widetilde{a}) + L\big)^{-1} - T(\widetilde{a}^{-1}) \in \mathcal{C}, \tag{5.5}$$

and $\|E_n\| \to 0$ as $n \to \infty$.

PROOF. Theorem 5.2 and Corollary 5.3 imply that A_n^{-1} is of the form (5.4) with $X, Y \in \mathcal{C}$ and $\{E_n\} \in \mathcal{N}$. Rewriting (5.4) as

$$I = A_n\big(T_n(a^{-1}) + P_n X P_n + W_n Y W_n + E_n\big),$$
$$I = W_n A_n W_n\big(T_n(\widetilde{a}^{-1}) + W_n X W_n + P_n Y P_n + W_n E_n W_n\big)$$

and passing to the strong limit $n \to \infty$, we get

$$I = \big(T(a) + K\big)\big(T(a^{-1}) + X\big), \quad I = \big(T(\widetilde{a}) + L\big)\big(T(\widetilde{a}^{-1}) + Y\big),$$

which gives (5.5). □

From (5.3) we infer that

$$\liminf_{n\to\infty} \|B_n\| \geq \|B\|,$$
$$\liminf_{n\to\infty} \|B_n\| = \liminf_{n\to\infty} \|W_n B_n W_n\| \geq \|\widetilde{B}\|.$$

Theorem 5.2 implies that

$$\limsup_{n\to\infty} \|B_n\| = \|\{B_n\}\|_{\mathcal{A}/\mathcal{N}} = \max(\|B\|, \|\widetilde{B}\|).$$

Consequently, $\lim_{n\to\infty} \|B_n\|$ exists and

$$\lim_{n\to\infty} \|B_n\| = \max(\|B\|, \|\widetilde{B}\|).$$

Letting $B_n = A_n$ and $B_n = A_n^{-1}$ and taking into account Corollary 5.4, we arrive at the following.

COROLLARY 5.5. *Let A_n be as in Corollary 5.3. Then*

$$\lim_{n\to\infty} \|A_n\| = \max(\|T(a) + K\|, \|T(\widetilde{a}) + L\|),$$
$$\lim_{n\to\infty} \|A_n^{-1}\| = \max(\|(T(a) + K)^{-1}\|, \|(T(\widetilde{a}) + L)^{-1}\|).$$
□

We note that Corollary 5.3 is Silbermann's [**54**], that Corollary 5.4 was first obtained by Widom [**62**], [**63**] (by different methods), and that Corollary 5.5 appeared explicitly in [**5**] for the first time. For far-reaching generalizations of the approach and the results of this section we recommend Hagen, Roch, and Silbermann's works [**26**], [**27**], [**45**]. The book [**14**] may also serve as an introduction to this topic.

6. Asymptotic inverses

In this section we show that if $a \in \mathcal{R}$ and $K, L \in \mathcal{K}$, then the norms of the operators E_n in (5.4) go to zero exponentially.

LEMMA 6.1. *If $R, S \in \mathcal{K}$ then*

$$\|P_n R P_n W_n S W_n\| = O(\tau^n) \text{ and } \|W_n S W_n P_n R P_n\| = O(\tau^n)$$

for some $\tau \in (0,1)$.

PROOF. Let $R, S \in \mathcal{K}_\sigma$ with $\sigma \in (0,1)$. We have

$$|(P_n R P_n W_n S W_n)_{jl}| = \left| \sum_{l=0}^{n-1} R_{jk} S_{n-k-1,n-l-1} \right|$$
$$= O\left(\sum_{l=0}^{n-1} \sigma^{j+l} \sigma^{n-k-1+n-l-1} \right) = O\left(n\sigma^{n-1+j}\sigma^{n-1-k} \right).$$

Hence,

$$\|P_nRP_nW_nSW_n\|^2 \le \|P_nRP_nW_nSW_n\|_F^2 = O\left(\sum_{j,k=0}^{n-1} n^2\sigma^{2(n-1+j)}\sigma^{2(n-1-k)}\right)$$

$$= O\left(n^2\sigma^{2(n-1)}\sum_{k=0}^{n-1}\sigma^{2(n-1-k)}\right) = O\left(n^2\sigma^{2(n-1)}\right) = O\left(n^2\sigma^{2n}\right) = O\left(\tau^n\right)$$

if only $\tau > \sigma^2$. Passage to adjoints gives the result for the operators $W_nSW_nP_nRP_n$. □

THEOREM 6.2. *Let $A_n = T_n(a) + P_nKP_n + W_nLW_n$ with $a \in \mathcal{R}$, $K \in \mathcal{K}$, $L \in \mathcal{K}$. If $T(a) + K$ and $T(\widetilde{a}) + L$ are invertible, then*

$$A_n^{-1} = T_n(a^{-1}) + P_nXP_n + W_nYW_n + E_n \tag{6.1}$$

with

(6.2) $X := (T(a) + K)^{-1} - T(a^{-1}) \in \mathcal{K}, \quad Y := (T(\widetilde{a}^{-1}) + l)^{-1} - T(\widetilde{a}^{-1}) \in \mathcal{K},$

(6.3) $\|E_n\| = O(\tau^n)$ *for some* $\tau \in (0, 1)$.

PROOF. Corollary 5.4 implies (6.1) with $X, Y \in \mathcal{C}$ and $\|E_n\| = o(1)$. From Lemma 4.5 we infer that in fact $X, Y \in \mathcal{K}$. We are left with (6.3).

Put

$$B_n = T_n(a^{-1}) + P_nXP_n + W_nYW_n.$$

Then

$$E_n = A_n^{-1} - B_n = A_n^{-1}(P_n - A_nB_n),$$

and because $\{A_n^{-1}\}$ is stable due to Corollary 5.3, it suffices to prove the estimate $\|A_nB_n - P_n\| = O(\tau^n)$. The matrix $A_nB_n - P_n$ equals

$$\begin{aligned} &P_n(T(a) + K)P_n(T(a^{-1}) + X)P_n - P_n \\ &\quad + P_n(T(a) + K)P_nW_nYW_n + W_nLW_nP_n(T(a^{-1}) + X)P_n \\ &\quad + W_nLW_nW_nYW_n. \end{aligned}$$

From Lemma 6.1 we know that

$$\|P_nKP_nW_nYW_n\| = O(\tau^n), \quad \|W_nLW_nP_nXP_n\| = O(\tau^n),$$

which shows that $A_nB_n - P_n$ is

$$\begin{aligned} &P_n(T(a) + K)P_n(T(a^{-1}) + X)P_n - P_n \\ &\quad + P_nT(a)W_nYW_n + W_nLW_nT(a^{-1})P_n + W_nLW_nW_nYW_n + \boldsymbol{O}(\tau^n) \\ &= P_n(T(a) + K)P_n(T(a^{-1}) + X)P_n - P_n \\ &\quad + W_n(T(\widetilde{a}) + L)P_n(T(\widetilde{a}^{-1}) + Y)W_n - W_nT(\widetilde{a})P_nT(\widetilde{a}^{-1})W_n + \boldsymbol{O}(\tau^n). \end{aligned}$$

Since $(T(a) + K)(T(a^{-1}) + X) = (T(\widetilde{a}) + L)(T(\widetilde{a}^{-1}) + Y) = I$, this is equal to

$$\begin{aligned} &P_n - P_n(T(a) + K)Q_n(T(a^{-1}) + X) - P_n \\ &\quad + P_n - W_n(T(\widetilde{a}) + L)Q_n(T(\widetilde{a}^{-1}) + Y)W_n \\ &\quad - W_nT(\widetilde{a})P_nT(\widetilde{a}^{-1})W_n + \boldsymbol{O}(\tau^n). \end{aligned}$$

Lemma 3.5 tells us that this expression is

$$\begin{aligned}
&P_n - P_nT(a)Q_nT(a^{-1}) + \boldsymbol{O}(\tau^n) - P_n \\
&\quad +P_n - W_nT(\widetilde{a})Q_nT(\widetilde{a}^{-1}W_n + \boldsymbol{O}(\tau^n) \\
&\quad -W_nT(\widetilde{a})P_nT(\widetilde{a}^{-1})W_n + \boldsymbol{O}(\tau^n) \\
&= P_n - P_nT(a)Q_nT(a^{-1})P_n - W_nT(\widetilde{a})T(\widetilde{a}^{-1})W_n + \boldsymbol{O}(\tau^n) \\
&= P_n - P_nT(a)T(a^{-1})P_n + P_nT(a)P_nT(a^{-1})P_n \\
&\quad -W_nT(\widetilde{a})T(\widetilde{a}^{-1})W_n + \boldsymbol{O}(\tau^n)
\end{aligned}$$

$$\begin{aligned}
&= P_n - P_n(I - H(a)H(\widetilde{a}^{-1}))P_n + T_n(a)T_n(a^{-1}) \\
&\quad -W_n(I - H(\widetilde{a})H(a^{-1}))W_n + \boldsymbol{O}(\tau^n) \quad \text{(by (4.1)} \\
&= T_n(a)T_n(a^{-1}) + P_nH(a)H(\widetilde{a}^{-1}) + W_nH(\widetilde{a})H(a^{-1})W_n - P_n + \boldsymbol{O}(\tau^n) \\
&= T_n(aa^{-1}) - P_n + \boldsymbol{O}(\tau^n) \quad \text{(by (2.3))} \\
&= P_n - P_n + \boldsymbol{O}(\tau^n) = \boldsymbol{O}(\tau^n).
\end{aligned}$$

□

7. Band matrices

For $s \in \{1, 2, 3, \ldots\}$, let $\mathcal{P}_s$ denote the collection of all trigonometric polynomials of degree at most s. Thus, $c \in \mathcal{P}_s$ if and only if

$$c(t) = \sum_{|j| \le s} c_j t^j \quad (t \in \mathbf{T}).$$

Also, let $\mathcal{F}_s$ stand for the matrices $R = (R_{jk})_{j,k=0}^{\infty}$ such that $R_{jk} = 0$ if $j > s$ or $k > s$. In other terms,

$$R \in \mathcal{F}_s \Longleftrightarrow P_sRP_s = R.$$

Clearly, $\mathcal{F}_s \subset \mathcal{K}$. Assume we are given

$$C_n = T_n(c) + P_nUP_n + W_nVW_n$$

with $c \in \mathcal{P}_s$, $c_sc_{-s} \neq 0$, $U \in \mathcal{F}_s$, $V \in \mathcal{F}_s$. Obviously, C_n is a band matrix with at most $2s + 1$ nonzero diagonals. Put

$$M := \max\left(\|T(c) + U\|, \|T(\widetilde{c}) + V\|\right), \quad M_0 := \|T(b)\|.$$

Here are our main tools.

THEOREM 7.1. *Suppose $M > M_0$ (generic case). If $s \geq 1$ and $n \geq 8s + 4$, then*

$$\big|\,\|C_n\| - M\big| \le 2sM\left(\frac{M}{M_0}\right)^{\frac{5}{2}}\left(\frac{M_0}{M}\right)^{\frac{n}{1s}}. \tag{7.1}$$

THEOREM 7.2. *Suppose $M \geq M_0$ (general case). If $s \geq 1$ and $n \geq 288s$, then*

$$M_0\left(1 - 146\frac{s}{n}\right) \le \|C_n\| \le M\left(1 + 4\frac{s}{n}\right). \tag{7.2}$$

In particular, if $M = M_0$ (exceptional case), then

$$\big|\,\|C_n\| - M\big| \le 146M\frac{s}{n}. \tag{7.3}$$

Theorem 7.1 is Theorem 1.3 of [**10**], and Theorem 7.2 is a combination of (the proof of) Theorem 1.5 and Proposition 3.2 of [**10**].

8. Approximation by band matrices

In this section we approximate the matrices we are interested in by band matrices and make use of Theorems 7.1 and 7.2 for the approximating band matrices.

LEMMA 8.1. *If $R \in \mathcal{K}_\sigma$ then $\|R - P_s R P_s\| = O(\sigma^s)$.*

PROOF. Because $R - P_s R P_s = P_s R Q_s + Q_s R P_s + Q_s R Q_s$, this is straightforward from Lemma 3.5. □

THEOREM 8.2. *Let*

$$B_n = T_n(b) + P_n U P_n + W_n V W_n, \quad b \in \mathcal{R},\ U \in \mathcal{K},\ V \in \mathcal{K},$$

and put

$$M := \max\big(\|T(b) + U\|,\ \|T(\widetilde{b}) + V\|\big), \quad M_0 := \|T(b)\|.$$

If $M > M_0$ (generic case), then there is a $\gamma > 0$ depending only on b, U, V such that

$$\big|\,\|B_n\| - M\big| = O(e^{-\gamma\sqrt{n}}). \tag{8.1}$$

If $M = M_0$ (exceptional case), then

$$\big|\,\|B_n\| - M\big| = O\Big(\frac{\log n}{n}\Big). \tag{8.2}$$

PROOF. Let $s(n)$ be a sequence of natural numbers such that

$$s(n) \to \infty, \quad \frac{n}{s(n)} \to \infty. \tag{8.3}$$

Put $U_{s(n)} = P_{s(n)} U P_{s(n)}$, $V_{s(n)} = P_{s(n)} V P_{s(n)}$ and let $c_n \in \mathcal{P}_{s(n)}$ denote the $s(n)$th partial sum of the Fourier (Laurent) series of b. From (1.2) with b in place of a we know that

$$\|b - c_n\|_\infty = O(\varrho^{s(n)}). \tag{8.4}$$

We have

$$\begin{aligned} &\Big|\,\|B_n\| - \|T_n(c_n) + P_n U_{s(n)} P_n + W_n V_{s(n)} W_n\|\,\Big| \\ &\quad \le \|B_n - T_n(c_n) - P_n U_{s(n)} P_n - W_n V_{s(n)} W_n\| \\ &\quad \le \|T_n(b - c_n)\| + \|P_n(U - U_{s(n)})P_n\| + \|W_n(V - V_{s(n)})W_n\| \\ &\quad \le \|b - c_n\|_\infty + \|U - U_{s(n)}\| + \|V - V_{s(n)}\|. \end{aligned} \tag{8.5}$$

Suppose $U, V \in \mathcal{K}_s$. From (8.4) and Lemma 8.1 we then deduce that (8.5) is

$$O(\varrho^{s(n)}) + O(\sigma^{s(n)}) = O(\tau^{s(n)}),$$

where $\tau \in (0,1)$ is any number such that $\varrho < \tau, \sigma < \tau$.

Let

$$M(s(n)) := \max\Big(\|T(c_n) + U_{s(n)}\|,\ \|(\widetilde{c}_n) + V_{s(n)}\|\Big), \quad M_0(s(n)) := \|T(c_n)\|.$$

Then $\big|M(s(n)) - M\big|$ does not exceed

$$\begin{aligned} &\max\Big(\Big|\,\|T(c_n) + U_{s(n)}\| - \|T(b) + U\|\,\Big|,\ \Big|\,\|T(\widetilde{c}_n) + V_{s(n)}\| - \|T(\widetilde{b}) + V\|\,\Big|\Big) \\ &\quad \le \max\Big(\|T(c_n - b) + U_{s(n)} - U\|,\ \|T(\widetilde{c}_n - \widetilde{b}) + V_{s(n)} - V\|\Big), \end{aligned}$$

and again taking into account (8.4) and Lemma 8.1, we see that this is

$$O(\varrho^{s(n)}) + O(\sigma^{s(n)}) = O(\tau^{s(n)}).$$

Furthermore, also by (8.4),

$$\big|M_0\big(s(n)\big) - M_0\big| = \Big| \|T(c_n)\| - \|T(b)\| \Big| \le \|T(c_n - b)\| = O(\varrho^{s(n)}) = O(\tau^{s(n)}).$$

In summary, at the present moment we have shown that

$$\begin{aligned} &\big| \|B_n\| - M\big| \\ &\quad\le \Big| \|T_n(c_n) + P_n U_{s(n)} V_n + W_n V_{s(n)} W_n\| - M\big(s(n)\big)\Big| + O(\tau^{s(n)}) \end{aligned} \tag{8.6}$$

and that

$$M\big(s(n)\big) = M + O(\tau^{s(n)}), \quad M_0\big(s(n)\big) = M_0 + O(\tau^{s(n)}). \tag{8.7}$$

Now suppose $M > M_0$. Then, by (8.7), $M(s(n)) > M_0(s(n))$ for all sufficiently large n, and (7.1) therefore gives

$$\begin{aligned} &\Big| \|T(c_n) + P_n U_{s(n)} V_n + W_n V_{s(n)} W_n\| - M\big(s(n)\big)\Big| \\ &\qquad \le 2s(n) M\big(s(n)\big) \left(\frac{M(s(n))}{M_0(s(n))}\right)^{\frac{5}{2}} \left(\frac{M_0(s(n))}{M(s(n))}\right)^{\frac{n}{4s(n)}}. \end{aligned} \tag{8.8}$$

Choose any $\varepsilon > 0$ so that $(M_0 + \varepsilon)/(M - \varepsilon) \in (0,1)$ and assume that $\tau \in (0,1)$ is larger than $(M_0 + \varepsilon)/(M - \varepsilon)$. If n is sufficiently large, then (8.8) is at most

$$2s(n)(M+\varepsilon)\left(\frac{M+\varepsilon}{M_0 - \varepsilon}\right)^{\frac{5}{2}} \left(\frac{M_0+\varepsilon}{M-\varepsilon}\right)^{\frac{n}{4s(n)}} = O\Big(s(n)\tau^{\frac{n}{4s(n)}}\Big).$$

This in conjunction with (8.6) yields the estimate

$$\big| \|B_n\| - M\big| = O\big(s(n)\tau^{n/(4s(n))}\big).$$

Letting $s(n) = [\sqrt{n}]$, we obtain (8.1) with any $\gamma > 0$ such that $\tau^{1/4} < e^{-\gamma}$.

Now assume $M = M_0$. Then $M(s(n)) \ge M_0(s(n))$ for all n by virtue of Lemma 4.1. From (7.2) and (8.7) we obtain

$$\begin{aligned} &\|T(c_n) + P_n U_{s(n)} P_n + W_n V_{s(n)} W_n\| - M\big(s(n)\big) \\ &\quad \le 4M\big(s(n)\big)\frac{s(n)}{n} = O\Big(\frac{s(n)}{n}\Big) \end{aligned}$$

and

$$\begin{aligned} &-\|T(c_n) + P_n U_{s(n)} P_n + W_n V_{s(n)} W_n\| + M\big(s(n)\big) \\ &\quad \le 146 M_0\big(s(n)\big)\frac{s(n)}{n} + \Big|M\big(s(n)\big) - M_0\big(s(n)\big)\Big| \\ &\quad = O\Big(\frac{s(n)}{n}\Big) + O(\tau^{s(n)}) \end{aligned}$$

(to get the $O(\tau^{s(n)})$ we used (8.7) and the equality $M = M_0$). Choose $\gamma > 0$ and $\alpha > 0$ so that $\tau < e^{-\gamma}$ and $\gamma\alpha > 1$. Then put $s(n) = [\alpha \log n]$. Clearly, (8.3) is satisfied. As

$$O\left(\frac{s(n)}{n}\right) = O\left(\frac{\log n}{n}\right),$$

$$O(\tau^{s(n)}) = O(e^{-\gamma\alpha \log n}) = O(n^{-\gamma\alpha}) = O\left(\frac{\log n}{n}\right),$$

we arrive at (8.2). □

9. Convergence of condition numbers

We are now in a position to prove Theorem 1.2. Thus, let

$$A_n = T_n(a) + P_n K P_n + W_n L W_n, \quad a \in \mathcal{R},\ K \in \mathcal{K},\ L \in \mathcal{K},$$

and put

$$M := \max\Big(\|T(a)+K\|,\ \|T(\widetilde{a})+L\|\Big), \quad M_0 := \|T(a)\|,$$

$$M^{(-1)} := \max\Big(\|(T(a)+K)^{-1}\|,\ \|(T(\widetilde{a})+L)^{-1}\|\Big), \quad M_0^{(-1)} := \|T(a^{-1})\|.$$

THEOREM 9.1. *If $M > M_0$ (generic case), then*

$$\big|\,\|A_n\| - M\big| = O(e^{-\gamma\sqrt{n}})$$

with some $\gamma > 0$ depending only on a, K, L, while if $M = M_0$ (exceptional case), then

$$\big|\,\|A_n\| - M\,\big| = O\left(\frac{\log n}{n}\right).$$

PROOF. This is nothing but Theorem 8.2 in other notation. □

THEOREM 9.2. *We have $\limsup_{n\to\infty} \|A_n^{-1}\| < \infty$ if and only if $T(a)+K$ and $T(\widetilde{a})+L$ are invertible. If $T(a)+K$ and $T(\widetilde{a})+L$ are invertible and $M^{(-1)} > M_0^{(-1)}$ (generic case), then*

$$\Big|\|A_n^{-1}\| - M^{(-1)}\Big| = O(e^{-\gamma\sqrt{n}})$$

with some $\gamma > 0$ depending only on a, K, L. If $T(a)+K$ and $T(\widetilde{a})+L$ are invertible and $M^{(-1)} = M_0^{(-1)}$ (exceptional case), then

$$\Big|\|A_n^{-1}\| - M^{(-1)}\Big| = O\left(\frac{\log n}{n}\right).$$

PROOF. The first assertion is Corollary 5.3. So assume $T(a)+K$ and $T(\widetilde{a})+L$ are invertible. Thus, by Theorem 6.2,

$$A_n^{-1} = T_n(a^{-1}) + P_n X P_n + W_n Y W_n + \boldsymbol{O}(\tau^n),$$

where X and Y are given by (6.2) and $\tau \in (0,1)$. Theorem 8.2 implies that $\|A_n^{-1}\|$ converges to

$$\max\Big(\|T(a^{-1})+X\|,\ \|T(\widetilde{a}^{-1})+Y\|\Big)$$

$$= \max\Big(\|\big(T(a)+K\big)^{-1}\|,\ \|\big(T(\widetilde{a})+L\big)^{-1}\|\Big) = M^{(-1)}$$

with the speed as asserted in the theorem. □

COROLLARY 9.3. *Suppose $T(a)+K$ and $T(\widetilde{a})+L$ are invertible. If $M > M_0$ and $M^{(-1)} > M_0^{(-1)}$ (generic case), then*

$$\left|\kappa(A_n) - MM^{(-1)}\right| = O(e^{-\gamma\sqrt{n}})$$

with some $\gamma > 0$ depending only on a, K, L. If $M = M_0$ or $M^{(-1)} = M_0^{(-1)}$ (exceptional case), then

$$\left|\kappa(A_n) - MM^{(-1)}\right| = O\left(\frac{\log n}{n}\right).$$ □

10. Pure Toeplitz matrices

We conclude with a few remarks on the condition numbers of pure Toeplitz matrices. Suppose $a \in \mathcal{R}$ does not vanish identically. If $T(a)$ is invertible, we know that

$$\kappa\big(T_n(a)\big) \to \|T(a)\|\,\|T^{-1}(a)\| = \kappa\big(T(a)\big),$$

and Corollary 9.3 tells us that the convergence speed is at least $O(\log n/n)$.

Now assume that $T(a)$ is not invertible. Then

$$\limsup_{n\to\infty} \kappa\big(T_n(a)\big) = \infty.$$

By Theorem 4.4, there are two reasons for $T(a)$ to be not invertible: the first is that a has no zeros on $\mathbf{T}$ but a nonvanishing winding number $\operatorname{wind} a$ about the origin and the second is that a has zeros on $\mathbf{T}$. The following results are from [**8**].

THEOREM 10.1. *If $0 \notin a(\mathbf{T})$ but $\operatorname{wind} a \neq 0$, then the condition numbers $\kappa(T_n(a))$ grow at least exponentially, that is, there are $D > 0$ and $\alpha > 0$ depending only on a such that*

$$\kappa\big(T_n(a)\big) \geq De^{\alpha n}$$

for all $n \geq 1$.

Getting upper bounds for $\kappa(T_n(a))$ is a more intricate problem. Let $\operatorname{sp} T_n(a)$ stand for the spectrum (= set of eigenvalues) of $T_n(a)$. The limiting set

$$\Lambda(a) = \limsup_{n\to\infty} \operatorname{sp} T_n(a)$$

is the set of all partial limits of the sequence $\{\operatorname{sp} T_n(a)\}_{n=1}^{\infty}$. In other words, $\lambda \in \Lambda(a)$ if and only if there exist $n_1 < n_2 < n_3 < \ldots$ and $\lambda_k \in \operatorname{sp} T_{n_k}(a)$ such that $\lambda_{n_k} \to \lambda$. The set $\Lambda(a)$ is known from the work of Schmidt and Spitzer [**48**], Hirschman [**32**], and Day [**15**]. It turns out that $\Lambda(a)$ is either a single point or the union of finitely many analytic arcs which have at most their endpoints on common. Thus, $\Lambda(a)$ is always a "thin" set. If $a(t) = t$ or $a(t) = t + t^{-1}/4$, then the spectrum of $T(a)$ is the closed unit disk and the closure of the interior of an ellipse with the foci -1 and 1, respectively, while $\Lambda(a)$ is the singleton $\{0\}$ and the segment $[-1, 1]$, respectively.

THEOREM 10.2. *If $0 \notin \Lambda(a) \cup a(\mathbf{T})$ but $\operatorname{wind} a \neq 0$, then there are constants $D \in (0,\infty)$ and $\alpha, \beta \in (0, \infty)$ depending only on a such that*

$$De^{\alpha n} \leq \kappa\big(T_n(a)\big) \leq De^{\beta n}$$

for all sufficiently large n.

THEOREM 10.3. *Let $a(t) = t + t^{-1}/4$ $(t \in \mathbf{T})$. Given any function $\varphi : \mathbf{N} \to (0, \infty)$, for example, $\varphi(n) = \exp(n^n)$, there exists a point $\lambda \in \Lambda(a)$ $(= \Lambda(a) \setminus a(\mathbf{T}))$ such that $\kappa(T_n(a - \lambda)) < \infty$ for all $n \geq 1$ but*

$$\kappa\big(T_{n_k}(a - \lambda)\big) > \varphi(n_k)$$

for infinitely many n_k.

Things are more difficult if a has zeros on $\mathbf{T}$. The following theorem provides us with a useful result in the case of only one zero.

THEOREM 10.4. *Suppose*

$$a(t) = (t - t_0)^{\beta} t^k c(t) \quad (t \in \mathbf{T})$$

where $\beta \in \{1, 2, \dots\}$, $k \in \mathbf{Z}$, $c \in \mathcal{R}$, $0 \notin c(\mathbf{T})$, $\operatorname{wind} c = 0$, $t_0 \in \mathbf{T}$. If $-\beta \leq k \leq 0$, there are constants $C_1, C_2 \in (0, \infty)$ such that

$$C_1 n^{\beta} \leq \kappa\big(T_n(a)\big) \leq C_2 n^{\beta}$$

for all sufficiently large n. If $k < -\beta$ or $k > 0$, then there are constants $D > 0$ and $\alpha > 0$ such that

$$\kappa\big(T_n(a)\big) \geq D e^{\alpha n}$$

for all sufficiently large n.

The following general lower estimate was proved in [**7**].

THEOREM 10.5. *Suppose $0 \in a(\mathbf{T})$ and let β be the maximal order of the zeros of a on $\mathbf{T}$. Then there is a constant $c > 0$ such that*

$$\kappa\big(T_n(a)\big) \geq C n^{\beta}.$$

In particular, we always have $\kappa(T_n(a)) \geq Cn$.

11. Additional notes

The remarks made in Section 2 focus on the problem considered here in its narrow sense. Of course, in a wider sense, the investigation of the condition numbers of Toeplitz-like matrices is in fact a big business.

Every result on the extreme eigenvalues of Hermitian Toeplitz matrices is a result on the condition numbers of these matrices. The pioneering work in this direction includes the papers Kac, Murdock, Szegö [**34**], Widom [**61**], Parter [**38**], and the book Grenander and Szegö [**24**]. For example, in [**24**, Section 5.4] it is shown that if $a : \mathbf{T} \to \mathbf{R}$ is a nonnegative smooth function which attains its minimum 0 on $\mathbf{T}$ at exactly one point $t_0 \in \mathbf{T}$ and if $a''(t_0) \neq 0$, then

$$\kappa(T_n(a)) \simeq n^2.$$

Only recently, Serra [**50**], [**51**] generalized this result to nonnegative L^1 functions with several minima of possibly higher orders. Extensions to positively semi-definite matrices are in [**7**]. Interesting results for matrices of the Kac-Murdock-Szegö type "without a symbol" are in Trench's paper [**58**].

It has been known for a long time that $\kappa(T_n(a))$ does increase exponentially if a is a nonnegative function on $\mathbf{T}$ with very strong zeros; see Rosenblatt [**47**] and Pourahmadi [**41**]. This is, for instance, the case if a vanishes identically on some subarc of $\mathbf{T}$. Conversely, Serra [**52**] observed that if a is a nonpathological nonnegative L^1 function on $\mathbf{T}$, then $\kappa(T_n(a))$ can never increase faster than $O(e^{\gamma n})$

with some $\gamma < \infty$. Here a nonnegative function a is said to be nonpathological if there is some subarc $\Gamma \subset \mathbf{T}$ such that $a(t) > 0$ for all $t \in \Gamma$. Tilli [**57**] informed us that he is now able to prove this result for every (and thus also pathological) nonnegative L^1 function. Notice in this connection that rational functions cannot have strong zeros unless they vanish identically. The exponentional growth of the condition numbers in Theorem 10.1 is caused by another reason.

The singular values of $T_n(a)$ are the eigenvalues of $(T_n(a)T_n(\overline{a}))^{1/2}$. Clearly, the minimal singular value of $T_n(a)$ is nothing but $\|T_n^{-1}(a)\|^{-1}$. Thus, the investigation of the asymptotic behavior of the condition numbers of Toeplitz matrices is equivalent to the study of the asymptotics of the smallest singular value. A well known theorem by Avram [**1**] and Parter [**39**], which was recently essentially generalized by Tyrtyshnikov and Zamarashkin [**59**], [**65**], states that the singular values $\{s_j(T_n(a)\}_{j=1}^n$ are asymptotically distributed like the values $\{|a(e^{2\pi ij/n})|\}_{j=1}^n$. Full proofs of the Avram-Parter theorem are also in [**14**] and [**60**]. This theorem *gives us the hint* that if, for example, $a(e^{i\theta}) = |\theta|^\alpha$ for θ in some neighborhood of zero and $|a(e^{i\theta})| \geq \varepsilon > 0$ for all other θ's, then $s_1(T_n(a))$ should decay as $1/n^\alpha$ (which would show that $\kappa(T_n(a))$ should increase as n^α), but the theorem *does not imply* that this really happens. This motivates the search for such results as Theorem 10.4.

The Avram-Parter theorem was extended by SeLegue [**49**] to the singular values of more general matrices; also see Section 5.7 of [**14**] and [**53**], [**56**]. In particular, we now have theorems on the asymptotic singular value distribution of P_nAP_n where A belongs to the smallest closed subalgebra $\mathbf{A}$ of $\mathcal{B}(l^2)$ which contains all Toeplitz operators generated by L^∞ functions. Clearly, all matrices of the form

$$A = \sum_{j=1}^{M} \prod_{k=1}^{M} T(a_{jk}), \quad a_{jk} \in L^\infty(\mathbf{T})$$

belong to $\mathbf{A}$ (and form a dense subset of $\mathbf{A}$). We remark that if all a_{jk} are in $\mathcal{R}$, then

$$A = T\left(\sum_{j=1}^{M} \prod_{k=1}^{M} a_{jk}\right) + K$$

with $K \in \mathcal{K}$. Furthermore, Fasino and Tilli [**17**] showed that if $\{D_n\}$ is any sequence of $n \times n$ matrices D_n such that $\|D_n\|_{\text{tr}}/n \to 0$ as $n \to \infty$, where $\|\cdot\|_{\text{tr}}$ is the trace norm, then the singular values of $A_n + D_n$ and A_n have the same asymptotic distribution. From Lemma 3.5 we deduce that the nth approximation number of $R \in \mathcal{K}_\sigma$ is at most $O(\sigma^n)$. This implies that operators in $\mathcal{K}$ are trace class operators, whence

$$\frac{\|P_nKP_n\|_{\text{tr}}}{n} \leq \frac{\|K\|_{\text{tr}}}{n}, \quad \frac{\|W_nLW_n\|_{\text{tr}}}{n} \leq \frac{\|L\|_{\text{tr}}}{n}$$

for $K, L \in \mathcal{K}$. Thus, if K and L are in $\mathcal{K}$, then the singular values of $T_n(a) + P_nKP_n + W_nLW_n$ are asymptotically distributed as the singular values of $T_n(a)$.

The limiting sets $\liminf_{n\to\infty} \Sigma(T_n(a))$ and $\limsup_{n\to\infty} \Sigma(T_n(a))$ of the set

$$\Sigma(T_n(a)) := \{s_1(T_n(a)), \dots, s_n(T_n(a))\}$$

of the singular values $s_1(T_n(a)) \le \ldots \le s_n(T_n(a))$ of $T_n(a)$) have also been studied in detail. See Widom [**64**], Roch and Silbermann [**44**], Silbermann [**55**], Zizler, Zuidwijk, Taylor, Arimoto [**66**], and see also [**5**] and [**14**].

A remarkable result on singular values of Toeplitz matrices is the splitting phenomenon discovered by Roch and Silbermann [**46**]. They showed that if $a : \mathbf{T} \to \mathbf{C}$ is piecewise continuous and $T(a)$ is Fredholm of index $k \in \mathbf{Z}$, then exactly $|k|$ singular values of $T_n(a)$ go to zero while the remaining $n - |k|$ singular values of $T_n(a)$ stay away from zero:

$$\lim_{n\to\infty} s_{|k|}(T_n(a)) = 0, \quad \liminf_{n\to\infty} s_{|k|+1}(T_n(a)) > 0.$$

An alternative proof of this result is in [**6**], and full proofs can also be found in [**14**]. Moreover, in [**8**] we proved that if $a \in \mathcal{R}$, then the $|k|$ lower singular values of $T_n(a)$ approach zero even with exponential speed. Applying the latter result to the smallest singular value, we obtain Theorem 10.1.

In this paper we considered perturbations of Toeplitz operators generated by rational functions. Much work has been done on Toeplitz operators with piecewise continuous generating functions. Notice that, for example, the Cauchy-Toeplitz matrix

$$\left(\frac{1}{j-k+\gamma}\right)_{j,k=0}^{\infty}, \quad \gamma \in \mathbf{C} \setminus \mathbf{Z},$$

is the Toeplitz matrix $T(a)$ with

$$a(e^{i\theta}) = \frac{\pi}{\sin \pi\gamma} e^{i\pi\gamma} e^{-i\gamma\theta}, \quad \theta \in [0, 2\pi),$$

and is therefore a Toeplitz matrix induced by a piecewise continuous function. Gohberg and Feldman [**20**] were the first to prove that $\kappa(T_n(a))$ remains bounded as $n \to \infty$ if a is piecewise continuous and $T(a)$ is invertible on l^2. Singular values and condition numbers of Toeplitz matrices with piecewise continuous generating functions are considered in [**5**], [**6**], [**7**], [**14**], [**44**], [**45**], [**46**], [**64**], for example.

Several results of this paper extend to block Toeplitz matrices. For instance, Corollaries 5.4 to 5.5 and Theorem 6.2 remain literally true in the block case. We think that Conjecture 1.1 and Theorem 1.2 are also valid for block matrices, but we have not checked the details. In connection with the problems studied here, the most powerful tool for tackling rationally generated block Toeplitz and block Toeplitz-like matrices is probably the state space method. Gohberg, Goldberg, and Kaashoek's two volumes [**21**] are an account and an inexhaustible source of this method. We also recommend Kaashoek's article [**33**]. Block Toeplitz matrices perturbed by matrices from the block analogue of $\mathcal{K}$ are considered by van der Mee, Seatzu, and Rodriguez [**36**].

Finally, we wish to emphasize that the class of matrices considered in this paper includes all rationally generated Toeplitz plus Hankel matrices. This is one of the most popular classes of matrices which admit a low rank reduction (Heinig and Rost [**28**]) or, equivalently, which have a displacement structure (Kailath and Sayed [**35**]). Fast and superfast algorithms for the solution of linear systems with such matrices and for matrix-vector multiplication with these matrices are an active research area; see, e.g., Bini and Pan [**4**], Gohberg, Kailath, Olshevsky [**22**], Gohberg and Koltracht [**23**], Heinig and Rost [**28**], [**29**], [**30**], [**31**], Olshevsky [**37**], Potts and

Steidl [**40**]. Asymptotic spectral properties of Toeplitz plus Hankel matrices were studied by Fasino [**16**], Fasino and Tilli [**17**], and Basor and Ehrhardt [**2**].

References

[1] F. Avram: On bilinear forms in Gaussian variables and Toeplitz matrices. *Probab. Theory and Related Fields* **79** (1988), 37-45.

[2] E. Basor and T. Ehrhardt: Asymptotic formulas for the determinants of a sum of Toeplitz and Hankel matrices. *Math. Nachr.*, to appear.

[3] G. Baxter: A norm inequality for a finite section Wiener-Hopf equation. *Illinois J. Math.* **7** (1963), 97-103.

[4] D. Bini and V. Pan: *Polynomial and Matrix Computations, I. Fundamental Algorithms.* Birkhäuser Verlag, Boston 1994.

[5] A. Böttcher: Pseudospectra and singular values of large convolution operators. *J. Integral Equations Appl.* **6** (1994), 267-301.

[6] A. Böttcher: On the approximation numbers of large Toeplitz matrices. *Documenta Mathematica* **2** (1997), 1-29.

[7] A. Böttcher and S. Grudsky: On the condition numbers of large semi-definite Toeplitz matrices. *Linear Algebra Appl.* **279** (1998), 285-301.

[8] A. Böttcher and S. Grudsky: Toeplitz band matrices with exponentially growing condition numbers. *Electronic J. Linear Algebra* **5** (1999), 104-125.

[9] A. Böttcher, S. Grudsky, and B. Silbermann: Norms of inverses, spectra, and pseudospectra of large truncated Wiener-Hopf operators and Toeplitz matrices. *New York J. Math.* **3** (1997), 1-31.

[10] A. Böttcher, S. Grudsky, A. Kozak, and B. Silbermann: Norms of large Toeplitz band matrices. *SIAM J. Matrix Analysis Appl.* **21** (1999), 547-561.

[11] A. Böttcher, S. Grudsky, A. Kozak, and B. Silbermann: Convergence speed estimates for the norms of the inverses of large truncated Toeplitz matrices. *Calcolo* **36** (1999), 103-122.

[12] A. Böttcher and B. Silbermann: The finite section method for Toeplitz operators on the quarter-plane with piecewise continuous symbols. *Math. Nachr.* **110** (1983), 279-291.

[13] A. Böttcher and B. Silbermann: *Analysis of Toeplitz Operators.* Akademie-Verlag, Berlin 1989 and Springer-Verlag, Berlin, Heidelberg, New York 1990.

[14] A. Böttcher and B. Silbermann: *Introduction to Large Truncated Toeplitz Matrices.* Springer-Verlag, New York 1998.

[15] K.M. Day: Measures associated with Toeplitz matrices generated by the Laurent expansion of rational functions. *Trans. Amer. Math. Soc.* **209** (1975), 175-183.

[16] D. Fasino: Spectral properties of Toeplitz plus Hankel matrices. *Calcolo* **33** (1996), 87-98.

[17] D. Fasino and P. Tilli: Spectral clustering properties of block multilevel Hankel matrices. *Linear Algebra. Appl.* **306** (2000), 155-163.

[18] I. Gohberg: On an application of the theory of normed rings to singular integral equations. *Uspekhi Matem. Nauk* **7** (1952), 149-156 [Russian].

[19] I. Gohberg and I.A. Feldman: On the finite section method for systems of the Wiener-Hopf type. *Dokl. Akad. Nauk SSSR* **165** (1965), 268-271 [Russian].

[20] I. Gohberg and I.A. Feldman: *Convolution Equations and Projection Methods for Their Solution.* Amer. Math. Soc., Providence, RI, 1974 [Russian original: Nauka, Moscow 1971].

[21] I. Gohberg, S. Goldberg, and M.A. Kaashoek: *Classes of Linear Operators.* Vol. I: Birkhäuser Verlag, Basel 1990; Vol. II: Birkhäuser Verlag, Basel 1993.

[22] I. Gohberg, T. Kailath, and V. Olshevsky: Fast Gaussian elemination with partial pivoting for matrices with displacement structure. *Math. Comput.* **64** (1995), 1557-1576.

[23] I. Gohberg and I. Koltracht: Efficient algorithm for Toeplitz plus Hankel matrices. *Integral Equations and Operator Theory* **12** (1989), 136-142.

[24] U. Grenander and G. Szegö: *Toeplitz Forms and Their Applications.* University of California Press, Berkeley and Los Angeles 1958.

[25] S. Grudsky and A. Kozak: On the convergence speed of the norms of the inverses of truncated Toeplitz operators. In: *Integro-Differential Equations and Appl.*, pp. 45-55, Rostov-on-Don University Press, Rostow-on-Don 1995 [Russian].

[26] R. Hagen, S. Roch, and B. Silbermann: *Spectral Theory of Approximation Methods for Convolution Equations.* Birkhäuser Verlag, Basel 1995.

[27] R. Hagen, S. Roch, and B. Silbermann: *C^*-Algebras and Numerical Analysis*. Book in preparation.

[28] G. Heinig and K. Rost: *Algebraic Methods for Toeplitz-like Matrices and Operators*. Akademie-Verlag, Berlin 1984 and Birkhäuser Verlag, Basel 1984.

[29] G. Heinig and K. Rost: On the inverses of Toeplitz plus Hankel matrices. *Linear Algebra. Appl.* **106** (1988), 39-52.

[30] G. Heinig and K. Rost: DFT representations of Toeplitz plus Hankel Bezoutians with application to fast matrix-vector multiplication. *Linear Algebra Appl.* **284** (1998), 157-175.

[31] G. Heinig and K. Rost: Hartley transform representations of inverses of real Toeplitz plus Hankel matrices. *Numer. Funct. Anal. Optim.* **21** (2000), 175-189.

[32] I.I. Hirschman, Jr.: The spectra of certain Toeplitz matrices. *Illinois J. Math.* **11** (1967), 145-159.

[33] M.A. Kaashoek: State space theory of rational matrix functions and applications. In: *Lectures on Operator Theory and Its Applications* (edited by P. Lancaster), pp. 233-333, Fields Institute Monographs, Vol. 3, Amer. Math. Soc., Providence, RI, 1995.

[34] M. Kac, W.L. Murdock, and G. Szegö: On the eigenvalues of certain Hermitian forms. *J. Rat. Mech. Analysis* **2** (1953), 767-800.

[35] T. Kailath and A.H. Sayed: Displacement structure: theory and applications. *SIAM Rev.* **37** (1985), 297-386.

[36] C.V.M. van der Mee, S. Seatzu, and G. Rodriguez: Solution methods for perturbed semi-infinite linear systems of block-Toeplitz type. Talk given at the Second Workshop "Large-Scale Scientific Computations", Sozopol, Bulgaria, June 5, 1999.

[37] V. Olshevsky: Unified superfast algorithm for passive tangential confluent rational boundary interpolation problems. Talk given at the Second Workshop "Large-Scale Scientific Computations", Sozopol, Bulgaria, June 3, 1999.

[38] S.V. Parter: Extreme eigenvalues of Toeplitz forms and applications to elliptic difference equations. *Trans. Amer. Math. Soc.* **99** (1966), 153-192.

[39] S.V. Parter: On the distribution of the singular values of Toeplitz matrices. *Linear Algebra Appl.* **80** (1986), 115-130.

[40] D. Potts and G. Steidl: Optimal trigonometric preconditioners for non-symmetric Toeplitz systems. *Linear Algebra Appl.* **281** (1998), 265-292.

[41] M. Pourahmadi: Remarks on the extreme eigenvalues of Toeplitz matrices. *Internat J. Math. & Math. Sci.* **11** (1988), 23-26.

[42] E. Reich: On non-Hermitian Toeplitz matrices. *Math. Scand.* **10** (1962), 145-152.

[43] L. Reichel and L.N. Trefethen: Eigenvalues and pseudo-eigenvalues of Toeplitz matrices. *Linear Algebra Appl.* **162** (1992), 153-185.

[44] S. Roch and B. Silbermann: Limiting sets of eigenvalues and singular values of Toeplitz matrices. *Asymptotic Analysis* **8** (1994), 293-309.

[45] S. Roch and B. Silbermann: C^*-algebra techniques in numerical analysis. *J. Operator Theory* **35** (1996), 241-280.

[46] S. Roch and B. Silbermann: Index calculus for approximation methods and singular value decomposition. *J. Math. Analysis Appl.* **225** (1998), 401-426.

[47] M. Rosenblatt: Some purely deterministic processes. *J. Math. Mech.* **6** (1957), 801-810.

[48] P. Schmidt and P. Spitzer: The Toeplitz matrices of an arbitrary Laurent polynomial. *Math. Scand.* **8** (1960), 15-38.

[49] D. SeLegue: A C^*-algebraic extension of the Szegö trace formula. Talk given at the GPOTS, Arizona State University, Tempe, May 22, 1996.

[50] S. Serra: On the extreme spectral properties of Toeplitz matrices generated by L^1 functions with several minima/maxima. *BIT* **34** (1996), 135-142.

[51] S. Serra: On the extreme eigenvalues of Hermitian (block) Toeplitz matrices. *Linear Algebra Appl.* **270** (1997), 109-129.

[52] S. Serra: How bad can positive definite Toeplitz matrices be? *Numer. Funct. Anal. Optim.* **21** (2000), 255-261.

[53] S. Serra and P. Tilli: From partial differential equations to generalized locally Toeplitz sequences. Technical report 12-99, Dept. Informatica, University of Pisa 1999.

[54] B. Silbermann: Lokale Theorie des Reduktionsverfahrens für Toeplitzoperatoren. *Math. Nachr.* **104** (1981), 137-146.

[55] B. Silbermann: On the limiting set of the singular values of Toeplitz matrices. *Linear Algebra Appl.* **182** (1993), 35-43.

[56] P. Tilli: Locally Toeplitz sequences - spectral properties and applications. *Linear Algebra Appl.* **278** (1998), 91-120.

[57] P. Tilli: private communication, April 1999.

[58] W.F. Trench: Asymptotic distribution of the spectra of a class of generalized Kac-Murdock-Szegö matrices. *Linear Algebra Appl.* **294** (1999), 181-192.

[59] E.E. Tyrtyshnikov: New theorems on the distribution of the eigenvalues and singular values of multilevel Toeplitz matrices. *Dokl. Akad. Nauk* **333** (1993), 300-303 [Russian].

[60] E.E. Tyrtyshnikov: A unifying approach to some old and new theorems on distribution and clustering. *Linear Algebra Appl.* **232** (1996), 1-43.

[61] H. Widom: On the eigenvalues of certain Hermitian operators. *Trans. Amer. Math. Soc.* **88** (1958), 491-522.

[62] H. Widom: Asymptotic inversion of convolution operators. *Publ. Math. I.H.E.S.* **44** (1975), 191-240.

[63] H. Widom: Asymptotic behavior of block Toeplitz matrices and determinants. II. *Adv. Math.* **21** (1976), 1-29.

[64] H. Widom: On the singular values of Toeplitz matrices. *Z. Analysis Anw.* **8** (1989), 221-229.

[65] N.L. Zamarashkin and E.E. Tyrtyshnikov: Distribution of the eigenvalues and singular values of Toeplitz matrices under weakened requirements on the generating function. *Matem. Sbornik* **188** (1997), 83-93 [Russian].

[66] P. Zizler, R.A. Zuidwijk, K.F. Taylor, and S. Arimoto: A finer aspect of eigenvalue distribution of selfadjoint band Toeplitz matrices. *SIAM J. Matrix Analysis Appl.*, to appear.

FACULTY OF MATHEMATICS, TU CHEMNITZ, D-09107 CHEMNITZ, GERMANY
E-mail address: `aboettch@mathematik.tu-chemnitz.de`

FACULTY OF MECHANICS AND MATHEMATICS, ROSTOV-ON-DON STATE UNIVERSITY, BOLSHAYA SADOVAYA 105, 344 711 ROSTOV-ON-DON, RUSSIAN FEDERATION
E-mail address: `grudsky@aaanet.ru`

Contemporary Mathematics
Volume **280**, 2001

How bad are symmetric Pick matrices?

Dario Fasino and Vadim Olshevsky

ABSTRACT. Let P be a symmetric positive definite Pick matrix of order n. The following facts will be proven here: **(a)** P is the Gram matrix of a set of rational functions, with respect to a inner product defined in terms of a "generating function" associated to P; **(b)** Its condition number is lower-bounded by a function growing exponentially in n; **(c)** P can be effectively preconditioned by the Pick matrix generated by the same nodes and a constant function.

1. Introduction

We consider here the class of real symmetric Pick matrices P defined as

$$P \equiv \left(\frac{d_i + d_j}{x_i + x_j} \right), \tag{1}$$

for $i, j = 1 \dots n$, where the *nodes* $x_1 \dots x_n$ are pairwise distinct and positive. For notational simplicity, the nodes are labeled in decreasing order, $x_1 > x_2 > \dots > x_n$. The latter hypothesis is of no restriction, since it can be fulfilled by a suitable row and column permutation on the original matrix. The class of Pick matrices is a notable example of displacement structured matrix class [**11**]. In fact, any matrix P as in (1) is the solution of a Sylvester equation having the form $D_x P + P D_x = de^* + ed^*$, where $e = (1 \dots 1)^*$, $d \equiv (d_i)$ and $D_x = \mathrm{Diag}(x_1 \dots x_n)$ is the diagonal matrix with x_i as ith diagonal entry. Matrices with definitions slightly different from (1) are also found in the literature under the same name, for example

$$\hat{P} \equiv \left(\frac{1 - c_i c_j}{x_i + x_j} \right), \tag{2}$$

where $-1 < c_i < 1$. These two definitions (and also other) are related by a diagonal congruence: Indeed, let $\Delta = \mathrm{Diag}(1 + c_1 \dots 1 + c_n)$. Then $P = \Delta^{-1} \hat{P} \Delta^{-1}/2$ is the matrix (1) with $d_i = (1 - c_i)/(1 + c_i)$.

Pick matrices arise in the classical Nevanlinna-Pick interpolation problem [**2, 5**]. In its simplest form, that problem amounts to find an analytic function mapping the unit disk into the closed unit disk, which fulfills a set of prescribed interpolation conditions. Variants of this problem are also considered, where the unit disk is replaced by a half plane, and these variants make up the differences among the

---

1991 *Mathematics Subject Classification.* Primary: 15A12, Secondary: 15A18 .

The work of the second author was supported by NSF grants CCR 9732355 and 0098222.

various definitions of Pick matrix. The solvability criterion for the Nevanlinna-Pick problem in all its forms is the positive semidefiniteness of a Pick matrix defined in terms of the given interpolation points.

Interest in the Nevanlinna-Pick interpolation problem and its confluent or degenerate cases [**5, 17**], as well as its generalizations to matrix- and operator-valued functions [**16**], and the the quest for practical solution algorithms [**9**], [**10**] (see also a survey [**12**]) as well as for superfast algorithms [**13**], [**14**], has become particularly intense recently, due to its deep connections with problems in time-invariant system theory, for example, in prediction theory and control theory [**2**]. Moreover, Pick matrices occur in the Ptàk-Young generalization of the Schur-Cohn theorem [**1**].

This paper focuses on the spectral conditioning of symmetric positive definite Pick matrices. The message to be gained is that, all matrices in this class are very ill-conditioned, hence checking their positive definiteness is a numerically difficult task. Indeed, it is well known that a very large condition number can spoil the positive definiteness in finite precision arithmetic.

In the next section, a "generating function" will be associated to any given Pick matrix. This function make it possible to consider a Pick matrix as the Gram matrix of a set of rational functions, with respect to a suitable inner product. This Gram structure is at the basis of the results in Section 3 and 4. There, firstly we derive a lower bound for the condition number of P that depends exponentially on the order n. Furthermore, it will be shown that particular Pick matrices, whose generating function is constant, turn out to be good preconditioners for all matrices in the class considered here.

The following notations are used throughout this paper: All matrices are n-by-n. Let $\lambda_{\max}(P) = \lambda_1(P) \geq \ldots \geq \lambda_n(P) = \lambda_{\min}(P)$ be the eigenvalues of P, and $\kappa_2(P)$ its spectral condition number. The purely imaginary numbers $\xi_i = \mathrm{i}x_i$ will be considered besides the real nodes x_i, the roman typeface i denoting the imaginary unit. Moreover, let

$$\pi(x) = \prod_{i=1}^{n}(x - \xi_i).$$

Finally, observe that if $p(x)$ is a polynomial of degree less than n then

$$\frac{p(x)}{\pi(x)} = \sum_{i=1}^{n} \frac{v_i}{x - \xi_i}, \qquad v_i = \frac{p(\xi_i)}{\pi'(\xi_i)}. \tag{3}$$

2. Generating functions for Pick matrices

Let $0 < a \leq +\infty$ be fixed, and let $w(x)$ be any real, even function, defined in the *reference interval* $(-a, a)$, such that

$$\int_{-a}^{a} \frac{1}{x^2 + x_i^2} w(x)\, \mathrm{d}x = \frac{d_i}{x_i}, \qquad i = 1 \ldots n. \tag{4}$$

Hereafter, we call any such function w a *generating function* for the matrix P. Its existence is not a problem, for if we choose a set of linearly independent real functions $w_1(x) \ldots w_n(x)$, with the property of being defined on $(-a, a)$ and even, then we can compute a function $w(x) = \sum_{i=1}^{n} c_i w_i(x)$ fulfilling conditions (4), by solving a linear system whose matrix is nonsingular. However, much more can be said. In what follows, a generating function $w(x)$ is said to be *good* if satisfies the constraints (4) with $a = +\infty$, and moreover there exist positive constants m, M

such that, for all $x \in \mathbf{R}$, $m \leq w(x) \leq M < +\infty$. The next theorem states precisely that every positive definite Pick matrix admits a good generating function.

THEOREM 2.1. *If P is positive definite, there exists a generating function for P, relative to the reference interval $\mathbf{R}$, which is bounded from above and below by positive constants. If P is positive semidefinite, and has rank k, then there exists a unique Borel measure generating P, whose spectrum consists precisely of k distinct points.*

The proof of the above theorem will be postponed to the end of this section, since it will be derived as a consequence of Theorem 2.3, which has an independent interest. To keep notation simple and coherent through the paper, $w(x)\,\mathrm{d}x$ sometimes will be used instead of the more appropriate $\mathrm{d}\mu(x)$, where $\mu(x)$ is a regular Borel measure.

As an example of the singular case mentioned in Theorem 2.1, consider the Pick matrix generated by a unit mass in zero, that is, $\int_{-a}^{a} f(x)w(x)\,\mathrm{d}x = f(0)$, for all continuous functions $f(x)$. Then from (4) we have $d_i = 1/x_i$, hence $P \equiv (x_i x_j)$, which clearly has rank one.

A more relevant example, which will play a special role in what follows, is the set of Pick matrices generated by a constant function in $\mathbf{R}$, $w(x) \equiv c$. Then, in this case, $d_i = c\pi$, independently on the nodes x_i. These matrices belong to the class of *symmetric Cauchy matrices*, see e.g., [**8, 13**]. A well known matrix in this class is the *Hilbert matrix*, $H \equiv (1/(i+j-1))$, which is generated by the function $w(x) \equiv 1/2\pi$ and the nodes $x_i = (2i-1)/2$. We will show in Section 4 that symmetric Cauchy matrices are spectrally close to all Pick matrices with the same nodes.

Remark that the symmetric Cauchy matrix $K \equiv (1/(x_i + x_j))$ fulfills the Lyapunov equation $D_x K + K D_x = ee^*$, hence it is positive definite, by a generalization of the Lyapunov theorem attributed to H. Schneider. Spectral properies of matrices defined as solutions of Lyapunov equations with a low rank right side are studied in [**15**], where it is shown that their eigenvalues have a fast decay. Moreover, Cauchy matrices are computationally very convenient: Since the coefficient matrices defining the above Lyapunov equation are diagonal, and the right side of that equation has rank one, there are stable and efficient algorithms for computing their triangular factors, see [**8**].

2.1. Pick matrices are Gram matrices.

THEOREM 2.2. *Consider the set of rational functions*

$$r_i(x) = \frac{1}{x - \xi_i} \qquad i = 1 \ldots n, \tag{5}$$

where $\xi_i = \mathrm{i}x_i$. Then, the matrix $P \equiv (p_{ij})$ is the Gram matrix of the functions $r_i(x)$ with respect to the inner product

$$\langle f, g \rangle_w = \int_{-a}^{a} f(x)\bar{g}(x)w(x)\,\mathrm{d}x,$$

i.e., $p_{ij} = \langle r_i, r_j \rangle_w$, where w is any generating function of P.

PROOF. In order to prove the equality $p_{ij} = \langle r_i, r_j \rangle_w$, consider separately the real part and the imaginary part, $\langle r_i, r_j \rangle_w = \Re\langle r_i, r_j \rangle_w + \mathrm{i}\Im\langle r_i, r_j \rangle_w$. For the real part, we have

$$\begin{aligned}
\Re\langle r_i, r_j\rangle_w &= \frac{1}{2}\left(\langle r_i, r_j\rangle_w + \langle r_j, r_i\rangle_w\right) \\
&= \frac{1}{2}\int_{-a}^{a}\left(\frac{1}{(x-\xi_i)(x+\xi_j)} + \frac{1}{(x+\xi_i)(x-\xi_j)}\right) w(x)\,\mathrm{d}x \\
&= \int_{-a}^{a}\left(\frac{1}{x^2+x_i^2}\frac{x_i}{x_i+x_j} + \frac{1}{x^2+x_j^2}\frac{x_j}{x_i+x_j}\right) w(x)\,\mathrm{d}x \\
&= \frac{x_i}{x_i+x_j}\frac{d_i}{x_i} + \frac{x_j}{x_i+x_j}\frac{d_j}{x_j} \\
&= \frac{d_i+d_j}{x_i+x_j}.
\end{aligned}$$

The imaginary part vanishes,

$$\begin{aligned}
2\mathrm{i}\Im\langle r_i, r_j\rangle_w &= \langle r_i, r_j\rangle_w - \langle r_j, r_i\rangle_w \\
&= \int_{-a}^{a}\left(\frac{1}{(x-\xi_i)(x+\xi_j)} - \frac{1}{(x+\xi_i)(x-\xi_j)}\right) w(x)\,\mathrm{d}x \\
&= 0,
\end{aligned}$$

since the last integrand is odd. □

Observe that a result similar to the one in the preceding theorem, but for a different definition of Pick matrix, can be found in [**17**], where the range of integration is the unit circle and the basis functions are polynomials.

As an immediate consequence, if P admits a nonnegative generating function, then P is positive definite. Now, in order to make a step toward the opposite conclusion, let $\omega(x) = \prod_{i=1}^{n}(x^2+x_i^2)$, and define the inner product

$$\langle p, q\rangle_{\hat{w}} = \int_{-a}^{a} p(x)\bar{q}(x)\hat{w}(x)\,\mathrm{d}x, \qquad \hat{w}(x) = \frac{w(x)}{\omega(x)}. \tag{6}$$

That inner product is well defined whenever p, q are polynomials of degree less than n. Indeed, if $f(x) = p(x)/\pi(x)$ and $g(x) = q(x)/\pi(x)$, then from (3) we see that $f(x)$ and $g(x)$ are in the linear span of the functions defined in (5). Moreover, $\langle f, g\rangle_w = \langle p, q\rangle_{\hat{w}}$. Let $p(x) = p_0 + \dots p_{n-1}x^{n-1}$, and denote by p and v the n-vectors with entries p_i and $v_i = p(\xi_i)\pi'(\xi_i)$ respectively. Then from (3) we have:

$$\begin{aligned}
v^*Pv &= \int_{-a}^{a} p(x)\overline{p}(x)\hat{w}(x)\,\mathrm{d}x \\
&= \sum_{i=0}^{n-1}\sum_{j=0}^{n-1} p_i\bar{p}_j \int_{-a}^{a} x^{i+j}\hat{w}(x)\,\mathrm{d}x \\
&= p^*Hp, \qquad (7)
\end{aligned}$$

where H is a Hankel matrix whose entries are the algebraic moments of $\hat{w}(x)$. This fact helps us to prove the following fact:

THEOREM 2.3. *Let $V \equiv (\xi_i^{j-1})$ be the Vandermonde matrix with nodes ξ_i, and D be the diagonal matrix with entries $\pi'(\xi_1)\dots\pi'(\xi_n)$. Then*

$$V^*D^*PDV = H, \qquad H \equiv \left(\int_{-a}^{a} x^{i+j}\hat{w}(x)\,\mathrm{d}x\right).$$

PROOF. On the basis of (7), it suffices to observe that $DVp = v$. □

2.2. Proof of Theorem 2.1. Let P be positive definite, and let $H \equiv (h_{i-j+2})$ be the Hankel matrix given by Theorem 2.3. Since H is congruent to P, it is positive definite, hence by classical results on the finite Hamburger moment problem [**6, 18**] there exists a nonnegative function $\hat{w}(x) \in L_1(\mathbf{R})$ such that

$$\int_{\mathbf{R}} x^i \hat{w}(x)\,\mathrm{d}x = h_i, \qquad i = 0 \dots 2n-2. \tag{8}$$

A close look at the definition of H reveals that $h_{2k-1} = 0$, hence we can safely suppose that $\hat{w}(x)$ is even. Now, we see from (6) that $w(x) = \omega(x)\hat{w}(x)$ fulfills the equalities

$$\int_{\mathbf{R}} \frac{1}{x^2 + x_i^2} w(x)\,\mathrm{d}x = \mu_i, \qquad i = 1 \dots n,$$

where $\mu_i = d_i / x_i$. We have, equivalently,

$$\int_{\mathbf{R}} a_i(x) \frac{w(x)}{x^2+1}\,\mathrm{d}x = \mu_i, \qquad a_i(x) = \frac{x^2+1}{x^2+x_i^2} \in L_\infty(\mathbf{R}).$$

Moreover, $(\mathbf{R}, (x^2+1)^{-1}\,\mathrm{d}x)$ is a finite measure space and, by construction, $w(x) \in L_1(\mathbf{R}, (x^2+1)^{-1}\,\mathrm{d}x)$. Hence, we are in the hypotheses of Theorem 2.9 of [**3**]. The consequence of that theorem is that we can assume $m \leq w(x) \in L_\infty(\mathbf{R})$, for some $m > 0$. The first part of the claim follows.

If P is semidefinite and has rank k, then the same holds also for H, hence the constraints (8) individuate a unique Borel measure $\hat{w}(x)\,\mathrm{d}x$ whose spectrum consists of precisely k points, see [**18**, Thm. 1.2]. The conclusion follows again from the relation $w(x) = \omega(x)\hat{w}(x)$. □

The preceding discussion gives us the basis for a procedure to compute a good generating function of a given P. Indeed, consider the convex functional

$$E(f) = \int_{\mathbf{R}} f(x)[\log f(x) - 1] \frac{\mathrm{d}x}{x^2+1},$$

which is a "nonautonomous" variant of the Shannon entropy [**3, 6**]. Observe that $E(f)$ is finite for every nonnegative $f \in L_\infty(\mathbf{R})$. Hence, we may consider the minimization of $E(f)$ among all bounded functions which are nonnegative and fulfil the constraints (4). The minimizer exists, since the admissible set is closed and not empty, is unique, since $E(f)$ is strictly convex, and can be expressed by Lagrange multiplier theorem as

$$w(x) = \exp\left(\sum_{i=1}^{n} \lambda_i \frac{x^2+1}{x^2+x_i^2} \right),$$

for some scalars $\lambda_1 \dots \lambda_n$, see [**3**, Thm. 4.8]. This function is, obviously, a good generating function for P. A similar argument, based on the minimization of the variant of the Burg entropy [**3**] given by

$$F(f) = -\int_{\mathbf{R}} \log f(x) \frac{\mathrm{d}x}{x^2+1},$$

allows us to derive the existence of a good generating function having the rational form

$$w(x) = \left(\sum_{i=1}^{n} \lambda_i \frac{x^2+1}{x^2+x_i^2} \right)^{-1}.$$

3. Exponential ill-conditioning

The purpose of this section is to prove that, for any symmetric positive definite Pick matrix P, $\kappa_2(P)$ is bounded from below by an exponential function in n, regardless of the generating function. This result is quite similar in spirit to other results available for Vandermonde [**7**], Krylov and Hankel matrices [**19, 20**], which are other examples of displacement structured matrices.

The starting point is a very simple technique: One chooses suitable vectors v, w and gets immediately the lower bound

$$\kappa_2(P) \geq \frac{v^*Pv}{v^*v}\frac{w^*w}{w^*Pw}.$$

In our case, the most convenient choice for v is a canonical basis vector, for example, $v = (1, 0 \dots 0)^*$, so that $v^*v = 1$ and

$$v^*Pv = \int_{-a}^{a} \frac{1}{x^2 + x_1^2} w(x)\, \mathrm{d}x.$$

To proceed further, consider the vector w whose entries are given by

$$\begin{aligned} w_1 &= \prod_{j\neq 1} \frac{\xi_1 + \xi_j}{\xi_1 - \xi_j} \\ w_i &= \frac{2\xi_i}{\xi_i - \xi_1} \prod_{j\neq 1,i} \frac{\xi_i + \xi_j}{\xi_i - \xi_j} \qquad i = 2 \dots n. \end{aligned}$$

From (3) we have

$$\sum_{i=1}^{n} \frac{w_i}{x - \xi_i} = \frac{1}{\pi(x)} \prod_{i=2}^{n} (x + \xi_i),$$

hence

$$w^*Pw = \int_{-a}^{a} \frac{1}{x^2 + x_1^2} w(x)\, \mathrm{d}x = v^*Pv.$$

So far, our estimate consists in the inequality $\kappa_2(P) \geq w^*w$ which, remarkably, does no longer depend on the generating function. It remains to estimate from below the squared norm of w. A rather crude bound is

$$w^*w > |w_1|^2 > \frac{(x_1 + x_n)^{2n-2}}{(x_1 - x_n)^{2n-2}}.$$

We obtain:

THEOREM 3.1. *Let P be given by (1) and positive definite, and let $r = (x_1 + x_n)^2/(x_1 - x_n)^2 > 1$. Then $\kappa_2(P) > r^{n-1}$.*

The conclusion in the above result looks weak, if compared to the above mentioned results concerning Vandermonde and positive definite Hankel matrices, since r can become arbitrarily close to 1. In fact, the condition numbers of the latter matrices can be bounded from below by exponentials like 2^n and 4^n respectively, apart of slowly varying factors, with no further assumptions.

Certainly, the preceding expression of r deserves some improvement. Nevertheless, the above theorem can be regarded as qualitatively optimal. Indeed, in Figure 1 we show some results concerning optimally conditioned symmetric Cauchy matrices. For $n = 2 \dots 20$, we have computed the minimum of the function $\kappa_2(P_n)$, where

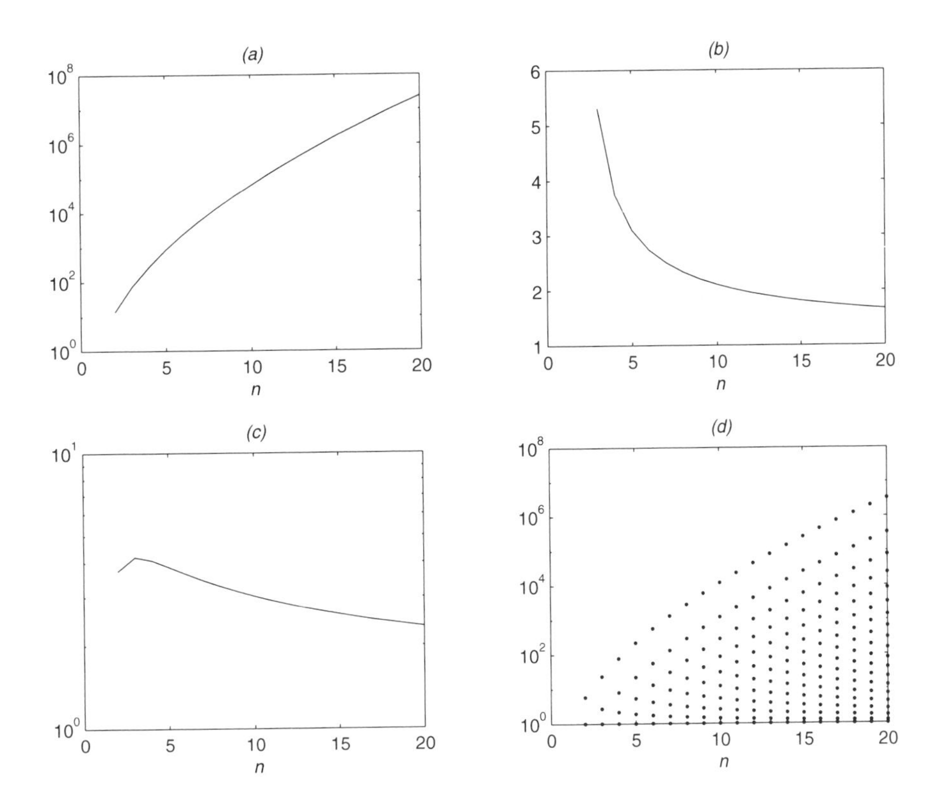

FIGURE 1. Optimally conditioned symmetric Cauchy matrices

P_n is the symmetric Cauchy matrix with nodes $x_1 \dots x_n$, assumed as free parameters, under the constraints $x_i > 0$. In Figure 1a, the computed minimum value of $\kappa_2(P_n)$ is plotted against n. Figures 1b and 1c show the plots of $\kappa_2(P_n)/\kappa_2(P_{n-1})$ and $(\kappa_2(P_n))^{1/n}$ respectively. These quantities are asymptotically equal, hence these numerical examples seem to indicate that $(\kappa_2(P_n))^{1/n}$ becomes close to one, if the nodes are unbounded and chosen in an optimal way. For completeness, we show in Figure 1d the optimal nodes, normalized so that the least node is equal to 1.

Moreover, the conclusion of the above theorem is in agreement with the main result of [**15**], which shows that the bound for the eigenvalue decay of symmetric Cauchy matrices may become arbitrarily close to 1, when the ratio x_1/x_n diverges.

Finally, remark that a lower bound for the conditioning of a Pick matrix defined as in (2) can be obtained from the inequality $\kappa_2(\hat{P}) \geq \kappa_2(\Delta)^{-2}\kappa_2(P)$, where $\Delta = \mathrm{Diag}(1 + c_1 \dots 1 + c_n)$ and P is in the form (1), since $\hat{P} = 2\Delta P \Delta$.

4. A preconditioning technique

The main result in this section is Corollary 4.3, stating that the symmetric Cauchy matrices introduced in Section 2 are good preconditioners for Pick matrices, since the conditioning of the preconditioned matrices does not depend on n. This result is the Pick counterpart of a similar result obtained in [**4**] for Hankel matrices.

The statement in the following lemma is a basic result in linear algebra, whose simple proof is omitted:

LEMMA 4.1. *Let P_1, P_2 be hermitian positive definite matrices, and suppose that for all vectors v it holds*

$$m \leq \frac{v^* P_1 v}{v^* P_2 v} \leq M.$$

Then, $\lambda_i(P_2^{-1}P_1) \in [m, M]$, and

$$\kappa_2(P_2^{-1}P_1) \leq \frac{M}{m}.$$

THEOREM 4.2. *Let P_1, P_2 be Pick matrices generated by nodes $x_1 \ldots x_n$ and functions w_1, w_2 in the same reference interval $(-a, a)$, and suppose that for $|x| \leq a$ it holds*

$$m \leq \frac{w_1(x)}{w_2(x)} \leq M.$$

Then $\lambda_i(P_2^{-1}P_1) \in [m, M]$ and $\kappa_2(P_2^{-1}P_1) \leq M/m$, with equality if and only if $m = M$.

PROOF. Let $\omega(x) = \prod_{i=1}^n (x^2 + x_i^2)$. For $v \equiv (v_i)$ let $p(x)$ be the polynomial defined by the interpolation conditions $p(\xi_i) = v_i \pi'(\xi_i)$. ¿From (3) and Theorem 2.2 we have

$$mv^* P_2 v = m \int_{-a}^{a} \frac{|p(x)|^2}{\omega(x)} w_2(x) \, \mathrm{d}x \leq \int_{-a}^{a} \frac{|p(x)|^2}{\omega(x)} w_1(x) \, \mathrm{d}x = v^* P_1 v,$$

since all integrands are nonnegative. Analogously, one proves

$$v^* P_1 v = \int_{-a}^{a} \frac{|p(x)|^2}{\omega(x)} w_1(x) \, \mathrm{d}x \leq M \int_{-a}^{a} \frac{|p(x)|^2}{\omega(x)} w_2(x) \, \mathrm{d}x = M v^* P_2 v.$$

Thesis follows from Lemma 4.1. □

COROLLARY 4.3. *If $w(x)$ is a good generating function for P, $m \leq w(x) \leq M$ for $x \in \mathbf{R}$, and K is a symmetric Cauchy matrix with the same nodes as P, then $\lambda_i(K^{-1}P) \in [m, M]$ and $\kappa_2(K^{-1}P) \leq M/m$.*

Finally, we would like to gain some insight about the distribution of the eigenvalues of $K^{-1}P$. This is the subject of the following proposition and the subsequent example.

THEOREM 4.4. *Introduce the set*

$$\mathcal{P}_n = \left\{ f(x) = \sum_{i=1}^{n} v_i r_i(x), \quad \int_{\mathbf{R}} |f(x)|^2 \, \mathrm{d}x = 1 \right\},$$

where the functions $r_i(x)$ are defined in (5). In the notations of Corollary 4.3 we have

$$\lambda_{\min}(K^{-1}P) = \min_{f \in \mathcal{P}_n} \int_{\mathbf{R}} |f(x)|^2 w(x) \, \mathrm{d}x,$$

and analogously for $\lambda_{\max}(K^{-1}P)$.

PROOF. Let $L \equiv (l_{ij})$ be the inverse of the lower triangular Cholesky factor of K, and let

$$s_i(x) = \sum_{j=1}^{n} l_{ij} \frac{1}{x - \xi_i}.$$

From the relation $LKL^* = I$, the identity matrix, and Theorem 2.2, we see that the functions $s_i(x)$ are orthonormal with respect to the inner product $\langle \cdot, \cdot \rangle_w$. Indeed, the above construction of $s_i(x)$ is equivalent to applying the Gram-Schmidt process to the set of rational functions $r_i(x)$. Now, for any vector $v \equiv (v_i)$, let

$$f(x) = \sum_{j=1}^{n} v_i s_i(x).$$

We have $\langle f, f \rangle_w = \|v\|_2^2$ by Parseval's identity, and

$$\frac{v^* LPL^* v}{v^* v} = \frac{\int_{\mathbf{R}} |f(x)|^2 w(x)\, \mathrm{d}x}{\int_{\mathbf{R}} |f(x)|^2\, \mathrm{d}x}.$$

Since LPL^* and $K^{-1}P$ are similar, we have the claim. □

Hence the distribution of $\lambda_i(K^{-1}P)$ depends on the capability of the functions $r_i(x)$ to approximate specific functions in $L_2(\mathbf{R})$. This topic seems not to have been dealt with in the literature, and it goes beyond the scope of the present paper. However, the following example shows that we cannot expect the values of $\lambda_i(K^{-1}P)$ to be fairly distributed in $[m, M]$, under general hypotheses.

In Figure 2 we plot the eigenvalues of preconditioned matrices $K_n^{-1}P_n$, of order n, where P_n is generated by $w(x) = 1 + (x^2 + 1)^{-1}$ and K_n is generated by the constant 1, under various configurations of the nodes. Remark that $1 < w(x) \le 2$. In Figure 2a the nodes are the number 2^{i-1}, for $i = 1 \ldots n$, and in Figure 2b their reciprocals. The nodes relative to Figure 2c are 2^{2i-n-1}, and those in Figure 2d are the optimal nodes depicted in Figure 1d.

References

[1] Ackner, R. and Kailath, T., On the Ptàk-Young generalization of the Schur-Cohn theorem. *IEEE Trans. Automat. Control*, 37 (1992), 1601–1604.

[2] Ball, J.A.; Gohberg, I. and Rodman, L., *Interpolation of rational matrix functions*. Operator Theory: Advances and Applications, 45. Birkhäuser Verlag, Basel, 1990.

[3] Borwein, J.M. and Lewis, A.S., Duality relationships for entropy-like minimization problems. *SIAM J. Control and Optimization*, 29 (1991), 325–338.

[4] Fasino, D., Spectral properties of Hankel matrices and numerical solutions of finite moment problems. *J. Comput. Appl. Math.*, 65 (1995), 145–155.

[5] Foias, C.; Frazho, A.E.; Gohberg, I. and Kaashoek, M.A., *Metric constrained interpolation, commutant lifting and systems*. Operator Theory: Advances and Applications, 100. Birkhäuser Verlag, Basel, 1998.

[6] Frontini, M. and Tagliani, A., Maximum entropy in the finite Stieltjes and Hamburger moment problem. *J. Math. Physics*, 35 (1994), 6748–6756.

[7] Gautschi, W. and Inglese, G., Lower bounds for the condition number of Vandermonde matrices. *Numer. Math.*, 52 (1988), 241–250.

[8] Gohberg, I.; Kailath, T. and Olshevsky, V., Fast Gaussian elimination with partial pivoting for matrices with displacement structure. *Math. Comp.*, 64 (1995), 1557–1576.

[9] Gohberg, I. and Olshevsky, V., Fast state space algorithms for matrix Nehari and Nehari-Takagi interpolation problems, *Integral Equations and Operator Theory*, **20, No. 1** (1994), 44-83.

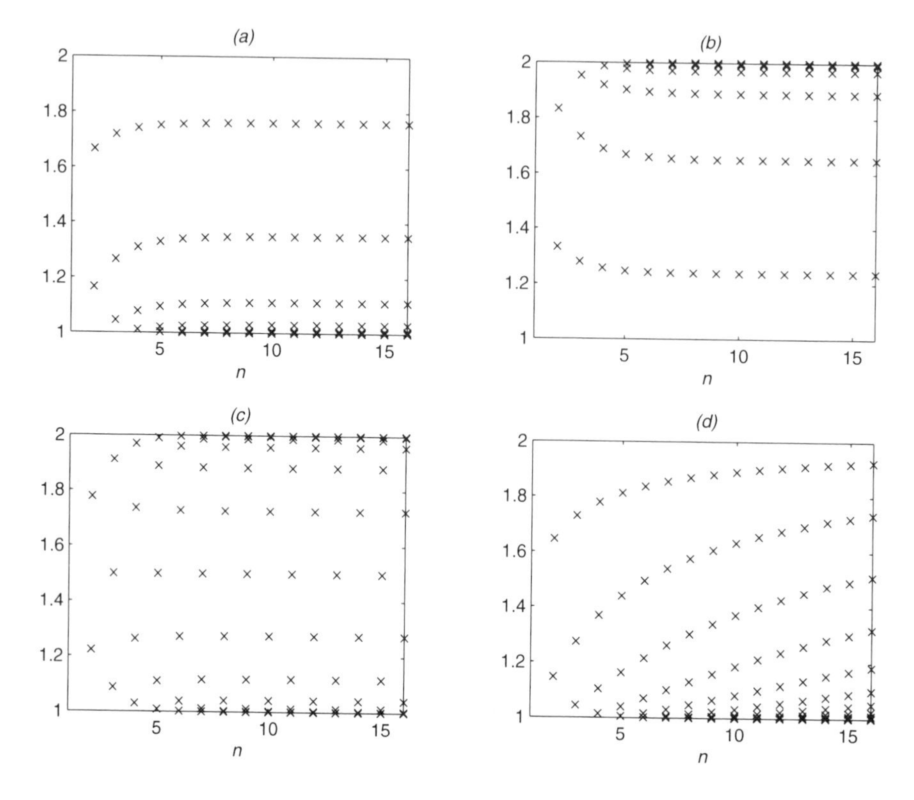

FIGURE 2. Eigenvalues of preconditioned Pick matrices

[10] Kailath, T. and Olshevsky, V. Diagonal Pivoting for Partially Reconstructible Cauchy-like Matrices, With Applications to Toeplitz-like Linear Equations and to Boundary Rational Matrix Interpolation Problems, *Linear Algebra and Its Applications*, **254** (1997), 251-302.

[11] Kailath, T. and Sayed, A.H., Displacement structure: Theory and applications. *SIAM Rev.*, 37 (1985), 297–386.

[12] Olshevsky, V., Pivoting for structured matrices, with applications, to appear in *Linear Algebra and Its Applications*, 53 p.
http://www.cs.gsu.edu/~matrvo

[13] Olshevsky, V. and Pan, V., *A unified superfast algorithm for boundary rational tangential interpolation problems and for inversion and factorization of dense structured matrices.* Proceedings of the 39th Annual Symposium on Foundations of Computer Science, pp. 192–201, IEEE Computer Society Press, Los Alamitos, 1998.

[14] Olshevsky, V. and Schokrollahi, A., A Unified Superfast Algorithm for Confluent Tangential interpolation problems, in *Advanced Signal Processing Algorithms, Architectures, and Implementations, SPIE publications*, 1999, 312-323.

[15] Penzl, T., Eigenvalue decay bounds for solutions of Lyapunov equations: The symmetric case. *Systems and Control Letters* (submitted), 1999.

[16] Sakhnovich, L.A., *Interpolation theory and its applications.* Mathematics and its Applications, 428. Kluwer Academic Publishers, Dordrecht, 1997.

[17] Sarason, D., Nevanlinna-Pick interpolation with boundary data. *Integral Equations Operator Theory*, 30 (1998), 231–250.

[18] Shohat, J.A. and Tamarkin, J.D., The problem of moments. *Amer. Math. Soc.*, New York 1943.

[19] Taylor, J.M., The condition of Gram matrices and related problems. *Proc. of the Royal Soc. of Edinburgh*, 80a (1978), 45–56.

[20] Tyrtyshnikov, E., How bad are Hankel matrices? *Numer. Math.*, 67 (1994), 261–269.

Dipartimento di Matematica e Informatica, University of Udine, Udine, Italy
E-mail address: `fasino@dimi.uniud.it`

Department of Mathematics and Computer Science, Georgia State University, Atlanta, GA 30303
E-mail address: `volshevsky@gsu.edu`

Contemporary Mathematics
Volume **280**, 2001

Spectral properties of real Hankel matrices

Miroslav Fiedler

Dedicated to the memory of Vlastimil Pták.

Abstract. We present two necessary conditions for the n-tuples of eigenvalues of real $n \times n$ Hankel matrices and completely characterize spectral properties of two subclasses of real Hankel matrices.

1. Introduction

K. R. Driessel [1] posed the following problem: Characterize all n-tuples of real numbers which can serve as n-tuples of eigenvalues of some $n \times n$ real Hankel matrix. We shall present two necessary conditions for such spectra and show that the second condition is also sufficient for the case $n = 3$. In the final part, we shall investigate and completely describe spectral properties of H-circulant and H-skew-circulant Hankel matrices.

Let us recall that a Hankel matrix is a matrix H of the form $H = (h_{i+j})$, $i, j = 0, \ldots, n-1$, i.e.

$$H = \begin{pmatrix} h_0 & h_1 & h_2 & \ldots & h_{n-1} \\ h_1 & h_2 & \ldots & h_{n-1} & h_n \\ h_2 & \ldots & & & \\ \ldots & & \ldots & & \ldots \\ h_{n-1} & & \ldots & h_{2n-3} & h_{2n-2} \end{pmatrix}$$

Thus it is always symmetric and in the case that it is real, its eigenvalues are real.

2. Results

Observe first that the identity matrix is for $n \geq 3$ not Hankel so that not all n-tuples $\lambda_1 \geq \lambda_2 \geq \cdots \geq \lambda_n$ can be n-tuples of $n \times n$ real Hankel matrices. This also suggests that there could be a positive constant bound from below for the "relative variance" of a non–zero real $n \times n$ Hankel matrix, i.e. for the number $\sum_{i<j}(\lambda_i - \lambda_j)^2 / \sum \lambda_i^2$. The best such bound is given in the first theorem.

1991 *Mathematics Subject Classification.* Primary: 15A57; Secondary: 15A18.
Key words and phrases. Hankel matrix, eigenvalue.
This research was supported by grant GAAV A1030701.

THEOREM 2.1. *An n-tuple of real numbers $\lambda_1, \dots, \lambda_n$ can serve as the n-tuple of eigenvalues of an $n \times n$ real Hankel matrix only if*

$$\sum_{1 \le i < j \le n} (\lambda_i - \lambda_j)^2 \ge K_n \sum_{i=1}^{n} \lambda_i^2, \tag{1}$$

where

$$\begin{aligned} K_n &= 2(\tfrac{2}{3} + \tfrac{4}{5} + \tfrac{6}{7} + \dots + \tfrac{n-2}{n-1}) \text{ if } n \text{ is even,} \\ K_n &= 2(\tfrac{2}{3} + \tfrac{4}{5} + \dots + \tfrac{n-3}{n-2} + \tfrac{n-1}{2n}) \text{ if } n \text{ is odd,} \end{aligned} \tag{2}$$

(thus $K_3 = \frac{2}{3}$).

The multiplicative constant K_n is the best possible for each $n \ge 2$ and the equality in (1) is attained if and only if the Hankel matrix is a multiple of the $n \times n$ matrix

$$H = \begin{pmatrix} 1 & 0 & \frac{1}{3} & 0 & \frac{1}{5} & \dots & \\ 0 & \frac{1}{3} & 0 & \frac{1}{5} & \dots & & \\ \frac{1}{3} & 0 & \frac{1}{5} & & \dots & & \\ 0 & \frac{1}{5} & & & \dots & & \\ \frac{1}{5} & & & & \dots & \frac{1}{3} & \\ \dots & & & & & & \\ & & & & \frac{1}{3} & 0 & \\ & & & \frac{1}{3} & 0 & 1 & \end{pmatrix},$$

i.e. $H = (h_{i+j})$, $i, j = 0, \dots, n-1$,

$$h_i = 0 \text{ if } t \text{ is odd}, \quad h_i = h_{2n-2-t} = \frac{1}{t+1}$$

if t is even and $t \le n-1$.

The condition (1) is, however, not sufficient even for $n = 3$.

Proof : Let $H = (h_{i+j})$, $i, j = 0, \dots, n-1$, be a real Hankel matrix, let $\lambda_1, \lambda_2, \dots, \lambda_n$ be its eigenvalues. Then

$$\sum \lambda_i = h_0 + h_2 + \dots + h_{2n-2}. \tag{3}$$

Since $\sum \lambda_i^2 = trHH^T$, we have

$$\sum \lambda_i^2 = h_0^2 + 2h_1^2 + 3h_2^2 + \dots + (n-1)h_{n-2}^2 + nh_n^2 + \dots + h_{2n-2}^2. \tag{4}$$

Denote by v_n, w_n respectively, the column vectors with n coordinates

$$\begin{aligned} v_n &= (h_0, h_2\sqrt{3}, h_4\sqrt{5}, \dots, h_{n-2}\sqrt{n-1}, h_n\sqrt{n-1}, h_{n+2}\sqrt{n-3}, \dots, h_{2n-2})^T, \\ w_n &= (1, \frac{1}{\sqrt{3}}, \frac{1}{\sqrt{5}}, \frac{1}{\sqrt{n-1}}, \frac{1}{\sqrt{n-1}}, \dots, 1)^T \end{aligned}$$

if n is even,

$$\begin{aligned} v_n &= (h_0, h_2\sqrt{3}, h_4\sqrt{5}, \dots, h_{n-3}\sqrt{n-2}, h_{n-1}\sqrt{n}, h_{n+1}\sqrt{n-2}, \dots, h_{2n-2})^T, \\ w_n &= (1, \frac{1}{\sqrt{3}}, \frac{1}{\sqrt{5}}, \frac{1}{\sqrt{n-2}}, \frac{1}{\sqrt{n}}, \frac{1}{\sqrt{n-2}}, \dots, 1)^T \end{aligned}$$

if n is odd.

Observe that by (2), (3) and (4) for the inner products,

$$(v_n, w_n) = \sum \lambda_i,$$
$$(w_n, w_n) = n - K_n,$$
$$(v_n, v_n) \leq \sum \lambda_i^2,$$

with equality if and only if $h_{2k+1} = 0$ for $k = 0, \dots, n-2$.

Thus, by the Cauchy-Schwarz inequality,

$$\begin{aligned}
\sum_{1 \leq i < j \leq n} (\lambda_i - \lambda_j)^2 &= n \sum \lambda_i^2 - (\sum \lambda_i)^2 \\
&= n \sum \lambda_i^2 - (v_n, w_n)^2 \\
&\geq n \sum \lambda_i^2 - (w_n, w_n)(v_n, v_n) \\
&\geq n \sum \lambda_i^2 - (n - K_n) \sum \lambda_i^2 \\
&= K_n \sum \lambda_i^2.
\end{aligned}$$

The equality conditions are easily established.

An example of a triple satisfying (1) but not corresponding to any 3×3 Hankel matrix will be given in Remark 2.3. ■

THEOREM 2.2. *An n-tuple $\lambda_1 \geq \lambda_2 \geq \dots \geq \lambda_n$ of real numbers can serve as the n-tuple of eigenvalues of an $n \times n$ Hankel matrix only if both conditions*

$$\lambda_1 - \lambda_{n-1} - 2\lambda_n \geq 0, \tag{5}$$
$$2\lambda_1 + \lambda_2 - \lambda_n \geq 0 \tag{6}$$

are fulfilled.

For $n = 3$, these conditions are also sufficient and a Hankel matrix with the eigenvalues $\lambda_1 \geq \lambda_2 \geq \lambda_3$ satisfying (5) and (6) is

$$\begin{pmatrix} \alpha & \beta & \gamma \\ \beta & \gamma & \beta \\ \gamma & \beta & \alpha \end{pmatrix}, \tag{7}$$

where

$$\begin{aligned}
\alpha &= \frac{1}{3}(\lambda_1 + 2\lambda_2 + \lambda_3), \\
\beta &= \sqrt{\frac{1}{18}(2\lambda_1 + \lambda_2 - \lambda_3)(\lambda_1 - \lambda_2 - 2\lambda_3)}, \\
\gamma &= \frac{1}{3}(\lambda_1 - \lambda_2 + \lambda_3).
\end{aligned}$$

Proof : Let H be Hankel with eigenvalues $\lambda_1 \geq \dots \geq \lambda_n$, let $Z = (z_{ik})$ be the symmetric matrix

$$Z = \begin{pmatrix} 0 & 0 & 1 & \cdots & 0 \\ 0 & -2 & 0 & \cdots & 0 \\ 1 & 0 & 0 & \cdots & 0 \\ . & . & . & \cdots & . \\ 0 & 0 & 0 & \cdots & 0 \end{pmatrix}.$$

The ordered eigenvalues of Z are $\beta_1 = 1$, $\beta_2 = \cdots = \beta_{n-2} = 0$, $\beta_{n-1} = -1, \beta_n = -2$.

By the v. Neumann inequalities,

$$\sum_{i=1}^{n} \lambda_i \beta_{n+1-i} \leq trHZ \leq \sum_{i=1}^{n} \lambda_i \beta_i.$$

Since $trHZ = 0$, we obtain

$$-(2\lambda_1 + \lambda_2 - \lambda_n) \leq 0 \leq \lambda_1 - \lambda_{n-1} - 2\lambda_n,$$

i.e. (5) and (6).

Since (7) has an eigenvalue $\alpha - \gamma = \lambda_2$ corresponding to the eigenvector $(1, 0, -1)^T$, the remaining two eigenvalues are easily seen to be λ_1 and λ_3. ■

REMARK 2.3. *The condition (1) is a consequence of (5) and (6), as follows immediately from the identity*

$$(\lambda_1 - \lambda_2)^2 + (\lambda_1 - \lambda_3)^2 + (\lambda_2 - \lambda_3)^2 - \frac{2}{3}(\lambda_1^2 + \lambda_2^2 + \lambda_3^2) =$$

$$\frac{2}{27}(7(\lambda_1 - \lambda_2 - 2\lambda_3)(2\lambda_1 + \lambda_2 - \lambda_3) + (2\lambda_1 - 5\lambda_2 + 2\lambda_3)^2).$$

It follows that (1) can be strenghtened to

$$(\lambda_1 - \lambda_2)^2 + (\lambda_1 - \lambda_3)^2 + (\lambda_2 - \lambda_3)^2 \geq \frac{2}{3}(\lambda_1^2 + \lambda_2^2 + \lambda_3^2) + (2\lambda_1 - 5\lambda_2 + 2\lambda_3)^2$$

(for $\lambda_1 \geq \lambda_2 \geq \lambda_3$), which is then , for $n = 3$, a necessary and sufficient condition.

The triple 2.9, 1, 1 shows that (1) does not suffice since (5) is not satisfied.

3. H-circulant matrices

In this section, we shall completely find spectral properties of a subclass of real Hankel matrices.

An $n \times n$ Hankel matrix (a_{i+k}), $n \geq 2$, will be called an H- circulant matrix, resp. H-skew-circulant matrix, if for all i, k, $i + k \geq n$, $a_{i+k} = a_{i+k-n}$, resp. $a_{i+k} = -a_{i+k-n}$.

THEOREM 3.1. *Let $A = (a_{i+k})$ be an $n \times n$ H-circulant matrix. Let ω denote the primitive n-th root of unity $\omega = e^{2\pi i/n}$.*

If n is even, then:

(a_1): *A has the eigenvalue $\sum_{j=0}^{n-1} a_j$ corresponding to the eigenvector $(1, 1, \ldots, 1)^T$;*

(a_2): *A has the eigenvalue $\sum_{j=0}^{n-1} (-1)^j a_j$ corresponding to the eigenvector $(1, -1, 1, -1, \ldots, -1)^T$;*

(a_3): *$\frac{n}{2} - 1$ pairs of eigenvalues*

$$\pm|a_0 + a_1\omega^j + \ldots + a_{n-1}\omega^{(n-1)j}|, \; j = 1, \ldots, \frac{n}{2} - 1.$$

If n is odd, then:

(b_1): *A has the eigenvalue $\sum_{j=0}^{n-1} a_j$ corresponding to the eigenvector $(1, 1, \ldots, 1)^T$;*

(b_2): $\frac{n-1}{2} - 1$ *pairs of eigenvalues*

$$\pm|a_0 + a_1\omega^j + \ldots + a_{n-1}\omega^{(n-1)j}|,\ j = 1, \ldots, \frac{n-1}{2} - 1.$$

Proof : Define by J_n the $n \times n$ counteridentity matrix $(\delta_{k,n+1-k})$, by Z_n the circulant $n \times n$ matrix

$$Z_n = \begin{pmatrix} 0 & 1 & & & \\ & & \cdot & & \\ & & & \cdot & \\ & & & & \cdot \\ & & & & 1 \\ 1 & & & & \end{pmatrix},$$

and further for $j = 0, 1, \ldots, n-1$ the vectors $u_{(j)} = (1, \omega^j, \omega^{2j}, \ldots, \omega^{(n-1)j})^T$. Let $\phi(x)$, $\psi(x)$ be the polynomials

$$\phi(x) \equiv a_0 + a_1 x + \ldots + a_{n-1}x^{n-1},$$

$$\psi(x) \equiv a_{n-1} + a_{n-2}x + \ldots + a_0 x^{n-1}.$$

Thus

$$A = \psi(Z_n)J_n,$$

$$Z_n u_{(j)} = \omega^j u_{(j)} \text{ for } j = 0, \ldots, n-1$$

which implies

$$AJ_n u_{(j)} = \psi(\omega^j)u_{(j)},$$

as well as for the complex conjugate

$$A\overline{J_n u_{(j)}} = \psi(\omega^{-j})\overline{u_{(j)}}.$$

Since

$$\omega^{(n-1)j}\overline{J_n u_{(j)}} = u_{(j)},$$

we obtain

$$Au_{(j)} = \omega^{(n-1)j}\psi(\omega^{-j})\overline{u_{(j)}},$$

i.e.

$$Au_{(j)} = \phi(\omega^j)\overline{u_{(j)}}.$$

For ω^j (and then also $u_{(j)}$) real, this yields (a_1), (a_2) and (b_1). Otherwise, we have for the complex conjugates,

$$A\overline{u_{(j)}} = \overline{\phi(\omega^j)}u_{(j)}.$$

It follows that if for such j, w_j is a fixed square root of $\phi(\omega^j)$ and the vectors $v^+_{(j)}$ and $v^-_{(j)}$ are defined by

$$v^\pm_{(j)} = \overline{w_j}u_{(j)} \pm w_j\overline{u_{(j)}},$$

then

$$Av^\pm_{(j)} = \pm|\phi(\omega^j)|v^\pm_{(j)}.$$

It is immediate that the vectors $u_{(j)}$, $j = 0, \ldots, n-1$, are all non-zero and mutually orthogonal. In addition, $u_{(j)} = \overline{u_{(n-j)}}$ if $j > \frac{n}{2}$. This implies that the vectors $v^\pm_{(j)}$, $1 \le j < \frac{n}{2}$, together with the real vectors $u_{(j)}$, form an orthogonal set of vectors. The proof is complete. ∎

THEOREM 3.2. *Let $A = (a_{i+k})$ be an $n \times n$ H-skew-circulant matrix. Let ω denote the primitive $2n$-th root of unity $\omega = e^{\pi i/n}$.*

If n is even, then:

(a_1): *A has $\frac{n}{2} - 1$ pairs of eigenvalues*

$$\pm|a_0 + a_1\omega^{2j+1} + \ldots + a_{n-1}\omega^{(n-1)(2j+1)}|,\ j = 0, 1, \ldots, \frac{n}{2} - 1.$$

If n is odd, then:

(b_1): *A has the eigenvalue $\sum_{j=0}^{n-1}(-1)^j a_j$ corresponding to the eigenvector $(1, -1, 1, -1, \ldots, 1)^T$, and*

(b_2): *A has $\frac{n-1}{2} - 1$ pairs of eigenvalues*

$$\pm|a_0 + a_1\omega^{2j+1} + \ldots + a_{n-1}\omega^{(n-1)(2j+1)}|,\ j = 0, \ldots, \frac{n-1}{2} - 2.$$

Proof : Let J_n be the $n \times n$ counteridentity matrix, Z_n the $n \times n$ matrix

$$Z_n = \begin{pmatrix} 0 & 1 & & & \\ & & \cdot & & \\ & & & \cdot & \\ & & & & \cdot \\ & & & & 1 \\ -1 & & & & \end{pmatrix}.$$

Define further the vectors $u_{(j)} = (1, \omega^{2j+1}, \omega^{2(2j+1)}, \ldots, \omega^{(n-1)(2j+1)})^T$ for $j = 0, \ldots, n-1$, and the polynomials

$$\phi(x) \equiv a_0 + a_1 x + \ldots + a_{n-2}x^{n-2},$$

$$\psi(x) \equiv a_{n-2}x + a_{n-3}x^2 + \ldots + a_0 x^{n-1}.$$

Thus

$$A = \psi(Z_n)J_n,$$

$$Z_n u_{(j)} = \omega^{2j+1}u_{(j)} \text{ for } j = 0, \ldots, n-1$$

which implies

$$AJ_n u_{(j)} = \psi(\omega^{2j+1})u_{(j)},$$

as well as

$$AJ_n\overline{u_{(j)}} = \psi(\omega^{-(2j+1)})\overline{u_{(j)}}.$$

Since

$$\begin{aligned} \psi(\overline{\omega^{2j+1}}) &= \psi(\omega^{-(2j+1)}) \\ &= -\omega^{2j+1}\phi(\omega^{2j+1}), \end{aligned}$$

and

$$J_n\overline{u_{(j)}} = -\omega^{2j+1}u_{(j)},$$

we obtain

$$Au_{(j)} = \phi(\omega^{2j+1})\overline{u_{(j)}}.$$

For ω^{2j+1} (and then also $u_{(j)}$) real, this yields (b_1). Otherwise, we have for the complex conjugates,

$$A\overline{u_{(j)}} = \phi(\overline{\omega^{2j+1}})u_{(j)}.$$

It follows that if for such j, w_j is a fixed square root of $\phi(\omega^{2j+1})$ and the vectors $v^+_{(j)}$ and $v^-_{(j)}$ are defined by

$$v^{\pm}_{(j)} = \overline{w_j} u_{(j)} \pm w_j \overline{u_{(j)}},$$

then

$$Av^{\pm}_{(j)} = \pm|\phi(\omega^{2j+1})| v^{\pm}_{(j)}.$$

It is immediate that the vectors $u_{(j)}$, $j = 0, \ldots, n-1$, are all non-zero and mutually orthogonal. In addition, $u_{(j)} = \overline{u_{(n-j)}}$ if $j > \frac{n}{2}$. This implies that the vectors $v^{\pm}_{(j)}$, $1 \leq j < \frac{n}{2}$, together with the real vector $u_{(n-1/2)}$ in the case that n is odd, form an orthogonal set of vectors. The proof is complete. ∎

4. Concluding remarks

We did not succeed to solve the eigenvalue problem for general Hankel matrices completely. One might try to use the method in the proof of Theorem 2.2. for some other appropriately chosen matrices Z to find other necessary conditions.

Of course, every (even not real) Hankel matrix is a (unique) sum of an H-circulant and an H-skew-circulant matrix. Similarly, if we define *H-symmetric* and *H-skew-symmetric* matrices as Hankel (square) matrices which are symmetric, resp. skew-symmetric with respect to the skew diagonal, every Hankel matrix is a unique sum of an H-symmetric and an H-skew-symmetric matrix.

Let us observe that the $n \times n$ H-symmetric matrix A satisfies

$$A = J_n A J_n \tag{8}$$

where J_n is the counteridentity matrix. Similarly,

$$A = -J_n A J_n$$

holds for every $n \times n$ H-skew-symmetric matrix. An easy consequence is :

THEOREM 4.1. *The spectrum of an H-skew-symmetric matrix is symmetric with respect to zero.*

Proof : Since

$$A - \lambda I = -J_n(A + \lambda I)J_n,$$

to every eigenvalue λ corresponds an eigenvalue $-\lambda$ with the same multiplicity. ∎

Since there are n eigenvalues but $2n-1$ parameters, the spectrum usually does not determine the Hankel matrix. There are, of course, cospectral Hankel matrices. For instance, the matrices

$$\begin{pmatrix} a & b & 0 \\ b & 0 & -b \\ 0 & -b & -a \end{pmatrix} \quad \text{and} \quad \begin{pmatrix} 0 & w & 0 \\ w & 0 & w \\ 0 & w & 0 \end{pmatrix}$$

where

$$w = \frac{1}{3}\sqrt{a^2 + b^2}$$

are cospectral.

On the other hand, there are spectra which correspond to one Hankel matrix only. An example is the spectrum of the matrix H from Theorem 2.1. It follows from (8) that in such case, the corresponding matrix is always H-symmetric. This suggests the following

CONJECTURE. *Every Hankel matrix is cospectral with an H-symmetric Hankel matrix.*

Observe that by Theorem 2.2, this is true for $n = 3$. It is also true for $n = 2$ (and $n = 1$) since every 2×2 real symmetric matrix (a_{ij}) (thus Hankel) is cospectral with the H-symmetric matrix

$$\begin{pmatrix} \frac{1}{2}(a_{11} + a_{22}) & \sqrt{a_{12}^2 + \frac{1}{4}(a_{11} - a_{22})^2} \\ \sqrt{a_{12}^2 + \frac{1}{4}(a_{11} - a_{22})^2} & \frac{1}{2}(a_{11} + a_{22}) \end{pmatrix}.$$

THEOREM 4.2. *If a Hankel matrix $H = (h_{i+k})$, $i, k = 0, \ldots, n-1$, is uniquely determined by its spectrum, then it is not only H-symmetric but also satisfies $h_j = 0$ for every odd j.*

Proof : This follows immediately from the fact that H and SHS have the same spectra and are both Hankel if $S = \operatorname{diag}(1, -1, 1, \ldots)$. ∎

REMARK 4.3. *This implies that for n even, such matrix has all eigenvalues with even multiplicity since it is permutation similar to a direct sum of two identical Hankel matrices.*

One could also formulate conditions which follow from general theorems about eigenvalues of a sum of symmetric matrices.

In addition to the problem of eigenvalues, the problem of orthogonal matrices which transform a real Hankel matrix into a diagonal matrix is interesting and open.

References

[1] K. Driessel, Private communication. 1991.

[2] M. Fiedler, Remarks on eigenvalues of Hankel matrices. *IMA Preprint Series, # 903.* Minneapolis 1991.

INSTITUTE OF COMPUTER SCIENCE, CZECH ACADEMY OF SCIENCES, POD VODÁRENSKOU VĚŽÍ 2, 182 07 PRAHA 8, THE CZECH REPUBLIC

E-mail address: fiedler@math.cas.cz

Contemporary Mathematics
Volume **280**, 2001

Conjectures and remarks on the limit of the spectral radius of nonnegative and block Toeplitz Matrices

Ludwig Elsner and S. Friedland

ABSTRACT. A finite sequence $\tau = \{t_{-m_-}, t_{-1}, t_0, t_1, t_2, ..., t_{m_+}\}$ of nonnegative $p \times p$ matrices defines a sequence of $m_- + m_+ + 1$ banded block Toeplitz matrices $T_n = (t_{ik}), n = 1, 2, ...,$, where $t_{ik} = t_{k-i}, i, k = 1, \ldots, n$ and $t_j = 0$ for $j \notin [-m_-, m_+]$. Let $\mu(\tau)$ be the limit of the spectral radius of T_n, as n tends to infinity. We state a conjecture which gives a formula for $\mu(\tau)$. In addition we discuss some related problems.

1. Introduction

Let $t = \{t_i\}_{\in \mathbf{Z}}$ be a bi-infinite sequence of $p \times p$ complex valued matrices. It defines a sequence of block Toeplitz matrices $T_n := (t_{ik})_{i,k=1}^n, n = 1, 2, \ldots,$ where $t_{ik} = t_{k-i}, i, k = 1, \ldots, n$. Let m_+, m_- be two nonnegative integers and assume that

$$t_j = 0, \quad \text{either } j > m_+, \quad \text{or } j < -m_-. \tag{1.1}$$

Let $m = m_+ + m_-$. Then for $n > m$ each T_n is $m+1$ banded block Toeplitz matrix. For $p \geq 1$ let $\mathbf{R}_+^{pp}$ denote the cone of $p \times p$ nonnegative matrices. In what follows we shall always assume that $\tau = \{t_{-m_-}, ..., t_{m_+}\} \in (\mathbf{R}_+^{pp})^{m+1}$. Write $T_n = T_n(\tau)$ and let $\mu_n(\tau) = \rho(T_n)$, where $\rho(A)$ is the spectral radius of a square matrix A. As T_n is a principal submatrix of the nonnegative matrix T_{n+1} we deduce that the sequence $\{\mu_n(\tau)\}_1^\infty$ is monotonously increasing. Let

$$\mu(\tau) = \lim_{n \to \infty} \mu_n(\tau). \tag{1.2}$$

As τ is a finite set it is straightforward to see that $\mu(\tau) \in [0, \infty)$. Let

$$\mu_0(\tau) = \inf_{\xi \in (0, \infty)} \rho\Big(\sum_{i=-m_-}^{m_+} \xi^i t_i\Big). \tag{1.3}$$

Then $\mu_0(\tau) \in [0, \infty)$ is a continuous function on $(\mathbf{R}_+^{pp})^{m+1}$. In [E-F] we showed that

$$\mu(\tau) \leq \mu_0(\tau) \tag{1.4}$$

1991 *Mathematics Subject Classification.* Primary 15A42, 15A48, 47B35, 47B65.

Supported by SFB 343 "Diskrete Strukturen in der Mathematik", Universität Bielefeld .

Furthermore

$$\mu(\tau) = \mu_0(\tau), \tag{1.5}$$

if t_{-m_-} and t_{m_+} are positive matrices. (For more general conditions see [E-F] or §2.) These conditions yield the result of Schmidt and Spitzer [S-S] that for $p = 1$ the equality (1.5) always holds. Thus for $p = 1$, the function $\mu(\tau)$ is a continuous function on $\mathbf{R}_+^{m+1}$. In [E-F] we gave a simple example of $\tilde{\tau} = \{\tilde{t}_{-1}, 0, \tilde{t}_1\}$, where $\tilde{t}_{-1}$ and $\tilde{t}_1$ are 2×2 positive definite diagonal matrices for which (1.5) fails. Hence $\mu(\tau)$ is not continuous at $\tilde{\tau}$.

The purpose of this note is to discuss some problems related to $\mu(\tau)$. In §2 we state a precise conjecture for the formula of $\mu(\tau)$. Furthermore, we show that for a given $\tilde{\tau}$, the function $\mu(\tau)$, which is restricted to all τ having the same $0-1$ pattern as $\tilde{\tau}$, is continuous at $\tilde{\tau}$. In §3 we discuss a matrix Riccati equation, which has an analytic solution $X(z)$ in a neighborhood of the origin. We show that the radius of convergence of the analytic series for $X(z)$ is bounded above by $\mu(\tau)^{-1}$, for a corresponding $\tau = (t_{-1}, t_0, t_1)$. It is an open question if this bound is sharp.

2. Conjectures and remarks

Let $A \in \mathbf{R}_+^{pp}$. Then A is called combinatorially singular if $\det(B) = 0$ for any $B \in \mathbf{R}^{pp}$ having zero entries at locations where A has. Otherwise A is called combinatorially nonsingular. In [E-F] we proved the equality (1.5) if one of the following conditions hold:

(a) t_{-m_-} is combinatorially nonsingular and t_{m_+} is irreducible;

(b) t_{m_+} is combinatorially nonsingular and t_{-m_-} is irreducible.

Assume that $\tilde{\tau}$ satisfies either (a) or (b). Then there exists a neighborhood O of $\tilde{\tau}$ in $\mathbf{R}_+^{pp}$ so that each $\tau \in O$ satisfies (a) or (b) respectively. The equality (1.5) yields that $\mu(\tau)$ is continuous at $\tilde{\tau}$.

PROPOSITION 2.1. *Let $\tilde{\tau} = (\tilde{t}_{-m_-}, ..., \tilde{t}_{m_+}) \in (\mathbf{R}_+^{pp})^{m+1}$. Then the function $\mu : (\mathbf{R}_+^{pp})^{m+1} \to \mathbf{R}_+$ is continuous at $\tilde{\tau}$ if and only if $\mu(\tilde{\tau}) = \mu_0(\tilde{\tau})$.*

Proof. Let int $(\mathbf{R}_+^{pp})^{m+1}$ be the set of all $m+1$ tuples of positive $p \times p$ matrices. For $\tau \in$ int $(\mathbf{R}_+^{pp})^{m+1}$ (1.5) holds. Hence μ is continous on int $(\mathbf{R}_+^{pp})^{m+1}$. Assume that $\tilde{\tau} \in \partial$int $(\mathbf{R}_+^{pp})^{m+1}$. Then for each sequence $\{\tau_i\}_1^\infty \subset$ int $(\mathbf{R}_+^{pp})^{m+1}$, which conveges to $\tilde{\tau}$, we have the equality

$$\lim_{i\to\infty} \mu(\tau_i) = \lim_{i\to\infty} \mu_0(\tau_i) = \mu_0(\tilde{\tau}).$$

Hence, if μ is continuous at $\tilde{\tau}$ then $\mu(\tilde{\tau}) = \mu_0(\tilde{\tau})$.

Let $\{\tau_i\}_1^\infty \subset (\mathbf{R}_+^{pp})^{m+1}$ be a sequence which converges to $\tilde{\tau}$. (1.4) yields

$$\limsup \mu(\tau_i) \le \limsup \mu_0(\tau_i) = \mu_0(\tilde{\tau}). \tag{2.1}$$

Given $\epsilon \in (0, 1)$ there exists $N(\epsilon)$ so that

$$(1 - \epsilon)\tilde{\tau} \le \tau_i, \quad i > N(\epsilon).$$

Hence

$$(1 - \epsilon)T_n(\tilde{\tau}) \le T_n(\tau_i), \quad i > N(\epsilon).$$

Therefore

$$(1 - \epsilon)\mu(\tilde{\tau}) \le \mu(\tau_i), \quad i > N(\epsilon),$$

and

$$\mu(\tilde{\tau}) \leq \liminf \mu(\tau_i). \tag{2.2}$$

If $\mu(\tilde{\tau}) = \mu_0(\tilde{\tau})$ then μ is continuous at $\tilde{\tau}$. □

Let

$$s(\tau, \xi) = \sum_{i=-m_-}^{m_+} \xi^i t_i \quad \xi \in (0, \infty) \tag{2.3}$$

be the symbol attached to biinfinite Toeplitz matrix $T = (t_{ik})_{i,k \in \mathbf{Z}}$. We say that $s(\tau, \cdot)$ is irreducible if $s(\tau, 1)$ is an irreducible matrix. Clearly, $s(\tau, 1)$ is irreducible if and only if $s(\tau, \xi)$ is irreducible for any $\xi \in (0, \infty)$.

Conjecture 2.2. *Let m_+, m_- be nonnegative integers and let $m = m_+ + m_-$. Suppose that $\tau \in (\mathbf{R}_+^{pp})^{m+1}$. Assume furthermore that $s(\tau, 1)$ is irreducible. Then (1.5) holds.*

Assume that Conjecture 2.2 holds. Suppose that $s(\tilde{\tau}, \cdot)$ is irreducible. Then $\mu(\tau)$ is continuous at $\tilde{\tau}$.

Assume that $s(\tau, 1)$ is reducible. Then there exists a permutation matrix $P \in \mathbf{R}_+^{pp}$ so that $Ps(\tau, 1)P^T$ is an $l \times l$ block upper triangular form, where each block is either an irreducible matrix or 1×1 zero matrix. ($Ps(\tau, 1)P^T$ is in the Frobenius normal form.) Each diagonal block of $Ps(\tau, 1)P^T$ induces symbols $s_k(\tau, \xi)$ for $k = 1, ..., l$. Each $s_k(\tau, 1)$ is a $p_k \times p_k$ irreducible matrix. (1×1 zero matrix is an irreducible matrix.) In that case we say that $s(\tau, \cdot)$ reducible and we will refer to

$$s_1(\tau, \cdot) \oplus \cdots \oplus s_l(\tau, \cdot) \tag{2.4}$$

as the principal part of the symbol $s(\tau, \cdot)$.

Conjecture 2.3. *Let m_+, m_- be nonnegative integers and let $m = m_+ + m_-$. Suppose that $\tau \in (\mathbf{R}_+^{pp})^{m+1}$. Assume that $s(\tau, 1)$ is reducible and (2.4) is the principal part of $s(\tau, \cdot)$. Then*

$$\mu(\tau) = \max_{1 \leq k \leq l} \inf_{\xi \in (0, \infty)} \rho(s_k(\tau, \xi)).$$

The following proposition is straightforward:

Proposition 2.4. *Let the assumptions of Conjecture 2.3 hold. Then*

$$\mu_0(\tau) = \inf_{\xi \in (0, \infty)} \max_{1 \leq k \leq m} \rho(s_k(\tau, \xi)).$$

The observation that μ is continuous on int $(\mathbf{R}_+^{pp})^{m+1}$ can be generalized as follows. Let $\tilde{\tau} = (\tilde{t}_{-m_-}, ..., \tilde{t}_{m_+}) \in (\mathbf{R}_+^{pp})^{m+1}$. Denote by $\mathcal{S}(\tilde{\tau})$ the set of $\tau = (t_{-m_-}, ..., t_{m_+}) \in (\mathbf{R}_+^{pp})^{m+1}$ so that the (i, j) entry of t_k is zero if the (i, j) entry of $\tilde{t}_k$ is zero for $i, j = 1, ..., p$ and $k = -m_-, ..., m_+$.

Proposition 2.5. *Let $\tilde{\tau} = (\tilde{t}_{-m_-}, ..., \tilde{t}_{m_+}) \in (\mathbf{R}_+^{pp})^{m+1}$. Then the function $\mu : \mathcal{S}(\tilde{\tau}) \to \mathbf{R}_+$ is continuous at $\tilde{\tau}$.*

Proof. Let $\{\tau_i\}_1^\infty \subset \mathcal{S}(\tilde{\tau})$ be a sequence which converges to $\tilde{\tau}$. Given $\epsilon \in (0, 1)$ there exists $N(\epsilon)$ so that

$$\tau_i \leq (1 + \epsilon)\tilde{\tau}, \quad i > N(\epsilon).$$

Hence

$$T_n(\tau_i) \leq (1 + \epsilon) T_n(\tilde{\tau}), \quad i > N(\epsilon).$$

Therefore

$$\mu(\tau_i) \le (1+\epsilon)\mu(\tilde{\tau}), \quad i > N(\epsilon).$$

Combine the above inequality with (2.2) to deduce the Proposition. □

The above Proposition can be strengthened as follows. Let $\mathcal{D} \subset \mathbf{R}^q$ be a convex domain. A function $f : \mathcal{D} \to \mathbf{R}_+$ is called logconvex if

$$f(ax + (1-a)y) \le f(x)^a f(y)^{1-a}, \quad x, y \in \mathcal{D}, \quad a \in (0,1).$$

Note that a logconvex function is either strictly positive on $\mathcal{D}$ or is equal to a zero function. It has been essentially shown by Artin in [Art], see also [Kin], that the set of all logconvex functions on $\mathcal{D}$ is a cone, which is closed with respect to: multiplication, raising to a nonnegative power and lim sup. Let $A(x) = (a_{ij}(x))_1^n, x \in \mathcal{D}$. Assume that each $a_{ij}(x)$ is a continuous logconvex function. Kingman [Kin] deducts from the above results that $\rho(A(x))$ is a continuous logconvex function. (See [Fri] for related results.) Combine Proposition 2.5 with Kingman's result to deduce:

PROPOSITION 2.6. *Let $\mathcal{D} \subset \mathbf{R}^q$ be a convex domain. Assume that $t_i(x)$ is a continuous logconvex function for $i = -m_-, ..., m_+$. Then $\mu(\tau(x))$ is a continuous logconvex function on $\mathcal{D}$.*

Observe that $\rho(s_k(\tau, e^t))$ is a logconvex function on $\mathbf{R}$. If $\rho(s_k(\tau,\xi))$ is a nonconstant function it achieves its infimum at the unique point $\hat{\xi}_k \in [0,\infty]$ for $k = 1, ..., l$. Assume the validity of Conjecture 2.3. Then Propositions 2.1 and 2.4 yield the exact conditions on points $\tilde{\tau}$ where μ is continuous.

3. Riccati equation

Consider the sequence of banded block Toeplitz matrices $\{T_n(\tau)\}_1^\infty$. By increasing the size of the blocks, if necessary, we can always assume that the above sequence is a sequence of tridiagonal block Toeplitz matrices. In this section we show that the limit spectral radius of an infinite tridiagonal nonnegative block Toeplitz matrix can bounded from above by using some special solutions of a certain matrix Riccati equation.

For $X \in \mathbf{C}^{pp}$ we denote by $||X||$ the l_2 operator norm of X, which is equal to $\sqrt{\rho(XX^*)}$. Let $A, B, C \in \mathbf{C}^{pp}$. In this section we consider the following Riccati equation

$$(I - zA - z^2CX)X = B, \quad X = X(z) \in \mathbf{C}^{pp}, \quad z \in \mathbf{C} \tag{3.1}$$

THEOREM 3.1. *Let $A, B, C \in \mathbf{C}^{pp}$. Then the Ricatti equation* (3.1) *has a unique power series solution in z of the form*

$$\begin{aligned} X(z) &= \sum_{k=0}^{\infty} z^k X_k, \quad X_k \in \mathbf{C}^{pp}, \quad k = 0, ..., \\ X_0 &= B, \quad X_1 = AB, \\ X_k &= AX_{k-1} + \sum_{j=0}^{k-2} CX_jX_{k-2-j}, \quad k = 2, ...,. \end{aligned} \tag{3.2}$$

Moreover, the above series and the majorant series

$$\sum_{0}^{\infty} |z|^k ||X_k||, \tag{3.3}$$

converge in the disk

$$|z| < \frac{1}{||A|| + 2\sqrt{||B||||C||}}. \tag{3.4}$$

Assume furthermore that $A, B, C \in \mathbf{R}_+^{pp}$. *Let* $\rho \in (0, \infty]$ *be the convergence radius of* $X(z)$. *Define recursively the following rational matrix functions in* z:

$$\begin{aligned} P_1(z) &= (I - zA)^{-1}B, \\ P_i(z) &= (I - zA - z^2CP_{i-1}(z))^{-1}B, \quad i = 2, ..., \end{aligned} \tag{3.5}$$

Then each $P_i(z)$ *is holomorphic in the disk* $|z| < \rho$ *for* $i = 1, ...,$. *Moreover*

$$\begin{aligned} &P_i(x) \leq X(x), \quad x \in [0, \rho), \quad i = 1, ..., \\ &\lim_{i \to \infty} P_i(z) = X(z), \quad |z| < \rho. \end{aligned} \tag{3.6}$$

Proof. Consider the recursive formula for $X_k = 0, ...,$ given by (3.2). Then the norm estimates for $||X_k||$ are

$$\begin{aligned} &||X_0|| \leq ||B||, \quad ||X_1|| \leq ||A|| \, ||B||, \\ &||X_k|| \leq ||A|| \, ||X_{k-1}|| + \sum_{j=0}^{k-2} ||C|| \, ||X_j|| \, ||X_{k-2-j}||, \quad k = 2, ...,. \end{aligned}$$

These estimates will be maximal if we replace all the inequalities by the equalities. Hence the upper estimates for $||X_k||, k = 0, ...,$ are obtained by the power series of the scalar version of (3.1) given by

$$(1 - z||A|| - z^2||C||x)x = ||B||.$$

The radius of convergence of the scalar series is given by the right-hand side of (3.4). Assume now that A, B, C are nonnegative. (3.2) it implies that $X_k \geq 0, k = 0, ...,$. Clearly, (3.5) yield that each $P_i(z)$ is analytic at $z = 0$. Let

$$P_i(z) = \sum_{k=0}^{\infty} z^k P_{i,k}, \quad i = 1, ...,. \tag{3.7}$$

We claim that

$$\begin{aligned} &P_{i,k} = X_k, \quad ,k = 0, ..., i, \\ &0 \leq P_{i,k} \leq X_k, \quad k = i + 1, ..., \\ &i = 1, ...,. \end{aligned} \tag{3.8}$$

We prove (3.8) by induction on i. For $i = 1$ we have

$$P_{1,k} = A^k B, \quad k = 0, ...,.$$

Compare the above equalities with the recursive definition of X_k in (3.2) to deduce (3.8) for $i = 1$. Assume the validity of (3.8) for $i = l - 1 \geq 1$. As

$$(I - zA - z^2 CP_{l-1}(z))P_l(z) = B, \quad l = 2, ...,$$

we obtain

$$P_{l,0} = B, \quad P_{l,1} = AB,$$
$$P_{l,k} = AP_{l,k-1} + \sum_{j=0}^{k-2} CP_{l-1,j}P_{l,k-2-j}, \quad k = 2, ...,. \tag{3.9}$$

Use the induction hypothesis together with above equalities and (3.2) to deduce the validity of (3.8) for $i = l$. It is straightforward to show that (3.8) yields (3.6). □

THEOREM 3.2. *Let $A, B, C \in \mathbf{R}_+^{pp}$. Set $t_{-1} = C, t_0 = A, t_1 = B$. Let μ be the limit spectral radius of the tridiagonal block Toeplitz matrices $T_n, n = 1, ...,$ induced by $\tau = (t_{-1}, t_0, t_1)$. Suppose that B is invertible and $T_n, n > n_0$, is irreducible. Then $\mu \leq \frac{1}{\rho}$.*

Proof. Fix $n > n_0$. Let $u^T = (0, ..., 0, v^T) \in \mathbf{R}_+^{np}, v > 0$. Assume that $\eta > \frac{1}{\rho}$. Let $z = \eta^{-1}$. Set $x^T = (x_1^T, ..., x_n^T) \in \mathbf{R}^{pn}$. Consider the equation

$$T_n x = \eta x - u. \tag{3.10}$$

Write down (3.10) as a system of block equations. The first equation of this block is $x_1 = zP_1(z)x_2$. Using the recursive deinition (3.5) we deduce that the next $n-2$ equations are

$$x_i = zP_i(z)x_{i+1}, \quad i = 2, ..., n-1.$$

The last equation of (3.10) is $x_n = zP_n(z)B^{-1}v$. Theorem 3.1 yields that (3.10) is solvable uniquely for every $\eta > \frac{1}{\rho}$. As T_n is irreducible for $n > n_0$ it follows that $\mu_n \leq \frac{1}{\rho}$. Hence $\mu \leq \frac{1}{\rho}$. □

In the scalar case $p = 1$ the assumptions of Theorem 3.2 yield $B, C > 0, A \geq 0$. We then get that $\rho^{-1} = A + 2\sqrt{BC}$ which is equal to the limit spectral radius of the corresponding tridiagonal Toeplitz matrix. It is an interesting question if the inequality given by Theorem 3.2 is sharp.

We conclude this paper with the following observation. Let

$$\tau(\xi) = (C(\xi), A(\xi), B(\xi)) = (\xi^{-1}C, A, \xi B), \quad A, B, C \in \mathbf{R}_+^{pp}, \quad \xi \in (0, \infty).$$

Let $\mu(\tau(\xi))$ be the limit spectral radius of the corresponding block tridiagonal matrices induced by the triple $\tau(\xi)$. Then $\mu(\tau(\xi)) = \mu(\tau(1))$. Consider the Ricatti equation (3.1) induced by the triple $\tau(\xi)$. Let $X(z, \xi)$ be the power series solution of this Ricatti equation. Let $\rho(\xi)$ be the radius of convergence of series $X(z, \xi)$. It is straightforward to see that $X(z, \xi) = \xi X(z, 1)$. Hence $\rho(\xi) = \rho(1)$.

References

[Art] E. Artin, Einführung in die Theorie der Gammafunktion, Teubner, Leipzig, 1931.

[E-F] L. Elsner and S. Friedland, The limit of the spectral radius of block Toeplitz matrices with nonnegative entries, *J. Integral Equations Operator Theory*, 36 (2000), 193-200.

[Fri] S. Friedland, Convex spectral functions, *Linear Multilin. Algebra* 9 (1981), 299-316.

[Kin] J.F.C. Kingman, A convexity property of positive matrices, *Quart. J. Math. Oxford Ser.* 12 (1961), 283-284.

[S-S] P. Schmidt and F. Spitzer, The Toeplitz matrices of an arbitrary Laurent polynomial, *Math. Scand.* 8, 15-38 (1960).

Fakultät für Mathematik,, Universität Bielefeld,, Postfach 100131, D-33501 Bielefeld, Germany

E-mail address: elsner@mathematik.uni-bielefeld.de

Department of Mathematics, Statistics and Computer Science,, University of Illinois at Chicago, Chicago, Illinois 60607-7045, USA

E-mail address: friedlan@uic.edu

Selected Titles in This Series

(Continued from the front of this publication)

For a complete list of titles in this series, visit the
AMS Bookstore at **www.ams.org/bookstore/**.